Differential Geometry

(Based on the Latest Syllabus as per NEP-2020)

For B.A. /B.Sc. Semester-V

Poonam Kumari
Associate Professor
Department of Mathematics
Magadh Mahila College
Patna University, Patna

SCIENTIFIC®
PUBLISHERS

SCIENTIFIC® PUBLISHERS

Published by:

Jodhpur
5 A, New Pali Road, P.O. Box 91
Jodhpur - 342 001 Rajasthan (India)
E-mail: info@scientificpub.com
Website: www.scientificpubonline.com

New Delhi
4806/24, Ansari Road, Daryaganj,
New Delhi - 110 002 (India)

ISBN: 978-81-961530-5-2 (H/B)

Printed in India

PREFACE

The book is intended to serve as a textbook for 5th semester students of B.A./B.Sc. (Hons.) Mathematics of Indian universities. It is written in accordance with the Latest Syllabus as per NEP-2020. It will also be found useful by the postgraduate students. The aim is to explain the subject matter in a simple way.

Each topic in the book is treated in a lucid way using the methods of differential calculus, integral calculus and vector algebra. A large number of solved examples followed by a carefully selected set of exercises are given at the end of each section. The answers and hints for the solution of unsolved problems are also given.

The book consists of five chapters. Chapter 1 begins with the theory of curves in three-dimensional Euclidean space. The curves are represented in a parametric form as a function of one variable. The geometric properties and various quantities associated with the curves such as the curvature and torsion are expressed via derivatives. At each point of a curve, there exists a set of three mutually perpendicular unit vectors **t**, **n**, **b** which form a right-handed system. The motion of this moving triad (**t**, **n**, **b**) along the curve is described by the Serret-Frenet formulae. The properties of a wide class of space curves called helices are also discussed briefly. The fact that a curve in space can be determined completely from the knowledge of its curvature and torsion is expressed in the form of the fundamental theorem of space curves. Other topics covered in the chapter include osculating circle, osculating sphere, evolutes, involutes and Bertrand curves.

Chapter 2 provides the representation of a surface in parametric form as well as in Monge's form. The concepts of characteristics, envelopes and developable surfaces are introduced in this chapter. Some properties of developable associated with space curves are also established.

In Chapter 3, the first fundamental form (or metric) and the second fundamental form of a surface and their properties are discussed. The fundamental magnitudes for some important surfaces such as the surface in Monge's form, the surface of revolution and the right helicoids are derived here. The formulae for the angle between two directions at a point on a surface are also derived in terms of direction coefficients and direction ratios.

Chapter 4 deals with the properties of curves on a surface. A formula for the normal curvature κ_n is derived in terms of second fundamental coefficients. Additionally, the maximum and minimum curvature along a given direction is found. This results in the definitions of principal directions, principal curvatures and lines of curvature. Various properties of the lines of curvature such as Rodrigue's formula and a few other theorems are discussed. Several other topics such as Dupin's indicatrix, third fundamental form, conjugate directions, asymptotic lines, ruled surfaces, isometric lines, null lines and the fundamental equations of the surface theory are also covered.

Chapter 5 addresses geodesic on a surface and geodesic curvature vector. The differential equation of a geodesic in various forms is derived and the formulae for the curvature and torsion of a geodesic are obtained. Some properties of geodesics and existence theorem are also established.

Suggestions for improvement of the book will be gratefully acknowledged.

Poonam Kumari

CONTENTS

SYMBOLS USED IN THE BOOK

s	Arc length of a curve
κ	Curvature of a curve
ρ	Radius of curvature of a curve
τ	Torsion of a curve
σ	Radius of torsion of a curve
$\mathbf{t}$	Unit tangent vector to a curve
$\mathbf{n}$	Unit principal normal vector to a curve
$\mathbf{b}$	Unit binormal vector to a curve
$\mathbf{N}$	Unit normal vector to a surface
κ_n	Normal curvature of a surface
κ_g	Geodesic curvature
K	Gaussian curvature

1

Curves in Space

1.1 INTRODUCTION

A curve is the locus of a point whose Cartesian coordinates x, y, z are the functions of a single parameter. This is equivalent to saying that a curve is the locus of a point whose position vector $\mathbf{r}$ relative to a fixed origin is a function of a single parameter. Thus a curve can be represented in parametric form by taking any real parameter t. It is usually convenient to choose as the scalar parameter the arc length s of the curve measured from a fixed point A on it.

When the curve is not a plane curve, it is said to be skew, tortuous or twisted.

1.2 PARAMETRIC REPRESENTATION OF SPACE CURVES

The parametric equations of a space curve are

$$x = x(t), \ y = y(t), \ z = z(t), \tag{1}$$

where x, y, z are real-valued functions of a single real parameter t ranging over a set of values $a \le t \le b$.

A space curve can also be represented by an equation of the form

$$\mathbf{r} = \mathbf{r}(t), \tag{2}$$

where $\mathbf{r}$ is the position vector of a current point on the curve.

The vector representation (2) is equivalent to the Cartesian representation (1), since the Cartesian coordinates x, y, z of the point $P(\mathbf{r})$ are components of $\mathbf{r}$. Thus, we have

$$\mathbf{r}(t) = x(t)\,\mathbf{i} + y(t)\mathbf{j} + z(t)\,\mathbf{k}$$

or $$\mathbf{r}(t) = (x(t), y(t), z(t)).$$

Remark. A curve can also be represented as the intersection of two surfaces represented by two equations of the form

$$F_1(x, y, z) = 0, \ F_2(x, y, z) = 0. \tag{3}$$

But the parametric representation given by (1) or (2) above is more suitable for the purpose of differential geometry.

1.3 ARC LENGTH OF A CURVE BETWEEN TWO POINTS

Let the equation of a curve C be given by

$$\mathbf{r} = \mathbf{r}(t), \tag{1}$$

where $\mathbf{r}$ is the position vector of a current point on the curve.

Suppose it is required to find the arc length between two points P and Q on the curve (1) corresponding to the values a and b of the parameter t. Let Δ denote a subdivision of $[a, b]$ by points $a = t_0, t_1, t_2, ..., t_n = b$, where

$$a = t_0 < t_1 < t_2 < ... < t_n = b.$$

Corresponding to this subdivision, the length L_Δ of the polygon inscribed to the arc joining the points P and Q is given by

$$L_\Delta = \sum_{i=1}^{n} P_{i-1}P_i = \sum_{i=1}^{n} |\,\mathbf{r}(t_i) - \mathbf{r}(t_{i-1})\,|. \tag{2}$$

It is obvious from (2) that if the number of points of subdivision is increased, then L_Δ will also increase (since the sum of two sides of a triangle is greater than the third side). Therefore the arc length is defined to be the supremum of L_Δ taken over all possible subdivisions of $[a, b]$.

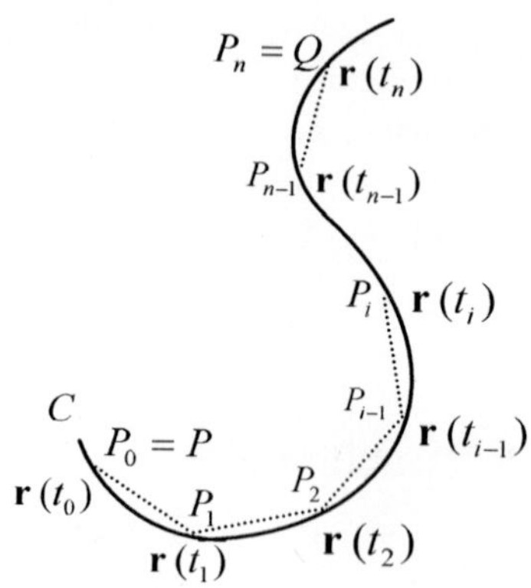

Since $\left| \int_{t_{i-1}}^{t_i} \dot{\mathbf{r}}(t)\, dt \right| = |\,\mathbf{r}(t_i) - \mathbf{r}(t_{i-1})\,|$, it follows from (2) that

$$L_\Delta = \sum_{i=1}^{n} \left| \int_{t_{i-1}}^{t_i} \dot{\mathbf{r}}(t)\, dt \right|. \tag{3}$$

Also, by Schwarz inequality, we have

$$\sum_{i=1}^{n} \left| \int_{t_{i-1}}^{t_i} \dot{\mathbf{r}}(t)\, dt \right| \le \sum_{i=1}^{n} \int_{t_{i-1}}^{t_i} |\,\dot{\mathbf{r}}(t)\,|\, dt = \int_{a}^{b} |\,\dot{\mathbf{r}}(t)\,|\, dt. \tag{4}$$

From (3) and (4), we get

$$L_\Delta \le \int_{a}^{b} |\,\dot{\mathbf{r}}(t)\,|\, dt. \tag{5}$$

Since the right hand member of (5) is finite and independent of Δ, the supremum of L_Δ is always finite. We shall now show that the supremum of L_Δ is actually

$$\int_{a}^{b} |\,\dot{\mathbf{r}}(t)\,|\, dt.$$

If $s = s(t)$ denotes the arc length from a to any point t, then the arc length from T to t is $s(t) - s(T)$, where $a \le T < t \le b$. Therefore, from (5), we have

$$s(t) - s(T) \le \int_T^t |\dot{\mathbf{r}}(t)|\, dt. \tag{6}$$

Also, by definition of arc length, we have

$$|\mathbf{r}(t) - \mathbf{r}(T)| \le s(t) - s(T). \tag{7}$$

From (6) and (7), we have

$$\left|\frac{\mathbf{r}(t) - \mathbf{r}(T)}{t - T}\right| \le \frac{s(t) - s(T)}{t - T} \le \frac{1}{t - T} \int_T^t |\dot{\mathbf{r}}(t)|\, dt. \tag{8}$$

Taking limit in (8) when $t \to T$, we get

$$|\dot{\mathbf{r}}(T)| \le \dot{s}(T) \le |\dot{\mathbf{r}}(T)|$$

$$\Rightarrow \quad \dot{s}(T) = |\dot{\mathbf{r}}(T)|.$$

Since this is true for any T in the range of t, we have

$$s = s(t) = \int_a^t |\dot{\mathbf{r}}(t)|\, dt. \tag{9}$$

Cartesian Equivalent. If (x, y, z) are the coordinates of a current point on the curve referred to fixed rectangular axes, then

$$\mathbf{r} = x\,\mathbf{i} + y\,\mathbf{j} + z\,\mathbf{k}.$$

$$\therefore \quad \dot{\mathbf{r}} = \dot{x}\,\mathbf{i} + \dot{y}\,\mathbf{j} + \dot{z}\,\mathbf{k} \text{ and } |\dot{\mathbf{r}}| = \sqrt{\dot{x}^2 + \dot{y}^2 + \dot{z}^2}.$$

Therefore, from (9), we have

$$s = s(t) = \int_a^t \sqrt{\dot{x}^2 + \dot{y}^2 + \dot{z}^2}\, dt. \tag{10}$$

1.4 TANGENT LINE

Definition. The tangent line to a curve C at a point P of C is defined as the limiting position of a straight line through P and a neighbouring point Q on C as Q approaches P along the curve.

Unit Tangent Vector

Let C be a space curve and let O be the vector origin. A is a fixed point on C from which the various arcual distances along C are measured.

Let P and Q be any two contiguous points on C whose position vectors are $\mathbf{r}$ and $\mathbf{r} + \delta\mathbf{r}$ respectively.

Again, let arc $AP = s$ and arc $AQ = s + \delta s$. Then arc $PQ = \delta s$.

From triangle law of addition of vectors,

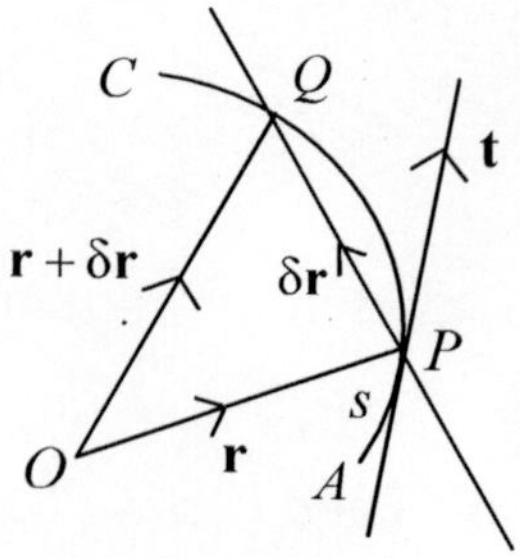

$$\overrightarrow{OP} + \overrightarrow{PQ} = \overrightarrow{OQ}$$

i.e., $\mathbf{r} + \overrightarrow{PQ} = \mathbf{r} + \delta\mathbf{r}$

i.e., $\overrightarrow{PQ} = \delta\mathbf{r}$.

$\therefore \quad \dfrac{\delta\mathbf{r}}{\delta s} =$ a vector parallel to $\overrightarrow{PQ}$.

Keep P fixed and allow $Q \to P$ along C. Then, in the limiting case, the chord PQ becomes the tangent to C at P.

$\therefore \quad \lim\limits_{Q\to P} \dfrac{\delta\mathbf{r}}{\delta s} =$ a vector parallel to the tangent at P

i.e., $\lim\limits_{\delta s\to 0} \dfrac{\delta\mathbf{r}}{\delta s} =$ a vector parallel to the tangent at P

i.e., $\dfrac{d\mathbf{r}}{ds} =$ a vector parallel to the tangent at P. (1)

Now,

$$\left|\frac{d\mathbf{r}}{ds}\right| = \lim_{Q\to P}\left|\frac{\delta\mathbf{r}}{\delta s}\right| = \lim_{Q\to P}\left|\frac{\overrightarrow{PQ}}{\text{arc } PQ}\right| = \lim_{Q\to P}\frac{\left|\overrightarrow{PQ}\right|}{\text{arc } PQ}$$

$$= \lim_{Q\to P}\frac{\text{chord } PQ}{\text{arc } PQ} = 1$$

i.e., $\left|\dfrac{d\mathbf{r}}{ds}\right| = 1$ (unity). (2)

From (1) and (2), it follows that $\dfrac{d\mathbf{r}}{ds}$ is a unit vector along the tangent at P. This unit vector along the tangent at P is denoted by $\mathbf{t}$ and is known as the unit tangent to C at P. We represent $\dfrac{d\mathbf{r}}{ds}$ by $\mathbf{r}'$ as well.

Equation of Tangent Line

Let the equation of a curve C be given by

$$\mathbf{r} = \mathbf{r}(s), \tag{1}$$

where $\mathbf{r}$ is the position vector of a current point P on the curve and s denotes the arc length of P from a fixed point A on it.

Suppose it is required to find the equation of the tangent line to the curve C at P.

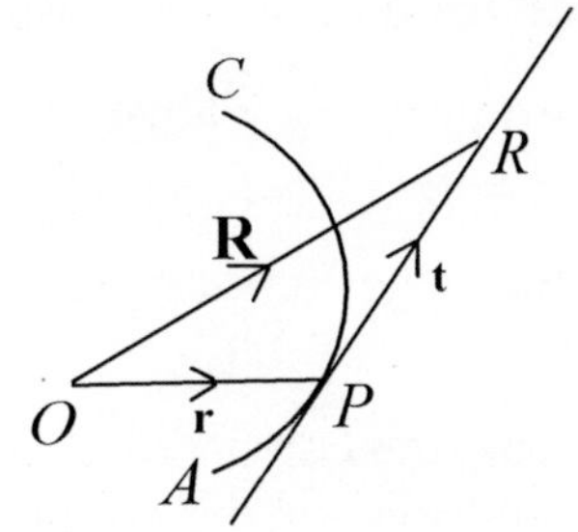

If $\mathbf{R}$ is the position vector of a current point R on the tangent to C at P, then

$$\vec{OR} = \vec{OP} + \vec{PR}$$

i.e., $\mathbf{R} = \mathbf{r} + c\mathbf{t}$

or $\mathbf{R} = \mathbf{r} + c\mathbf{r}'$, (2)

where c is a scalar parameter.

Equation (2) is the equation of the tangent to curve (1) at P.

If the equation of curve C is given in terms of parameter t, then $\dot{\mathbf{r}} = \dfrac{d\mathbf{r}}{dt}$ is parallel to the vector $\mathbf{t}$. Hence, in this case, equation (2) may be written as

$$\mathbf{R} = \mathbf{r} + \lambda\dot{\mathbf{r}}, \tag{3}$$

where λ is a scalar parameter.

Cartesian Form. If (x, y, z) are the coordinates of P referred to fixed rectangular axes and (X, Y, Z) are the coordinates of the current point R, then equation (2) is equivalent to

$$X\mathbf{i} + Y\mathbf{j} + Z\mathbf{k} = (x\mathbf{i} + y\mathbf{j} + z\mathbf{k}) + c\,(x'\mathbf{i} + y'\mathbf{j} + z'\mathbf{k}).$$

Equating the coefficients of **i**, **j**, **k** on both sides, we get

$$X = x + cx',\ Y = y + cy',\ Z = z + cz'$$

or
$$\frac{X - x}{x'} = \frac{Y - y}{y'} = \frac{Z - z}{z'} (= c). \tag{4}$$

It follows from (4) that the direction cosines of the tangent are x', y', z'.

Similarly, the Cartesian equivalent to (3) is

$$\frac{X - x}{\dot{x}} = \frac{Y - y}{\dot{y}} = \frac{Z - z}{\dot{z}}\ (= \lambda). \tag{5}$$

It follows from (5) that the direction cosines of the tangent are proportional to $\dot{x}, \dot{y}, \dot{z}$.

1.5 ORDER OF CONTACT

Definition. Let $P, P_1, P_2, \ldots, P_n$ be points on a given curve. If the points $P, P_1, P_2, \ldots, P_n$ lie on a surface and $P_1, P_2, \ldots, P_n \to P$ along the curve, then the curve and the surface are said to have the contact of nth order at the point P.

Condition for a Curve and a Surface to have a Contact of nth Order

Let the equation of a curve C be given by

$$\mathbf{r} = (x(t), y(t), z(t)) \tag{1}$$

and let the equation of a surface S be given by

$$F(x, y, z) = 0. \tag{2}$$

The values of t corresponding to the points which are common to C and S are the solutions of the equation

$$F(x(t), y(t), z(t)) = 0. \tag{3}$$

Let t_0 be one of the solutions of (3). Then,

$$F(t_0) = 0. \tag{4}$$

Now, expanding $F(t)$ by Taylor's theorem about t_0, we get

$$F(t) = F(t - t_0 + t_0)$$

$$= F(t_0) + (t - t_0)\, F'(t_0) + \frac{(t - t_0)^2}{2!} F''(t_0) + \dots + \frac{(t - t_0)^n}{n!} F^n(t_0) + \dots$$

Substituting the value of $F(t_0)$ from (4) in this equation, we get

$$F(t) = (t - t_0)\, F'(t_0) + \frac{(t - t_0)^2}{2!} F''(t_0) + \dots + \frac{(t - t_0)^n}{n!} F^n(t_0) + \dots \tag{5}$$

If $F'(t_0) \neq 0$, then t_0 is called a simple zero of $F(t)$. In this case, C and S are said to have a simple intersection at $\mathbf{r}(t_0)$.

If $F'(t_0) = 0$, but $F''(t_0) \neq 0$, then $F(t)$ is of second order of $t - t_0$. In this case, we say that t_0 is a double zero of $F(t)$, and C and S have two-point contact (or contact of first order) at $\mathbf{r}(t_0)$.

If $F'(t_0) = F''(t_0) = 0$, but $F'''(t_0) \neq 0$, then $F(t)$ is of third order of $t - t_0$. In this case, we say that t_0 is a triple zero of $F(t)$, and C and S have three-point contact (or contact of second order) at $\mathbf{r}(t_0)$.

In general, if $F'(t_0) = F''(t_0) = \dots = F^{n-1}(t_0) = 0$, but $F^n(t_0) \neq 0$, then C and S are said to have n-point contact (or contact of $(n-1)^{\text{th}}$ order) at $\mathbf{r}(t_0)$.

Solved Examples

Example 1. Find the length of the curve

$$\frac{x^2}{a^2} - \frac{y^2}{b^2} = 1, \; x = a\cosh\frac{z}{a}$$

from the point $(a, 0, 0)$ to the point (x, y, z).

Solution. The equation of the given curve in parametric form is

$$x = a\cosh t, \; y = b\sinh t, \; z = at.$$

$$\therefore \quad \dot{x} = a\sinh t, \; \dot{y} = b\cosh t, \; \dot{z} = a.$$

The point $(a, 0, 0)$ corresponds to the value $t = 0$. Hence the required length of the curve is given by

$$s = \int_0^t \sqrt{(\dot{x}^2 + \dot{y}^2 + \dot{z}^2)}\, dt$$

$$= \int_0^t \sqrt{(a^2\sinh^2 t + b^2\cosh^2 t + a^2)}\, dt$$

$$= \int_0^t \sqrt{a^2(\sinh^2 t + 1) + b^2\cosh^2 t}\, dt$$

$$= \int_0^t \sqrt{a^2\cosh^2 t + b^2\cosh^2 t}\, dt \qquad [\because \cosh^2\theta - \sinh^2\theta = 1]$$

$$= \int_0^t \sqrt{(a^2 + b^2)\cosh^2 t}\, dt$$

$$= \int_0^t \sqrt{(a^2 + b^2)}\, \cosh t\, dt$$

$$= \sqrt{(a^2 + b^2)}\, \sinh t$$

$$= \frac{y}{b}\sqrt{(a^2 + b^2)}.$$

Example 2. Find the unit tangent to the circular helix

$$\mathbf{r} = (a\cos t, a\sin t, bt),$$

where $-\infty < t < \infty$, $a > 0$ and write down equation of tangent at any point. Also, find the length of one complete turn of the helix.

Solution. **First Part.** The equation of the circular helix is

$$\mathbf{r} = (a\cos t, a\sin t, bt).$$

$\therefore$ $$\dot{\mathbf{r}} = (-a\sin t, a\cos t, b)$$

and $$|\dot{\mathbf{r}}| = \sqrt{(a^2\sin^2 t + a^2\cos^2 t + b^2)} = \sqrt{(a^2+b^2)}.$$

Hence the unit tangent vector $\mathbf{t}$ is given by

$$\mathbf{t} = \frac{\dot{\mathbf{r}}}{|\dot{\mathbf{r}}|} = \frac{1}{\sqrt{(a^2+b^2)}}(-a\sin t,\ a\cos t,\ b).$$

The equation of tangent at any point is

$$\mathbf{R} = \mathbf{r} + \lambda\dot{\mathbf{r}}, \tag{1}$$

where λ is a scalar parameter.

Substituting the values of $\mathbf{r}$ and $\dot{\mathbf{r}}$ in (1), we get

$$\mathbf{R} = (a\cos t,\ a\sin t, bt) + \lambda(-a\sin t, a\cos t,\ b).$$

or $$\mathbf{R} = (a\cos t - \lambda a\sin t, a\sin t + \lambda a\cos t, bt + \lambda b). \tag{2}$$

Since $\mathbf{R} = X\mathbf{i} + Y\mathbf{j} + Z\mathbf{k}$, equation (2) of the tangent can be written in Cartesian form as

$$\frac{X - a\cos t}{-a\sin t} = \frac{Y - a\sin t}{a\cos t} = \frac{Z - bt}{b}\ (=\lambda). \tag{3}$$

Second Part. The range of parameter t corresponding to one complete turn of the helix is $t_0 \le t \le t_0 + 2\pi$. Therefore the length s of one complete turn of the helix is given by

$$s = \int_{t_0}^{t_0+2\pi} |\dot{\mathbf{r}}|\,dt = \int_{t_0}^{t_0+2\pi} \sqrt{(a^2+b^2)}\,dt = 2\pi\sqrt{(a^2+b^2)}.$$

Example 3. Find the tangent at a point of the curve of intersection of the ellipsoid

$$\frac{x^2}{a^2} + \frac{y^2}{b^2} + \frac{z^2}{c^2} = 1$$

and the confocal

$$\frac{x^2}{a^2-\lambda} + \frac{y^2}{b^2-\lambda} + \frac{z^2}{c^2-\lambda} = 1,$$

where λ is a parameter.

Solution. We know that the equation of the tangent is

$$\frac{X-x}{\dot{x}} = \frac{Y-y}{\dot{y}} = \frac{Z-z}{\dot{z}}. \tag{1}$$

The equation of the given ellipsoid is

$$\frac{x^2}{a^2}+\frac{y^2}{b^2}+\frac{z^2}{c^2}-1=0. \qquad (2)$$

The equation of the confocal to (2) is

$$\frac{x^2}{a^2-\lambda}+\frac{y^2}{b^2-\lambda}+\frac{z^2}{c^2-\lambda}-1=0. \qquad (3)$$

Let the curve of intersection of the surfaces (2) and (3) be $\mathbf{r}\,(t)$.

Differentiating (2) with respect to t, we get

$$\frac{2x\dot{x}}{a^2}+\frac{2y\dot{y}}{b^2}+\frac{2z\dot{z}}{c^2}=0$$

or $$\frac{x\dot{x}}{a^2}+\frac{y\dot{y}}{b^2}+\frac{z\dot{z}}{c^2}=0. \qquad (4)$$

Differentiating (3) with respect to t, we get

$$\frac{2x\dot{x}}{a^2-\lambda}+\frac{2y\dot{y}}{b^2-\lambda}+\frac{2z\dot{z}}{c^2-\lambda}=0$$

or $$\frac{x\dot{x}}{a^2-\lambda}+\frac{y\dot{y}}{b^2-\lambda}+\frac{z\dot{z}}{c^2-\lambda}=0. \qquad (5)$$

Solving (4) and (5) for $\dot{x}, \dot{y}, \dot{z}$, we get

$$\frac{\dot{x}}{\dfrac{yz}{b^2(c^2-\lambda)}-\dfrac{yz}{c^2(b^2-\lambda)}}=\frac{\dot{y}}{\dfrac{zx}{c^2(a^2-\lambda)}-\dfrac{zx}{a^2(c^2-\lambda)}}$$

$$=\frac{\dot{z}}{\dfrac{xy}{a^2(b^2-\lambda)}-\dfrac{xy}{b^2(a^2-\lambda)}}$$

or $$\frac{\dot{x}}{\dfrac{yz\lambda\,(b^2-c^2)}{b^2c^2(c^2-\lambda)\,(b^2-\lambda)}}=\frac{\dot{y}}{\dfrac{zx\lambda\,(c^2-a^2)}{c^2a^2(a^2-\lambda)\,(c^2-\lambda)}}$$

$$=\frac{\dot{z}}{\dfrac{xy\lambda\,(a^2-b^2)}{a^2b^2(b^2-\lambda)\,(a^2-\lambda)}}$$

or $$\frac{\dot{x}}{\dfrac{a^2\,(b^2-c^2)\,(a^2-\lambda)}{x}}=\frac{\dot{y}}{\dfrac{b^2\,(c^2-a^2)\,(b^2-\lambda)}{y}}$$

$$=\frac{\dot{z}}{\dfrac{c^2\,(a^2-b^2)\,(c^2-\lambda)}{z}}=k\,(\text{let}).$$

$$\therefore \quad \left.\begin{aligned} \dot{x} &= \frac{ka^2(b^2-c^2)(a^2-\lambda)}{x}, \\ \dot{y} &= \frac{kb^2(c^2-a^2)(b^2-\lambda)}{y}, \\ \dot{z} &= \frac{kc^2(a^2-b^2)(c^2-\lambda)}{z}. \end{aligned}\right\} \qquad (6)$$

Substituting the values of $\dot{x}, \dot{y}, \dot{z}$ from (6) in (1), we get

$$\frac{X-x}{\dfrac{ka^2(b^2-c^2)(a^2-\lambda)}{x}} = \frac{Y-y}{\dfrac{kb^2(c^2-a^2)(b^2-\lambda)}{y}} = \frac{Z-z}{\dfrac{kc^2(a^2-b^2)(c^2-\lambda)}{z}}$$

or

$$\frac{x(X-x)}{a^2(b^2-c^2)(a^2-\lambda)} = \frac{y(Y-y)}{b^2(c^2-a^2)(b^2-\lambda)} = \frac{z(Z-z)}{c^2(a^2-b^2)(c^2-\lambda)}.$$

This is the required equation of the tangent.

Example 4. Find the equation of the plane that has three-point contact at the origin with the curve $x = t^4 - 1,\ y = t^3 - 1,\ z = t^2 - 1$.

Solution. Let the equation of the required plane be

$$ax + by + cz = 0. \qquad (1)$$

The equation of the given curve is

$$x = t^4 - 1,\ y = t^3 - 1,\ z = t^2 - 1. \qquad (2)$$

At the origin,

$$x = 0, y = 0, z = 0$$

i.e., $t^4 - 1 = 0,\ t^3 - 1 = 0,\ t^2 - 1 = 0$ [using (2)]

i.e., $t = 1$.

The points of intersection of the curve (2) and plane (1) are the zeros of the function

$$F(t) = a(t^4 - 1) + b(t^3 - 1) + c(t^2 - 1). \qquad (3)$$

From (3), we have

$$F'(t) = 4at^3 + 3bt^2 + 2ct,$$
$$F''(t) = 12at^2 + 6bt + 2c.$$

Since the plane (1) has three-point contact with the curve (2) at the origin, i.e., at $t=1$, we have

$$F'(1) = 4a + 3b + 2c = 0, \tag{4}$$

$$F''(1) = 12a + 6b + 2c = 0. \tag{5}$$

Solving (4) and (5) for a, b, c, we get

$$\frac{a}{6-12} = \frac{b}{24-8} = \frac{c}{24-36}$$

or $$\frac{a}{-6} = \frac{b}{16} = \frac{c}{-12}$$

or $$\frac{a}{3} = \frac{b}{-8} = \frac{c}{6} = k \text{ (let)}.$$

$\therefore$ $$a = 3k,\ b = -8k,\ c = 6k. \tag{6}$$

Substituting the values of a, b, c from (6) in (1), we get

$$3x - 8y + 6z = 0.$$

This is the required equation of the plane.

Example 5. Determine a, h, b so that the paraboloid

$$2z = ax^2 + 2hxy + by^2$$

may have the closest contact at the origin with the curve

$$x = t^3 - 2t^2 + 1,\ y = t^3 - 1,\ z = t^2 - 2t + 1.$$

Find also the order of contact.

Solution. The equation of the given paraboloid is

$$2z = ax^2 + 2hxy + by^2. \tag{1}$$

The equation of the given curve is

$$x = t^3 - 2t^2 + 1,\ y = t^3 - 1,\ z = t^2 - 2t + 1. \tag{2}$$

At the origin,

$$x = 0,\ y = 0,\ z = 0$$

i.e., $$t^3 - 2t^2 + 1 = 0,\ t^3 - 1 = 0,\ t^2 - 2t + 1 = 0 \qquad \text{[using (2)]}$$

i.e., $$t = 1.$$

The points of intersection of the curve (2) and paraboloid (1) are the zeros of the function

$$F(t) = 2(t^2 - 2t + 1) - a(t^3 - 2t^2 + 1)^2 - 2h(t^3 - 2t^2 + 1)(t^3 - 1) - b(t^3 - 1)^2. \tag{3}$$

From (3), we have

$$F'(t) = 4t - 4 - 2a(t^3 - 2t^2 + 1)(3t^2 - 4t) - 2h(6t^5 - 10t^4 + 4t) - 2b(t^3 - 1)(3t^2), \tag{4}$$

which is clearly zero at $t = 1$, i.e., at the origin.

From (4), we have

$$F''(t) = 4 - 2a(t^3 - 2t^2 + 1)(6t - 4) - 2a(3t^2 - 4t)^2 - 2h(30t^4 - 40t^3 + 4) - 2b(t^3 - 1)(6t) - 18bt^4$$

or
$$F''(t) = 4 - 2a(15t^4 - 40t^3 + 24t^2 + 6t - 4) - 2h(30t^4 - 40t^3 + 4) - 6b(5t^4 - 2t). \quad (5)$$

For contact of second order at the origin, i.e., at $t = 1$, we have

$$F''(1) = 4 - 2a + 12h - 18b = 0$$

or
$$a - 6h + 9b = 2. \quad (6)$$

Now, from (5), we have

$$F'''(t) = -2a(60t^3 - 120t^2 + 48t + 6) - 2h(120t^3 - 120t^2) - 6b(20t^3 - 2). \quad (7)$$

For contact of third order at the origin, i.e., at $t = 1$, we have

$$F'''(1) = 12a - 108b = 0$$

or
$$a - 9b = 0. \quad (8)$$

Again, from (7), we have

$$F^{iv}(t) = -2a(180t^2 - 240t + 48) - 2h(360t^2 - 240t) - 6b(60t^2). \quad (9)$$

For contact of fourth order at the origin, i.e., at $t = 1$, we have

$$F^{iv}(1) = 24a - 240h - 360b = 0$$

or
$$a - 10h - 15b = 0. \quad (10)$$

Solving (6), (8) and (10), we get

$$a = \frac{5}{6},\ b = \frac{5}{54},\ h = -\frac{1}{18}. \quad (11)$$

From (9), we have

$$F^{v}(t) = -2a(360t - 240) - 2h(720t - 240) - 6b(120t).$$

$$\therefore \quad F^{v}(1) = -240a - 960h - 720b.$$

Obviously, $F^{v}(1) \neq 0$ for the values of a, b and h given by (11). Hence the closest possible contact of the paraboloid (1) at the origin with the curve (2) is a contact of fourth order at the origin, and the values of a, b and h are given by (11).

Example 6. Prove that if the circle

$$lx + my + nz = 0,\ x^2 + y^2 + z^2 = 2cz,$$

has three-point contact with the paraboloid $ax^2 + by^2 = 2z$, then

$$c = \frac{l^2 + m^2}{bl^2 + am^2}.$$

Solution. The equation of the given circle is

$$lx + my + nz = 0,\ x^2 + y^2 + z^2 = 2cz. \tag{1}$$

The equation of the given paraboloid is

$$ax^2 + by^2 - 2z = 0. \tag{2}$$

Differentiating (1) with respect to t, we get

$$l\dot{x} + m\dot{y} + n\dot{z} = 0,$$

$$2x\dot{x} + 2y\dot{y} + 2z\dot{z} - 2c\dot{z} = 0$$

i.e., $\quad l\dot{x} + m\dot{y} + n\dot{z} = 0,$

$$x\dot{x} + y\dot{y} + (z - c)\dot{z} = 0.$$

Solving these for $\dot{x}, \dot{y}, \dot{z}$ we get

$$\frac{\dot{x}}{m(z-c) - ny} = \frac{\dot{y}}{nx - l(z-c)} = \frac{\dot{z}}{ly - mx} = \lambda, \text{ say}.$$

$\therefore \quad \dot{x} = \lambda\,(mz - ny - mc),\ \dot{y} = \lambda\,(nx - lz + lc),\ \dot{z} = \lambda\,(ly - mx). \quad (3)$

At the origin, these reduce to

$$\dot{x} = -\lambda mc,\ \dot{y} = \lambda lc,\ \dot{z} = 0. \tag{4}$$

Differentiating (3) again with respect to t, we get

$$\ddot{x} = \lambda\,(m\dot{z} - n\dot{y}),\ \ddot{y} = \lambda\,(n\dot{x} - l\dot{z}),\ \ddot{z} = \lambda\,(l\dot{y} - m\dot{x}).$$

At the origin, these reduce, with the help of (4), to

$$\ddot{x} = -\lambda^2 nlc,\ \ddot{y} = -\lambda^2 mnc,\ \ddot{z} = \lambda^2 (l^2 + m^2)\,c. \tag{5}$$

If the circle (1) has three-point contact with the paraboloid (2), then we must have

$$2ax\dot{x} + 2by\dot{y} - 2\dot{z} = 0$$

i.e., $\quad ax\dot{x} + by\dot{y} - \dot{z} = 0 \quad (6)$

and $\quad a\dot{x}^2 + ax\ddot{x} + b\dot{y}^2 + by\ddot{y} - \ddot{z} = 0. \quad (7)$

At the origin, (7) reduces to

$$a\dot{x}^2 + b\dot{y}^2 - \ddot{z} = 0. \tag{8}$$

Using (4) and (5) in (8), we get

$$a\lambda^2 m^2 c^2 + b\lambda^2 l^2 c^2 - \lambda^2 (l^2 + m^2)\,c = 0$$

or $\quad \lambda^2 c\,[c\,(am^2 + bl^2) - (l^2 + m^2)] = 0$

or $\quad c\,(am^2 + bl^2) - (l^2 + m^2) = 0$

or $\quad c = \dfrac{l^2 + m^2}{bl^2 + am^2}.$

Exercises

1. Prove that the length of the curve

$$x = 2a\left[\sin^{-1} t + t\sqrt{1-t^2}\right], y = 2at^2, z = 4at$$

between the points $t = t_1$ and $t = t_2$ is $4a\sqrt{2}(t_2 - t_1)$.

2. Show that the tangent at any point of the curve whose equations, referred to rectangular axes, are

$$x = 3t,\ y = 3t^2,\ z = 2t^3$$

makes a constant angle with the line $y = z - x = 0$.

Find the lines that have four point contact at $(0, 0, 1)$ with the surface

$$x^4 + 3xyz + x^2 - y^2 - z^2 + 2yz - 3xy - 2y + 2z = 1.$$

$$\left[\textbf{Ans. } \frac{x}{0} = \frac{y}{1} = \frac{z-1}{1},\ \frac{x}{1} = \frac{y}{0} = \frac{z-1}{\pm 1},\ \frac{x}{1} = \frac{y}{\pm 1} = \frac{z-1}{0}.\right]$$

3. Find the inflexional tangent at (x_1, y_1, z_1) on the surface $y^2 z = 4ax$.

$$\left[\textbf{Ans. } \frac{x - x_1}{3x_1} = \frac{y - y_1}{2y_1} = \frac{z - z_1}{-z_1}.\right]$$

1.6 OSCULATING PLANE OR PLANE OF CURVATURE

Definition. Let C be a curve and let P and Q be two neighbouring points on C. Then the osculating plane of C at P is the limiting position of the plane which contains the tangent line at P and also contains the point Q as $Q \to P$.

Alternative Definition. Let P, Q, R be three consecutive points on a curve C. Then the osculating plane of C at P is the limiting position of the plane PQR as Q and R tend to P.

Equation of Osculating Plane

Let the equation of a curve C be given by

$$\mathbf{r} = \mathbf{r}(s), \tag{1}$$

where $\mathbf{r}$ is the position vector of a current point on the curve and s is the arc length of this point from a fixed point on the curve.

Let P and Q be two neighbouring points on C with position vectors $\mathbf{r}(s)$ and $\mathbf{r}(s + \delta s)$ respectively. Let the position vector of a current point R on the osculating plane of C at P be $\mathbf{R}$. Then the vectors

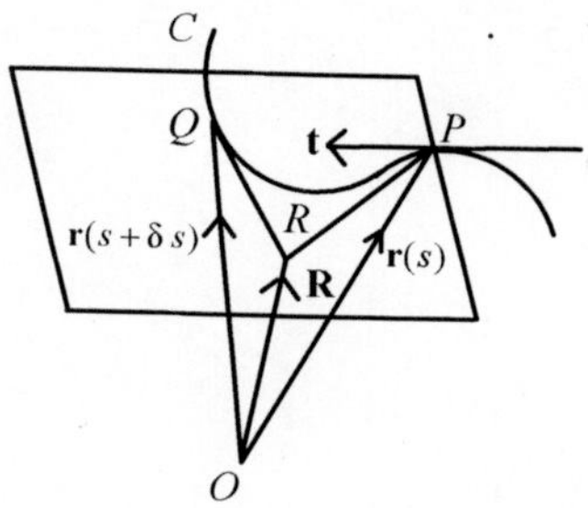

$$\overrightarrow{PR} = \mathbf{R} - \mathbf{r}(s),$$

$$\mathbf{t} = \mathbf{r}'(s)$$

and

$$\overrightarrow{PQ} = \mathbf{r}(s+\delta s) - \mathbf{r}(s)$$

lie in the plane RPQ. Hence their scalar triple product must be zero, i.e.,

$$[\mathbf{R} - \mathbf{r}(s)] \cdot \mathbf{r}'(s) \times [\mathbf{r}(s+\delta s) - \mathbf{r}(s)] = 0. \tag{2}$$

Now, expanding $\mathbf{r}(s+\delta s)$ by Taylor's theorem in ascending powers of δs, we have

$$\mathbf{r}(s+\delta s) = \mathbf{r}(s) + \delta s\, \mathbf{r}'(s) + \frac{(\delta s)^2}{2!}\mathbf{r}''(s) + O\{(\delta s)^3\}. \tag{3}$$

Using (3), (2) becomes

$$[\mathbf{R} - \mathbf{r}(s)] \cdot \mathbf{r}'(s) \times \left[\delta s\, \mathbf{r}'(s) + \frac{(\delta s)^2}{2!}\mathbf{r}''(s) + O\{(\delta s)^3\}\right] = 0$$

or $$[\mathbf{R} - \mathbf{r}(s)] \cdot \mathbf{r}'(s) \times \left[\frac{(\delta s)^2}{2!}\mathbf{r}''(s) + O\{(\delta s)^3\}\right] = 0$$

or $$[\mathbf{R} - \mathbf{r}(s)] \cdot \mathbf{r}'(s) \times [\mathbf{r}''(s) + O\{(\delta s)^3\}] = 0$$

or $$[\mathbf{R} - \mathbf{r}(s)] \cdot \mathbf{r}'(s) \times \mathbf{r}''(s) = 0$$

or $$[\mathbf{R} - \mathbf{r}(s),\ \mathbf{r}'(s),\ \mathbf{r}''(s)] = 0. \tag{4}$$

This is the equation of the osculating plane of C at P.

Equation of Osculating Plane in terms of General Parameter

Let the equation of a curve C be given by

$$\mathbf{r} = \mathbf{r}(t). \tag{5}$$

Then

$$\mathbf{r}' = \frac{d\mathbf{r}}{ds} = \frac{d\mathbf{r}}{dt}\Big/\frac{ds}{dt} = \frac{\dot{\mathbf{r}}}{\dot{s}},$$

$$\mathbf{r}'' = \frac{d}{ds}\left(\frac{d\mathbf{r}}{ds}\right) = \frac{d}{dt}\left(\frac{d\mathbf{r}}{ds}\right)\frac{dt}{ds} = \left[\frac{d}{dt}\left(\frac{\dot{\mathbf{r}}}{\dot{s}}\right)\right]\frac{1}{\dot{s}} = \frac{\ddot{\mathbf{r}}}{\dot{s}^2} - \frac{\dot{\mathbf{r}}\ddot{s}}{\dot{s}^3}.$$

Substituting the above values in (4), we get

$$\left[\mathbf{R} - \mathbf{r},\ \frac{\dot{\mathbf{r}}}{\dot{s}},\ \frac{\ddot{\mathbf{r}}}{\dot{s}^2} - \frac{\dot{\mathbf{r}}\ddot{s}}{\dot{s}^3}\right] = 0$$

or $$(\mathbf{R}-\mathbf{r}) \cdot \frac{\dot{\mathbf{r}}}{\dot{s}} \times \left(\frac{\ddot{\mathbf{r}}}{\dot{s}^2} - \frac{\dot{\mathbf{r}}\ddot{s}}{\dot{s}^3} \right) = 0$$

or $$(\mathbf{R}-\mathbf{r}) \cdot \frac{\dot{\mathbf{r}}}{\dot{s}} \times \frac{\ddot{\mathbf{r}}}{\dot{s}^2} = 0$$

or $$(\mathbf{R}-\mathbf{r}) \cdot \dot{\mathbf{r}} \times \ddot{\mathbf{r}} = 0$$

or $$[\mathbf{R}-\mathbf{r},\ \dot{\mathbf{r}},\ \ddot{\mathbf{r}}] = 0. \tag{6}$$

This is the equation of the osculating plane for the curve (5) at P.

Cartesian Form. If (x, y, z) are the coordinates of P referred to fixed rectangular axes and (X, Y, Z) are the coordinates of the current point R, then

$$\mathbf{r} = x\mathbf{i} + y\mathbf{j} + z\mathbf{k} \quad \text{and} \quad \mathbf{R} = X\mathbf{i} + Y\mathbf{j} + Z\mathbf{k}.$$

$$\therefore \quad \mathbf{R}-\mathbf{r} = (X-x)\,\mathbf{i} + (Y-y)\,\mathbf{j} + (Z-z)\,\mathbf{k}.$$

Also, $\dot{\mathbf{r}} = \dot{x}\mathbf{i} + \dot{y}\mathbf{j} + \dot{z}\mathbf{k}$ and $\ddot{\mathbf{r}} = \ddot{x}\mathbf{i} + \ddot{y}\mathbf{j} + \ddot{z}\mathbf{k}$.

Hence equation (6) is equivalent to

$$\begin{vmatrix} X-x & Y-y & Z-z \\ \dot{x} & \dot{y} & \dot{z} \\ \ddot{x} & \ddot{y} & \ddot{z} \end{vmatrix} = 0.$$

Osculating Plane at the Point of Inflexion

Definition. A point P is called the point of inflexion if $\mathbf{r}'' = \mathbf{0}$ at P. In this case, the tangent line at P is called the inflexional tangent.

Theorem. When the curve is analytic, there exists a definite osculating plane at a point of inflexion P, provided the curve is not a straight line.

Proof. Since $\mathbf{r}' (= \mathbf{t})$ is a unit tangent vector, we have

$$\mathbf{r}'^2 = 1.$$

Differentiating with respect to s, we get

$$2\mathbf{r}' \cdot \mathbf{r}'' = 0$$

or $$\mathbf{r}' \cdot \mathbf{r}'' = 0.$$

Differentiating this again with respect to s, we get

$$\mathbf{r}'' \cdot \mathbf{r}'' + \mathbf{r}' \cdot \mathbf{r}''' = 0$$

or $$\mathbf{r}' \cdot \mathbf{r}''' = 0. \qquad [\because \mathbf{r}'' = \mathbf{0} \text{ at the point of inflexion}]$$

This shows that $\mathbf{r}'$ is linearly independent of $\mathbf{r}'''$, except when $\mathbf{r}''' = \mathbf{0}$. Continuing this argument, we shall arrive at the result

$$\mathbf{r}' \cdot \mathbf{r}^k = 0,$$

where $\mathbf{r}^k\,(k \geq 2)$ is the first non zero derivative of $\mathbf{r}$ at P. Therefore, from (3), we have

$$\mathbf{r}\,(s+\delta s) = \mathbf{r}\,(s) + \frac{(\delta s)^k}{k\,!}\mathbf{r}^k\,(s) + O\{\,(\delta s)^{k+1}\}. \tag{7}$$

Using (7), (2) becomes

$$[\mathbf{R} - \mathbf{r}\,(s),\;\; \mathbf{r}'\,(s),\;\; \mathbf{r}^k\,(s)] = 0. \tag{8}$$

If $\mathbf{r}^k = \mathbf{0}$ for all $k \geq 2$, then $\mathbf{r}'\,(=\mathbf{t})$ is constant (since the curve under consideration is analytic), i.e., the tangent vector is same at each point of the curve. Hence the curve is a straight line.

Therefore equation (8) is the equation of the osculating plane at a point of inflexion P if the curve is not a straight line.

1.7 OSCULATING PLANE AT A POINT OF THE CURVE OF INTERSECTION OF TWO SURFACES

Let the equations of the surfaces be

$$f(\mathbf{r}) = 0,\;\; \phi(\mathbf{r}) = 0. \tag{1}$$

We know that the tangent vector $\dot{\mathbf{r}}$ to the curve of intersection of the surfaces given by (1) at $P\,(\mathbf{r})$ is at right angles to their respective normals $\nabla f,\; \nabla\phi.$ Therefore

$$\dot{\mathbf{r}}.\,\nabla f = 0,\;\; \dot{\mathbf{r}}.\,\nabla\phi = 0 \tag{2}$$

$\Rightarrow$ $\dot{\mathbf{r}}$ is perpendicular to each of ∇f and $\nabla\phi$.

$\Rightarrow$ $\dot{\mathbf{r}}$ is parallel to $\nabla f \times \nabla\phi$.

$$\Rightarrow \quad \lambda\dot{\mathbf{r}} = \nabla f \times \nabla\phi. \tag{3}$$

Differentiating (3) with respect to t, we get

$$\lambda\ddot{\mathbf{r}} + \dot{\lambda}\dot{\mathbf{r}} = (\dot{\nabla f}) \times \nabla\phi + \nabla f \times (\dot{\nabla\phi}).$$

Taking the cross product of both sides with $\dot{\mathbf{r}},$ we get

$$\dot{\mathbf{r}} \times (\lambda\ddot{\mathbf{r}} + \dot{\lambda}\dot{\mathbf{r}}) = \dot{\mathbf{r}} \times [(\dot{\nabla f}) \times \nabla\phi + \nabla f \times (\dot{\nabla\phi})]$$

$$\begin{aligned}\Rightarrow \quad \lambda\dot{\mathbf{r}} \times \ddot{\mathbf{r}} &= \dot{\mathbf{r}} \times [(\dot{\nabla f}) \times (\nabla\phi)] + \dot{\mathbf{r}} \times [(\nabla f) \times (\dot{\nabla\phi})] \\ &= [(\dot{\mathbf{r}}.\,\nabla\phi)\,(\dot{\nabla f}) - (\dot{\mathbf{r}}.\,(\dot{\nabla f}))\nabla\phi] \\ &\qquad + [(\dot{\mathbf{r}}.\,(\dot{\nabla\phi}))\nabla f - (\dot{\mathbf{r}}.\,\nabla f)\,(\dot{\nabla\phi})] \\ &= -(\dot{\mathbf{r}}.\,(\dot{\nabla f}))\nabla\phi + (\dot{\mathbf{r}}.\,(\dot{\nabla\phi}))\nabla f \qquad \text{[using (2)]}\end{aligned}$$

$$\Rightarrow \quad \dot{\mathbf{r}} \times \ddot{\mathbf{r}} = \frac{(\dot{\mathbf{r}}.\,(\dot{\nabla\phi}))\nabla f - (\dot{\mathbf{r}}.\,(\dot{\nabla f}))\nabla\phi}{\lambda} \tag{4}$$

Now, the equation of the osculating plane at $P\,(\mathbf{r})$ is

$$(\mathbf{R}-\mathbf{r})\,.\,\dot{\mathbf{r}}\times\ddot{\mathbf{r}}=0$$

or $$(\mathbf{R}-\mathbf{r})\,.\,[(\dot{\mathbf{r}}\,.\,(\dot{\nabla\phi}))\nabla f-(\dot{\mathbf{r}}\,.\,(\dot{\nabla f}))\nabla\phi]=0 \qquad \text{[using (4)]}$$

or $$(\mathbf{R}-\mathbf{r})\,.\,[(\dot{\mathbf{r}}\,.\,(\dot{\nabla\phi}))\nabla f]=(\mathbf{R}-\mathbf{r})\,.\,[(\dot{\mathbf{r}}\,.\,(\dot{\nabla f}))\nabla\phi]$$

or $$\frac{(\mathbf{R}-\mathbf{r})\,.\,\nabla f}{\dot{\mathbf{r}}\,.\,(\dot{\nabla f})}=\frac{(\mathbf{R}-\mathbf{r})\,.\,\nabla\phi}{\dot{\mathbf{r}}\,.\,(\dot{\nabla\phi})}. \qquad (5)$$

Cartesian Form. If we write

$$f(\mathbf{r})=f(x,y,z),\ \ \phi(\mathbf{r})=\phi(x,y,z),$$

$$\mathbf{r}=x\mathbf{i}+y\mathbf{j}+z\mathbf{k},\quad \mathbf{R}=X\mathbf{i}+Y\mathbf{j}+Z\mathbf{k},$$

$$\nabla f=\frac{\partial f}{\partial x}\mathbf{i}+\frac{\partial f}{\partial y}\mathbf{j}+\frac{\partial f}{\partial z}\mathbf{k},\ \ \nabla\phi=\frac{\partial\phi}{\partial x}\mathbf{i}+\frac{\partial\phi}{\partial y}\mathbf{j}+\frac{\partial\phi}{\partial z}\mathbf{k},$$

then

$$\dot{\mathbf{r}}=\dot{x}\mathbf{i}+\dot{y}\mathbf{j}+\dot{z}\mathbf{k},$$

$$(\dot{\nabla f})=\sum\left(\frac{\partial^2 f}{\partial x^2}\dot{x}+\frac{\partial^2 f}{\partial y\,\partial x}\dot{y}+\frac{\partial^2 f}{\partial z\,\partial x}\dot{z}\right)\mathbf{i},$$

$$(\dot{\nabla\phi})=\sum\left(\frac{\partial^2 \phi}{\partial x^2}\dot{x}+\frac{\partial^2 \phi}{\partial y\,\partial x}\dot{y}+\frac{\partial^2 \phi}{\partial z\,\partial x}\dot{z}\right)\mathbf{i}.$$

Substituting the above values in (5), we get

$$\frac{(X-x)\dfrac{\partial f}{\partial x}+(Y-y)\dfrac{\partial f}{\partial y}+(Z-z)\dfrac{\partial f}{\partial z}}{\dot{x}^2\dfrac{\partial^2 f}{\partial x^2}+\ldots+2\dot{y}\dot{z}\dfrac{\partial^2 f}{\partial y\,\partial z}+\ldots}$$

$$=\frac{(X-x)\dfrac{\partial\phi}{\partial x}+(Y-y)\dfrac{\partial\phi}{\partial y}+(Z-z)\dfrac{\partial\phi}{\partial z}}{\dot{x}^2\dfrac{\partial^2\phi}{\partial x^2}+\ldots+2\dot{y}\dot{z}\dfrac{\partial^2\phi}{\partial y\,\partial z}+\ldots}. \qquad (6)$$

1.8 NORMAL PLANE

Definition. Let C be a space curve and let P be any point on it. Let $\mathbf{t}$ be the unit tangent to C at P.

The infinitely many lines passing through P and each perpendicular to $\mathbf{t}$ are known as the normals to C at P.

All the normals to C at P lie in a plane passing through P and perpendicular to $\mathbf{t}$. This plane is known as the normal plane of C at P.

Equation of Normal Plane

Let the equation of a curve C be given by

$$\mathbf{r} = \mathbf{r}(s), \tag{1}$$

where $\mathbf{r}$ is the position vector of a current point P on the curve and s denotes the arc length of P from a fixed point on the curve.

If $\mathbf{R}$ is the position vector of a current point R on the normal plane of C at P, then the vector $\mathbf{R} - \mathbf{r}$ lies in the normal plane. But the unit tangent $\mathbf{t}$ is perpendicular to the normal plane. Therefore, we have

$$(\mathbf{R} - \mathbf{r}) \cdot \mathbf{t} = 0. \tag{2}$$

This is the equation of the normal plane of C at P.

1.9 PRINCIPAL NORMAL AND BINORMAL

Principal Normal. Definition.

The normal which lies in the osculating plane at a point P of a curve C is called the principal normal to C at P. Obviously, the principal normal is the line of intersection of the osculating plane and the normal plane to C at P.

Since the principal normal lies in the osculating plane, it follows that the principal normal is perpendicular to the unit tangent $\mathbf{t}$ which lies in the osculating plane.

We denote the unit vector along the principal normal by $\mathbf{n}$.

Equation of Principal Normal.

Let the equation of a curve C be given by

$$\mathbf{r} = \mathbf{r}(s), \tag{1}$$

where the letters $\mathbf{r}$ and s have their usual meaning.

If $\mathbf{R}$ is the position vector of a current point R on the principal normal of C at P, then

$$\vec{OR} = \vec{OP} + \vec{PR}$$

i.e.,
$$\mathbf{R} = \mathbf{r} + \lambda \mathbf{n}, \tag{2}$$

where λ is some scalar.

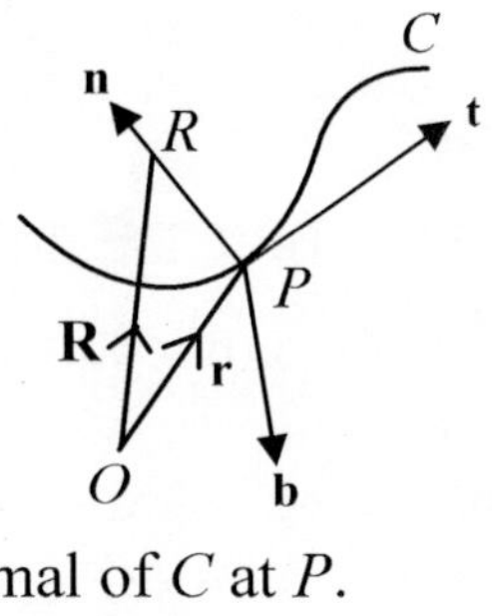

Equation (2) is the equation of the principal normal of C at P.

Binormal. Definition.

The normal which is perpendicular to the osculating plane at a point P of a curve C is called the binormal to C at P. The unit vector along the binormal is denoted by $\mathbf{b}$.

The sense of $\mathbf{b}$ is chosen in such a manner that the triad $\mathbf{t}, \mathbf{n}, \mathbf{b}$ form a right handed system of axes, i.e.,

$$\mathbf{t} \times \mathbf{n} = \mathbf{b}, \mathbf{n} \times \mathbf{b} = \mathbf{t}, \mathbf{b} \times \mathbf{t} = \mathbf{n};$$

$$\mathbf{t} \cdot \mathbf{n} = 0, \ \mathbf{n} \cdot \mathbf{b} = 0, \ \mathbf{b} \cdot \mathbf{t} = 0.$$

Equation of Binormal.

Let the equation of a curve C be given by

$$\mathbf{r} = \mathbf{r}(s), \tag{1}$$

where the letters $\mathbf{r}$ and s have their usual meaning.

If $\mathbf{R}$ is the position vector of a current point R on the binormal of C at P, then

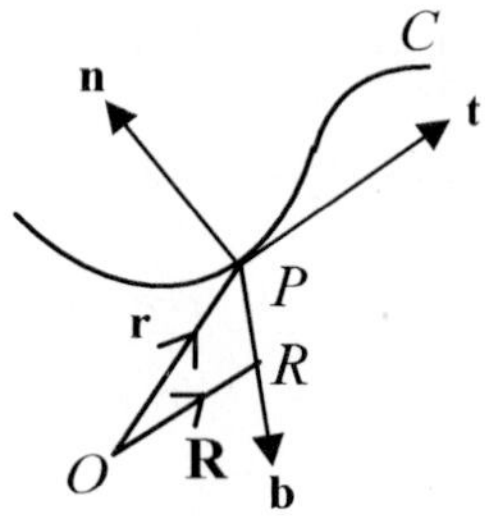

$$\vec{OR} = \vec{OP} + \vec{PR}$$

i.e., $\mathbf{R} = \mathbf{r} + \lambda \mathbf{b}$, (2)

where λ is some scalar.

Equation (2) is the equation of the binormal of C at P.

1.10 RECTIFYING PLANE

Definition. The plane containing the tangent and the binormal to a curve C at a point P is called the rectifying plane of C at P.

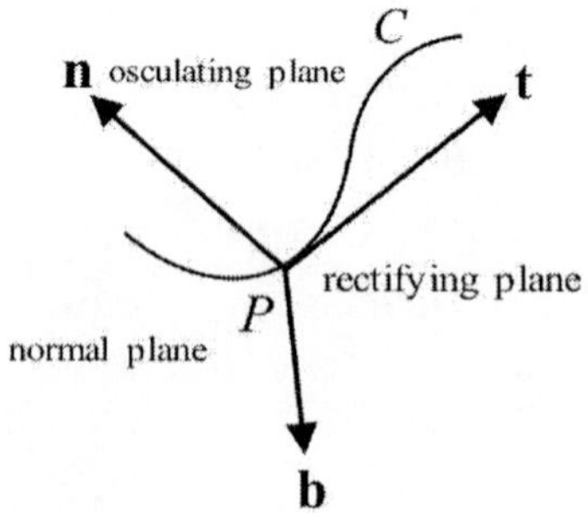

Equation of Rectifying Plane

Let the equation of a curve C be given by

$$\mathbf{r} = \mathbf{r}(s), \tag{1}$$

where the letters $\mathbf{r}$ and s have their usual meaning.

If $\mathbf{R}$ is the position vector of a current point R on the rectifying plane of C at P, then the vector $\mathbf{R} - \mathbf{r}$ lies in the rectifying plane. But the unit principal normal $\mathbf{n}$ is perpendicular to the rectifying plane. Therefore, we have

$$(\mathbf{R} - \mathbf{r}) \cdot \mathbf{n} = 0. \tag{2}$$

This is the equation of the rectifying plane of C at P.

Remark. At each point of a curve, the moving triad of unit vectors $\mathbf{t}, \mathbf{n}, \mathbf{b}$ determines three mutually perpendicular planes. The equations of these planes may be written as

$$(\mathbf{R}-\mathbf{r}) \cdot \mathbf{b} = 0, \qquad \text{(osculating plane)}$$
$$(\mathbf{R}-\mathbf{r}) \cdot \mathbf{t} = 0, \qquad \text{(normal plane)}$$
$$(\mathbf{R}-\mathbf{r}) \cdot \mathbf{n} = 0. \qquad \text{(rectifying plane)}$$

These planes are also called fundamental planes.

1.11 CURVATURE, TORSION AND SCREW CURVATURE

Curvature.

The arc rate of rotation of the tangent is known as the curvature or the circular curvature.

If the unit tangent corresponding to the arcual distance s at a point P on a space curve C be $\mathbf{t}$, then the curvature at P is given by

$$\left| \frac{d\mathbf{t}}{ds} \right| = \kappa.$$

The vector $\dfrac{d\mathbf{t}}{ds}$ is known as the curvature vector.

Let P and Q be any two neighbouring points on a space curve C, and let $\mathbf{t}$ and $\mathbf{t}+\delta\mathbf{t}$ denote unit tangents at P and Q respectively. Also, let s and $s+\delta s$ denote the arcual distances of these points respectively from a fixed point on C.

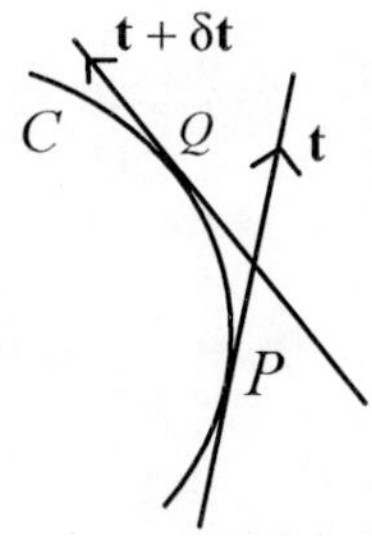

In $\Delta\, EFG$, we have

$$FG^2 = EF^2 + EG^2 - 2(EF)(EG)\cos\delta\theta$$
$$\Rightarrow \quad |\delta\mathbf{t}|^2 = |\mathbf{t}|^2 + |\mathbf{t}+\delta\mathbf{t}|^2 - 2|\mathbf{t}|\,|\mathbf{t}+\delta\mathbf{t}|\cos\delta\theta$$
$$\Rightarrow \quad |\delta\mathbf{t}|^2 = 1+1-2\cos\delta\theta$$
$$= 2(1-\cos\delta\theta)$$
$$= 2\left(2\sin^2\frac{\delta\theta}{2}\right)$$
$$= 4\sin^2\frac{\delta\theta}{2}$$

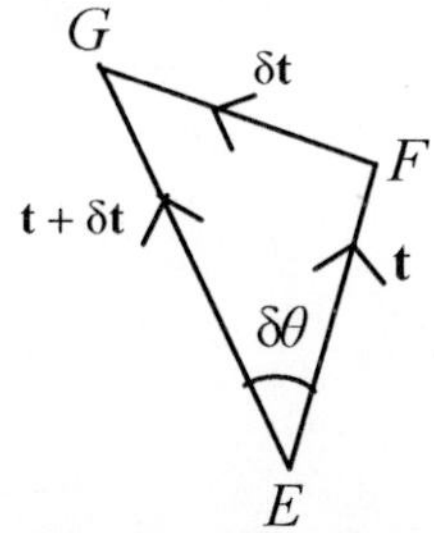

$$\Rightarrow \quad \left|\frac{\delta \mathbf{t}}{\delta s}\right|^2 = \left(\frac{\sin\frac{\delta\theta}{2}}{\frac{\delta\theta}{2}}\,\frac{\delta\theta}{\delta s}\right)^2$$

$$\Rightarrow \quad \left|\frac{\delta \mathbf{t}}{\delta s}\right| = \frac{\sin\frac{\delta\theta}{2}}{\frac{\delta\theta}{2}}\,\frac{\delta\theta}{\delta s}$$

$$\Rightarrow \quad \lim_{\delta s\to 0}\left|\frac{\delta \mathbf{t}}{\delta s}\right| = \lim_{\delta\theta\to 0}\left(\frac{\sin\frac{\delta\theta}{2}}{\frac{\delta\theta}{2}}\right)\lim_{\delta s\to 0}\left(\frac{\delta\theta}{\delta s}\right)$$

$$\Rightarrow \quad \left|\frac{d\mathbf{t}}{ds}\right| = \frac{d\theta}{ds}$$

$$\Rightarrow \quad \kappa = \frac{d\theta}{ds}.$$

The reciprocal of the curvature is known as the radius of curvature and it is denoted by ρ, i.e.,

$$\rho = \frac{1}{\kappa}.$$

Theorem 1. A necessary and sufficient condition for a curve to be a straight line is that the curvature $\kappa = 0$ at all points of the curve.

Proof. Necessary condition. Let the curve be a straight line.

To prove : $\kappa = 0$ at all points of the curve

We know that the vector equation of a straight line passing through a point whose position vector is **a** and parallel to a vector **c** is

$$\mathbf{r} = \mathbf{a} + s\mathbf{c}.$$

$$\Rightarrow \quad \frac{d\mathbf{r}}{ds} = \mathbf{c} \qquad [\because \mathbf{a} \text{ and } \mathbf{c} \text{ are constant vectors}]$$

$$\Rightarrow \quad \mathbf{t} = \mathbf{c}$$

$$\Rightarrow \quad \frac{d\mathbf{t}}{ds} = \mathbf{0}$$

$$\Rightarrow \quad \left|\frac{d\mathbf{t}}{ds}\right| = 0$$

$$\Rightarrow \quad \kappa = 0. \qquad \left[\because \left|\frac{d\mathbf{t}}{ds}\right| = \kappa\right]$$

Sufficient condition. Let $\kappa = 0$ at all points of the curve.
To prove : The curve is a straight line.
Since $\kappa = 0$, we have

$$\left|\frac{d\mathbf{t}}{ds}\right| = 0 \qquad \left[\because \left|\frac{d\mathbf{t}}{ds}\right| = \kappa\right]$$

$\Rightarrow \quad \dfrac{d\mathbf{t}}{ds} = \mathbf{0}$

$\Rightarrow \quad \displaystyle\int \frac{d\mathbf{t}}{ds} ds = \mathbf{c}$, where $\mathbf{c}$ is the constant of integration.

$\Rightarrow \quad \mathbf{t} = \mathbf{c}$

$\Rightarrow \quad \dfrac{d\mathbf{r}}{ds} = \mathbf{c}$

$\Rightarrow \quad \displaystyle\int \frac{d\mathbf{r}}{ds} ds = \int \mathbf{c}\, ds + \mathbf{a}$, where $\mathbf{a} =$ constant of integration

$\Rightarrow \quad \mathbf{r} = \mathbf{c}s + \mathbf{a}$

$\Rightarrow \quad \mathbf{r} = \mathbf{a} + s\mathbf{c}$,

which represents a straight line passing through a point whose position vector is **a** and parallel to a vector **c**.

Torsion

The arc rate of rotation of the binormal is known as the torsion.
If P be a point on a space curve C corresponding to the arc length s and if **b** denotes the unit binormal at P, then the torsion at P is given by

$$\left|\frac{d\mathbf{b}}{ds}\right| = \tau.$$

The reciprocal of the torsion is known as the radius of torsion and it is denoted by σ, i.e.,

$$\sigma = \frac{1}{\tau}.$$

Let P and Q be any two contiguous points on a space curve C corresponding to the arcual distances s and $s + \delta s$ respectively, and let **b** and $\mathbf{b} + \delta\mathbf{b}$ denote unit binormals at P and Q respectively.

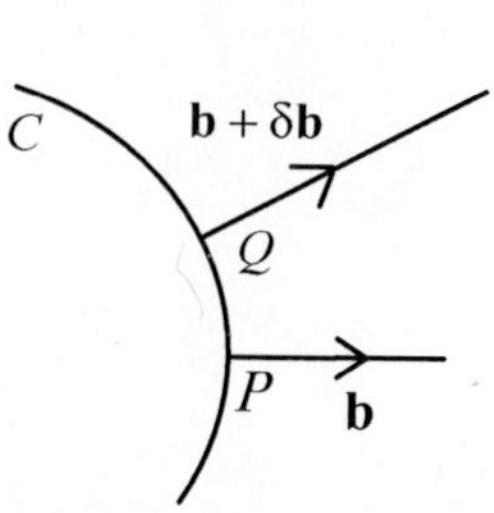

Also, let $\delta\phi$ = angle between the binormals at P and Q.

Then, as earlier, it can be proved that

$$\tau = \frac{d\phi}{ds}.$$

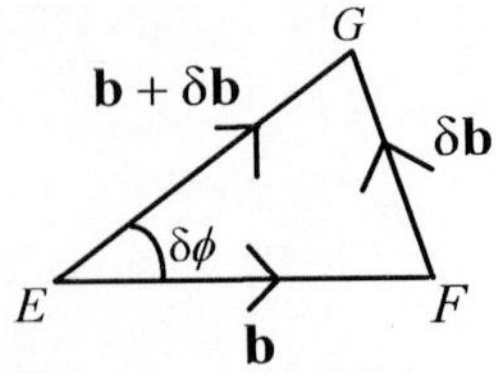

Theorem 2. A necessary and sufficient condition for a curve to be a plane curve is that the torsion $\tau = 0$ at all points of the curve.

Proof. Necessary condition. Let the curve be a plane curve.

To prove : $\tau = 0$ at all points of the curve.

Since the given curve is a plane curve, it follows that the plane containing the curve is the osculating plane for all the points of the curve. Therefore, we have

$$\mathbf{b} = \text{a constant vector}$$

$$\Rightarrow \quad \frac{d\mathbf{b}}{ds} = \mathbf{0}$$

$$\Rightarrow \quad \left|\frac{d\mathbf{b}}{ds}\right| = 0$$

$$\Rightarrow \quad \tau = 0. \qquad \left[\because \left|\frac{d\mathbf{b}}{ds}\right| = \tau\right]$$

Sufficient condition. Let $\tau = 0$ at all points of the curve.

To prove : The curve is a plane curve.

Since $\tau = 0$, we have

$$\left|\frac{d\mathbf{b}}{ds}\right| = 0 \qquad \left[\because \left|\frac{d\mathbf{b}}{ds}\right| = \tau\right]$$

$$\Rightarrow \quad \frac{d\mathbf{b}}{ds} = \mathbf{0}$$

$$\Rightarrow \quad \int \frac{d\mathbf{b}}{ds}\, ds = \text{constant of integration}$$

$$\Rightarrow \quad \mathbf{b} = \text{a constant vector.}$$

Now,

$$\frac{d}{ds}(\mathbf{r}.\mathbf{b}) = \frac{d\mathbf{r}}{ds}.\mathbf{b} + \mathbf{r}.\frac{d\mathbf{b}}{ds}$$

$$= \mathbf{t}.\mathbf{b} + \mathbf{r}.\mathbf{0} \qquad \left[\because \frac{d\mathbf{b}}{ds} = \mathbf{0} \text{ and } \frac{d\mathbf{r}}{ds} = \mathbf{t}\right]$$

$$= 0 \qquad [\because \mathbf{t} \text{ and } \mathbf{b} \text{ are mutually perpendicular}]$$

$\Rightarrow \quad \mathbf{r}.\mathbf{b} = \text{constant.} \qquad (1)$

Since **b** is a constant vector of magnitude unity, it follows that $\mathbf{r}.\mathbf{b}$ is the projection of the position vector **r** on **b** and is therefore the same at all points of the curve given by (1). This implies that the curve must lie in a plane.

Screw Curvature

The arc rate of rotation of the principal normal is known as the screw curvature.

If P be a point on a space curve C corresponding to the arc length s and if **n** denotes the unit principal normal at P, then the screw curvature at P is given by

$$\left|\frac{d\mathbf{n}}{ds}\right| = \sqrt{\kappa^2 + \tau^2}.$$

1.12 SERRET-FRENET FORMULAE

Let C be a space curve and let P be any point on it whose arcual distance from a fixed point A on C is s. Let **t**, **n**, **b** be the unit tangent, unit principal normal and unit binormal respectively of C at P. Also, let κ and τ be the curvature and torsion respectively of C at P.

Now, we have

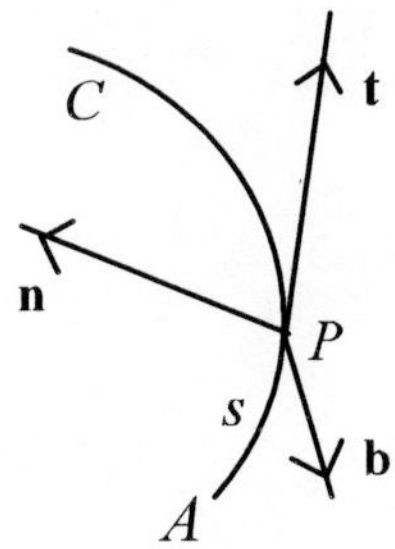

$$\mathbf{t}.\mathbf{t} = |\mathbf{t}|^2 = 1$$

$$\Rightarrow \quad \frac{d}{ds}(\mathbf{t}.\mathbf{t}) = 0$$

$$\Rightarrow \quad \frac{d\mathbf{t}}{ds}.\mathbf{t} + \mathbf{t}.\frac{d\mathbf{t}}{ds} = 0$$

$$\Rightarrow \quad \frac{d\mathbf{t}}{ds}.\mathbf{t} + \frac{d\mathbf{t}}{ds}.\mathbf{t} = 0 \quad [\because \text{Scalar product of two vectors is commutative}]$$

$$\Rightarrow \quad 2\frac{d\mathbf{t}}{ds}.\mathbf{t} = 0$$

$$\Rightarrow \quad \frac{d\mathbf{t}}{ds}.\mathbf{t} = 0$$

$$\Rightarrow \quad \frac{d\mathbf{t}}{ds} = \mathbf{t}' \text{ is perpendicular to } \mathbf{t}. \qquad (1)$$

Also, we know that

$$\frac{d\mathbf{t}}{ds} \text{ lies in the osculating plane of } C \text{ at } P. \tag{2}$$

$$(1) \text{ and } (2) \Rightarrow \frac{d\mathbf{t}}{ds} \text{ is parallel to } \mathbf{n}. \tag{3}$$

Also, by definition,

$$\left|\frac{d\mathbf{t}}{ds}\right| = \kappa. \tag{4}$$

$$(3) \text{ and } (4) \Rightarrow \frac{d\mathbf{t}}{ds} = \pm\, \kappa\mathbf{n}.$$

We make a convention that

$$\frac{d\mathbf{t}}{ds} = \kappa\mathbf{n}. \tag{5}$$

Again,

$$\mathbf{b}\,.\,\mathbf{b} = |\,\mathbf{b}\,|^2 = 1$$

$$\Rightarrow \quad \frac{d}{ds}(\mathbf{b}\,.\,\mathbf{b}\,) = 0$$

$$\Rightarrow \quad \frac{d\mathbf{b}}{ds}\,.\,\mathbf{b} + \mathbf{b}\,.\,\frac{d\mathbf{b}}{ds} = 0$$

$$\Rightarrow \quad \frac{d\mathbf{b}}{ds}\,.\,\mathbf{b} + \frac{d\mathbf{b}}{ds}\,.\,\mathbf{b} = 0$$

[∵ Scalar product of two vectors is commutative]

$$\Rightarrow \quad 2\frac{d\mathbf{b}}{ds}\,.\,\mathbf{b} = 0$$

$$\Rightarrow \quad \frac{d\mathbf{b}}{ds}\,.\,\mathbf{b} = 0$$

$$\Rightarrow \quad \frac{d\mathbf{b}}{ds} = \mathbf{b}' \text{ is perpendicular to } \mathbf{b}. \tag{6}$$

Also,

$$\mathbf{b}\,.\,\mathbf{t} = 0$$

$$\Rightarrow \quad \frac{d}{ds}(\mathbf{b}\,.\,\mathbf{t}\,) = 0$$

$$\Rightarrow \quad \frac{d\mathbf{b}}{ds}\,.\,\mathbf{t} + \mathbf{b}\,.\,\frac{d\mathbf{t}}{ds} = 0$$

$$\Rightarrow \quad \frac{d\mathbf{b}}{ds}\,.\,\mathbf{t} + \mathbf{b}\,.\,\kappa\mathbf{n} = 0$$

$$\Rightarrow \quad \frac{d\mathbf{b}}{ds}\,.\,\mathbf{t} + \kappa\,(\mathbf{b}\,.\,\mathbf{n}) = 0$$

$$\Rightarrow \quad \frac{d\mathbf{b}}{ds}\,.\,\mathbf{t} = 0 \qquad [\because \mathbf{b} \text{ is perpendicular to } \mathbf{n}]$$

$$\Rightarrow \quad \frac{d\mathbf{b}}{ds} \text{ is perpendicular to } \mathbf{t}. \tag{7}$$

$$(6) \text{ and } (7) \Rightarrow \frac{d\mathbf{b}}{ds} \text{ is parallel to } \mathbf{n}. \tag{8}$$

Also, by definition,

$$\left|\frac{d\mathbf{b}}{ds}\right| = \tau. \tag{9}$$

$$(8) \text{ and } (9) \Rightarrow \frac{d\mathbf{b}}{ds} = \pm\,\tau\mathbf{n}.$$

By convention, we write

$$\frac{d\mathbf{b}}{ds} = -\,\tau\,\mathbf{n}. \tag{10}$$

Again, we have

$$\mathbf{n} = \mathbf{b} \times \mathbf{t}$$

$$\begin{aligned}
\Rightarrow \quad \frac{d\mathbf{n}}{ds} &= \frac{d}{ds}(\mathbf{b} \times \mathbf{t}) \\
&= \frac{d\mathbf{b}}{ds} \times \mathbf{t} + \mathbf{b} \times \frac{d\mathbf{t}}{ds} \\
&= (-\,\tau\mathbf{n}) \times \mathbf{t} + \mathbf{b} \times \kappa\mathbf{n} \qquad [\text{using (5) and (10)}] \\
&= -\,\tau\,(\mathbf{n} \times \mathbf{t}) + \kappa\,(\mathbf{b} \times \mathbf{n}) \\
&= -\,\tau\,(-\,\mathbf{b}) + \kappa\,(-\,\mathbf{t})
\end{aligned}$$

$$\Rightarrow \quad \frac{d\mathbf{n}}{ds} = \tau\,\mathbf{b} - \kappa\mathbf{t}. \tag{11}$$

The results (5), (10) and (11) are known as Serret-Frenet formulae.

1.13 CURVATURE AND TORSION OF THE CURVE $\mathbf{r} = \mathbf{r}(t)$

We have

$$\dot{\mathbf{r}} = \frac{d\mathbf{r}}{dt} = \frac{d\mathbf{r}}{ds}\frac{ds}{dt} = \mathbf{t}\,\dot{s} \tag{1}$$

$$\Rightarrow \quad |\dot{\mathbf{r}}| = |\mathbf{t}\dot{s}| = |\mathbf{t}|\,\dot{s} = \dot{s}. \tag{2}$$

Differentiating (1) w. r. to t, we get

$$\ddot{\mathbf{r}} = \frac{d}{dt}(\mathbf{t}\dot{s})$$
$$= \frac{d\mathbf{t}}{dt}\dot{s} + \mathbf{t}\ddot{s}$$
$$= \frac{d\mathbf{t}}{ds}\frac{ds}{dt}\dot{s} + \mathbf{t}\ddot{s}$$
$$= \dot{s}^2\frac{d\mathbf{t}}{ds} + \mathbf{t}\ddot{s}$$

i.e., $$\ddot{\mathbf{r}} = \dot{s}^2\kappa\mathbf{n} + \mathbf{t}\ddot{s}. \quad (3)$$

Taking the cross product of (1) and (3), we get

$$\dot{\mathbf{r}} \times \ddot{\mathbf{r}} = \kappa\dot{s}^3\mathbf{b} \quad (4)$$
$$\Rightarrow \quad |\dot{\mathbf{r}} \times \ddot{\mathbf{r}}| = \kappa\dot{s}^3. \quad (5)$$

Differentiating (4) w. r. to t, we get

$$\frac{d}{dt}(\dot{\mathbf{r}} \times \ddot{\mathbf{r}}) = \frac{d}{dt}(\kappa\dot{s}^3\mathbf{b})$$

or $$\ddot{\mathbf{r}} \times \ddot{\mathbf{r}} + \dot{\mathbf{r}} \times \dddot{\mathbf{r}} = \left[\frac{d}{dt}(\kappa\dot{s}^3)\right]\mathbf{b} + \kappa\dot{s}^3\frac{d\mathbf{b}}{dt}$$

or $$\dot{\mathbf{r}} \times \dddot{\mathbf{r}} = \left[\frac{d}{dt}(\kappa\dot{s}^3)\right]\mathbf{b} + \kappa\dot{s}^3(-\tau\mathbf{n})\,\dot{s}$$

or $$\dot{\mathbf{r}} \times \dddot{\mathbf{r}} = \left[\frac{d}{dt}(\kappa\dot{s}^3)\right]\mathbf{b} - \kappa\tau\dot{s}^4\mathbf{n}. \quad (6)$$

Taking the dot product of (3) and (6), we have

$$\ddot{\mathbf{r}}\,.\,(\dot{\mathbf{r}} \times \dddot{\mathbf{r}}) = (\dot{s}^2\kappa\mathbf{n} + \mathbf{t}\ddot{s})\,.\left[\left(\frac{d}{dt}(\kappa\dot{s}^3)\right)\mathbf{b} - \kappa\tau\dot{s}^4\mathbf{n}\right]$$

or $$[\ddot{\mathbf{r}},\ \dot{\mathbf{r}},\ \dddot{\mathbf{r}}] = -\kappa^2\tau\dot{s}^6$$

or $$-[\dot{\mathbf{r}},\ \ddot{\mathbf{r}},\ \dddot{\mathbf{r}}] = -\kappa^2\tau\dot{s}^6$$

or $$[\dot{\mathbf{r}},\ \ddot{\mathbf{r}},\ \dddot{\mathbf{r}}] = \kappa^2\tau\dot{s}^6. \quad (7)$$

From (2) and (5), we get

$$\kappa = \frac{|\dot{\mathbf{r}} \times \ddot{\mathbf{r}}|}{|\dot{\mathbf{r}}|^3}. \quad (8)$$

From (5) and (7), we get

$$\tau = \frac{[\dot{\mathbf{r}},\ \ddot{\mathbf{r}},\ \dddot{\mathbf{r}}]}{|\dot{\mathbf{r}} \times \ddot{\mathbf{r}}|^2}. \quad (9)$$

1.14 HELICES

Definition. A curve drawn on a cylinder in such a way that it cuts the generators of the cylinder at a constant angle is said to be a helix or a cylindrical helix.

Since the generators of a cylinder are parallel to a fixed direction, it follows that the tangent (i.e., **t**) at each point of the helix makes a constant angle with a fixed direction.

Theorem 1. A necessary and sufficient condition for a curve to be a helix is that its curvature and torsion are in a constant ratio.

Proof. Necessity. Let the given curve C be a helix.

To prove : $\frac{\kappa}{\tau} = \text{constant}.$

Let **a** denote the unit vector in the direction of the axis.

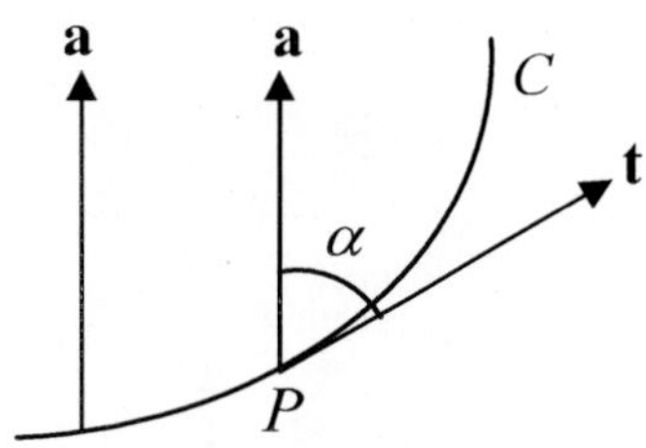

Since a helix cuts the generators of the cylinder at a constant angle, we have

$$\mathbf{t} \cdot \mathbf{a} = \cos\alpha = \text{constant}$$

$$\Rightarrow \quad \frac{d}{ds}(\mathbf{t} \cdot \mathbf{a}) = 0$$

$$\Rightarrow \quad \kappa\mathbf{n} \cdot \mathbf{a} = 0 \qquad \left[\because \frac{d\mathbf{a}}{ds} = \mathbf{0}, \text{ as } \mathbf{a} \text{ is a constant vector.}\right]$$

$$\Rightarrow \quad \mathbf{n} \cdot \mathbf{a} = 0 \qquad [\because \kappa \neq 0]$$

$\Rightarrow$ **n** is perpendicular to **a**.

$\Rightarrow$ **a** lies in the rectifying plane.

$$\Rightarrow \quad \mathbf{a} = l\mathbf{t} + m\mathbf{b} \qquad (1)$$

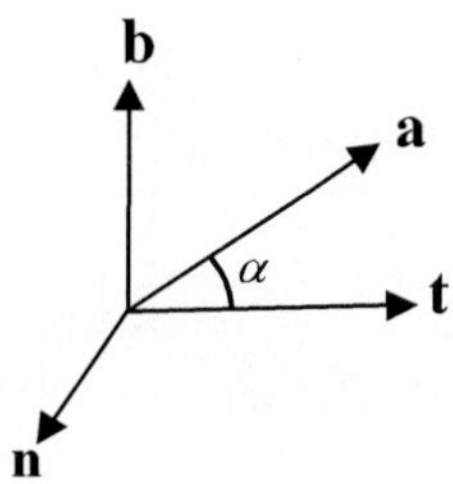

$$\Rightarrow \quad \left.\begin{array}{l} \mathbf{a} \cdot \mathbf{t} = l, \\ \mathbf{a} \cdot \mathbf{b} = m. \end{array}\right\}$$

$$\Rightarrow \quad \left.\begin{array}{l} \cos\alpha = l, \\ \sin\alpha = m. \end{array}\right\} \qquad (2)$$

Substituting the values of l and m from (2) in (1), we get

$$\mathbf{a} = \mathbf{t}\cos\alpha + \mathbf{b}\sin\alpha$$

$$\Rightarrow \quad \frac{d\mathbf{a}}{ds} = \frac{d}{ds}(\mathbf{t}\cos\alpha + \mathbf{b}\sin\alpha)$$

$$\Rightarrow \quad \mathbf{0} = \frac{d\mathbf{t}}{ds}\cos\alpha + \frac{d\mathbf{b}}{ds}\sin\alpha \qquad \left[\because \frac{d\mathbf{a}}{ds} = \mathbf{0}, \text{as } \mathbf{a} \text{ is a constant vector.}\right]$$

$$\Rightarrow \quad \mathbf{0} = (\kappa\mathbf{n})\cos\alpha + (-\tau\mathbf{n})\sin\alpha$$

$$\Rightarrow \quad \mathbf{0} = (\kappa\cos\alpha - \tau\sin\alpha)\,\mathbf{n}$$

$$\Rightarrow \quad 0 = \kappa\cos\alpha - \tau\sin\alpha \qquad [\because \mathbf{n} \neq \mathbf{0}]$$

$$\Rightarrow \quad \kappa\cos\alpha = \tau\sin\alpha$$

$$\Rightarrow \quad \frac{\kappa}{\tau} = \tan\alpha = \text{constant.} \qquad [\because \alpha \text{ is constant}]$$

Sufficiency. Now, we suppose that

$$\frac{\kappa}{\tau} = \text{constant} = \tan\alpha, \text{ where } -\infty < \alpha < \infty.$$

To prove : The given curve is a helix.

Since $\frac{\kappa}{\tau} = \tan\alpha$, we have

$$\kappa\cos\alpha - \tau\sin\alpha = 0$$

$$\Rightarrow \quad (\kappa\cos\alpha - \tau\sin\alpha)\,\mathbf{n} = \mathbf{0}$$

$$\Rightarrow \quad (\kappa\mathbf{n})\cos\alpha + (-\tau\mathbf{n})\sin\alpha = \mathbf{0}$$

$$\Rightarrow \quad \frac{d\mathbf{t}}{ds}\cos\alpha + \frac{d\mathbf{b}}{ds}\sin\alpha = \mathbf{0}$$

$$\Rightarrow \quad \frac{d}{ds}(\mathbf{t}\cos\alpha + \mathbf{b}\sin\alpha) = \mathbf{0}.$$

Integrating, we get

$$\mathbf{t}\cos\alpha + \mathbf{b}\sin\alpha = \text{constant of integration}$$

$$\Rightarrow \quad \mathbf{t}\cos\alpha + \mathbf{b}\sin\alpha = \mathbf{a} \text{ (unit constant vector)}$$

$$\Rightarrow \quad \mathbf{t}\,.\,\mathbf{a} = \cos\alpha$$

$$\Rightarrow \quad \mathbf{t}\,.\,\mathbf{a} = \text{constant} \qquad [\because \alpha \text{ is constant}]$$

⇒ **t** makes a constant angle with **a**.

⇒ The given curve is a helix.

Theorem 2. Curvature κ and torsion τ of a helix are in a constant ratio to the curvature κ_1 of the plane curve C_1 obtained by projecting C on a plane perpendicular to the axis of the helix and are connected by the relations

$$\kappa = \kappa_1\sin^2\alpha \text{ and } \tau = \kappa_1\sin\alpha\cos\alpha,$$

where α is the constant angle at which the helix cuts the generators.

Proof. Let the generators be parallel to z-axis and let the arc length of C be measured from A, the point of intersection of the helix with xy-plane.

Let s_1 be the arc length of the projection C_1 of C on the xy-plane measured from A to the generators through the current point $P\,(x, y, z)$.

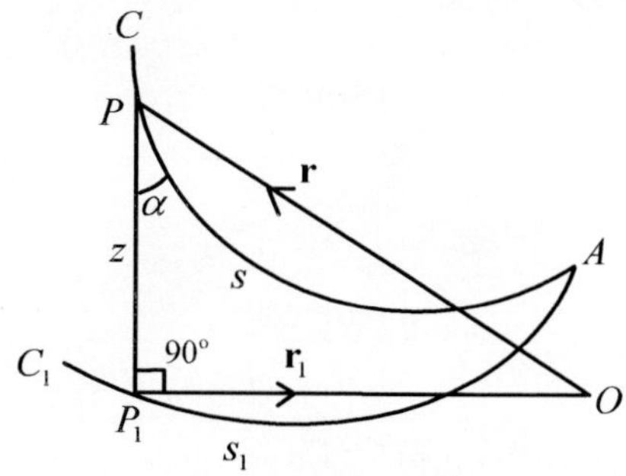

If α is the constant angle at which the helix cuts the generators, then

$$s_1 = s \sin\alpha.$$

Differentiating this with respect to s, we get

$$\frac{ds_1}{ds} = \sin\alpha. \tag{1}$$

Now, the coordinates (x, y) of P are functions of s_1 and $z = s\cos\alpha$. Hence if $\mathbf{r}$ is the position vector of P, then

$$\mathbf{r} = (x, y, s\cos\alpha).$$

$$\therefore \quad \mathbf{t} = \frac{d\mathbf{r}}{ds} = \left(\frac{dx}{ds_1}\frac{ds_1}{ds}, \frac{dy}{ds_1}\frac{ds_1}{ds}, \cos\alpha\right)$$

$$= \left(\frac{dx}{ds_1}\sin\alpha, \frac{dy}{ds_1}\sin\alpha, \cos\alpha\right). \qquad \text{[using (1)]}$$

$$\therefore \quad \mathbf{t}' = \left(\frac{d^2x}{ds_1^{\,2}}\frac{ds_1}{ds}\sin\alpha, \frac{d^2y}{ds_1^{\,2}}\frac{ds_1}{ds}\sin\alpha, 0\right)$$

or
$$\kappa\mathbf{n} = \left(\frac{d^2x}{ds_1^{\,2}}\sin^2\alpha, \frac{d^2y}{ds_1^{\,2}}\sin^2\alpha, 0\right).$$

Squaring, we get

$$\kappa^2 = \left[\left(\frac{d^2x}{ds_1^{\,2}}\right)^2 + \left(\frac{d^2y}{ds_1^{\,2}}\right)^2\right]\sin^4\alpha. \tag{2}$$

Now, we find the curvature κ_1 of the projection C_1. If $\mathbf{r}_1$ is the position vector of the projection P_1 of P, then

$$\mathbf{r}_1 = (x, y, 0).$$

$$\therefore \quad \mathbf{t}_1 = \frac{d\mathbf{r}_1}{ds_1} = \left(\frac{dx}{ds_1}, \frac{dy}{ds_1}, 0\right)$$

and
$$\mathbf{t}_1' = \left(\frac{d^2x}{ds_1^{\,2}}, \frac{d^2y}{ds_1^{\,2}}, 0\right)$$

or $$\kappa_1 \mathbf{n}_1 = \left(\frac{d^2 x}{ds_1^{\,2}}, \frac{d^2 y}{ds_1^{\,2}}, 0 \right).$$

Squaring, we get

$$\kappa_1^{\,2} = \left(\frac{d^2 x}{ds_1^{\,2}} \right)^2 + \left(\frac{d^2 y}{ds_1^{\,2}} \right)^2. \qquad (3)$$

From (2) and (3), we have

$$\kappa^2 = \kappa_1^{\,2} \sin^4 \alpha$$

i.e., $$\kappa = \kappa_1 \sin^2 \alpha. \qquad (4)$$

Also, for the helix, we know that

$$\frac{\kappa}{\tau} = \tan \alpha.$$

∴ $$\tau = \kappa \cot \alpha$$

or $$\tau = \kappa_1 \sin^2 \alpha \cot \alpha \qquad \text{[using (4)]}$$

or $$\tau = \kappa_1 \sin \alpha \cos \alpha. \qquad (5)$$

Equations (4) and (5) give the required results.

Circular Helix

A helix drawn on a circular cylinder is known as a circular helix.

Let O be the centre of the circular base of the right circular cylinder.

Let the circular helix cut the circular base at A.

Take O for the origin, OA produced for the x-axis and the axis of the right circular cylinder for the z-axis.

Now, take any point $P\,(x, y, z)$ on the helix.

PQ is perpendicular from P to xy-plane. QN is perpendicular from Q to x-axis. Let $\angle QON = \theta$ and let α be the angle between $\mathbf{t}$ and $\mathbf{a}$, where $\mathbf{a}$ is the unit vector along the axis of the cylinder.

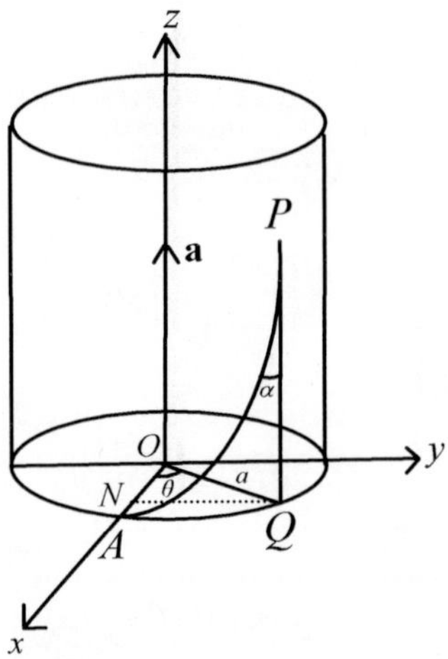

Let a be the radius of the right circular cylinder. Then,

$$x = ON = \frac{ON}{OQ} OQ = a \cos \theta,$$

$$y = NQ = \frac{NQ}{OQ} OQ = a \sin\theta,$$

$$z = PQ = \frac{PQ}{\text{arc}AP} \; \frac{\text{arc}AP}{\text{arc}AQ} \; \frac{\text{arc}AQ}{a} a$$

$$= \cos\alpha \frac{1}{\sin\alpha} \theta \, a = a\theta \cot\alpha.$$

Hence, the equation of a circular helix may be written in the form

$$\mathbf{r} = (a\cos\theta, a\sin\theta, a\theta\cot\alpha).$$

Solved Examples

Example 1. Prove that

$$\mathbf{r}''' = \kappa'\mathbf{n} - \kappa^2\mathbf{t} + \kappa\tau\mathbf{b}$$

and hence that

$$\mathbf{r}'''' = (\kappa'' - \kappa^3 - \kappa\tau^2)\,\mathbf{n} - 3\kappa'\kappa\mathbf{t} + (2\kappa'\tau + \tau'\kappa)\,\mathbf{b}.$$

Solution. We know that

$$\mathbf{r}' = \mathbf{t}$$

$$\Rightarrow \quad \mathbf{r}'' = \mathbf{t}' = \kappa\mathbf{n}$$

$$\Rightarrow \quad \mathbf{r}''' = \kappa'\mathbf{n} + \kappa\mathbf{n}' = \kappa'\mathbf{n} + \kappa\,(\tau\mathbf{b} - \kappa\mathbf{t}) = \kappa'\mathbf{n} - \kappa^2\mathbf{t} + \kappa\tau\mathbf{b}.$$

Hence

$$\mathbf{r}'''' = (\kappa''\mathbf{n} + \kappa'\mathbf{n}') - (2\kappa\kappa'\mathbf{t} + \kappa^2\mathbf{t}') + (\kappa'\tau\mathbf{b} + \kappa\tau'\mathbf{b} + \kappa\tau\mathbf{b}')$$

$$= [\kappa''\mathbf{n} + \kappa'\,(\tau\mathbf{b} - \kappa\mathbf{t})] - [2\kappa\kappa'\mathbf{t} + \kappa^2(\kappa\mathbf{n})]$$

$$+ [\kappa'\tau\mathbf{b} + \kappa\tau'\mathbf{b} + \kappa\tau\,(-\tau\mathbf{n})]$$

$$= (\kappa'' - \kappa^3 - \kappa\tau^2)\,\mathbf{n} - 3\kappa'\kappa\,\mathbf{t} + (2\kappa'\tau + \tau'\kappa)\,\mathbf{b}.$$

Example 2. Prove the following relations :

(i) $\mathbf{r}'.\mathbf{r}'' = 0$

(ii) $\mathbf{r}'.\mathbf{r}''' = -\kappa^2$

(iii) $\mathbf{r}'.\mathbf{r}'''' = -3\kappa\kappa'$

(iv) $\mathbf{r}''.\mathbf{r}''' = \kappa\kappa'$

(v) $\mathbf{r}''.\mathbf{r}'''' = \kappa\,(\kappa'' - \kappa^3 - \kappa\tau^2)$

(vi) $\mathbf{r}'''.\mathbf{r}'''' = \kappa'\kappa'' + 2\kappa^3\kappa' + \kappa^2\tau\tau' + \kappa\kappa'\tau^2.$

Solution. (i) We know that

$$\mathbf{r}' = \mathbf{t} \text{ and } \mathbf{r}'' = \mathbf{t}' = \kappa\mathbf{n}.$$

$$\therefore \quad \mathbf{r}'.\mathbf{r}'' = \mathbf{t}\,.\,\kappa\mathbf{n}$$

$$= \kappa\,(\mathbf{t}\,.\,\mathbf{n})$$

$$= 0. \qquad [\because \mathbf{t}\,.\,\mathbf{n} = 0]$$

(ii) Proceeding as in Example 1, we have

$$\mathbf{r}'=\mathbf{t},\ \mathbf{r}''=\mathbf{t}'=\kappa\mathbf{n} \text{ and } \mathbf{r}'''=\kappa'\mathbf{n}-\kappa^2\mathbf{t}+\kappa\tau\mathbf{b}.$$

Therefore

$$\begin{aligned}\mathbf{r}'.\mathbf{r}''' &= \mathbf{t}.(\kappa'\mathbf{n}-\kappa^2\mathbf{t}+\kappa\tau\mathbf{b})\\ &= \kappa'(\mathbf{t}.\mathbf{n})-\kappa^2(\mathbf{t}.\mathbf{t})+\kappa\tau(\mathbf{t}.\mathbf{b})\\ &= -\kappa^2. \qquad [\because \mathbf{t}.\mathbf{n}=0,\ \mathbf{t}.\mathbf{b}=0,\ \mathbf{t}.\mathbf{t}=1]\end{aligned}$$

(iii) Proceeding as in Example 1, we have

$$\mathbf{r}'=\mathbf{t},\ \mathbf{r}''=\mathbf{t}'=\kappa\mathbf{n},\ \mathbf{r}'''=\kappa'\mathbf{n}-\kappa^2\mathbf{t}+\kappa\tau\mathbf{b},$$

$$\mathbf{r}''''=(\kappa''-\kappa^3-\kappa\tau^2)\,\mathbf{n}-3\kappa'\kappa\,\mathbf{t}+(2\kappa'\tau+\tau'\kappa)\,\mathbf{b}.$$

Therefore

$$\begin{aligned}\mathbf{r}'.\mathbf{r}'''' &= \mathbf{t}.[(\kappa''-\kappa^3-\kappa\tau^2)\,\mathbf{n}-3\kappa'\kappa\,\mathbf{t}+(2\kappa'\tau+\tau'\kappa)\,\mathbf{b}]\\ &= (\kappa''-\kappa^3-\kappa\tau^2)\,\mathbf{t}.\mathbf{n}-3\kappa'\kappa\,\mathbf{t}.\mathbf{t}+(2\kappa'\tau+\tau'\kappa)\,\mathbf{t}.\mathbf{b}\\ &= -3\kappa'\kappa. \qquad [\because \mathbf{t}.\mathbf{n}=0,\ \mathbf{t}.\mathbf{b}=0,\ \mathbf{t}.\mathbf{t}=1]\end{aligned}$$

(iv) We have

$$\begin{aligned}\mathbf{r}''.\mathbf{r}''' &= \kappa\mathbf{n}.(\kappa'\mathbf{n}-\kappa^2\mathbf{t}+\kappa\tau\mathbf{b})\\ &= \kappa\kappa'\mathbf{n}.\mathbf{n}-\kappa^3\mathbf{n}.\mathbf{t}+\kappa^2\tau\mathbf{n}.\mathbf{b}\\ &= \kappa\kappa'. \qquad [\because \mathbf{n}.\mathbf{t}=0,\ \mathbf{n}.\mathbf{b}=0,\ \mathbf{n}.\mathbf{n}=1]\end{aligned}$$

(v) We have

$$\begin{aligned}\mathbf{r}''.\mathbf{r}'''' &= \kappa\mathbf{n}.[(\kappa''-\kappa^3-\kappa\tau^2)\,\mathbf{n}-3\kappa'\kappa\mathbf{t}+(2\kappa'\tau+\tau'\kappa)\,\mathbf{b}]\\ &= \kappa\,(\kappa''-\kappa^3-\kappa\tau^2)\,\mathbf{n}.\mathbf{n}-3\kappa'\kappa^2\mathbf{n}.\mathbf{t}+(2\kappa\kappa'\tau+\tau'\kappa^2)\,\mathbf{n}.\mathbf{b}\\ &= \kappa\,(\kappa''-\kappa^3-\kappa\tau^2). \qquad [\because \mathbf{n}.\mathbf{t}=0, \mathbf{n}.\mathbf{b}=0, \mathbf{n}.\mathbf{n}=1]\end{aligned}$$

(vi) We have

$$\begin{aligned}\mathbf{r}'''.\mathbf{r}'''' &= (\kappa'\mathbf{n}-\kappa^2\mathbf{t}+\kappa\tau\mathbf{b})\,.\\ &\qquad [(\kappa''-\kappa^3-\kappa\tau^2)\,\mathbf{n}-3\kappa'\kappa\mathbf{t}+(2\kappa'\tau+\tau'\kappa)\,\mathbf{b}]\\ &= \kappa'\,(\kappa''-\kappa^3-\kappa\tau^2)+3\kappa^3\kappa'+\kappa\tau\,(2\kappa'\tau+\tau'\kappa).\\ &\qquad [\because \mathbf{t}.\mathbf{n}=\mathbf{t}.\mathbf{b}=\mathbf{n}.\mathbf{b}=0, \mathbf{t}.\mathbf{t}=\mathbf{n}.\mathbf{n}=\mathbf{b}.\mathbf{b}=1]\\ &= \kappa'\kappa''+2\kappa^3\kappa'+\kappa^2\tau\tau'+\kappa\kappa'\tau^2.\end{aligned}$$

Example 3. Prove that $[\mathbf{r}',\ \mathbf{r}'',\ \mathbf{r}''']=\kappa^2\tau$.

Solution. Proceeding as in Example 1, we have

$$\mathbf{r}'=\mathbf{t},\ \mathbf{r}''=\mathbf{t}'=\kappa\mathbf{n} \text{ and } \mathbf{r}'''=\kappa'\mathbf{n}-\kappa^2\mathbf{t}+\kappa\tau\mathbf{b}.$$

$$\begin{aligned}\therefore\quad [\mathbf{r}',\ \mathbf{r}'',\ \mathbf{r}'''] &= \mathbf{r}'.(\mathbf{r}''\times\mathbf{r}''')\\ &= \mathbf{t}.[\kappa\mathbf{n}\times(\kappa'\mathbf{n}-\kappa^2\mathbf{t}+\kappa\tau\mathbf{b})]\\ &= \mathbf{t}.[\kappa^3\mathbf{b}+\kappa^2\tau\mathbf{t}]\end{aligned}$$

$$= \kappa^3 \mathbf{t}\,.\,\mathbf{b} + \kappa^2 \tau \mathbf{t}\,.\,\mathbf{t}$$
$$= \kappa^2 \tau. \qquad [\because \mathbf{t}\,.\,\mathbf{b} = 0,\ \mathbf{t}\,.\,\mathbf{t} = 1]$$

Example 4. Prove that the Serret-Frenet formulae can be written in the form

$$\mathbf{t}' = \mathbf{d} \times \mathbf{t}, \quad \mathbf{n}' = \mathbf{d} \times \mathbf{n}, \quad \mathbf{b}' = \mathbf{d} \times \mathbf{b}$$

and determine **d** (the Darboux vector of the curve).

Solution. We have

$$\begin{aligned}\mathbf{t}' &= \kappa \mathbf{n}\\ &= \kappa(\mathbf{b} \times \mathbf{t})\\ &= \tau(\mathbf{t} \times \mathbf{t}) + \kappa(\mathbf{b} \times \mathbf{t}) \qquad [\because \mathbf{t} \times \mathbf{t} = \mathbf{0}]\\ &= (\tau \mathbf{t} \times \mathbf{t}) + (\kappa \mathbf{b} \times \mathbf{t})\\ &= (\tau \mathbf{t} + \kappa \mathbf{b}) \times \mathbf{t}\\ &= \mathbf{d} \times \mathbf{t}, \text{ where } \mathbf{d} = \tau \mathbf{t} + \kappa \mathbf{b}.\end{aligned}$$

Similarly,

$$\begin{aligned}\mathbf{n}' &= \tau \mathbf{b} - \kappa \mathbf{t}\\ &= \tau(\mathbf{t} \times \mathbf{n}) + \kappa(\mathbf{b} \times \mathbf{n})\\ &= (\tau \mathbf{t} \times \mathbf{n}) + (\kappa \mathbf{b} \times \mathbf{n})\\ &= (\tau \mathbf{t} + \kappa \mathbf{b}) \times \mathbf{n}\\ &= \mathbf{d} \times \mathbf{n}\end{aligned}$$

and

$$\begin{aligned}\mathbf{b}' &= -\tau \mathbf{n}\\ &= -\tau(\mathbf{b} \times \mathbf{t})\\ &= \tau(\mathbf{t} \times \mathbf{b})\\ &= \tau(\mathbf{t} \times \mathbf{b}) + \kappa(\mathbf{b} \times \mathbf{b}) \qquad [\because \mathbf{b} \times \mathbf{b} = \mathbf{0}]\\ &= (\tau \mathbf{t} \times \mathbf{b}) + (\kappa \mathbf{b} \times \mathbf{b})\\ &= (\tau \mathbf{t} + \kappa \mathbf{b}) \times \mathbf{b}\\ &= \mathbf{d} \times \mathbf{b}.\end{aligned}$$

Example 5. Prove, with the usual notations, that

(i) $[\mathbf{t}', \mathbf{t}'', \mathbf{t}'''] = [\mathbf{r}'', \mathbf{r}''', \mathbf{r}''''] = \kappa^3(\kappa\tau' - \kappa'\tau) = \kappa^5 \dfrac{d}{ds}\left(\dfrac{\tau}{\kappa}\right)$

(ii) $[\mathbf{b}', \mathbf{b}'', \mathbf{b}'''] = \tau^3(\kappa\tau' - \kappa'\tau) = \tau^5 \dfrac{d}{ds}\left(\dfrac{\kappa}{\tau}\right)$.

Solution. (i) Using Serret-Frenet formulae, we have

$$\mathbf{t}' = \kappa \mathbf{n}$$

$$\Rightarrow \quad \mathbf{t}'' = \kappa'\mathbf{n} + \kappa\mathbf{n}' = \kappa'\mathbf{n} + \kappa(\tau\mathbf{b} - \kappa\mathbf{t}) = -\kappa^2\mathbf{t} + \kappa'\mathbf{n} + \kappa\tau\mathbf{b}$$

$$\Rightarrow \quad \mathbf{t}''' = -(2\kappa\kappa'\mathbf{t} + \kappa^2\mathbf{t}') + (\kappa''\mathbf{n} + \kappa'\mathbf{n}') + (\kappa'\tau\mathbf{b} + \kappa\tau'\mathbf{b} + \kappa\tau\mathbf{b}')$$

$$= -[2\kappa\kappa'\mathbf{t} + \kappa^2(\kappa\mathbf{n})] + [\kappa''\mathbf{n} + \kappa'(\tau\mathbf{b} - \kappa\mathbf{t})] + [\kappa'\tau\mathbf{b} + \kappa\tau'\mathbf{b} + \kappa\tau(-\tau\mathbf{n})]$$

$$= -3\kappa'\kappa\mathbf{t} + (\kappa'' - \kappa^3 - \kappa\tau^2)\mathbf{n} + (2\kappa'\tau + \tau'\kappa)\mathbf{b}.$$

Therefore

$$\mathbf{t}' \times \mathbf{t}'' = \kappa\mathbf{n} \times (-\kappa^2\mathbf{t} + \kappa'\mathbf{n} + \kappa\tau\mathbf{b})$$

$$= \kappa^3\mathbf{b} + \kappa^2\tau\mathbf{t} \qquad [\because \mathbf{n}\times\mathbf{t} = -\mathbf{b},\ \mathbf{n}\times\mathbf{n} = \mathbf{0}, \mathbf{n}\times\mathbf{b} = \mathbf{t}]$$

$$= \kappa^2\tau\mathbf{t} + \kappa^3\mathbf{b}.$$

Now, we have

$$[\mathbf{t}',\ \mathbf{t}'',\ \mathbf{t}'''] = (\mathbf{t}' \times \mathbf{t}'') . \mathbf{t}'''$$

$$= (\kappa^2\tau\mathbf{t} + \kappa^3\mathbf{b}). [-3\kappa'\kappa\mathbf{t} + (\kappa'' - \kappa^3 - \kappa\tau^2)\mathbf{n} + (2\kappa'\tau + \tau'\kappa)\mathbf{b}]$$

$$= -3\kappa^3\kappa'\tau + \kappa^3(2\kappa'\tau + \tau'\kappa)$$

$$[\because \mathbf{t}.\mathbf{n} = \mathbf{t}.\mathbf{b} = \mathbf{n}.\mathbf{b} = 0, \mathbf{t}.\mathbf{t} = \mathbf{n}.\mathbf{n} = \mathbf{b}.\mathbf{b} = 1]$$

$$= -\kappa^3\kappa'\tau + \kappa^4\tau'$$

$$= \kappa^3(\kappa\tau' - \kappa'\tau)$$

$$= \kappa^5\left(\frac{\kappa\tau' - \kappa'\tau}{\kappa^2}\right)$$

$$= \kappa^5 \frac{d}{ds}\left(\frac{\tau}{\kappa}\right).$$

(ii) Using Serret-Frenet formulae, we have

$$\mathbf{b}' = -\tau\mathbf{n}$$

$$\Rightarrow \quad \mathbf{b}'' = -\tau'\mathbf{n} - \tau\mathbf{n}' = -\tau'\mathbf{n} - \tau(\tau\mathbf{b} - \kappa\mathbf{t}) = \tau\kappa\mathbf{t} - \tau'\mathbf{n} - \tau^2\mathbf{b}$$

$$\Rightarrow \quad \mathbf{b}''' = (\tau'\kappa\mathbf{t} + \tau\kappa'\mathbf{t} + \tau\kappa\mathbf{t}') - (\tau''\mathbf{n} + \tau'\mathbf{n}') - (2\tau\tau'\mathbf{b} + \tau^2\mathbf{b}')$$

$$= [\tau'\kappa\mathbf{t} + \tau\kappa'\mathbf{t} + \tau\kappa(\kappa\mathbf{n})] - [\tau''\mathbf{n} + \tau'(\tau\mathbf{b} - \kappa\mathbf{t})] - [2\tau\tau'\mathbf{b} + \tau^2(-\tau\mathbf{n})]$$

$$\Rightarrow \quad \mathbf{b}''' = (2\tau'\kappa + \tau\kappa')\mathbf{t} + (\tau^3 + \tau\kappa^2 - \tau'')\mathbf{n} - 3\tau\tau'\mathbf{b}.$$

Therefore

$$\mathbf{b}' \times \mathbf{b}'' = (-\tau\mathbf{n}) \times (\tau\kappa\mathbf{t} - \tau'\mathbf{n} - \tau^2\mathbf{b})$$

$$= \tau^2\kappa\mathbf{b} + \tau^3\mathbf{t} \qquad [\because \mathbf{n}\times\mathbf{t} = -\mathbf{b},\ \mathbf{n}\times\mathbf{n} = \mathbf{0},\ \mathbf{n}\times\mathbf{b} = \mathbf{t}]$$

$$= \tau^3\mathbf{t} + \tau^2\kappa\mathbf{b}.$$

Now, we have

$$
\begin{aligned}
[\mathbf{b}',\ \mathbf{b}'',\ \mathbf{b}'''] &= (\mathbf{b}'\times\mathbf{b}'')\,.\,\mathbf{b}''' \\
&= (\tau^3\mathbf{t}+\tau^2\kappa\mathbf{b}). \\
&\qquad [(2\tau'\kappa+\tau\kappa')\,\mathbf{t}+(\tau^3+\tau\kappa^2-\tau'')\,\mathbf{n}-3\tau\tau'\mathbf{b}] \\
&= \tau^3(2\tau'\kappa+\tau\kappa')-3\tau^3\tau'\kappa \\
&\qquad [\because \mathbf{t}\,.\,\mathbf{n}=\mathbf{t}\,.\,\mathbf{b}=\mathbf{n}\,.\,\mathbf{b}=0, \mathbf{t}\,.\,\mathbf{t}=\mathbf{n}\,.\,\mathbf{n}=\mathbf{b}\,.\,\mathbf{b}=1] \\
&= -\tau^3\tau'\kappa+\tau^4\kappa' \\
&= \tau^3(\tau\kappa'-\tau'\kappa) \\
&= \tau^5\left(\frac{\tau\kappa'-\tau'\kappa}{\tau^2}\right) \\
&= \tau^5\frac{d}{ds}\left(\frac{\kappa}{\tau}\right).
\end{aligned}
$$

Example 6. Prove that the position vector $\mathbf{r}(s)$ of any point on a space curve satisfies the differential equation

$$
\mathbf{r}^{iv}-\left(\frac{2\kappa'}{\kappa}+\frac{\tau'}{\tau}\right)\mathbf{r}'''+\left(\kappa^2+\tau^2+\frac{\kappa'\tau'}{\tau\kappa}+\frac{2\kappa'^2-\kappa\kappa''}{\kappa^2}\right)\mathbf{r}''
+\kappa^2\left(\frac{\kappa'}{\kappa}-\frac{\tau'}{\tau}\right)\mathbf{r}'=\mathbf{0}.
$$

Solution. The given differential equation is

$$
\mathbf{r}^{iv}-\left(\frac{2\kappa'}{\kappa}+\frac{\tau'}{\tau}\right)\mathbf{r}'''+\left(\kappa^2+\tau^2+\frac{\kappa'\tau'}{\tau\kappa}+\frac{2\kappa'^2-\kappa\kappa''}{\kappa^2}\right)\mathbf{r}''
+\kappa^2\left(\frac{\kappa'}{\kappa}-\frac{\tau'}{\tau}\right)\mathbf{r}'=\mathbf{0}. \qquad (1)
$$

Proceeding as in Example 1, we have

$$\mathbf{r}'=\mathbf{t},\ \mathbf{r}''=\mathbf{t}'=\kappa\mathbf{n},\ \mathbf{r}'''=\kappa'\mathbf{n}-\kappa^2\mathbf{t}+\kappa\tau\mathbf{b},$$

$$\mathbf{r}^{iv}=(\kappa''-\kappa^3-\kappa\tau^2)\,\mathbf{n}-3\kappa'\kappa\,\mathbf{t}+(2\kappa'\tau+\tau'\kappa)\,\mathbf{b}.$$

Hence, L.H.S. of (1)

$$
\begin{aligned}
&= (\kappa''-\kappa^3-\kappa\tau^2)\,\mathbf{n}-3\kappa'\kappa\,\mathbf{t}+(2\kappa'\tau+\tau'\kappa)\,\mathbf{b} \\
&\qquad -\left(\frac{2\kappa'}{\kappa}+\frac{\tau'}{\tau}\right)(\kappa'\mathbf{n}-\kappa^2\mathbf{t}+\kappa\tau\mathbf{b}) \\
&\qquad +\left(\kappa^2+\tau^2+\frac{\kappa'\tau'}{\tau\kappa}+\frac{2\kappa'^2-\kappa\kappa''}{\kappa^2}\right)\kappa\mathbf{n}+\kappa^2\left(\frac{\kappa'}{\kappa}-\frac{\tau'}{\tau}\right)\mathbf{t}
\end{aligned}
$$

$$=\left[-3\kappa'\kappa+\kappa^2\left(\frac{2\kappa'}{\kappa}+\frac{\tau'}{\tau}\right)+\kappa^2\left(\frac{\kappa'}{\kappa}-\frac{\tau'}{\tau}\right)\right]\mathbf{t}$$

$$+\left[(\kappa''-\kappa^3-\kappa\tau^2)-\kappa'\left(\frac{2\kappa'}{\kappa}+\frac{\tau'}{\tau}\right)\right.$$

$$\left.+\kappa\left(\kappa^2+\tau^2+\frac{\kappa'\tau'}{\tau\kappa}+\frac{2\kappa'^2-\kappa\kappa''}{\kappa^2}\right)\right]\mathbf{n}$$

$$+\left[(2\kappa'\tau+\tau'\kappa)-\kappa\tau\left(\frac{2\kappa'}{\kappa}+\frac{\tau'}{\tau}\right)\right]\mathbf{b}$$

$$=\left[-3\kappa'\kappa+\left(2\kappa'\kappa+\frac{\kappa^2\tau'}{\tau}\right)+\left(\kappa'\kappa-\frac{\kappa^2\tau'}{\tau}\right)\right]\mathbf{t}$$

$$+\left[(\kappa''-\kappa^3-\kappa\tau^2)-\left(\frac{2\kappa'^2}{\kappa}+\frac{\kappa'\tau'}{\tau}\right)\right.$$

$$\left.+\left(\kappa^3+\kappa\tau^2+\frac{\kappa'\tau'}{\tau}+\frac{2\kappa'^2}{\kappa}-\kappa''\right)\right]\mathbf{n}$$

$$+\left[(2\kappa'\tau+\tau'\kappa)-(2\kappa'\tau+\tau'\kappa)\right]\mathbf{b}$$

$$=0\mathbf{t}+0\mathbf{n}+0\mathbf{b}$$
$$=\mathbf{0}.$$

Therefore, the position vector $\mathbf{r}(s)$ of the current point on a space curve satisfies the differential equation (1).

Example 7. Prove that the position vector $\mathbf{r}(s)$ of the current point on a space curve satisfies the differential equation

$$\frac{d}{ds}\left[\sigma\frac{d}{ds}\left(\rho\frac{d^2\mathbf{r}}{ds^2}\right)\right]+\frac{d}{ds}\left(\frac{\sigma}{\rho}\frac{d\mathbf{r}}{ds}\right)+\frac{\rho}{\sigma}\frac{d^2\mathbf{r}}{ds^2}=\mathbf{0}.$$

Solution. The given differential equation is

$$\frac{d}{ds}\left[\sigma\frac{d}{ds}\left(\rho\frac{d^2\mathbf{r}}{ds^2}\right)\right]+\frac{d}{ds}\left(\frac{\sigma}{\rho}\frac{d\mathbf{r}}{ds}\right)+\frac{\rho}{\sigma}\frac{d^2\mathbf{r}}{ds^2}=\mathbf{0}. \tag{1}$$

We know that

$$\frac{d\mathbf{r}}{ds}=\mathbf{r}'=\mathbf{t}$$

$$\Rightarrow \quad \frac{d^2\mathbf{r}}{ds^2}=\mathbf{r}''=\mathbf{t}'=\kappa\mathbf{n}.$$

Also, by definition,

$$\rho = \frac{1}{\kappa} \text{ and } \sigma = \frac{1}{\tau}.$$

Therefore

$$\frac{d}{ds}\left[\sigma \frac{d}{ds}\left(\rho \frac{d^2\mathbf{r}}{ds^2}\right)\right] + \frac{d}{ds}\left(\frac{\sigma}{\rho}\frac{d\mathbf{r}}{ds}\right) + \frac{\rho}{\sigma}\frac{d^2\mathbf{r}}{ds^2}$$
$$= \frac{d}{ds}\left[\frac{1}{\tau}\frac{d}{ds}\left(\frac{1}{\kappa}\kappa\mathbf{n}\right)\right] + \frac{d}{ds}\left(\frac{\kappa}{\tau}\mathbf{t}\right) + \frac{\tau}{\kappa}(\kappa\mathbf{n})$$
$$= \frac{d}{ds}\left[\frac{1}{\tau}\frac{d\mathbf{n}}{ds}\right] + \frac{d}{ds}\left(\frac{\kappa}{\tau}\mathbf{t}\right) + \tau\mathbf{n}$$
$$= \frac{d}{ds}\left[\frac{1}{\tau}(\tau\mathbf{b} - \kappa\mathbf{t})\right] + \frac{d}{ds}\left(\frac{\kappa}{\tau}\mathbf{t}\right) + \tau\mathbf{n}$$
$$= \frac{d}{ds}\left[\mathbf{b} - \frac{\kappa}{\tau}\mathbf{t}\right] + \frac{d}{ds}\left(\frac{\kappa}{\tau}\mathbf{t}\right) + \tau\mathbf{n}$$
$$= \left[\frac{d\mathbf{b}}{ds} - \frac{d}{ds}\left(\frac{\kappa}{\tau}\mathbf{t}\right)\right] + \frac{d}{ds}\left(\frac{\kappa}{\tau}\mathbf{t}\right) + \tau\mathbf{n}$$
$$= \frac{d\mathbf{b}}{ds} + \tau\mathbf{n}$$
$$= -\tau\mathbf{n} + \tau\mathbf{n}$$
$$= \mathbf{0}.$$

Therefore, the position vector $\mathbf{r}(s)$ of the current point on a space curve satisfies the differential equation (1).

Example 8. Prove that a necessary and sufficient condition that a curve is plane is that $[\mathbf{r}', \mathbf{r}'', \mathbf{r}'''] = 0$.

Solution. Proceeding as in Example 3, we have

$$[\mathbf{r}', \mathbf{r}'', \mathbf{r}'''] = \kappa^2\tau. \tag{1}$$

Now, we shall prove that a curve is plane if and only if $[\mathbf{r}', \mathbf{r}'', \mathbf{r}'''] = 0$.

Necessary condition. Let the curve be a plane curve.

To prove : $[\mathbf{r}', \mathbf{r}'', \mathbf{r}'''] = 0$.

Since the given curve is a plane curve, it follows that the plane containing the curve is the osculating plane for all the points of the curve. Therefore, we have

$$\mathbf{b} = \text{a constant vector}$$

$$\Rightarrow \quad \frac{d\mathbf{b}}{ds} = \mathbf{0}$$

$\Rightarrow \quad \left|\dfrac{d\mathbf{b}}{ds}\right| = 0$

$\Rightarrow \quad \tau = 0$ $\quad\left[\because \left|\dfrac{d\mathbf{b}}{ds}\right| = \tau\right]$

$\Rightarrow \quad \kappa^2\tau = 0$

$\Rightarrow \quad [\mathbf{r}', \mathbf{r}'', \mathbf{r}'''] = 0.$ [using (1)]

Sufficient condition. Let $[\mathbf{r}', \mathbf{r}'', \mathbf{r}'''] = 0$.
To prove : The curve is a plane curve.
By supposition,

$$[\mathbf{r}', \mathbf{r}'', \mathbf{r}'''] = 0.$$

This $\Rightarrow \kappa^2\tau = 0$ [using (1)]

$\Rightarrow$ either $\kappa = 0$ or $\tau = 0$. (2)

We show that $\tau = 0$ always.
Suppose, if possible, that $\tau \neq 0$ at some point of the curve.
This $\Rightarrow$ There exists a neighbourhood of this point where $\tau \neq 0$.

$\Rightarrow \kappa = 0$ in this neighbourhood. [using (2)]

$\Rightarrow$ The arc is a straight line.

$\Rightarrow \tau = 0$ on this line.

But, this is a contradiction to our supposition that $\tau \neq 0$ at some point of the curve.

Hence our supposition that $\tau \neq 0$ at some point of the curve is wrong.

Therefore, we must have $\tau = 0$ at all points of the curve and consequently, the curve is a plane curve.

Example 9. Prove that a necessary and sufficient condition that a curve is plane is that $[\dot{\mathbf{r}}, \ddot{\mathbf{r}}, \dddot{\mathbf{r}}] = 0$.

Solution. We know that

$$\tau = \frac{[\dot{\mathbf{r}}, \ddot{\mathbf{r}}, \dddot{\mathbf{r}}]}{|\dot{\mathbf{r}} \times \ddot{\mathbf{r}}|^2}. \tag{1}$$

Now, we shall prove that a curve is plane if and only if $[\dot{\mathbf{r}}, \ddot{\mathbf{r}}, \dddot{\mathbf{r}}] = 0$.

Necessary condition. Let the curve be a plane curve.

To prove : $[\dot{\mathbf{r}}, \ddot{\mathbf{r}}, \dddot{\mathbf{r}}] = 0$.

Since the given curve is a plane curve, it follows that the plane containing the curve is the osculating plane for all the points of the curve. Therefore, we have

$$\mathbf{b} = \text{a constant vector}$$

$$\Rightarrow \quad \frac{d\mathbf{b}}{ds} = \mathbf{0}$$

$$\Rightarrow \quad \left| \frac{d\mathbf{b}}{ds} \right| = 0$$

$$\Rightarrow \quad \tau = 0 \qquad \left[\because \left| \frac{d\mathbf{b}}{ds} \right| = \tau \right]$$

$$\Rightarrow \quad [\dot{\mathbf{r}},\ \ddot{\mathbf{r}},\ \dddot{\mathbf{r}}] = 0. \qquad \text{[using (1)]}$$

Sufficient condition. Let $[\dot{\mathbf{r}},\ \ddot{\mathbf{r}},\ \dddot{\mathbf{r}}] = 0$.

To prove : The curve is a plane curve.

By supposition,

$$[\dot{\mathbf{r}},\ \ddot{\mathbf{r}},\ \dddot{\mathbf{r}}] = 0$$

$$\Rightarrow \quad \frac{[\dot{\mathbf{r}},\ \ddot{\mathbf{r}},\ \dddot{\mathbf{r}}]}{|\dot{\mathbf{r}} \times \ddot{\mathbf{r}}|^2} = 0$$

$$\Rightarrow \quad \tau = 0. \qquad \text{[using (1)]}$$

$\Rightarrow$ The curve is a plane curve.

Example 10. If the tangent and the binormal at a point of curve makes angles $\theta,\ \phi$ respectively with a fixed direction, prove that

$$\frac{\sin\theta}{\sin\phi}\frac{d\theta}{d\phi} = -\frac{\kappa}{\tau}.$$

Solution. Let **a** denote the unit vector along the fixed direction. Then, by the given conditions, we have

$$\mathbf{t} \,.\, \mathbf{a} = \cos\theta \qquad (1)$$

and $$\mathbf{b} \,.\, \mathbf{a} = \cos\phi. \qquad (2)$$

Differentiating (1) with respect to s, we get

$$\frac{d}{ds}(\mathbf{t} \,.\, \mathbf{a}) = \frac{d}{ds}(\cos\theta)$$

or $$\frac{d\mathbf{t}}{ds} \,.\, \mathbf{a} + \mathbf{t} \,.\, \frac{d\mathbf{a}}{ds} = -\sin\theta \frac{d\theta}{ds}$$

or $$(\kappa\mathbf{n}) \,.\, \mathbf{a} + \mathbf{t} \,.\, \mathbf{0} = -\sin\theta \frac{d\theta}{ds} \qquad \left[\because \frac{d\mathbf{a}}{ds} = \mathbf{0}, \text{ as } \mathbf{a} \text{ is a constant vector.} \right]$$

or $$\kappa(\mathbf{n} \,.\, \mathbf{a}) = -\sin\theta \frac{d\theta}{ds}. \qquad (3)$$

Similarly, differentiating (2) with respect to s, we get

$$\frac{d}{ds}(\mathbf{b}\,.\,\mathbf{a}) = \frac{d}{ds}(\cos\phi)$$

or $$\frac{d\mathbf{b}}{ds}\,.\,\mathbf{a} + \mathbf{b}\,.\,\frac{d\mathbf{a}}{ds} = -\sin\phi\frac{d\phi}{ds}$$

or $$(-\tau\mathbf{n})\,.\,\mathbf{a} + \mathbf{b}\,.\,\mathbf{0} = -\sin\phi\frac{d\phi}{ds} \quad \left[\because \frac{d\mathbf{a}}{ds} = \mathbf{0}, \text{ as } \mathbf{a} \text{ is a constant vector.}\right]$$

or $$-\tau\,(\mathbf{n}\,.\,\mathbf{a}) = -\sin\phi\frac{d\phi}{ds}$$

or $$\tau\,(\mathbf{n}\,.\,\mathbf{a}) = \sin\phi\frac{d\phi}{ds}. \tag{4}$$

From (3) and (4), we get

$$\frac{\kappa\,(\mathbf{n}\,.\,\mathbf{a})}{\tau\,(\mathbf{n}\,.\,\mathbf{a})} = -\frac{\sin\theta}{\sin\phi}\frac{d\theta}{d\phi}$$

or $$\frac{\sin\theta}{\sin\phi}\frac{d\theta}{d\phi} = -\frac{\kappa}{\tau}.$$

Example 11. Find the curvature and torsion for the circular helix

$$x = a\cos\theta,\ y = a\sin\theta,\ z = a\theta\cot\alpha.$$

Solution. First Method

The position vector **r** of any point on the circular helix is given by

$$\mathbf{r} = a\,(\cos\theta, \sin\theta, \theta\cot\alpha)$$

$$\Rightarrow \quad \frac{d\mathbf{r}}{ds} = a\,(-\sin\theta, \cos\theta,\ \cot\alpha)\frac{d\theta}{ds}$$

$$\Rightarrow \quad \mathbf{t} = a\,(-\sin\theta, \cos\theta,\ \cot\alpha)\frac{d\theta}{ds}. \tag{1}$$

Squaring both sides of (1), we get

$$1 = a^2\left(\frac{d\theta}{ds}\right)^2(\sin^2\theta + \cos^2\theta + \cot^2\alpha)$$

$$= a^2\left(\frac{d\theta}{ds}\right)^2(1 + \cot^2\alpha)$$

$$\Rightarrow \quad 1 = a^2\left(\frac{d\theta}{ds}\right)^2\operatorname{cosec}^2\alpha$$

$$\Rightarrow \quad \left(\frac{d\theta}{ds}\right)^2 = \frac{1}{a^2 \operatorname{cosec}^2\alpha}$$

$$\Rightarrow \quad \frac{d\theta}{ds} = \frac{1}{a \operatorname{cosec}\alpha} = \frac{\sin\alpha}{a}. \tag{2}$$

Substituting the value of $\frac{d\theta}{ds}$ from (2) in (1), we get

$$\mathbf{t} = (-\sin\theta, \cos\theta, \cot\alpha)\sin\alpha \tag{3}$$

$$\Rightarrow \quad \frac{d\mathbf{t}}{ds} = (-\cos\theta, -\sin\theta, 0)\sin\alpha\frac{d\theta}{ds}$$

$$\Rightarrow \quad \kappa\mathbf{n} = (-\cos\theta, -\sin\theta, 0)\frac{\sin^2\alpha}{a}. \tag{4}$$

[using (2)]

Squaring both sides of (4), we get

$$\kappa^2 = (\cos^2\theta + \sin^2\theta)\left(\frac{\sin^2\alpha}{a}\right)^2$$

or $\quad \kappa^2 = \left(\frac{\sin^2\alpha}{a}\right)^2$

or $\quad \kappa = \frac{\sin^2\alpha}{a}. \qquad (5)$

Substituting the value of κ from (5) in (4), we get

$$\mathbf{n} = (-\cos\theta, -\sin\theta, 0). \tag{6}$$

Now, we have

$$\mathbf{b} = \mathbf{t}\times\mathbf{n}$$

$$= (-\sin\theta, \cos\theta, \cot\alpha)\sin\alpha \times (-\cos\theta, -\sin\theta, 0)$$

[using (3) and (6)]

$$\Rightarrow \quad \mathbf{b} = (\cot\alpha\sin\theta, -\cot\alpha\cos\theta, 1)\sin\alpha$$

$$\Rightarrow \quad \frac{d\mathbf{b}}{ds} = (\cot\alpha\cos\theta, \cot\alpha\sin\theta, 0)\sin\alpha\frac{d\theta}{ds}$$

$$\Rightarrow \quad -\tau\mathbf{n} = (\cot\alpha\cos\theta, \cot\alpha\sin\theta, 0)\frac{\sin^2\alpha}{a}. \tag{7}$$

[using (2)]

Squaring both sides of (7), we get

$$\tau^2 = \cot^2\alpha\,(\cos^2\theta + \sin^2\theta)\left(\frac{\sin^2\alpha}{a}\right)^2$$

or $$\tau^2 = \cot^2\alpha\left(\frac{\sin^2\alpha}{a}\right)^2$$

or $$\tau = \cot\alpha\frac{\sin^2\alpha}{a} = \frac{\cos\alpha\sin\alpha}{a}. \quad (8)$$

The curvature and torsion for the circular helix are given by (5) and (8) respectively.

Alternative Method

We have

$$\mathbf{r} = a(\cos\theta, \sin\theta, \theta\cot\alpha).$$

$$\therefore \quad \dot{\mathbf{r}} = a(-\sin\theta, \cos\theta, \cot\alpha), \quad (1)$$

$$\ddot{\mathbf{r}} = a(-\cos\theta, -\sin\theta, 0), \quad (2)$$

$$\dddot{\mathbf{r}} = a(\sin\theta, -\cos\theta, 0). \quad (3)$$

From (1), we have

$$|\dot{\mathbf{r}}| = \sqrt{a^2(\sin^2\theta + \cos^2\theta + \cot^2\alpha)}$$

$$= \sqrt{a^2(1+\cot^2\alpha)}$$

$$= \sqrt{a^2\operatorname{cosec}^2\alpha}$$

$$= a\operatorname{cosec}\alpha$$

i.e., $$|\dot{\mathbf{r}}| = \frac{a}{\sin\alpha}. \quad (4)$$

Taking the cross product of (1) and (2), we get

$$\dot{\mathbf{r}}\times\ddot{\mathbf{r}} = a^2(-\sin\theta, \cos\theta, \cot\alpha)\times(-\cos\theta, -\sin\theta, 0)$$

i.e., $$\dot{\mathbf{r}}\times\ddot{\mathbf{r}} = a^2(\cot\alpha\sin\theta, -\cot\alpha\cos\theta, 1). \quad (5)$$

$$\therefore \quad |\dot{\mathbf{r}}\times\ddot{\mathbf{r}}| = \sqrt{a^4(\cot^2\alpha\sin^2\theta + \cot^2\alpha\cos^2\theta + 1)}$$

$$= \sqrt{a^4[\cot^2\alpha(\sin^2\theta + \cos^2\theta) + 1]}$$

$$= \sqrt{a^4[\cot^2\alpha + 1]}$$

$$= \sqrt{a^4\operatorname{cosec}^2\alpha}$$

$$= a^2\operatorname{cosec}\alpha$$

i.e., $$|\dot{\mathbf{r}}\times\ddot{\mathbf{r}}| = \frac{a^2}{\sin\alpha}. \quad (6)$$

Again, from (3) and (5), we have

$$[\dot{\mathbf{r}}, \ddot{\mathbf{r}}, \dddot{\mathbf{r}}] = (\dot{\mathbf{r}}\times\ddot{\mathbf{r}})\cdot\dddot{\mathbf{r}}$$

$$= a^3(\cot\alpha\sin\theta, -\cot\alpha\cos\theta, 1)\cdot(\sin\theta, -\cos\theta, 0)$$

or $\quad [\dot{\mathbf{r}},\ \ddot{\mathbf{r}},\ \dddot{\mathbf{r}}\,] = a^3 \cot\alpha. \qquad (7)$

Now, the curvature κ is given by

$$\kappa = \frac{|\,\dot{\mathbf{r}} \times \ddot{\mathbf{r}}\,|}{|\,\dot{\mathbf{r}}\,|^3}$$

$$= \frac{a^2/\sin\alpha}{(a/\sin\alpha)^3} \qquad \text{[using (4) and (6)]}$$

$$= \frac{\sin^2\alpha}{a}$$

and the torsion τ is given by

$$\tau = \frac{[\dot{\mathbf{r}},\ \ddot{\mathbf{r}},\ \dddot{\mathbf{r}}\,]}{|\,\dot{\mathbf{r}} \times \ddot{\mathbf{r}}\,|^2}$$

$$= \frac{a^3 \cot\alpha}{(a^2/\sin\alpha)^2} \qquad \text{[using (6) and (7)]}$$

$$= \frac{\cos\alpha \sin\alpha}{a}.$$

Example 12. For the curve

$$x = a\,(3u - u^3),\ y = 3au^2,\ z = a\,(3u + u^3),$$

prove that

$$\kappa = \tau = \frac{1}{3a\,(1+u^2)^2}.$$

Solution. The position vector $\mathbf{r}$ of any point on the given curve is

$$\mathbf{r} = a\,(3u - u^3, 3u^2,\ 3u + u^3).$$

$$\therefore \quad \dot{\mathbf{r}} = a\,(3 - 3u^2, 6u,\ 3 + 3u^2), \qquad (1)$$

$$\ddot{\mathbf{r}} = a\,(-6u, 6,\ 6u), \qquad (2)$$

$$\dddot{\mathbf{r}} = a\,(-6, 0,\ 6). \qquad (3)$$

From (1), we have

$$|\,\dot{\mathbf{r}}\,| = \sqrt{a^2\,(3-3u^2)^2 + 36\,a^2u^2 + a^2\,(3+3u^2)^2}$$

$$= 3a\sqrt{(1-u^2)^2 + 4u^2 + (1+u^2)^2}$$

$$= 3a\sqrt{2\,(u^4 + 2u^2 + 1)}$$

$$= 3a\sqrt{2\,(u^2+1)^2}$$

i.e., $|\dot{\mathbf{r}}| = 3\sqrt{2}\, a\,(u^2+1).$ (4)

Taking the cross product of (1) and (2), we get

$$\dot{\mathbf{r}} \times \ddot{\mathbf{r}} = a^2\,(3-3u^2, 6u,\ 3+3u^2) \times (-6u, 6,\ 6u)$$
$$= a^2\,(18u^2-18, -36u,\ 18u^2+18)$$

i.e., $\dot{\mathbf{r}} \times \ddot{\mathbf{r}} = 18\,a^2(u^2-1, -2u,\ u^2+1).$ (5)

$\therefore$ $|\dot{\mathbf{r}} \times \ddot{\mathbf{r}}| = 18\,a^2\sqrt{(u^2-1)^2+4u^2+(u^2+1)^2} = 18\,a^2\sqrt{2\,(u^2+1)^2}$

i.e., $|\dot{\mathbf{r}} \times \ddot{\mathbf{r}}| = 18\sqrt{2}\,a^2(u^2+1).$ (6)

Again, we have

$$[\dot{\mathbf{r}}, \ddot{\mathbf{r}}, \dddot{\mathbf{r}}] = (\dot{\mathbf{r}} \times \ddot{\mathbf{r}}) \cdot \dddot{\mathbf{r}}$$
$$= 18\,a^3(u^2-1, -2u,\ u^2+1) \cdot (-6, 0, 6)$$

[using (3) and (5)]

or $[\dot{\mathbf{r}}, \ddot{\mathbf{r}}, \dddot{\mathbf{r}}] = 216\,a^3.$ (7)

Now, the curvature κ is given by

$$\kappa = \frac{|\dot{\mathbf{r}} \times \ddot{\mathbf{r}}|}{|\dot{\mathbf{r}}|^3}$$

$$= \frac{18\sqrt{2}\,a^2(u^2+1)}{\{3\sqrt{2}\,a(u^2+1)\}^3}$$ [using (4) and (6)]

i.e., $\kappa = \dfrac{1}{3a\,(1+u^2)^2}$ (8)

and the torsion τ is given by

$$\tau = \frac{[\dot{\mathbf{r}}, \ddot{\mathbf{r}}, \dddot{\mathbf{r}}]}{|\dot{\mathbf{r}} \times \ddot{\mathbf{r}}|^2}$$

$$= \frac{216\,a^3}{\{18\sqrt{2}\,a^2(u^2+1)\}^2}$$ [using (6) and (7)]

i.e., $\tau = \dfrac{1}{3a\,(1+u^2)^2}.$ (9)

From (8) and (9), we get

$$\kappa = \tau = \frac{1}{3a\,(1+u^2)^2}.$$

Example 13. For the curve $\mathbf{r} = (u, u^2, u^3)$, prove that

$$\kappa^2 = \frac{4(9u^4 + 9u^2 + 1)}{(9u^4 + 4u^2 + 1)^3}$$

and $$\tau = \frac{3}{9u^4 + 9u^2 + 1}.$$

Solution. We have

$$\mathbf{r} = (u, u^2, u^3).$$

$$\therefore \quad \dot{\mathbf{r}} = (1, 2u, 3u^2), \tag{1}$$

$$\ddot{\mathbf{r}} = (0, 2, 6u), \tag{2}$$

$$\dddot{\mathbf{r}} = (0, 0, 6). \tag{3}$$

From (1), we have

$$|\dot{\mathbf{r}}| = \sqrt{1 + 4u^2 + 9u^4}. \tag{4}$$

Taking the cross product of (1) and (2), we get

$$\dot{\mathbf{r}} \times \ddot{\mathbf{r}} = (1, 2u, 3u^2) \times (0, 2, 6u) = (12u^2 - 6u^2, -6u, 2)$$

i.e., $$\dot{\mathbf{r}} \times \ddot{\mathbf{r}} = (6u^2, -6u, 2). \tag{5}$$

$$\therefore \quad |\dot{\mathbf{r}} \times \ddot{\mathbf{r}}| = \sqrt{36u^4 + 36u^2 + 4}$$

i.e., $$|\dot{\mathbf{r}} \times \ddot{\mathbf{r}}| = 2\sqrt{9u^4 + 9u^2 + 1}. \tag{6}$$

Again, we have

$$[\dot{\mathbf{r}}, \ddot{\mathbf{r}}, \dddot{\mathbf{r}}] = (\dot{\mathbf{r}} \times \ddot{\mathbf{r}}) \cdot \dddot{\mathbf{r}}$$

$$= (6u^2, -6u, 2) \cdot (0, 0, 6) \qquad \text{[using (3) and (5)]}$$

or $$[\dot{\mathbf{r}}, \ddot{\mathbf{r}}, \dddot{\mathbf{r}}] = 12. \tag{7}$$

Now, the curvature κ is given by

$$\kappa = \frac{|\dot{\mathbf{r}} \times \ddot{\mathbf{r}}|}{|\dot{\mathbf{r}}|^3}.$$

$$\therefore \quad \kappa^2 = \frac{|\dot{\mathbf{r}} \times \ddot{\mathbf{r}}|^2}{|\dot{\mathbf{r}}|^6} = \frac{4(9u^4 + 9u^2 + 1)}{(9u^4 + 4u^2 + 1)^3} \tag{8}$$

[using (4) and (6)]

and the torsion τ is given by

$$\tau = \frac{[\dot{\mathbf{r}}, \ddot{\mathbf{r}}, \dddot{\mathbf{r}}]}{|\dot{\mathbf{r}} \times \ddot{\mathbf{r}}|^2} = \frac{3}{9u^4 + 9u^2 + 1}. \tag{9}$$

[using (6) and (7)]

Equations (8) and (9) give the required results.

Example 14. For the curve

$$x = 4a\cos^3 u,\ \ y = 4a\sin^3 u,\ \ \ z = 3c\cos 2u,$$

prove that

$$\mathbf{n} = (\sin u, \cos u, 0)$$

and

$$\kappa = \frac{a}{6(a^2 + c^2)\sin 2u}.$$

Solution. The position vector **r** of any point on the given curve is

$$\mathbf{r} = (4a\cos^3 u, 4a\sin^3 u, 3c\cos 2u).$$

Differentiating both sides with respect to s, we get

$$\frac{d\mathbf{r}}{ds} = (-12a\cos^2 u\sin u, 12a\sin^2 u\cos u,\ -6c\sin 2u)\frac{du}{ds}$$

or $$\mathbf{t} = (-12a\cos^2 u\sin u, 12a\sin^2 u\cos u, -12c\sin u\cos u)\frac{du}{ds}. \tag{1}$$

Squaring both sides, we get

$$1 = 144\left(\frac{du}{ds}\right)^2 (a^2\cos^4 u\sin^2 u + a^2\sin^4 u\cos^2 u + c^2\sin^2 u\cos^2 u)$$

$$= 144\left(\frac{du}{ds}\right)^2 \{a^2\sin^2 u\cos^2 u(\cos^2 u + \sin^2 u) + c^2\sin^2 u\cos^2 u\}$$

$$= 144\left(\frac{du}{ds}\right)^2 (a^2\sin^2 u\cos^2 u + c^2\sin^2 u\cos^2 u)$$

$$= 144\left(\frac{du}{ds}\right)^2 (a^2 + c^2)\sin^2 u\cos^2 u$$

$$\Rightarrow \qquad \left(\frac{du}{ds}\right)^2 = \frac{1}{144(a^2 + c^2)\sin^2 u\cos^2 u}$$

$$\Rightarrow \qquad \frac{du}{ds} = \frac{1}{12\sqrt{(a^2 + c^2)}\ \sin u\cos u}. \tag{2}$$

Substituting the value of $\frac{du}{ds}$ from (2) in (1), we get

$$\mathbf{t} = \frac{1}{\sqrt{(a^2 + c^2)}}(-a\cos u, a\sin u, -c). \tag{3}$$

$$\therefore \qquad \frac{d\mathbf{t}}{ds} = \frac{1}{\sqrt{(a^2+c^2)}}(a\sin u, a\cos u, 0)\frac{du}{ds}$$

or $$\kappa\mathbf{n} = \frac{a}{12\sin u\cos u\,(a^2+c^2)}(\sin u, \cos u, 0) \qquad \text{[using (2)]}$$

or $$\kappa\mathbf{n} = \frac{a}{6\sin 2u\,(a^2+c^2)}(\sin u, \cos u, 0). \qquad (4)$$

Squaring both sides of (4), we get

$$\kappa^2 = \frac{a^2}{36\,(a^2+c^2)^2\sin^2 2u}(\sin^2 u + \cos^2 u)$$

$$= \frac{a^2}{36\,(a^2+c^2)^2\sin^2 2u}$$

$$\Rightarrow \qquad \kappa = \frac{a}{6\,(a^2+c^2)\sin 2u}. \qquad (5)$$

Substituting the value of κ from (5) in (4), we get

$$\mathbf{n} = (\sin u, \cos u, 0). \qquad (6)$$

Equations (5) and (6) give the required results.

Example 15. Find the curvature and torsion of the curve

$$\mathbf{r} = (au - a\sin u, a - a\cos u, bu).$$

Solution. We have

$$\mathbf{r} = (au - a\sin u, a - a\cos u, bu).$$

$$\therefore \qquad \dot{\mathbf{r}} = (a - a\cos u, a\sin u,\ b), \qquad (1)$$

$$\ddot{\mathbf{r}} = (a\sin u, a\cos u,\ 0), \qquad (2)$$

$$\dddot{\mathbf{r}} = (a\cos u, -a\sin u,\ 0). \qquad (3)$$

From (1), we have

$$|\,\dot{\mathbf{r}}\,| = \sqrt{(a - a\cos u)^2 + (a\sin u)^2 + b^2}$$

or $$|\,\dot{\mathbf{r}}\,| = \sqrt{2a^2(1-\cos u) + b^2}. \qquad (4)$$

Taking the cross product of (1) and (2), we get

$$\dot{\mathbf{r}}\times\ddot{\mathbf{r}} = (a - a\cos u, a\sin u,\ b)\times(a\sin u, a\cos u,\ 0)$$

i.e., $$\dot{\mathbf{r}}\times\ddot{\mathbf{r}} = (-ab\cos u, ab\sin u, a^2(\cos u - 1)). \qquad (5)$$

$$\therefore \qquad |\,\dot{\mathbf{r}}\times\ddot{\mathbf{r}}\,| = \sqrt{(-ab\cos u)^2 + (ab\sin u)^2 + a^4(\cos u - 1)^2}$$

i.e., $$|\,\dot{\mathbf{r}}\times\ddot{\mathbf{r}}\,| = a\sqrt{b^2 + a^2(1-\cos u)^2}. \qquad (6)$$

Again, we have

$$[\dot{\mathbf{r}}, \ddot{\mathbf{r}}, \dddot{\mathbf{r}}] = (\dot{\mathbf{r}} \times \ddot{\mathbf{r}}) . \dddot{\mathbf{r}}$$

$$= (-ab\cos u, ab\sin u, a^2(\cos u - 1)) \ .(a\cos u, -a\sin u, 0)$$

[using (3) and (5)]

$$= -a^2 b\cos^2 u - a^2 b\sin^2 u$$

$$= -a^2 b(\cos^2 u + \sin^2 u)$$

or $$[\dot{\mathbf{r}}, \ddot{\mathbf{r}}, \dddot{\mathbf{r}}] = -a^2 b. \tag{7}$$

Now, the curvature κ is given by

$$\kappa = \frac{|\dot{\mathbf{r}} \times \ddot{\mathbf{r}}|}{|\dot{\mathbf{r}}|^3} = \frac{a\sqrt{b^2 + a^2(1-\cos u)^2}}{\{2a^2(1-\cos u) + b^2\}^{3/2}} \tag{8}$$

[using (4) and (6)]

and the torsion τ is given by

$$\tau = \frac{[\dot{\mathbf{r}}, \ddot{\mathbf{r}}, \dddot{\mathbf{r}}]}{|\dot{\mathbf{r}} \times \ddot{\mathbf{r}}|^2} = -\frac{b}{b^2 + a^2(1-\cos u)^2}. \tag{9}$$

[using (6) and (7)]

Example 16. Find the curvature and torsion of the curve

$$\mathbf{r} = (a + a\cos u, a\sin u, 2a\sin\tfrac{1}{2}u).$$

Solution. We have

$$\mathbf{r} = a(1 + \cos u, \sin u, 2\sin\tfrac{1}{2}u).$$

$\therefore$ $$\dot{\mathbf{r}} = a(-\sin u, \cos u, \ \cos\tfrac{1}{2}u), \tag{1}$$

$$\ddot{\mathbf{r}} = a(-\cos u, -\sin u, \ -\tfrac{1}{2}\sin\tfrac{1}{2}u), \tag{2}$$

$$\dddot{\mathbf{r}} = a(\sin u, -\cos u, \ -\tfrac{1}{4}\cos\tfrac{1}{2}u). \tag{3}$$

From (1), we have

$$|\dot{\mathbf{r}}| = a\sqrt{\sin^2 u + \cos^2 u + \cos^2\tfrac{1}{2}u} = a\sqrt{1 + \frac{1+\cos u}{2}}$$

or $$|\dot{\mathbf{r}}| = a\sqrt{\frac{3+\cos u}{2}}. \tag{4}$$

Taking the cross product of (1) and (2), we get

$$\dot{\mathbf{r}} \times \ddot{\mathbf{r}} = a^2(-\sin u, \cos u, \cos\tfrac{1}{2}u) \times (-\cos u, -\sin u, -\tfrac{1}{2}\sin\tfrac{1}{2}u)$$

or $$\dot{\mathbf{r}} \times \ddot{\mathbf{r}} = a^2(\sin u\cos\tfrac{1}{2}u - \tfrac{1}{2}\cos u\sin\tfrac{1}{2}u,$$

$$-\cos u\cos\tfrac{1}{2}u - \tfrac{1}{2}\sin u\sin\tfrac{1}{2}u, 1). \tag{5}$$

$$\therefore \qquad |\dot{\mathbf{r}}\times\ddot{\mathbf{r}}| = a^2\sqrt{\frac{13+3\cos u}{8}}. \tag{6}$$

Again, from (3) and (5), we have

$$[\dot{\mathbf{r}},\ \ddot{\mathbf{r}},\ \dddot{\mathbf{r}}] = (\dot{\mathbf{r}}\times\ddot{\mathbf{r}})\,.\,\dddot{\mathbf{r}} = \frac{3}{4}a^3\cos\frac{u}{2}. \tag{7}$$

Now, the curvature κ is given by

$$\kappa = \frac{|\dot{\mathbf{r}}\times\ddot{\mathbf{r}}|}{|\dot{\mathbf{r}}|^3} = \frac{\sqrt{13+3\cos u}}{a\,(3+\cos u)^{3/2}} \tag{8}$$

[using (4) and (6)]

and the torsion τ is given by

$$\tau = \frac{[\dot{\mathbf{r}},\ \ddot{\mathbf{r}},\ \dddot{\mathbf{r}}]}{|\dot{\mathbf{r}}\times\ddot{\mathbf{r}}|^2} = \frac{6\cos\frac{1}{2}u}{a\,(13+3\cos u)}. \tag{9}$$

[using (6) and (7)]

Example 17. For the curve $x = 3t,\ y = 3t^2,\ z = 2t^3$, show that

$$\rho = -\sigma = \frac{3}{2}(1+2t^2)^2\,.$$

Solution. The position vector $\mathbf{r}$ of any point on the given curve is

$$\mathbf{r} = (3t, 3t^2, 2t^3).$$

$$\therefore \qquad \dot{\mathbf{r}} = (3, 6t, 6t^2), \tag{1}$$

$$\ddot{\mathbf{r}} = (0, 6, 12t), \tag{2}$$

$$\dddot{\mathbf{r}} = (0, 0, 12). \tag{3}$$

From (1), we have

$$|\dot{\mathbf{r}}| = \sqrt{(3)^2+(6t)^2+(6t^2)^2} = \sqrt{9+36t^2+36t^4}$$

$$= 3\sqrt{1+4t^2+4t^4} = 3\sqrt{(1+2t^2)^2}$$

i.e., $$|\dot{\mathbf{r}}| = 3\,(1+2t^2). \tag{4}$$

Taking the cross product of (1) and (2), we get

$$\dot{\mathbf{r}}\times\ddot{\mathbf{r}} = (3,\ 6t,\ 6t^2)\times(0,6,12t) = (36t^2, -36t, 18)$$

i.e., $$\dot{\mathbf{r}}\times\ddot{\mathbf{r}} = 18\,(2t^2, -2t, 1). \tag{5}$$

$$\therefore \qquad |\dot{\mathbf{r}}\times\ddot{\mathbf{r}}| = 18\sqrt{(2t^2)^2+(-2t)^2+1} = 18\sqrt{(1+2t^2)^2}$$

i.e., $$|\dot{\mathbf{r}}\times\ddot{\mathbf{r}}| = 18\,(1+2t^2). \tag{6}$$

Again, we have

$$[\dot{\mathbf{r}}, \ddot{\mathbf{r}}, \dddot{\mathbf{r}}] = (\dot{\mathbf{r}} \times \ddot{\mathbf{r}}) \cdot \dddot{\mathbf{r}}$$

$$= 18(2t^2, -2t, 1) \cdot (0, 0, 12) \qquad \text{[using (3) and (5)]}$$

$$= 18 \times 12$$

i.e., $\left[\dot{\mathbf{r}}, \ddot{\mathbf{r}}, \dddot{\mathbf{r}}\right] = 216.$ (7)

Now, we have

$$\rho = \frac{1}{\kappa} = \frac{|\dot{\mathbf{r}}|^3}{|\dot{\mathbf{r}} \times \ddot{\mathbf{r}}|} = \frac{27(1+2t^2)^3}{18(1+2t^2)} = \frac{3}{2}(1+2t^2)^2 \qquad (8)$$

[using (4) and (6)]

and

$$\sigma = \frac{1}{\tau} = \frac{|\dot{\mathbf{r}} \times \ddot{\mathbf{r}}|^2}{\left[\dot{\mathbf{r}}, \ddot{\mathbf{r}}, \dddot{\mathbf{r}}\right]} = \frac{324(1+2t^2)^2}{216} = \frac{3}{2}(1+2t^2)^2. \qquad (9)$$

[using (6) and (7)]

From (8) and (9), we get

$$\rho = \sigma = \frac{3}{2}(1+2t^2)^2.$$

For a left handed system, we have

$$\sigma = -\frac{3}{2}(1+2t^2)^2.$$

Hence, in this case,

$$\rho = -\sigma = \frac{3}{2}(1+2t^2)^2.$$

Example 18. For the curve

$$x = a\tan\theta,\ y = a\cot\theta,\ z = a\sqrt{2}\log\tan\theta,$$

show that

$$\rho = -\sigma = \frac{2\sqrt{2}a}{\sin^2 2\theta}.$$

Solution. First Method

The position vector **r** of any point on the given curve is

$$\mathbf{r} = a(\tan\theta,\ \cot\theta,\ \sqrt{2}\log\tan\theta)$$

$$\Rightarrow \quad \frac{d\mathbf{r}}{ds} = a\left(\sec^2\theta, -\operatorname{cosec}^2\theta,\ \sqrt{2}\frac{\sec^2\theta}{\tan\theta}\right)\frac{d\theta}{ds}$$

$$\Rightarrow \quad \mathbf{t} = a(\sec^2\theta, -\operatorname{cosec}^2\theta,\ \sqrt{2}\sec\theta\operatorname{cosec}\theta)\frac{d\theta}{ds}. \qquad (1)$$

Squaring both sides of (1), we get

$$1 = a^2\left(\frac{d\theta}{ds}\right)^2\left(\sec^4\theta + \operatorname{co}\sec^4\theta + 2\sec^2\theta\operatorname{cosec}^2\theta\right)$$

$$= a^2\left(\frac{d\theta}{ds}\right)^2(\sec^2\theta + \operatorname{cosec}^2\theta)^2$$

$$\Rightarrow \quad 1 = a\left(\frac{d\theta}{ds}\right)(\sec^2\theta + \operatorname{cosec}^2\theta)$$

$$= a\left(\frac{d\theta}{ds}\right)\left(\frac{1}{\cos^2\theta} + \frac{1}{\sin^2\theta}\right)$$

$$= a\left(\frac{d\theta}{ds}\right)\left(\frac{\sin^2\theta + \cos^2\theta}{\cos^2\theta\sin^2\theta}\right)$$

$$= a\left(\frac{d\theta}{ds}\right)\left(\frac{1}{\cos^2\theta\sin^2\theta}\right)$$

$$\Rightarrow \quad \left(\frac{d\theta}{ds}\right) = \frac{\cos^2\theta\sin^2\theta}{a}. \tag{2}$$

Substituting the value of $\frac{d\theta}{ds}$ from (2) in (1), we get

$$\mathbf{t} = \left(\sin^2\theta, -\cos^2\theta, \frac{1}{\sqrt{2}}\sin 2\theta\right). \tag{3}$$

$$\therefore \quad \frac{d\mathbf{t}}{ds} = (\sin 2\theta, \sin 2\theta, \sqrt{2}\cos 2\theta)\frac{d\theta}{ds}$$

or $$\frac{d\mathbf{t}}{ds} = (\sin 2\theta, \sin 2\theta, \sqrt{2}\cos 2\theta)\frac{\cos^2\theta\sin^2\theta}{a} \qquad \text{[using (2)]}$$

or $$\kappa\mathbf{n} = (\sin 2\theta, \sin 2\theta, \sqrt{2}\cos 2\theta)\frac{\cos^2\theta\sin^2\theta}{a}. \tag{4}$$

Squaring both sides of (4), we get

$$\kappa^2 = (\sin^2 2\theta + \sin^2 2\theta + 2\cos^2 2\theta)\frac{\cos^4\theta\sin^4\theta}{a^2}$$

$$= 2(\sin^2 2\theta + \cos^2 2\theta)\frac{\cos^4\theta\sin^4\theta}{a^2}$$

$$= \frac{2\cos^4\theta\sin^4\theta}{a^2}$$

$$\Rightarrow \quad \kappa = \frac{\sqrt{2}\cos^2\theta\sin^2\theta}{a} = \frac{\sin^2 2\theta}{2\sqrt{2}a}. \tag{5}$$

From (5), we have

$$\rho = \frac{1}{\kappa} = \frac{2\sqrt{2}a}{\sin^2 2\theta}. \tag{6}$$

Substituting the value of κ from (5) in (4), we get

$$\mathbf{n} = \frac{1}{\sqrt{2}}(\sin 2\theta, \sin 2\theta, \sqrt{2}\cos 2\theta) \tag{7}$$

$$\Rightarrow \quad \frac{d\mathbf{n}}{ds} = \frac{1}{\sqrt{2}}(2\cos 2\theta, 2\cos 2\theta, -2\sqrt{2}\sin 2\theta)\frac{d\theta}{ds}$$

$$\Rightarrow \quad \tau\mathbf{b} - \kappa\mathbf{t} = \frac{1}{\sqrt{2}}(2\cos 2\theta, 2\cos 2\theta, -2\sqrt{2}\sin 2\theta)\frac{d\theta}{ds}$$

$$\Rightarrow \quad \tau\mathbf{b} = \frac{1}{\sqrt{2}}(2\cos 2\theta, 2\cos 2\theta, -2\sqrt{2}\sin 2\theta)\frac{d\theta}{ds} + \kappa\mathbf{t}$$

$$= \frac{1}{\sqrt{2}}(2\cos 2\theta, 2\cos 2\theta, -2\sqrt{2}\sin 2\theta)\frac{\cos^2\theta\sin^2\theta}{a}$$

$$+ \frac{\sqrt{2}\cos^2\theta\sin^2\theta}{a}\left(\sin^2\theta, -\cos^2\theta, \frac{1}{\sqrt{2}}\sin 2\theta\right)$$

[using (3) and (5)]

i.e.,
$$\tau\mathbf{b} = \frac{\sqrt{2}\cos^2\theta\sin^2\theta}{a}\Big(\cos 2\theta + \sin^2\theta,\ \cos 2\theta - \cos^2\theta,$$

$$-\frac{1}{\sqrt{2}}\sin 2\theta\Big). \tag{8}$$

Squaring both sides of (8), we get

$$\tau^2 = \frac{2\cos^4\theta\sin^4\theta}{a^2}\Big\{(\cos 2\theta + \sin^2\theta)^2 + (\cos 2\theta - \cos^2\theta)^2$$

$$+ \frac{1}{2}\sin^2 2\theta\Big\}$$

$$= \frac{2\cos^4\theta\sin^4\theta}{a^2}\Big\{2\cos^2 2\theta + (\sin^4\theta + \cos^4\theta)$$

$$-2\cos 2\theta\,(\cos^2\theta - \sin^2\theta) + \frac{1}{2}\sin^2 2\theta\Big\}$$

$$= \frac{2\cos^4\theta\sin^4\theta}{a^2}\Big\{2\cos^2 2\theta + (\sin^4\theta + \cos^4\theta)$$

$$-2\cos^2 2\theta + \frac{1}{2}\sin^2 2\theta\Big\}$$

$$= \frac{2\cos^4\theta\sin^4\theta}{a^2}\left\{(\sin^4\theta+\cos^4\theta)+\frac{1}{2}\sin^2 2\theta\right\}$$

$$= \frac{2\cos^4\theta\sin^4\theta}{a^2}\left\{(\sin^4\theta+\cos^4\theta)+2\sin^2\theta\cos^2\theta\right\}$$

$$= \frac{2\cos^4\theta\sin^4\theta}{a^2}\left\{(\sin^2\theta+\cos^2\theta)^2\right\}$$

$$= \frac{2\cos^4\theta\sin^4\theta}{a^2}.$$

$$\therefore \quad \tau = \frac{\sqrt{2}\cos^2\theta\sin^2\theta}{a} = \frac{\sin^2 2\theta}{2\sqrt{2}a}. \tag{9}$$

From (9), we have

$$\sigma = \frac{1}{\tau} = \frac{2\sqrt{2}a}{\sin^2 2\theta}. \tag{10}$$

From (6) and (10), we get

$$\rho = \sigma = \frac{2\sqrt{2}a}{\sin^2 2\theta}.$$

For a left handed system, we have

$$\sigma = -\frac{2\sqrt{2}a}{\sin^2 2\theta}.$$

Hence, in this case,

$$\rho = -\sigma = \frac{2\sqrt{2}a}{\sin^2 2\theta}.$$

Alternative Method

We have

$$\mathbf{r} = a(\tan\theta,\ \cot\theta,\ \sqrt{2}\log\tan\theta)$$

$$\Rightarrow \quad \frac{d\mathbf{r}}{ds}\frac{ds}{d\theta} = a\left(\sec^2\theta, -\operatorname{cosec}^2\theta,\ \sqrt{2}\frac{\sec^2\theta}{\tan\theta}\right)$$

$$\Rightarrow \quad \mathbf{t}\dot{s} = \frac{a}{\sin^2\theta\cos^2\theta}(\sin^2\theta, -\cos^2\theta, \sqrt{2}\sin\theta\cos\theta). \tag{1}$$

Squaring both sides of (1), we get

$$\dot{s}^2 = \frac{a^2}{\sin^4\theta\cos^4\theta}(\sin^4\theta+\cos^4\theta+2\sin^2\theta\cos^2\theta)$$

$$\text{or} \quad \dot{s}^2 = \frac{a^2}{\sin^4\theta\cos^4\theta}(\sin^2\theta+\cos^2\theta)^2 = \frac{a^2}{\sin^4\theta\cos^4\theta}$$

or $$\dot{s} = \frac{a}{\sin^2\theta\cos^2\theta}. \tag{2}$$

Multiplying both sides of (1) by $\sin^2\theta\cos^2\theta$, we get

$$(\dot{s}\sin^2\theta\cos^2\theta)\,\mathbf{t} = a\,(\sin^2\theta, -\cos^2\theta, \sqrt{2}\sin\theta\cos\theta).$$

Differentiating this with respect to θ, we get

$$\left[\frac{d}{d\theta}(\dot{s}\sin^2\theta\cos^2\theta)\right]\mathbf{t} + (\dot{s}\sin^2\theta\cos^2\theta)\frac{d\mathbf{t}}{ds}\frac{ds}{d\theta}$$
$$= a\,(\sin 2\theta, \sin 2\theta, \sqrt{2}\cos 2\theta)$$

or $$\left[\frac{d}{d\theta}(\dot{s}\sin^2\theta\cos^2\theta)\right]\mathbf{t} + (\dot{s}\sin^2\theta\cos^2\theta)\kappa\mathbf{n}\dot{s}$$
$$= a\,(\sin 2\theta, \sin 2\theta, \sqrt{2}\cos 2\theta)$$

or $$\left[\frac{d}{d\theta}(\dot{s}\sin^2\theta\cos^2\theta)\right]\mathbf{t} + (\kappa\dot{s}^2\sin^2\theta\cos^2\theta)\mathbf{n}$$
$$= a\,(\sin 2\theta, \sin 2\theta, \sqrt{2}\cos 2\theta). \tag{3}$$

Taking the cross product of (1) and (3), we get

$$(\kappa\dot{s}^3\sin^2\theta\cos^2\theta)\,\mathbf{b}$$
$$= \frac{a^2}{\sin^2\theta\cos^2\theta}(-\sqrt{2}\cos^2\theta\cos 2\theta - \sqrt{2}\sin\theta\cos\theta\sin 2\theta,$$
$$\sqrt{2}\sin\theta\cos\theta\sin 2\theta - \sqrt{2}\sin^2\theta\cos 2\theta,$$
$$\sin 2\theta\sin^2\theta + \sin 2\theta\cos^2\theta)$$

or $$(\kappa\dot{s}^3\sin^2\theta\cos^2\theta)\,\mathbf{b}$$
$$= \frac{a^2}{\sin^2\theta\cos^2\theta}\Big(-\sqrt{2}\cos\theta\,(\cos 2\theta\cos\theta + \sin 2\theta\sin\theta),$$
$$\sqrt{2}\sin\theta\,(\sin 2\theta\cos\theta - \cos 2\theta\sin\theta),$$
$$\sin 2\theta\,(\sin^2\theta + \cos^2\theta)\Big)$$

or $$(\kappa\dot{s}^3\sin^2\theta\cos^2\theta)\,\mathbf{b}$$
$$= \frac{a^2}{\sin^2\theta\cos^2\theta}\Big(-\sqrt{2}\cos\theta\cos(2\theta - \theta),$$
$$\sqrt{2}\sin\theta\sin(2\theta - \theta), \sin 2\theta\Big)$$

or $$(\kappa\dot{s}^3\sin^2\theta\cos^2\theta)\,\mathbf{b}$$
$$= \frac{a^2}{\sin^2\theta\cos^2\theta}(-\sqrt{2}\cos^2\theta, \sqrt{2}\sin^2\theta, 2\sin\theta\cos\theta). \tag{4}$$

Squaring both sides of (4), we get

$$\kappa^2\dot{s}^6\sin^4\theta\cos^4\theta$$
$$=\frac{a^4}{\sin^4\theta\cos^4\theta}(2\cos^4\theta+2\sin^4\theta+4\sin^2\theta\cos^2\theta)$$
$$=\frac{2a^4}{\sin^4\theta\cos^4\theta}(\cos^2\theta+\sin^2\theta)^2$$

i.e., $$\kappa^2\dot{s}^6\sin^4\theta\cos^4\theta=\frac{2a^4}{\sin^4\theta\cos^4\theta} \qquad (5)$$

$$\Rightarrow \quad \kappa^2=\frac{2a^4}{\dot{s}^6\sin^8\theta\cos^8\theta}$$
$$=\left(\frac{\sin^2\theta\cos^2\theta}{a}\right)^6\frac{2a^4}{\sin^8\theta\cos^8\theta} \qquad \text{[using (2)]}$$
$$=\frac{2\sin^4\theta\cos^4\theta}{a^2}$$
$$\Rightarrow \quad \kappa=\frac{\sqrt{2}\sin^2\theta\cos^2\theta}{a}=\frac{\sin^2 2\theta}{2\sqrt{2}a}$$
$$\Rightarrow \quad \rho=\frac{1}{\kappa}=\frac{2\sqrt{2}a}{\sin^2 2\theta}. \qquad (6)$$

We write (4) as follows :

$$(\kappa\dot{s}^3\sin^4\theta\cos^4\theta)\,\mathbf{b}=a^2(-\sqrt{2}\cos^2\theta,\sqrt{2}\sin^2\theta,\sin 2\theta).$$

Differentiating this with respect to θ, we get

$$\left[\frac{d}{d\theta}(\kappa\dot{s}^3\sin^4\theta\cos^4\theta)\right]\mathbf{b}+(\kappa\dot{s}^3\sin^4\theta\cos^4\theta)\frac{d\mathbf{b}}{ds}\frac{ds}{d\theta}$$
$$=a^2\,(2\sqrt{2}\cos\theta\sin\theta,2\sqrt{2}\sin\theta\cos\theta,2\cos 2\theta\,)$$

or $$\left[\frac{d}{d\theta}(\kappa\dot{s}^3\sin^4\theta\cos^4\theta)\right]\mathbf{b}+(\kappa\dot{s}^3\sin^4\theta\cos^4\theta\,)(-\tau\mathbf{n})\,\dot{s}$$
$$=\sqrt{2}a^2(\sin 2\theta,\sin 2\theta,\sqrt{2}\cos 2\theta)$$

or $$\left[\frac{d}{d\theta}(\kappa\dot{s}^3\sin^4\theta\cos^4\theta)\right]\mathbf{b}-(\tau\kappa\dot{s}^4\sin^4\theta\cos^4\theta\,)\,\mathbf{n}$$
$$=\sqrt{2}a^2(\sin 2\theta,\sin 2\theta,\sqrt{2}\cos 2\theta\,). \qquad (7)$$

Taking the dot product of (3) and (7), we get

$$-\tau\,\kappa^2\dot{s}^6\sin^6\theta\cos^6\theta=\sqrt{2}\,a^3(\sin^2 2\theta+\sin^2 2\theta+2\cos^2 2\theta\,)$$
$$=2\sqrt{2}\,a^3(\sin^2 2\theta+\cos^2 2\theta\,)=2\sqrt{2}\,a^3$$

$$\Rightarrow \qquad \tau = -\frac{2\sqrt{2}\, a^3}{\kappa^2\, \dot{s}^6 \sin^6\theta \cos^6\theta}$$

$$= -\frac{2\sqrt{2}\, a^3}{\sin^2\theta\cos^2\theta}\left(\frac{1}{\kappa^2\, \dot{s}^6 \sin^4\theta\cos^4\theta}\right)$$

$$= -\frac{2\sqrt{2}\, a^3}{\sin^2\theta\cos^2\theta}\left(\frac{\sin^4\theta\cos^4\theta}{2a^4}\right) \qquad \text{[using (5)]}$$

$$= -\frac{\sqrt{2}\sin^2\theta\cos^2\theta}{a}$$

$$= -\frac{\sin^2 2\theta}{2\sqrt{2}a}$$

$$\Rightarrow \qquad \sigma = \frac{1}{\tau} = -\frac{2\sqrt{2}a}{\sin^2 2\theta}. \tag{8}$$

From (6) and (8), we get

$$\rho = -\sigma = \frac{2\sqrt{2}a}{\sin^2 2\theta}.$$

Example 19. For a point of intersection of the surfaces

$$x^2 - y^2 = c^2,\ y = x\tanh\frac{z}{c},$$

prove that

$$\rho = -\sigma = \frac{2x^2}{c}.$$

Solution. The parametric equation of the curve may be written as

$$x = c\cosh\theta,\ y = c\sinh\theta,\ z = c\theta.$$

Therefore the position vector **r** of any point on the curve is given by

$$\mathbf{r} = (c\cosh\theta,\ c\sinh\theta,\ c\theta) = c\,(\cosh\theta,\ \sinh\theta, \theta).$$

$$\therefore \qquad \dot{\mathbf{r}} = c\,(\sinh\theta, \cosh\theta,\ 1), \tag{1}$$

$$\ddot{\mathbf{r}} = c\,(\cosh\theta, \sinh\theta, 0), \tag{2}$$

$$\dddot{\mathbf{r}} = c\,(\sinh\theta, \cosh\theta, 0). \tag{3}$$

From (1), we have

$$|\,\dot{\mathbf{r}}\,| = c\sqrt{\sinh^2\theta + \cosh^2\theta + 1}$$

or $\qquad |\,\dot{\mathbf{r}}\,| = c\sqrt{2\cosh^2\theta} \qquad [\because \cosh^2\theta - \sinh^2\theta = 1]$

or $\qquad |\,\dot{\mathbf{r}}\,| = \sqrt{2}\,c\cosh\theta. \qquad (4)$

Taking the cross product of (1) and (2), we get

$$\dot{\mathbf{r}} \times \ddot{\mathbf{r}} = c^2 (\sinh\theta, \cosh\theta, 1) \times (\cosh\theta, \sinh\theta, 0)$$

i.e., $$\dot{\mathbf{r}} \times \ddot{\mathbf{r}} = c^2 (-\sinh\theta, \cosh\theta, -1). \tag{5}$$

$\therefore$ $$|\dot{\mathbf{r}} \times \ddot{\mathbf{r}}| = c^2 \sqrt{\sinh^2\theta + \cosh^2\theta + 1} = c^2 \sqrt{2\cosh^2\theta}$$

or $$|\dot{\mathbf{r}} \times \ddot{\mathbf{r}}| = \sqrt{2}\, c^2 \cosh\theta. \tag{6}$$

Again, from (3) and (5), we have

$$[\dot{\mathbf{r}}, \ddot{\mathbf{r}}, \dddot{\mathbf{r}}] = (\dot{\mathbf{r}} \times \ddot{\mathbf{r}}) \cdot \dddot{\mathbf{r}} = c^3. \tag{7}$$

Now, the curvature κ is given by

$$\kappa = \frac{|\dot{\mathbf{r}} \times \ddot{\mathbf{r}}|}{|\dot{\mathbf{r}}|^3}$$

$$= \frac{\sqrt{2}\, c^2 \cosh\theta}{2\sqrt{2}\, c^3 \cosh^3\theta} \qquad \text{[using (4) and (6)]}$$

$$= \frac{1}{2c\cosh^2\theta}$$

$$= \frac{c}{2x^2}.$$

$\therefore$ $$\rho = \frac{1}{\kappa} = \frac{2x^2}{c} \tag{8}$$

and the torsion τ is given by

$$\tau = \frac{[\dot{\mathbf{r}}, \ddot{\mathbf{r}}, \dddot{\mathbf{r}}]}{|\dot{\mathbf{r}} \times \ddot{\mathbf{r}}|^2}$$

$$= \frac{c^3}{2c^4 \cosh^2\theta} \qquad \text{[using (6) and (7)]}$$

$$= \frac{1}{2c\cosh^2\theta}$$

$$= \frac{c}{2x^2}.$$

$\therefore$ $$\sigma = \frac{1}{\tau} = \frac{2x^2}{c}. \tag{9}$$

From (8) and (9), we get

$$\rho = \sigma = \frac{2x^2}{c}.$$

For a left handed system, we have

$$\sigma = -\frac{2x^2}{c}.$$

Hence, in this case,

$$\rho = -\sigma = \frac{2x^2}{c}.$$

Example 20. Find the radii of curvature and torsion at a point of the curve

$$x^2 + y^2 = a^2,\ x^2 - y^2 = az.$$

Solution. The parametric equation of the curve may be written as

$$x = a\cos\theta,\ y = a\sin\theta,\ z = a\cos 2\theta.$$

Therefore the position vector **r** of any point on the curve is given by

$$\mathbf{r} = (a\cos\theta, a\sin\theta, a\cos 2\theta) = a(\cos\theta, \sin\theta, \cos 2\theta).$$

$$\therefore \quad \dot{\mathbf{r}} = a(-\sin\theta, \cos\theta, -2\sin 2\theta), \tag{1}$$

$$\ddot{\mathbf{r}} = a(-\cos\theta, -\sin\theta, -4\cos 2\theta), \tag{2}$$

$$\dddot{\mathbf{r}} = a(\sin\theta, -\cos\theta, 8\sin 2\theta). \tag{3}$$

From (1), we have

$$|\dot{\mathbf{r}}| = a\sqrt{\sin^2\theta + \cos^2\theta + 4\sin^2 2\theta} = a\sqrt{1 + 4\sin^2 2\theta}$$

or $$|\dot{\mathbf{r}}| = a\sqrt{5 - 4\cos^2 2\theta}. \tag{4}$$

Taking the cross product of (1) and (2), we get

$$\dot{\mathbf{r}} \times \ddot{\mathbf{r}} = a^2(-\sin\theta, \cos\theta, -2\sin 2\theta) \times (-\cos\theta, -\sin\theta, -4\cos 2\theta)$$

$$= a^2(-4\cos 2\theta\cos\theta - 2\sin 2\theta\sin\theta,$$

$$2\sin 2\theta\cos\theta - 4\cos 2\theta\sin\theta, \sin^2\theta + \cos^2\theta)$$

$$= a^2(-2\cos\theta\cos 2\theta - 2\cos\theta, 2\sin\theta - 2\sin\theta\cos 2\theta, 1)$$

or $$\dot{\mathbf{r}} \times \ddot{\mathbf{r}} = a^2(-4\cos^3\theta, 4\sin^3\theta, 1). \tag{5}$$

$$\therefore \quad |\dot{\mathbf{r}} \times \ddot{\mathbf{r}}| = a^2\sqrt{16\cos^6\theta + 16\sin^6\theta + 1}$$

$$= a^2\sqrt{16(\cos^6\theta + \sin^6\theta) + 1}$$

$$= a^2\sqrt{16\{(\cos^2\theta + \sin^2\theta)^3 - 3\cos^2\theta\sin^2\theta\} + 1}$$

$$= a^2\sqrt{17 - 48\cos^2\theta\sin^2\theta}$$

$$= a^2\sqrt{17 - 12\sin^2 2\theta}$$

i.e., $$|\dot{\mathbf{r}} \times \ddot{\mathbf{r}}| = a^2\sqrt{5 + 12\cos^2 2\theta}. \tag{6}$$

Again, we have

$$[\dot{\mathbf{r}},\ \ddot{\mathbf{r}},\ \dddot{\mathbf{r}}] = (\dot{\mathbf{r}} \times \ddot{\mathbf{r}}) \cdot \dddot{\mathbf{r}}$$

$$= a^3(-4\cos^3\theta, 4\sin^3\theta, 1) \cdot (\sin\theta, -\cos\theta,\ 8\sin 2\theta)$$

[using (3) and (5)]

$$= a^3(-4\cos^3\theta\sin\theta - 4\sin^3\theta\cos\theta + \ 8\sin 2\theta)$$

$$= a^3\{-4\sin\theta\cos\theta(\cos^2\theta + \sin^2\theta) + \ 8\sin 2\theta\}$$

$$= a^3(-2\sin 2\theta + \ 8\sin 2\theta)$$

i.e., $$[\dot{\mathbf{r}},\ \ddot{\mathbf{r}},\ \dddot{\mathbf{r}}] = 6a^3\sin 2\theta. \tag{7}$$

Now,

$$\kappa = \frac{|\dot{\mathbf{r}} \times \ddot{\mathbf{r}}|}{|\dot{\mathbf{r}}|^3}$$

$$= \frac{\sqrt{5 + 12\cos^2 2\theta}}{a(5 - 4\cos^2 2\theta)^{3/2}}$$ [using (4) and (6)]

$$= \frac{\sqrt{5 + 12\left(\dfrac{z^2}{a^2}\right)}}{a\left\{5 - 4\left(\dfrac{z^2}{a^2}\right)\right\}^{3/2}}$$ $\left[\because \cos 2\theta = \dfrac{z}{a}\right]$

$$= \frac{a\sqrt{5a^2 + 12z^2}}{(5a^2 - 4z^2)^{3/2}}.$$

Therefore the radius of curvature ρ is given by

$$\rho = \frac{1}{\kappa} = \frac{(5a^2 - 4z^2)^{3/2}}{a\sqrt{5a^2 + 12z^2}}. \tag{8}$$

Again,

$$\tau = \frac{[\dot{\mathbf{r}},\ \ddot{\mathbf{r}},\ \dddot{\mathbf{r}}]}{|\dot{\mathbf{r}} \times \ddot{\mathbf{r}}|^2}$$

$$= \frac{6a^3\sin 2\theta}{a^4(5 + 12\cos^2 2\theta)}$$ [using (6) and (7)]

$$= \frac{6\sqrt{1 - \cos^2 2\theta}}{a(5 + 12\cos^2 2\theta)}$$

$$= \frac{6\sqrt{1-(z^2/a^2)}}{a\,\{5+12\,(z^2/a^2)\}} \qquad \left[\because \cos 2\theta = \frac{z}{a}\right]$$

$$= \frac{6\sqrt{a^2 - z^2}}{5a^2 + 12\,z^2}.$$

Therefore, the radius of torsion σ is given by

$$\sigma = \frac{1}{\tau} = \frac{5a^2 + 12\,z^2}{6\sqrt{a^2 - z^2}}. \tag{9}$$

Example 21. Determine the function $f\,(u)$ so that the curve given by $\mathbf{r} = (a\cos u, a\sin u, f\,(u))$ shall be plane.

Solution. The condition that the curve be plane is $\tau = 0$. So, we first calculate τ.

We have

$$\mathbf{r} = (a\cos u, a\sin u, f\,(u)).$$

$$\therefore \quad \dot{\mathbf{r}} = (-\,a\sin u, a\cos u, \dot{f}\,(u)), \tag{1}$$

$$\ddot{\mathbf{r}} = (-\,a\cos u, -\,a\sin u,\ \ddot{f}\,(u)), \tag{2}$$

$$\dddot{\mathbf{r}} = (a\sin u, -\,a\cos u,\ \dddot{f}\,(u)). \tag{3}$$

Taking the cross product of (1) and (2), we get

$$\dot{\mathbf{r}} \times \ddot{\mathbf{r}} = (-\,a\sin u, a\cos u, \dot{f}\,(u)) \times (-\,a\cos u, -\,a\sin u,\ \ddot{f}\,(u))$$

$$= \big(a\cos u\ \ddot{f}\,(u) + a\sin u\ \dot{f}\,(u),\ -\,a\cos u\ \dot{f}\,(u) + a\sin u\ \ddot{f}\,(u),$$

$$a^2\sin^2 u + a^2\cos^2 u\big)$$

i.e., $$\dot{\mathbf{r}} \times \ddot{\mathbf{r}} = a\,(\sin u\ \dot{f}\,(u) + \cos u\ \ddot{f}\,(u), \sin u\ \ddot{f}\,(u) - \cos u\ \dot{f}\,(u), a). \tag{4}$$

$$\therefore \quad |\,\dot{\mathbf{r}} \times \ddot{\mathbf{r}}\,| = a\sqrt{\{\dot{f}\,(u)\}^2 + \{\ddot{f}\,(u)\}^2 + a^2}. \tag{5}$$

Again, from (3) and (4), we have

$$(\dot{\mathbf{r}} \times \ddot{\mathbf{r}})\,.\,\dddot{\mathbf{r}}$$

$$= a\,(\sin u\ \dot{f}\,(u) + \cos u\ \ddot{f}\,(u), \sin u\ \ddot{f}\,(u) - \cos u\ \dot{f}\,(u), a)$$

$$.\,(a\sin u, -\,a\cos u,\ \dddot{f}\,(u))$$

or $$\left[\dot{\mathbf{r}}, \ddot{\mathbf{r}}, \dddot{\mathbf{r}}\right] = a^2(\dot{f}\,(u) + \dddot{f}\,(u)). \tag{6}$$

Now, the torsion τ is given by

$$\tau = \frac{[\dot{\mathbf{r}},\ \ddot{\mathbf{r}},\ \dddot{\mathbf{r}}]}{|\dot{\mathbf{r}} \times \ddot{\mathbf{r}}|^2} = \frac{\dot{f}(u) + \dddot{f}(u)}{\{\dot{f}(u)\}^2 + \{\ddot{f}(u)\}^2 + a^2}. \tag{7}$$

[using (5) and (6)]

The curve is plane if

$$\tau = 0$$

i.e., if $\dfrac{\dot{f}(u) + \dddot{f}(u)}{\{\dot{f}(u)\}^2 + \{\ddot{f}(u)\}^2 + a^2} = 0$ [using (7)]

i.e., if $\dot{f}(u) + \dddot{f}(u) = 0$

i.e., if $\dfrac{df}{du} + \dfrac{d^3 f}{du^3} = 0$

i.e., if $(D + D^3) f(u) = 0,$ (8)

where $D \equiv \dfrac{d}{du}.$

Now, it remains to solve the differential equation (8).

Obviously, the auxiliary equation is

$$m + m^3 = 0$$

or $m\,(1 + m^2) = 0.$

The roots of the above equation are $m = 0, \pm i.$

Therefore, the solution of the differential equation (8) is

$$f(u) = ae^{0u} + be^{0u} \sin(u + c)$$

i.e., $f(u) = a + b \sin(u + c),$

where a, b, c are arbitrary constants.

Example 22. Prove that in order that, the principal normals of a curve be binormals of another, the relation

$$a\ (\kappa^2 + \tau^2) = b\kappa$$

must hold, where a and b are constants.

Solution. Let C and C_1 be a pair of curves such that the principal normals of C are the binormals of C_1.

We distinguish the quantities of C_1 from those of C by using the suffix unity. Let P and P_1 be the corresponding points of C and C_1 respectively.

Let $\mathbf{r}$ and $\mathbf{r}_1$ denote the position vectors of the points P and Q respectively. Then

$$\mathbf{r}_1 = \mathbf{r} + c\mathbf{n}, \tag{1}$$

where c is a scalar quantity

and $\mathbf{b}_1 = \mathbf{n}$. (2)

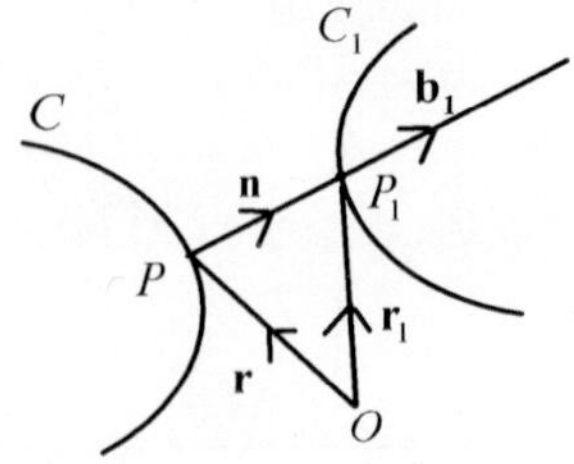

Differentiating (1) with respect to s, we get

$$\frac{d\mathbf{r}_1}{ds} = \frac{d\mathbf{r}}{ds} + c\frac{d\mathbf{n}}{ds}$$

or $$\frac{d\mathbf{r}_1}{ds_1}\frac{ds_1}{ds} = \mathbf{t} + c\,(\tau\mathbf{b} - \kappa\mathbf{t})$$

or $$\mathbf{t}_1\frac{ds_1}{ds} = (1 - c\kappa)\,\mathbf{t} + c\tau\mathbf{b}. \quad (3)$$

Squaring (3), we get

$$\left(\frac{ds_1}{ds}\right)^2 = (1 - c\kappa)^2 + c^2\tau^2. \quad (4)$$

Differentiating (3) again with respect to s, we get

$$\frac{d}{ds}\left(\mathbf{t}_1\frac{ds_1}{ds}\right) = \frac{d}{ds}[(1 - c\kappa)\,\mathbf{t}] + \frac{d}{ds}(c\tau\mathbf{b})$$

or $$\left(\frac{d\mathbf{t}_1}{ds_1}\frac{ds_1}{ds}\right)\frac{ds_1}{ds} + \mathbf{t}_1\frac{d^2s_1}{ds^2} = [(-c\kappa')\,\mathbf{t} + (1 - c\kappa)\,\kappa\mathbf{n}] + [c\tau'\mathbf{b} + c\tau\,(-\tau\mathbf{n})]$$

or $$\kappa_1\mathbf{n}_1\left(\frac{ds_1}{ds}\right)^2 + \mathbf{t}_1\frac{d^2s_1}{ds^2} = -c\kappa'\,\mathbf{t} + (\kappa - c\kappa^2 - c\tau^2)\,\mathbf{n} + c\tau'\mathbf{b}. \quad (5)$$

Taking the dot product of (2) and (5), we get

$$0 = \kappa - c\kappa^2 - c\tau^2$$

or $$\kappa = c\,(\kappa^2 + \tau^2).$$

Taking $c = \dfrac{a}{b}$, where a, b are constants, we get

$$\kappa = \frac{a}{b}(\kappa^2 + \tau^2)$$

or $$a\,(\kappa^2 + \tau^2) = b\kappa.$$

Example 23. On the tangent to a given curve C, a point Q is taken at a constant distance c from the point of contact. Prove that the curvature κ_1 of the locus C_1 of Q is given by

$$\kappa_1^2\,(1 + c^2\kappa^2)^3 = c^2\kappa^2\tau^2(1 + c^2\kappa^2) + (\kappa + c\kappa' + c^2\kappa^3)^2.$$

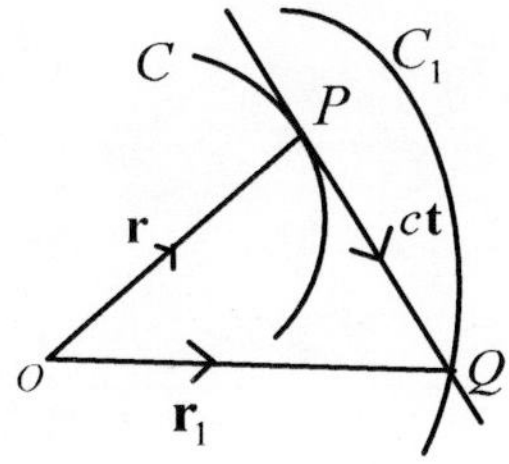

Solution. Let us distinguish the quantities of C_1 from those of C by using the suffix unity.

Let $\mathbf{r}$ and $\mathbf{r}_1$ denote the position vectors of the points P and Q respectively.

Then, we have

$$\mathbf{r}_1 = \mathbf{r} + c\mathbf{t}, \tag{1}$$

where c is a scalar quantity.

Differentiating (1) with respect to s, we get

$$\frac{d\mathbf{r}_1}{ds} = \frac{d\mathbf{r}}{ds} + c\frac{d\mathbf{t}}{ds}$$

or $$\frac{d\mathbf{r}_1}{ds_1}\frac{ds_1}{ds} = \mathbf{t} + c\,(\kappa\mathbf{n})$$

or $$\mathbf{t}_1\frac{ds_1}{ds} = \mathbf{t} + c\kappa\mathbf{n}. \tag{2}$$

Squaring (2), we get

$$\left(\frac{ds_1}{ds}\right)^2 = 1 + c^2\kappa^2. \tag{3}$$

Differentiating (2) again with respect to s, we get

$$\frac{d}{ds}\left(\mathbf{t}_1\frac{ds_1}{ds}\right) = \frac{d\mathbf{t}}{ds} + \frac{d}{ds}(c\kappa\mathbf{n})$$

or $$\left(\frac{d\mathbf{t}_1}{ds_1}\frac{ds_1}{ds}\right)\frac{ds_1}{ds} + \mathbf{t}_1\frac{d^2s_1}{ds^2} = \kappa\mathbf{n} + [c\kappa'\mathbf{n} + c\kappa\,(\tau\mathbf{b} - \kappa\mathbf{t})]$$

or $$\kappa_1\mathbf{n}_1\left(\frac{ds_1}{ds}\right)^2 + \mathbf{t}_1\frac{d^2s_1}{ds^2} = -\,c\kappa^2\mathbf{t} + (c\kappa' + \kappa)\,\mathbf{n} + c\kappa\tau\mathbf{b}. \tag{4}$$

Taking the cross product of (2) and (4), we get

$$\kappa_1\left(\frac{ds_1}{ds}\right)^3\mathbf{b}_1 = (c\kappa' + \kappa)\,\mathbf{b} - c\kappa\tau\mathbf{n} + c^2\kappa^3\mathbf{b} + c^2\kappa^2\tau\mathbf{t}$$

or $$\kappa_1\left(\frac{ds_1}{ds}\right)^3\mathbf{b}_1 = c^2\kappa^2\tau\,\mathbf{t} - c\kappa\tau\mathbf{n} + (c^2\kappa^3 + c\kappa' + \kappa)\,\mathbf{b}. \tag{5}$$

Squaring (5), we get

$$\kappa_1^{\,2}\left(\frac{ds_1}{ds}\right)^6 = c^4\kappa^4\tau^2 + c^2\kappa^2\tau^2 + (c^2\kappa^3 + c\kappa' + \kappa)^2$$

or $$\kappa_1^2\left(\frac{ds_1}{ds}\right)^6 = c^2\kappa^2\tau^2(1+c^2\kappa^2)+(\kappa+c\kappa'+c^2\kappa^3)^2. \qquad (6)$$

Substituting the value of $\frac{ds_1}{ds}$ from (3) in (6), we get

$$\kappa_1^2(1+c^2\kappa^2)^3 = c^2\kappa^2\tau^2(1+c^2\kappa^2)+(\kappa+c\kappa'+c^2\kappa^3)^2.$$

Example 24. On the principal normal to a given curve, a point Q is taken at a constant distance c from the curve. Show that the curvature κ_1 of the locus of Q is given by

$$\kappa_1^2[(1-c\kappa)^2+c^2\tau^2]^3 = [c\tau^2-\kappa(1-c\kappa)]^2[c^2\tau^2+(1-c\kappa)^2] + c^2[\tau'-c(\tau'\kappa-\tau\kappa')]^2.$$

Solution. Let us distinguish the quantities of C_1 from those of C by using the suffix unity.

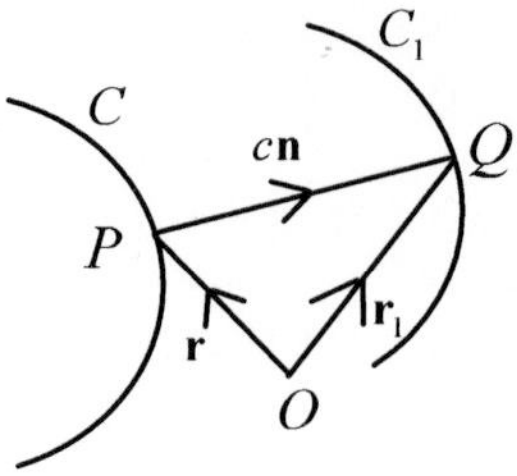

Let $\mathbf{r}$ and $\mathbf{r}_1$ denote the position vectors of the points P and Q respectively.

Then, we have

$$\mathbf{r}_1 = \mathbf{r} + c\mathbf{n}, \qquad (1)$$

where c is a scalar quantity.

Differentiating (1) with respect to s, we get

$$\frac{d\mathbf{r}_1}{ds} = \frac{d\mathbf{r}}{ds} + c\frac{d\mathbf{n}}{ds}$$

or $$\frac{d\mathbf{r}_1}{ds_1}\frac{ds_1}{ds} = \mathbf{t} + c(\tau\mathbf{b} - \kappa\mathbf{t})$$

or $$\mathbf{t}_1\frac{ds_1}{ds} = (1-c\kappa)\mathbf{t} + c\tau\mathbf{b}. \qquad (2)$$

Squaring (2), we get

$$\left(\frac{ds_1}{ds}\right)^2 = (1-c\kappa)^2 + c^2\tau^2. \qquad (3)$$

Differentiating (2) again with respect to s, we get

$$\frac{d}{ds}\left(\mathbf{t}_1\frac{ds_1}{ds}\right) = \frac{d}{ds}[(1-c\kappa)\mathbf{t}] + \frac{d}{ds}(c\tau\mathbf{b})$$

or $$\left(\frac{d\mathbf{t}_1}{ds_1}\frac{ds_1}{ds}\right)\frac{ds_1}{ds}+\mathbf{t}_1\frac{d^2s_1}{ds^2}=[(-c\kappa')\,\mathbf{t}+(1-c\kappa)\,\kappa\mathbf{n}]$$
$$+[c\tau'\mathbf{b}+c\tau\,(-\tau\mathbf{n})]$$

or $$\kappa_1\mathbf{n}_1\left(\frac{ds_1}{ds}\right)^2+\mathbf{t}_1\frac{d^2s_1}{ds^2}=-c\kappa'\,\mathbf{t}+(\kappa-c\kappa^2-c\tau^2)\,\mathbf{n}+c\tau'\mathbf{b}. \tag{4}$$

Taking the cross product of (2) and (4), we get

$$\kappa_1\left(\frac{ds_1}{ds}\right)^3\mathbf{b}_1=(1-c\kappa)(\kappa-c\kappa^2-c\tau^2)\,\mathbf{b}-(1-c\kappa)\,c\tau'\mathbf{n}$$
$$-c^2\tau\kappa'\mathbf{n}-(\kappa-c\kappa^2-c\tau^2)\,c\tau\mathbf{t}$$

or $$\kappa_1\left(\frac{ds_1}{ds}\right)^3\mathbf{b}_1=-c\tau\,(\kappa-c\kappa^2-c\tau^2)\,\mathbf{t}-(c\tau'-c^2\tau'\kappa+c^2\tau\kappa')\,\mathbf{n}$$
$$+(1-c\kappa)(\kappa-c\kappa^2-c\tau^2)\,\mathbf{b}. \tag{5}$$

Squaring (5), we get

$$\kappa_1^{\,2}\left(\frac{ds_1}{ds}\right)^6=c^2\tau^2(\kappa-c\kappa^2-c\tau^2)^2+(c\tau'-c^2\tau'\kappa+c^2\tau\kappa')^2$$
$$+(1-c\kappa)^2(\kappa-c\kappa^2-c\tau^2)^2$$

or $$\kappa_1^{\,2}\left(\frac{ds_1}{ds}\right)^6=[c\tau^2-\kappa\,(1-c\kappa)]^2\,[c^2\tau^2+(1-c\kappa)^2]$$
$$+c^2[\tau'-c\,(\tau'\kappa-\tau\kappa')]^2. \tag{6}$$

Substituting the value of $\dfrac{ds_1}{ds}$ from (3) in (6), we get

$$\kappa_1^{\,2}\,[(1-c\kappa)^2+c^2\tau^2]^3=[c\tau^2-\kappa\,(1-c\kappa)]^2\,[c^2\tau^2+(1-c\kappa)^2]$$
$$+c^2[\tau'-c\,(\tau'\kappa-\tau\kappa')]^2.$$

Example 25. On the binormal to a given curve, a point Q is taken at a constant distance c from the curve. Prove that the curvature κ_1 of the locus of Q is given by

$$\kappa_1^{\,2}\,(1+c^2\tau^2)^3=c^2\tau^4(1+c^2\tau^2)+(\kappa-c\tau'+c^2\kappa\tau^2)^2.$$

Solution. Let us distinguish the quantities of C_1 from those of C by using the suffix unity.

Let $\mathbf{r}$ and $\mathbf{r}_1$ denote the position vectors of the points P and Q respectively.

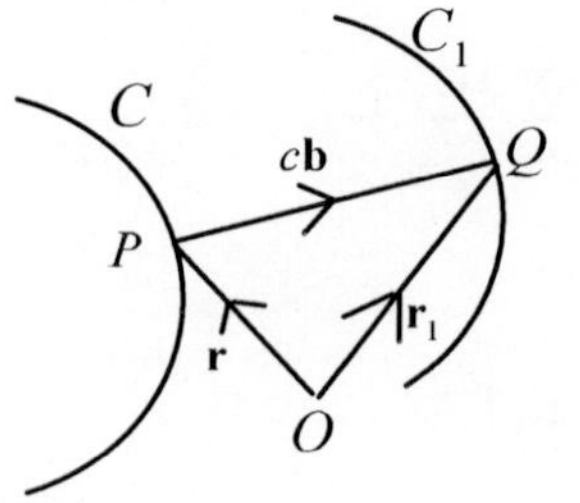

Then, we have

$$\mathbf{r}_1 = \mathbf{r} + c\mathbf{b}, \tag{1}$$

where c is a scalar quantity.

Differentiating (1) with respect to s, we get

$$\frac{d\mathbf{r}_1}{ds} = \frac{d\mathbf{r}}{ds} + c\frac{d\mathbf{b}}{ds}$$

or $$\frac{d\mathbf{r}_1}{ds_1}\frac{ds_1}{ds} = \mathbf{t} + c\,(-\tau\mathbf{n})$$

or $$\mathbf{t}_1\frac{ds_1}{ds} = \mathbf{t} - c\tau\mathbf{n}. \tag{2}$$

Squaring (2), we get

$$\left(\frac{ds_1}{ds}\right)^2 = 1 + c^2\tau^2. \tag{3}$$

Differentiating (2) again with respect to s, we get

$$\frac{d}{ds}\left(\mathbf{t}_1\frac{ds_1}{ds}\right) = \frac{d\mathbf{t}}{ds} - \frac{d}{ds}(c\tau\mathbf{n})$$

or $$\left(\frac{d\mathbf{t}_1}{ds_1}\frac{ds_1}{ds}\right)\frac{ds_1}{ds} + \mathbf{t}_1\frac{d^2s_1}{ds^2} = \kappa\mathbf{n} - [c\tau'\mathbf{n} + c\tau\,(\tau\mathbf{b} - \kappa\mathbf{t})]$$

or $$\kappa_1\mathbf{n}_1\left(\frac{ds_1}{ds}\right)^2 + \mathbf{t}_1\frac{d^2s_1}{ds^2} = c\kappa\tau\mathbf{t} + (\kappa - c\tau')\,\mathbf{n} - c\tau^2\mathbf{b}. \tag{4}$$

Taking the cross product of (2) and (4), we get

$$\kappa_1\left(\frac{ds_1}{ds}\right)^3\mathbf{b}_1 = (\kappa - c\tau')\,\mathbf{b} + c\tau^2\mathbf{n} + c^2\tau^2\kappa\mathbf{b} + c^2\tau^3\mathbf{t}$$

or $$\kappa_1\left(\frac{ds_1}{ds}\right)^3\mathbf{b}_1 = c^2\tau^3\mathbf{t} + c\tau^2\mathbf{n} + (\kappa - c\tau' + c^2\tau^2\kappa)\,\mathbf{b}. \tag{5}$$

Squaring (5), we get

$$\kappa_1^2\left(\frac{ds_1}{ds}\right)^6 = c^4\tau^6 + c^2\tau^4 + (\kappa - c\tau' + c^2\tau^2\kappa)^2$$

or $$\kappa_1^2\left(\frac{ds_1}{ds}\right)^6 = c^2\tau^4(1+c^2\tau^2)+(\kappa-c\tau'+c^2\tau^2\kappa)^2. \qquad (6)$$

Substituting the value of $\dfrac{ds_1}{ds}$ from (3) in (6), we get

$$\kappa_1^2\,(1+c^2\tau^2)^3 = c^2\tau^4(1+c^2\tau^2)+(\kappa-c\tau'+c^2\kappa\tau^2)^2.$$

Example 26. On the binormal of a curve of constant torsion τ, a point Q is taken at a constant distance c from the curve. Show that the binormal to the locus of Q is inclined to the binormal of the given curve at an angle $\tan^{-1}\left\{c\tau^2/\kappa\sqrt{(c^2\tau^2+1)}\right\}$.

Solution. Let us distinguish the quantities of C_1 from those of C by using the suffix unity.

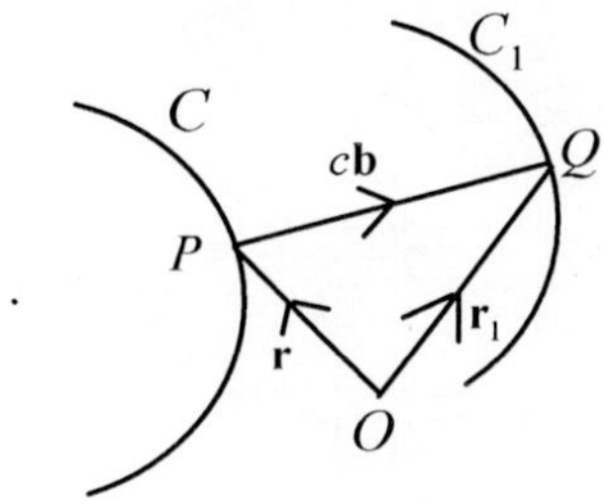

Let $\mathbf{r}$ and $\mathbf{r}_1$ denote the position vectors of the points P and Q respectively.

Then, we have

$$\mathbf{r}_1 = \mathbf{r} + c\mathbf{b}, \qquad (1)$$

where c is a scalar quantity.

Differentiating (1) with respect to s, we get

$$\frac{d\mathbf{r}_1}{ds} = \frac{d\mathbf{r}}{ds} + c\frac{d\mathbf{b}}{ds}$$

or $$\frac{d\mathbf{r}_1}{ds_1}\frac{ds_1}{ds} = \mathbf{t} + c\,(-\tau\mathbf{n})$$

or $$\mathbf{t}_1\frac{ds_1}{ds} = \mathbf{t} - c\tau\mathbf{n}. \qquad (2)$$

Squaring (2), we get

$$\left(\frac{ds_1}{ds}\right)^2 = 1 + c^2\tau^2. \qquad (3)$$

Differentiating (2) again with respect to s, we get

$$\frac{d}{ds}\left(\mathbf{t}_1\frac{ds_1}{ds}\right) = \frac{d\mathbf{t}}{ds} - \frac{d}{ds}(c\tau\mathbf{n})$$

or $$\left(\frac{d\mathbf{t}_1}{ds_1}\frac{ds_1}{ds}\right)\frac{ds_1}{ds} + \mathbf{t}_1\frac{d^2s_1}{ds^2} = \kappa\mathbf{n} - c\tau\,(\tau\mathbf{b} - \kappa\mathbf{t})$$

or $$\kappa_1 \mathbf{n}_1 \left(\frac{ds_1}{ds}\right)^2 + \mathbf{t}_1 \frac{d^2 s_1}{ds^2} = c\kappa\tau\mathbf{t} + \kappa\mathbf{n} - c\tau^2\mathbf{b}. \tag{4}$$

Taking the cross product of (2) and (4), we get

$$\kappa_1 \left(\frac{ds_1}{ds}\right)^3 \mathbf{b}_1 = \kappa\mathbf{b} + c\tau^2\mathbf{n} + c^2\tau^2\kappa\mathbf{b} + c^2\tau^3\mathbf{t}$$

or $$\kappa_1 \left(\frac{ds_1}{ds}\right)^3 \mathbf{b}_1 = c^2\tau^3\mathbf{t} + c\tau^2\mathbf{n} + \kappa\,(1 + c^2\tau^2)\,\mathbf{b}. \tag{5}$$

Squaring (5), we get

$$\kappa_1^2 \left(\frac{ds_1}{ds}\right)^6 = c^4\tau^6 + c^2\tau^4 + \kappa^2 (1 + c^2\tau^2)^2$$

or $$\kappa_1^2 \left(\frac{ds_1}{ds}\right)^6 = c^2\tau^4 (c^2\tau^2 + 1) + \kappa^2 (1 + c^2\tau^2)^2$$

or $$\kappa_1^2 \left(\frac{ds_1}{ds}\right)^6 = (1 + c^2\tau^2)\,[\,c^2\tau^4 + \kappa^2 (1 + c^2\tau^2)]. \tag{6}$$

Taking the dot product of both sides of (5) with **b**, we get

$$\kappa_1 \left(\frac{ds_1}{ds}\right)^3 \mathbf{b}_1 . \mathbf{b} = \kappa\,(1 + c^2\tau^2)$$

or $$\kappa_1 \left(\frac{ds_1}{ds}\right)^3 \cos\theta = \kappa\,(1 + c^2\tau^2), \tag{7}$$

where θ is the angle between **b** and $\mathbf{b}_1$.
Squaring (7), we get

$$\kappa_1^2 \left(\frac{ds_1}{ds}\right)^6 \cos^2\theta = \kappa^2 (1 + c^2\tau^2)^2 \tag{8}$$

From (6) and (8), we get

$$(1 + c^2\tau^2)\,[c^2\tau^4 + \kappa^2 (1 + c^2\tau^2)]\cos^2\theta = \kappa^2 (1 + c^2\tau^2)^2$$

or $$[c^2\tau^4 + \kappa^2 (1 + c^2\tau^2)]\cos^2\theta = \kappa^2 (1 + c^2\tau^2)$$

or $$\sec^2\theta = \frac{[c^2\tau^4 + \kappa^2 (1 + c^2\tau^2)]}{\kappa^2 (1 + c^2\tau^2)}$$

or $$1 + \tan^2\theta = \frac{[c^2\tau^4 + \kappa^2 (1 + c^2\tau^2)]}{\kappa^2 (1 + c^2\tau^2)}$$

or $$\tan^2\theta = \frac{[c^2\tau^4 + \kappa^2(1+c^2\tau^2)]}{\kappa^2(1+c^2\tau^2)} - 1 = \frac{c^2\tau^4}{\kappa^2(1+c^2\tau^2)}$$

or $$\tan\theta = \frac{c\tau^2}{\kappa\sqrt{(1+c^2\tau^2)}}.$$

$\therefore$ $$\theta = \tan^{-1}\frac{c\tau^2}{\kappa\sqrt{(1+c^2\tau^2)}}.$$

Example 27. If there is one-one correspondence between the points of two curves and the tangents at the corresponding points are parallel, show that the principal normals are parallel and therefore also the binormals. Also, prove that

$$\frac{\kappa_1}{\kappa} = \frac{ds}{ds_1} = \frac{\tau_1}{\tau}.$$

Solution. Let C and C_1 be two curves. We shall use the suffix unity to denote the quantities belonging to the curve C_1.

Since the tangents at the corresponding points of C and C_1 are parallel, we have

$$\mathbf{t}_1 = \pm\mathbf{t}. \tag{1}$$

Differentiating (1) with respect to s, we get

$$\frac{d\mathbf{t}_1}{ds_1}\frac{ds_1}{ds} = \pm\frac{d\mathbf{t}}{ds}.$$

or $$\kappa_1\mathbf{n}_1\frac{ds_1}{ds} = \pm\kappa\mathbf{n}. \tag{2}$$

It follows from equation (2) that the principal normals at the corresponding points of C and C_1 are parallel.

Squaring (2), we get

$$\kappa_1^2\left(\frac{ds_1}{ds}\right)^2 = \kappa^2$$

or $$\frac{ds_1}{ds} = \frac{\kappa}{\kappa_1}$$

or $$\frac{ds}{ds_1} = \frac{\kappa_1}{\kappa}. \tag{3}$$

Taking the cross product of (1) and (2), we get

$$\kappa_1\frac{ds_1}{ds}\mathbf{b}_1 = \kappa\mathbf{b}. \tag{4}$$

It follows from equation (2) that the binormals at the corresponding points of C and C_1 are parallel.

Again, differentiating (4) with respect to s, we get

$$\left[\frac{d}{ds}\left(\kappa_1\frac{ds_1}{ds}\right)\right]\mathbf{b}_1+\left(\kappa_1\frac{ds_1}{ds}\right)\left(-\tau_1\mathbf{n}_1\frac{ds_1}{ds}\right)=\frac{d\kappa}{ds}\mathbf{b}+\kappa(-\tau\mathbf{n}). \tag{5}$$

Taking the dot product of (2) and (5), we get

$$\kappa_1^{\ 2}\tau_1\left(\frac{ds_1}{ds}\right)^3=\pm\kappa^2\tau.$$

Taking positive sign, we have

$$\left(\frac{ds_1}{ds}\right)^3=\frac{\tau}{\tau_1}\frac{\kappa^2}{\kappa_1^{\ 2}}$$

$$=\frac{\tau}{\tau_1}\left(\frac{ds_1}{ds}\right)^2. \qquad \text{[using (3)]}$$

$$\therefore \quad \frac{ds}{ds_1}=\frac{\tau_1}{\tau}. \tag{6}$$

Combining (3) and (6), we get

$$\frac{\kappa_1}{\kappa}=\frac{ds}{ds_1}=\frac{\tau_1}{\tau}.$$

Exercises

1. Prove that $\left[\dot{\mathbf{r}},\ \ddot{\mathbf{r}},\ \dddot{\mathbf{r}}\right]=0$ is a necessary and sufficient condition that the curve be plane.
2. If the nth derivative of $\mathbf{r}$ with respect to s is given by
   $$\mathbf{r}^{(n)}=a_n\mathbf{t}+b_n\mathbf{n}+c_n\mathbf{b},$$
   prove the reduction formulae
   $$a_{n+1}=a_n{}'-\kappa b_n,$$
   $$b_{n+1}=b_n{}'+\kappa a_n-\tau c_n,$$
   $$c_{n+1}=c_n{}'+\tau b_n.$$
3. Prove that the principal normals at consecutive points do not intersect unless $\tau=0$.
4. Find the radii of curvature and torsion of the circular helix
   $$x=a\cos\theta,\ y=a\sin\theta,\ z=a\theta\tan\alpha.$$
   [**Ans.** $\rho=a\sec^2\alpha,\ \ \sigma=a\sec\alpha\operatorname{cosec}\alpha$]

5. Find the radii of curvature and torsion at the point $\theta = \pi/4$ of the curve given by

$$\mathbf{r} = (a\cos\theta,\ a\sin\theta,\ a\cos 2\theta). \qquad \left[\textbf{Ans. } \rho = 5a,\ \sigma = \frac{5}{6}a\right]$$

6. Find the curvature and torsion of the curves given by

(a) $\mathbf{r} = (e^t,\ -e^t,\ \sqrt{2}\,t)$.

$$[\textbf{ Ans. } \kappa = \sqrt{2}(e^t + e^{-t})^2,\ \tau = -\sqrt{2}(e^t + e^{-t})^2\ ]$$

(b) $\mathbf{r} = [2a\sin^{-1} t + 2at\sqrt{(1-t^2)},\ 2at^2, 4at)$.

$$\left[\textbf{Ans. } \kappa = \tau = \frac{1}{8a\sqrt{(1-t^2)}}\right]$$

7. A point moves on a sphere of radius a so that its latitude is equal to its longitude. Prove that at $(x,\ y,\ z)$,

$$\rho = \frac{(2a^2 - z^2)^{3/2}}{a\,(8a^2 - 3z^2)^{1/2}}, \quad \sigma = \frac{8a^2 - 3z^2}{6\,(a^2 - z^2)^{1/2}}.$$

[**Hint.** Any point on the sphere is given by

$x = a\cos\theta\cos\phi,\ y = a\cos\theta\sin\phi,\ z = a\sin\theta.$

In case, its latitude is equal to longitude, $\theta = \phi$. Therefore

$x = a\cos^2\theta,\ y = a\cos\theta\sin\theta,\ z = a\sin\theta.$]

8. Show that if the space curve C, $\mathbf{r} = \mathbf{r}\,(s)$ has constant torsion τ, then the curve C_1

$$\mathbf{r}_1 = -\frac{\mathbf{n}}{\tau} + \int \mathbf{b}\, ds$$

has constant torsion $\pm\tau$.

1.15 FUNDAMENTAL THEOREMS FOR SPACE CURVES

Theorem 1. Existence Theorem for Space Curves. If f and g are continuous functions of a real variable $s\,(s \geq 0)$, then there exists a space curve for which s is the arc length measured from some suitable base point and for which $f\,(s)$ and $g\,(s)$ are the curvature and the torsion respectively.

Proof. We consider the three linear differential equations

$$\frac{d\alpha}{ds} = f(s)\,\beta, \frac{d\beta}{ds} = g(s)\,\gamma - f(s)\,\alpha, \frac{d\gamma}{ds} = -g(s)\,\beta. \tag{1}$$

From existence theorem on linear differential equations, we know that if $\alpha_0, \beta_0, \gamma_0$ are given constants, then equations (1) admit a unique set of solutions $\alpha(s), \beta(s), \gamma(s)$, for which $\alpha(0) = \alpha_0, \beta(0) = \beta_0, \gamma(0) = \gamma_0$.

Hence, we can find a unique set of solutions $\alpha_1(s), \beta_1(s), \gamma_1(s)$ which assume values 1, 0, 0 at $s = 0$.

Similarly, there exists a unique set of solutions $\alpha_2(s), \beta_2(s), \gamma_2(s)$ which assume values 0, 1, 0 at $s = 0$ and a unique set of solutions $\alpha_3(s), \beta_3(s), \gamma_3(s)$ which assume values 0, 0, 1 at $s = 0$.

Now, we prove that for all values of s,

$$\alpha_1^2 + \beta_1^2 + \gamma_1^2 = \alpha_2^2 + \beta_2^2 + \gamma_2^2 = \alpha_3^2 + \beta_3^2 + \gamma_3^2 = 1$$

and $$\alpha_1\alpha_2 + \beta_1\beta_2 + \gamma_1\gamma_2 = \alpha_2\alpha_3 + \beta_2\beta_3 + \gamma_2\gamma_3$$

$$= \alpha_3\alpha_1 + \beta_3\beta_1 + \gamma_3\gamma_1 = 0.$$

We have

$$\frac{d}{ds}(\alpha_1^2 + \beta_1^2 + \gamma_1^2) = 2\left(\alpha_1\frac{d\alpha_1}{ds} + \beta_1\frac{d\beta_1}{ds} + \gamma_1\frac{d\gamma_1}{ds}\right)$$

$$= 2\left[\alpha_1(f\beta_1) + \beta_1(g\gamma_1 - f\alpha_1) + \gamma_1(-g\beta_1)\right]$$

[using (1)]

$$= 0.$$

Integrating, we get

$$\alpha_1^2 + \beta_1^2 + \gamma_1^2 = c_1,$$

where c_1 is an arbitrary constant.
At $s = 0$, we have

$$\alpha_1 = 1, \beta_1 = 0, \gamma_1 = 0.$$

Hence, $c_1 = 1$.
Thus, we have

$$\alpha_1^2 + \beta_1^2 + \gamma_1^2 = 1.$$

Similarly,

$$\alpha_2^2 + \beta_2^2 + \gamma_2^2 = 1$$

and $$\alpha_3^2 + \beta_3^2 + \gamma_3^2 = 1.$$

Therefore, we have

$$\left.\begin{aligned}\alpha_1^2+\beta_1^2+\gamma_1^2&=1,\\ \alpha_2^2+\beta_2^2+\gamma_2^2&=1,\\ \alpha_3^2+\beta_3^2+\gamma_3^2&=1.\end{aligned}\right\} \tag{2}$$

Again,

$$\frac{d}{ds}(\alpha_1\alpha_2+\beta_1\beta_2+\gamma_1\gamma_2)$$

$$=\left(\frac{d\alpha_1}{ds}\alpha_2+\alpha_1\frac{d\alpha_2}{ds}\right)+\left(\frac{d\beta_1}{ds}\beta_2+\beta_1\frac{d\beta_2}{ds}\right)+\left(\frac{d\gamma_1}{ds}\gamma_2+\gamma_1\frac{d\gamma_2}{ds}\right)$$

$$=[(f\beta_1)\alpha_2+\alpha_1(f\beta_2)]+[(g\gamma_1-f\alpha_1)\beta_2+\beta_1(g\gamma_2-f\alpha_2)]$$

$$+[(-g\beta_1)\gamma_2+\gamma_1(-g\beta_2)] \quad \text{[using (1)]}$$

$$=0$$

Integrating, we get

$$\alpha_1\alpha_2+\beta_1\beta_2+\gamma_1\gamma_2=d_1,$$

where d_1 is an arbitrary constant.
At $s=0$, we have

$$\alpha_1=1,\beta_1=0,\gamma_1=0 \text{ and } \alpha_2=0,\beta_2=1,\gamma_2=0.$$

Hence, $d_1=0$.
Thus, we have

$$\alpha_1\alpha_2+\beta_1\beta_2+\gamma_1\gamma_2=0.$$

Similarly,

$$\alpha_2\alpha_3+\beta_2\beta_3+\gamma_2\gamma_3=0$$

and $\alpha_3\alpha_1+\beta_3\beta_1+\gamma_3\gamma_1=0.$

Therefore

$$\left.\begin{aligned}\alpha_1\alpha_2+\beta_1\beta_2+\gamma_1\gamma_2&=0,\\ \alpha_2\alpha_3+\beta_2\beta_3+\gamma_2\gamma_3&=0,\\ \alpha_3\alpha_1+\beta_3\beta_1+\gamma_3\gamma_1&=0.\end{aligned}\right\} \tag{3}$$

Thus we have six equations given by (2) and (3).
If we write

$$A=\begin{bmatrix}\alpha_1&\beta_1&\gamma_1\\ \alpha_2&\beta_2&\gamma_2\\ \alpha_3&\beta_3&\gamma_3\end{bmatrix} \text{ and } A^T=\begin{bmatrix}\alpha_1&\alpha_2&\alpha_3\\ \beta_1&\beta_2&\beta_3\\ \gamma_1&\gamma_2&\gamma_3\end{bmatrix},$$

where A^T denotes the transpose of A, then the six equations given by (2) and (3) are equivalent to the single matrix equation

$AA^T = I$, where I is the identity matrix of order 3.

$\Rightarrow \quad A^{-1}(AA^T) = A^{-1}I$

$\Rightarrow \quad (A^{-1}A)A^T = A^{-1}$

$\Rightarrow \quad IA^T = A^{-1}$

$\Rightarrow \quad A^T = A^{-1}$

$\Rightarrow \quad A^T A = A^{-1}A$

$\Rightarrow \quad A^T A = I.$

This is equivalent to the six equations given by

$$\left.\begin{aligned} \alpha_1^2 + \alpha_2^2 + \alpha_3^2 &= 1, \\ \beta_1^2 + \beta_2^2 + \beta_3^2 &= 1, \\ \gamma_1^2 + \gamma_2^2 + \gamma_3^2 &= 1, \end{aligned}\right\} \tag{4}$$

and

$$\left.\begin{aligned} \alpha_1\beta_1 + \alpha_2\beta_2 + \alpha_3\beta_3 &= 0, \\ \beta_1\gamma_1 + \beta_2\gamma_2 + \beta_3\gamma_3 &= 0, \\ \gamma_1\alpha_1 + \gamma_2\alpha_2 + \gamma_3\alpha_3 &= 0. \end{aligned}\right\} \tag{5}$$

It follows from (4) and (5) that there are three mutually perpendicular unit vectors

$$\vec{\alpha} = (\alpha_1, \alpha_2, \alpha_3),\ \vec{\beta} = (\beta_1, \beta_2, \beta_3), \vec{\gamma} = (\gamma_1, \gamma_2, \gamma_3)$$

defined for each value of s.

Let $\vec{\alpha}, \vec{\beta}, \vec{\gamma}$ form a right handed system so that $\vec{\alpha} \times \vec{\beta} = \vec{\gamma}$, $\vec{\beta} \times \vec{\gamma} = \vec{\alpha}$ and $\vec{\gamma} \times \vec{\alpha} = \vec{\beta}$.

Now, we consider the curve

$$\mathbf{r} = \mathbf{r}(s) = \int_0^s \vec{\alpha}\, ds. \tag{6}$$

Differentiating (6) with respect to s, we get

$$\frac{d\mathbf{r}}{ds} = \vec{\alpha}$$

$\Rightarrow \quad s$ is the arc length and $\vec{\alpha}$ is the unit tangent vector of the curve (6).

$$\Rightarrow \quad \mathbf{t} = \vec{\alpha}. \tag{7}$$

Differentiating (7) with respect to s, we get

$$\frac{d\mathbf{t}}{ds} = \frac{d\vec{\alpha}}{ds}$$

$$= f(s)\vec{\beta} \qquad \text{[using (1)]}$$

$$\Rightarrow \quad \kappa\mathbf{n} = f(s)\vec{\beta} \qquad (8)$$

$\Rightarrow$ The unit principal normal $\mathbf{n}$ is parallel to $\vec{\beta}$.

$\Rightarrow$ We can take

$$\mathbf{n} = \vec{\beta}. \qquad (9)$$

[measuring s in the direction which makes $\mathbf{n}$ to have the same direction as $\vec{\beta}$]

Substituting the value of $\mathbf{n}$ from (9) in (8), we get

$$\kappa = f(s). \qquad (10)$$

Taking the cross product of (7) and (9), we get

$$\mathbf{t} \times \mathbf{n} = \vec{\alpha} \times \vec{\beta}$$

$$\Rightarrow \quad \mathbf{b} = \vec{\gamma}. \qquad (11)$$

Differentiating (11) with respect to s, we get

$$\frac{d\mathbf{b}}{ds} = \frac{d\vec{\gamma}}{ds}$$

$$= -g(s)\vec{\beta} \qquad \text{[using (1)]}$$

$$\Rightarrow \quad -\tau\mathbf{n} = -g(s)\vec{\beta}$$

$$\Rightarrow \quad \tau = g(s). \qquad (12)$$

[using (9)]

It follows from (10) and (12) that the curvature and torsion of the curve (6) are $f(s)$ and $g(s)$ respectively. Hence the theorem is proved.

Theorem 2. Uniqueness Theorem for Space Curves. A curve is uniquely determined except as to position in space when its curvature and torsion are given functions of its arc length s.

Proof. Let C and C_1 be any two curves such that their curvatures and torsions are same for the same arc length s.

Let $\mathbf{t}, \mathbf{n}, \mathbf{b}$ and $\mathbf{t}_1, \mathbf{n}_1, \mathbf{b}_1$ refer to C and C_1 respectively at the corresponding points. By the corresponding points, we mean the points on the two curves having the same arc length.

Now,

$$\frac{d}{ds}(\mathbf{t}\cdot\mathbf{t}_1)=\mathbf{t}\cdot\frac{d\mathbf{t}_1}{ds}+\frac{d\mathbf{t}}{ds}\cdot\mathbf{t}_1=\mathbf{t}\cdot\kappa\mathbf{n}_1+\kappa\mathbf{n}\cdot\mathbf{t}_1, \tag{1}$$

$$\frac{d}{ds}(\mathbf{n}\cdot\mathbf{n}_1)=\mathbf{n}\cdot\frac{d\mathbf{n}_1}{ds}+\frac{d\mathbf{n}}{ds}\cdot\mathbf{n}_1=\mathbf{n}\cdot(\tau\mathbf{b}_1-\kappa\mathbf{t}_1)+(\tau\mathbf{b}-\kappa\mathbf{t})\cdot\mathbf{n}_1, \tag{2}$$

$$\frac{d}{ds}(\mathbf{b}\cdot\mathbf{b}_1)=\mathbf{b}\cdot\frac{d\mathbf{b}_1}{ds}+\frac{d\mathbf{b}}{ds}\cdot\mathbf{b}_1=\mathbf{b}\cdot(-\tau\mathbf{n}_1)+(-\tau\mathbf{n})\cdot\mathbf{b}_1. \tag{3}$$

Adding (1), (2) and (3), we get

$$\frac{d}{ds}(\mathbf{t}\cdot\mathbf{t}_1+\mathbf{n}\cdot\mathbf{n}_1+\mathbf{b}\cdot\mathbf{b}_1)=0.$$

Integrating, we get

$$\mathbf{t}\cdot\mathbf{t}_1+\mathbf{n}\cdot\mathbf{n}_1+\mathbf{b}\cdot\mathbf{b}_1=h, \tag{4}$$

where h is an arbitrary constant.

Move C_1 and place it in such a way that the initial points of C and C_1 are coincident. Now, rotate C_1 without any deformation to it and place it in such a way that the corresponding principal planes at the coincident initial points of C and C_1 are coincident.

At $s=0$, we have $\mathbf{t}=\mathbf{t}_1, \mathbf{n}=\mathbf{n}_1, \mathbf{b}=\mathbf{b}_1$. Hence it follows that

$$\mathbf{t}\cdot\mathbf{t}_1=\mathbf{n}\cdot\mathbf{n}_1=\mathbf{b}\cdot\mathbf{b}_1=1 \text{ at } s=0. \tag{5}$$

From (4) and (5), we get $h = 3$.

Substituting this value of h in (4), we get

$$\mathbf{t}\cdot\mathbf{t}_1+\mathbf{n}\cdot\mathbf{n}_1+\mathbf{b}\cdot\mathbf{b}_1=3 \text{ for all values of } s$$

$$\Rightarrow \quad \mathbf{t}\cdot\mathbf{t}_1=\mathbf{n}\cdot\mathbf{n}_1=\mathbf{b}\cdot\mathbf{b}_1=1 \text{ for all values of } s$$

[∵ Sum of cosines of three angles can be equal to 3 only if cosine of each angle is equal to 1]

$\Rightarrow$ $\mathbf{t}=\mathbf{t}_1, \mathbf{n}=\mathbf{n}_1, \mathbf{b}=\mathbf{b}_1$ for all values of s

$\Rightarrow$ The principal planes of the two curves at the corresponding points are parallel.

Also, $\mathbf{t}=\mathbf{t}_1$ for all values of s

$$\Rightarrow \quad \frac{d\mathbf{r}}{ds}=\frac{d\mathbf{r}_1}{ds} \text{ for all values of } s$$

$$\Rightarrow \quad \frac{d}{ds}(\mathbf{r}-\mathbf{r}_1)=\mathbf{0} \text{ for all values of } s.$$

Integrating, we get

$$\mathbf{r}-\mathbf{r}_1=\mathbf{d} \tag{6}$$

for all values of s, where $\mathbf{d}$ is a constant vector.

Since (6) is true for all values of s, it is true for $s = 0$. Therefore, we have

$$\mathbf{d} = \mathbf{0} \qquad [\because \mathbf{r} = \mathbf{r}_1 \text{ at } s = 0]$$

$\Rightarrow \quad \mathbf{r} - \mathbf{r}_1 = \mathbf{0}$ for all values of s [from (6)]

$\Rightarrow \quad \mathbf{r} = \mathbf{r}_1$ for all values of s

$\Rightarrow$ The corresponding points of C and C_1 are coincident.

Hence the theorem is proved.

1.16 OSCULATING CIRCLE (OR CIRCLE OF CURVATURE)

Definition. If a circle has three points of contact with a curve Γ at a point, then the circle is known as the osculating circle (or circle of curvature) of the curve at the point.

Let Γ be a space curve and let P, Q, R be any three points on Γ.

The circle through P, Q, R lies in the plane through P, Q, R. Therefore the limiting case of the circle lies in the limiting case of the plane as $Q \to P$ and $R \to P$ along Γ.

From this, it follows that the osculating circle (or circle of curvature) of Γ at P lies in the osculating plane of Γ at P.

The centre of the circle through the three points P, Q, R is the point of intersection of the perpendicular bisectors of the chords PQ and QR. As Q and R tend to P along Γ, these perpendicular bisectors tend to be a normal to Γ at P.

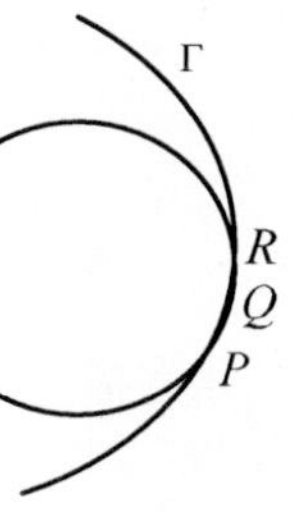

Since the limiting case of the plane through P, Q, R as $Q \to P$ and $R \to P$ along Γ is the osculating plane of Γ at P, it follows that the limiting case of the perpendicular bisectors of the chords PQ and QR lies in the osculating plane.

From the above arguments, we can conclude that the centre of osculating circle at P lies on the normal to Γ at P, i.e., on the principal normal to Γ at P.

The Centre and Radius of the Circle of Curvature

Let Γ be a space curve and let P be a point on it. Let C be the centre of the osculating circle and let ρ be the radius of curvature of Γ at P.

If $\mathbf{r}$ and $\mathbf{r}_1$ be the position vectors of P and C respectively, then

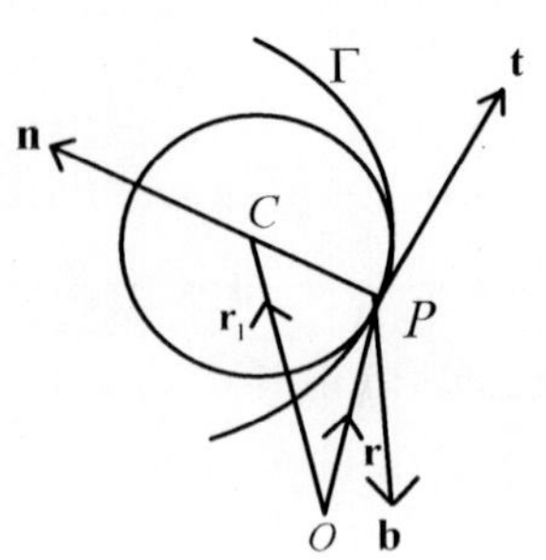

$$\mathbf{r}_1 = \mathbf{r} + \rho\mathbf{n}$$

$$\Rightarrow \quad \mathbf{r}_1 - \mathbf{r} = \rho\mathbf{n}$$

$$\Rightarrow \quad (\mathbf{r}_1 - \mathbf{r})^2 = \rho^2.$$

Let $F(s) = (\mathbf{r}_1 - \mathbf{r})^2 - \rho^2$,

where s is the arc length of Γ.

Since the circle has three points of contact with Γ at P, we have

$$F(s) = F'(s) = F''(s) = 0.$$

Now,

$$F'(s) = 0 \Rightarrow 2(\mathbf{r}_1 - \mathbf{r}) . (-\mathbf{t}) = 0$$

$$\Rightarrow (\mathbf{r}_1 - \mathbf{r}) . \mathbf{t} = 0 \qquad (1)$$

$\Rightarrow (\mathbf{r}_1 - \mathbf{r})$ is perpendicular to $\mathbf{t}$.

$\Rightarrow (\mathbf{r}_1 - \mathbf{r})$ lies in the normal plane of Γ at P.

Again,

$F''(s) = 0 \Rightarrow (\mathbf{r}_1 - \mathbf{r}) . \kappa\mathbf{n} - \mathbf{t} . \mathbf{t} = 0$, where κ is curvature of Γ at P. [using (1)]

$$\Rightarrow (\mathbf{r}_1 - \mathbf{r}) . \kappa\mathbf{n} - 1 = 0$$

$$\Rightarrow (\mathbf{r}_1 - \mathbf{r}) . \mathbf{n} = \frac{1}{\kappa}$$

$$\Rightarrow \rho\mathbf{n} . \mathbf{n} = \frac{1}{\kappa} \qquad [\because \mathbf{r}_1 - \mathbf{r} = \rho\mathbf{n}]$$

$$\Rightarrow \rho = \frac{1}{\kappa}.$$

Thus it follows that the radius of the osculating circle (or the circle of curvature) is nothing but the radius of curvature of the curve at the point.

Some Properties of the Locus of the Centre of Curvature

Let Γ be a space curve and let Γ_1 be the locus of the centre of curvature of Γ. Then Γ_1 has the following properties.

(i) The tangent to Γ_1 lies in the normal plane of Γ and makes an angle $\tan^{-1}(\rho/\sigma\rho')$ with the principal normal of Γ.

(ii) If curvature of Γ is constant, then curvature of Γ_1 is also constant and torsion of Γ_1 varies inversely as the torsion of Γ.

Proof. Let the quantities of Γ_1 be distinguished from those of Γ by using the suffix unity.

(i) The position vector $\mathbf{r}_1$ of the centre of curvature of Γ is given by

$$\mathbf{r}_1 = \mathbf{r} + \rho\mathbf{n}.$$

Differentiating this with respect to s, we get

$$\frac{d\mathbf{r}_1}{ds_1}\frac{ds_1}{ds} = \mathbf{t} + \rho'\mathbf{n} + \rho(\tau\mathbf{b} - \kappa\mathbf{t})$$

$$\Rightarrow \quad \mathbf{t}_1 \frac{ds_1}{ds} = \rho'\mathbf{n} + \rho\tau\mathbf{b} \tag{1}$$

$\Rightarrow$ $\mathbf{t}_1$ lies in the plane containing **n** and **b**, i.e., in the normal plane of Γ.

If $\mathbf{t}_1$ is inclined at an angle α to the principal normal **n**, then

$$\left.\begin{aligned} \cos\alpha &= \mathbf{n}\,.\,\mathbf{t}_1, \\ \sin\alpha &= \mathbf{b}\,.\,\mathbf{t}_1 \end{aligned}\right\} \tag{2}$$

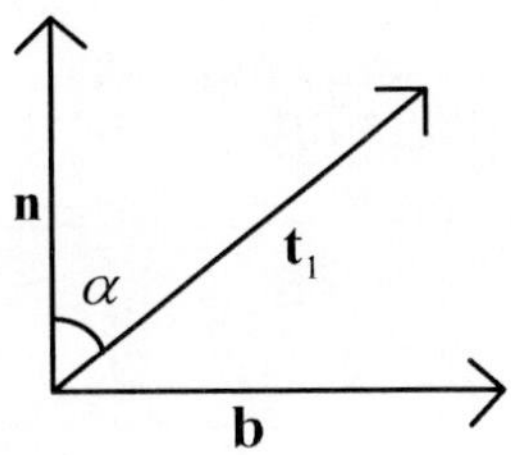

Substituting the value of $\mathbf{t}_1$ from (1) in (2), we get

$$\cos\alpha = \mathbf{n}\,.\,(\rho'\mathbf{n} + \rho\tau\mathbf{b})\frac{ds}{ds_1} = \rho'\frac{ds}{ds_1},$$

$$\sin\alpha = \mathbf{b}\,.\,(\rho'\mathbf{n} + \rho\tau\mathbf{b})\frac{ds}{ds_1} = \rho\tau\frac{ds}{ds_1}.$$

$$\therefore \quad \tan\alpha = \frac{\sin\alpha}{\cos\alpha} = \frac{\rho\tau}{\rho'} = \frac{\rho}{\rho'\sigma}$$

$$\Rightarrow \quad \alpha = \tan^{-1}\left(\frac{\rho}{\rho'\sigma}\right).$$

(ii) If κ is constant, i.e., ρ is constant, then $\rho' = 0$. Therefore, from (1), we have

$$\mathbf{t}_1\frac{ds_1}{ds} = \rho\tau\mathbf{b}. \tag{3}$$

Squaring both sides of (3), we get

$$\left(\frac{ds_1}{ds}\right)^2 = (\rho\tau)^2. \qquad [\because \mathbf{t}_1^{\,2} = \mathbf{b}^2 = 1]$$

$$\therefore \quad \frac{ds_1}{ds} = \rho\tau. \tag{4}$$

Substituting this value of $\frac{ds_1}{ds}$ in (3), we get

$$\mathbf{t}_1 = \mathbf{b}. \tag{5}$$

Differentiating (5) with respect to s, we get

$$\frac{d\mathbf{t}_1}{ds_1}\frac{ds_1}{ds} = -\tau\mathbf{n}$$

$$\Rightarrow \quad \kappa_1\mathbf{n}_1\frac{ds_1}{ds} = -\tau\mathbf{n}$$

$$\Rightarrow \quad \kappa_1\mathbf{n}_1(\rho\tau) = -\tau\mathbf{n} \qquad \text{[using (4)]}$$

$$\Rightarrow \quad \kappa_1\mathbf{n}_1 = -\frac{1}{\rho}\mathbf{n} = -\kappa\mathbf{n}. \tag{6}$$

$$[\because \rho = 1/\kappa]$$

$\Rightarrow$ The principal normal to Γ_1 is parallel to the principal normal to Γ.

We may choose the direction of $\mathbf{n}_1$ as opposite to that of $\mathbf{n}$, so that

$$\mathbf{n}_1 = -\mathbf{n}. \tag{7}$$

From (6) and (7), we get

$$\kappa_1 = \kappa$$

i.e., if curvature of Γ is constant, then curvature of Γ_1 is also constant.

Now,

$$\begin{aligned} \mathbf{b}_1 &= \mathbf{t}_1 \times \mathbf{n}_1 \\ &= \mathbf{b} \times (-\mathbf{n}) \qquad \text{[using (5) and (7)]} \\ &= \mathbf{n} \times \mathbf{b} \\ &= \mathbf{t} \end{aligned}$$

i.e., $\mathbf{b}_1 = \mathbf{t}.$ (8)

Differentiating (8) with respect to s, we get

$$\frac{d\mathbf{b}_1}{ds_1}\frac{ds_1}{ds} = \kappa\mathbf{n}$$

$$\Rightarrow \quad (-\tau_1\mathbf{n}_1)\frac{ds_1}{ds} = \kappa\mathbf{n}$$

$$\Rightarrow \quad (-\tau_1\mathbf{n}_1)(\rho\tau) = \kappa\mathbf{n} \qquad \text{[using (4)]}$$

$$\Rightarrow \quad \tau_1\rho\tau = \kappa \qquad \text{[using (7)]}$$

$$\Rightarrow \quad \tau_1 = \frac{\kappa}{\rho\tau} = \frac{\kappa^2}{\tau} \qquad [\because \rho = 1/\kappa]$$

i.e., torsion of Γ_1 varies inversely as the torsion of Γ.

1.17 OSCULATING SPHERE

Definition. The osculating sphere of a curve at a point is the sphere which has four points of contact with the curve at the point. The centre of the osculating sphere is called the centre of spherical curvature and its radius is called the radius of spherical curvature.

The Centre and Radius of the Osculating Sphere

Let Γ be a space curve and let P be a point on it. Let C be the centre and R the radius of the osculating sphere of Γ at P.

If $\mathbf{r}$ and $\mathbf{r}_1$ be the position vectors of P and C respectively, then

$$\mathbf{r}_1 = \mathbf{r} + \overrightarrow{PC}$$

$$\Rightarrow \quad \mathbf{r}_1 - \mathbf{r} = \overrightarrow{PC}$$

$$\Rightarrow \quad (\mathbf{r}_1 - \mathbf{r})^2 = R^2.$$

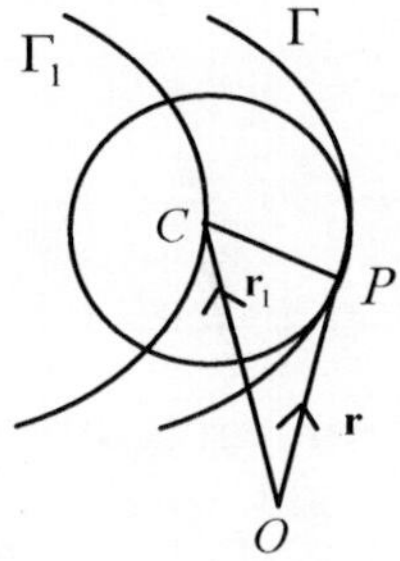

Let $F(s) = (\mathbf{r}_1 - \mathbf{r})^2 - R^2$,

where s is the arc length of Γ.

Since the sphere has four points of contact with Γ at P, we have

$$F(s) = F'(s) = F''(s) = F'''(s) = 0.$$

Now,

$$F'(s) = 0 \Rightarrow 2(\mathbf{r}_1 - \mathbf{r}).(-\mathbf{t}) = 0$$

$$\Rightarrow (\mathbf{r}_1 - \mathbf{r}).\ \mathbf{t} = 0 \tag{1}$$

$\Rightarrow (\mathbf{r}_1 - \mathbf{r})$ is perpendicular to $\mathbf{t}$.

$\Rightarrow (\mathbf{r}_1 - \mathbf{r})$ lies in the normal plane of Γ at P.

$$\Rightarrow \mathbf{r}_1 - \mathbf{r} = \lambda\mathbf{n} + \mu\mathbf{b}. \tag{2}$$

Again,

$$F''(s) = 0 \Rightarrow (\mathbf{r}_1 - \mathbf{r}).\kappa\mathbf{n} - \mathbf{t}.\mathbf{t} = 0$$

$$\Rightarrow (\mathbf{r}_1 - \mathbf{r}).\kappa\mathbf{n} - 1 = 0$$

$$\Rightarrow (\mathbf{r}_1 - \mathbf{r}).\mathbf{n} = \frac{1}{\kappa} = \rho \tag{3}$$

$$\Rightarrow (\lambda\mathbf{n} + \mu\mathbf{b}).\mathbf{n} = \rho \qquad \text{[using (2)]}$$

$$\Rightarrow \lambda = \rho. \tag{4}$$

Similarly,

$$F'''(s) = 0 \Rightarrow (\mathbf{r}_1 - \mathbf{r}).(\tau\mathbf{b} - \kappa\mathbf{t}) - \mathbf{t}.\mathbf{n} = \rho' \qquad \text{[using (3)]}$$

$$\Rightarrow (\lambda\mathbf{n} + \mu\mathbf{b}).(\tau\mathbf{b} - \kappa\mathbf{t}) - \mathbf{t}.\mathbf{n} = \rho' \qquad \text{[using (2)]}$$

$$\Rightarrow \mu\tau = \rho'$$

$$\Rightarrow \mu = \frac{\rho'}{\tau} = \rho'\sigma. \qquad (5)$$

Substituting the values of λ and μ from (4) and (5) respectively in (2), we get

$$\mathbf{r}_1 - \mathbf{r} = \rho\mathbf{n} + \rho'\sigma\mathbf{b}$$

$$\Rightarrow \quad \mathbf{r}_1 = \mathbf{r} + \rho\mathbf{n} + \rho'\sigma\mathbf{b}, \qquad (6)$$

which gives the position vectors of the centre of the osculating sphere.

Also, from (6), we have

$$(\mathbf{r}_1 - \mathbf{r})^2 = \rho^2 + \rho'^2\sigma^2$$

$$\Rightarrow \quad R^2 = \rho^2 + \rho'^2\sigma^2 \qquad [\because (\mathbf{r}_1 - \mathbf{r})^2 = R^2]$$

$$\Rightarrow \quad R = \sqrt{\rho^2 + \rho'^2\sigma^2}, \qquad (7)$$

which gives the radius of the osculating sphere.

Locus of the Centre of Spherical Curvature

Let Γ be a space curve and let Γ_1 be the locus of the centre of spherical curvature of Γ.

Let the quantities of Γ_1 be distinguished from those of Γ by using the suffix unity. Then Γ_1 has the following properties.

(i) $\mathbf{t}_1 = \mathbf{b}$

(ii) $\mathbf{n}_1 = -\mathbf{n}$

(iii) $\mathbf{b}_1 = \mathbf{t}$

(iv) $\rho_1 = \dfrac{1}{\kappa_1} = \rho + \sigma\dfrac{d}{ds}(\rho'\sigma)$

(v) $\sigma_1 = \dfrac{1}{\tau_1} = \dfrac{\rho^2}{\sigma} + \rho\dfrac{d}{ds}(\rho'\sigma)$

(vi) $\kappa\kappa_1 = \tau\tau_1$, i.e., $\rho\rho_1 = \sigma\sigma_1$.

Proof. (i) The position vector $\mathbf{r}_1$ of the centre of spherical curvature of Γ is given by

$$\mathbf{r}_1 = \mathbf{r} + \rho\mathbf{n} + \rho'\sigma\mathbf{b}.$$

Differentiating this with respect to s, we get

$$\frac{d\mathbf{r}_1}{ds_1}\frac{ds_1}{ds} = \mathbf{t} + \rho'\mathbf{n} + \rho(\tau\mathbf{b} - \kappa\mathbf{t}) + \mathbf{b}\frac{d}{ds}(\sigma\rho') + (\sigma\rho')(-\tau\mathbf{n})$$

or $$\mathbf{t}_1 \frac{ds_1}{ds} = \left[\frac{\rho}{\sigma} + \frac{d}{ds}(\sigma\rho')\right]\mathbf{b}. \tag{1}$$

[$\because \rho = 1/\kappa$ and $\sigma = 1/\tau$]

Squaring both sides of (1), we get

$$\left(\frac{ds_1}{ds}\right)^2 = \left[\frac{\rho}{\sigma} + \frac{d}{ds}(\sigma\rho')\right]^2.$$ [$\because \mathbf{t}_1^{\ 2} = \mathbf{b}^2 = 1$]

$$\therefore \quad \frac{ds_1}{ds} = \frac{\rho}{\sigma} + \frac{d}{ds}(\sigma\rho'). \tag{2}$$

Substituting this value of $\frac{ds_1}{ds}$ in (1), we get

$$\mathbf{t}_1 = \mathbf{b}. \tag{3}$$

(ii) Differentiating (3) with respect to s, we get

$$\frac{d\mathbf{t}_1}{ds_1}\frac{ds_1}{ds} = -\tau\mathbf{n}$$

$$\Rightarrow \quad \kappa_1 \mathbf{n}_1 \frac{ds_1}{ds} = -\tau\mathbf{n} \tag{4}$$

$\Rightarrow$ The principal normal to Γ_1 is parallel to the principal normal to Γ. We may choose the direction of $\mathbf{n}_1$ as opposite to that of $\mathbf{n}$, so that

$$\mathbf{n}_1 = -\mathbf{n}. \tag{5}$$

(iii) From (4) and (5), we get

$$\kappa_1 \frac{ds_1}{ds} = \tau. \tag{6}$$

Now,

$$\begin{aligned} \mathbf{b}_1 &= \mathbf{t}_1 \times \mathbf{n}_1 \\ &= \mathbf{b} \times (-\mathbf{n}) \qquad \text{[using (3) and (5)]} \\ &= \mathbf{n} \times \mathbf{b} \\ &= \mathbf{t} \end{aligned}$$

i.e., $$\mathbf{b}_1 = \mathbf{t}. \tag{7}$$

(iv) Differentiating (7) with respect to s, we get

$$\frac{d\mathbf{b}_1}{ds_1}\frac{ds_1}{ds} = \kappa\mathbf{n}$$

$$\Rightarrow \quad (-\tau_1\mathbf{n}_1)\frac{ds_1}{ds} = \kappa\mathbf{n}$$

$$\Rightarrow \qquad \tau_1 \frac{ds_1}{ds} = \kappa. \tag{8}$$

[using (5)]

From (6), we get

$$\frac{1}{\kappa_1}\frac{ds}{ds_1} = \frac{1}{\tau}$$

$$\Rightarrow \qquad \rho_1 \frac{ds}{ds_1} = \sigma$$

$$\Rightarrow \qquad \rho_1 = \sigma \frac{ds_1}{ds}$$

$$= \sigma \left[\frac{\rho}{\sigma} + \frac{d}{ds}(\sigma\rho') \right] \qquad \text{[using (2)]}$$

$$\Rightarrow \qquad \rho_1 = \rho + \sigma \frac{d}{ds}(\sigma\rho').$$

(v) From (8), we get

$$\frac{1}{\tau_1}\frac{ds}{ds_1} = \frac{1}{\kappa}.$$

$$\Rightarrow \qquad \sigma_1 \frac{ds}{ds_1} = \rho$$

$$\Rightarrow \qquad \sigma_1 = \rho \frac{ds_1}{ds}$$

$$= \rho \left[\frac{\rho}{\sigma} + \frac{d}{ds}(\sigma\rho') \right] \qquad \text{[using (2)]}$$

$$\Rightarrow \qquad \sigma_1 = \frac{\rho^2}{\sigma} + \rho \frac{d}{ds}(\sigma\rho').$$

(vi) From (6) and (8), we get

$$\frac{\kappa_1}{\tau_1} = \frac{\tau}{\kappa}.$$

$$\Rightarrow \qquad \kappa\kappa_1 = \tau\tau_1$$

$$\Rightarrow \qquad \rho\rho_1 = \sigma\sigma_1.$$

Solved Examples

Example 1. Show that the radius R of the sphere of curvature is given by

$$R^2 = \rho^4\sigma^2\mathbf{r}'''^2 - \sigma^2.$$

Solution. We know that

$$\mathbf{r}' = \mathbf{t}$$

$$\Rightarrow \quad \mathbf{r}'' = \mathbf{t}' = \kappa \mathbf{n} = \frac{1}{\rho}\mathbf{n}$$

$$\Rightarrow \quad \mathbf{r}''' = \frac{1}{\rho}\,(\tau \mathbf{b} - \kappa \mathbf{t}) - \frac{\rho'}{\rho^2}\mathbf{n}$$

i.e., $$\mathbf{r}''' = -\frac{1}{\rho^2}\mathbf{t} - \frac{\rho'}{\rho^2}\mathbf{n} + \frac{1}{\rho\sigma}\mathbf{b}. \qquad [\because \kappa = 1/\rho \text{ and } \tau = 1/\sigma]$$

Squaring, we get

$$\mathbf{r}'''^2 = \frac{1}{\rho^4} + \frac{\rho'^2}{\rho^4} + \frac{1}{\rho^2\sigma^2}$$

$$\Rightarrow \quad \rho^4\sigma^2\mathbf{r}'''^2 = \sigma^2 + \rho'^2\sigma^2 + \rho^2$$

$$= \sigma^2 + R^2 \qquad [\because R^2 = \rho^2 + \rho'^2\sigma^2]$$

$$\Rightarrow \quad R^2 = \rho^4\sigma^2\mathbf{r}'''^2 - \sigma^2.$$

Example 2. If R is the radius of spherical curvature, show that

$$R = \left| (\mathbf{t} \times \mathbf{t}'')/\kappa^2\tau \right|.$$

Solution. We know that

$$\mathbf{r}' = \mathbf{t}$$

$$\Rightarrow \quad \mathbf{t}' = \kappa \mathbf{n} = \frac{1}{\rho}\mathbf{n}$$

$$\Rightarrow \quad \mathbf{t}'' = \frac{1}{\rho}\,(\tau \mathbf{b} - \kappa \mathbf{t}) - \frac{\rho'}{\rho^2}\mathbf{n}$$

i.e., $$\mathbf{t}'' = -\frac{1}{\rho^2}\mathbf{t} - \frac{\rho'}{\rho^2}\mathbf{n} + \frac{1}{\rho\sigma}\mathbf{b}.$$

$$\therefore \quad \mathbf{t} \times \mathbf{t}'' = -\frac{\rho'}{\rho^2}\mathbf{b} - \frac{1}{\rho\sigma}\mathbf{n} \qquad [\because \mathbf{t} \times \mathbf{t} = \mathbf{0},\ \mathbf{t} \times \mathbf{n} = \mathbf{b},\ \mathbf{t} \times \mathbf{b} = -\mathbf{n}]$$

$$\Rightarrow \quad |\mathbf{t} \times \mathbf{t}''| = \sqrt{\left(\frac{\rho'^2}{\rho^4} + \frac{1}{\rho^2\sigma^2}\right)}$$

$$= \sqrt{\left(\frac{\rho'^2\sigma^2 + \rho^2}{\rho^4\sigma^2}\right)}$$

i.e., $$|\mathbf{t} \times \mathbf{t}''| = \sqrt{\frac{R^2}{\rho^4\sigma^2}} = \frac{R}{\rho^2\sigma}. \qquad [\because R^2 = \rho^2 + \rho'^2\sigma^2]$$

$\therefore \qquad R = \left| \rho^2\sigma\,(\mathbf{t}\times\mathbf{t}'') \right| = \left| (\mathbf{t}\times\mathbf{t}'')/\kappa^2\tau \right|.$

Example 3. Prove that

$$x'''^2 + y'''^2 + z'''^2 = \frac{1}{\rho^2\sigma^2} + \frac{1+\rho'^2}{\rho^4} = \frac{1}{\rho^4} + \frac{R^2}{\rho^4\sigma^2},$$

where dashes denote differentiation with respect to s.

Solution. We have

$$\mathbf{r} = (x,\ y,\ z)$$

$$\Rightarrow \qquad \mathbf{r}' = (x',\ y',\ z')$$

$$\Rightarrow \qquad \mathbf{r}'' = (x'',\ y'',\ z'')$$

$$\Rightarrow \qquad \mathbf{r}''' = (x''',\ y''',\ z''').$$

Squaring, we get

$$\mathbf{r}'''^2 = x'''^2 + y'''^2 + z'''^2. \tag{1}$$

Again,

$$\mathbf{r}'' = \mathbf{t}' = \kappa\mathbf{n} = \frac{1}{\rho}\mathbf{n}$$

$$\Rightarrow \qquad \mathbf{r}''' = \frac{1}{\rho}\,(\tau\mathbf{b} - \kappa\mathbf{t}) - \frac{\rho'}{\rho^2}\mathbf{n}$$

i.e., $\qquad \mathbf{r}''' = -\dfrac{1}{\rho^2}\mathbf{t} - \dfrac{\rho'}{\rho^2}\mathbf{n} + \dfrac{1}{\rho\sigma}\mathbf{b}. \qquad [\because \kappa = 1/\rho \text{ and } \tau = 1/\sigma]$

Squaring, we get

$$\mathbf{r}'''^2 = \frac{1}{\rho^4} + \frac{\rho'^2}{\rho^4} + \frac{1}{\rho^2\sigma^2}$$

or $\qquad \mathbf{r}'''^2 = \dfrac{1}{\rho^2\sigma^2} + \dfrac{1+\rho'^2}{\rho^4}. \qquad (2)$

From (1) and (2), we get

$$x'''^2 + y'''^2 + z'''^2 = \frac{1}{\rho^2\sigma^2} + \frac{1+\rho'^2}{\rho^4}. \tag{3}$$

Also, we know that

$$R^2 = \rho^2 + \rho'^2\sigma^2$$

$$\Rightarrow \qquad \frac{R^2}{\rho^4\sigma^2} = \frac{1}{\rho^2\sigma^2} + \frac{\rho'^2}{\rho^4}$$

$$\Rightarrow \qquad \frac{R^2}{\rho^4\sigma^2} + \frac{1}{\rho^4} = \frac{1}{\rho^2\sigma^2} + \frac{\rho'^2}{\rho^4} + \frac{1}{\rho^4}$$

$$\Rightarrow \quad \frac{1}{\rho^4} + \frac{R^2}{\rho^4\sigma^2} = \frac{1}{\rho^2\sigma^2} + \frac{1+\rho'^2}{\rho^4}. \tag{4}$$

From (3) and (4), we get

$$x'''^2 + y'''^2 + z'''^2 = \frac{1}{\rho^2\sigma^2} + \frac{1+\rho'^2}{\rho^4} = \frac{1}{\rho^4} + \frac{R^2}{\rho^4\sigma^2}.$$

Example 4. Show that the principal normal to a curve is normal to the locus of the centre of curvature at points for which the value of κ is stationary.

Solution. We know that the position vector $\mathbf{r}_1$ of the centre of curvature is given by

$$\mathbf{r}_1 = \mathbf{r} + \rho\mathbf{n}.$$

Differentiating this with respect to s, we get

$$\frac{d\mathbf{r}_1}{ds_1}\frac{ds_1}{ds} = \mathbf{t} + \rho'\mathbf{n} + \rho(\tau\mathbf{b} - \kappa\mathbf{t})$$

$$\Rightarrow \quad \mathbf{t}_1\frac{ds_1}{ds} = \rho'\mathbf{n} + \rho\tau\mathbf{b} \qquad [\because \rho = 1/\kappa]$$

$$\Rightarrow \quad \mathbf{t}_1 = (\rho'\mathbf{n} + \rho\tau\mathbf{b})\frac{ds}{ds_1}$$

$$\Rightarrow \quad \mathbf{n}\,.\mathbf{t}_1 = \rho'\frac{ds}{ds_1}. \tag{1}$$

$$[\because \mathbf{n}\,.\,\mathbf{n} = 1,\ \mathbf{n}\,.\,\mathbf{b} = 0]$$

For stationary values of κ, we have

$$\kappa' = 0$$

$$\Rightarrow \quad \rho' = 0.$$

Hence, from (1), we get

$$\mathbf{n}\,.\,\mathbf{t}_1 = 0$$

$\Rightarrow$ $\mathbf{n}$ is orthogonal to $\mathbf{t}_1$, i.e., the principal normal to the given curve is normal to the locus of the centre of curvature.

Example 5. If s_1 is the arc length of the locus of the centre of curvature, show that

$$\frac{ds_1}{ds} = \frac{\sqrt{(\kappa^2\tau^2 + \kappa'^2)}}{\kappa^2} = \sqrt{\left[\left(\frac{\rho}{\sigma}\right)^2 + \rho'^2\right]}.$$

Solution. We know that if $\mathbf{r}_1$ denotes the position vector of the centre of curvature, then

$$\mathbf{r}_1 = \mathbf{r} + \rho\mathbf{n}$$

or $$\mathbf{r}_1 = \mathbf{r} + \frac{1}{\kappa}\mathbf{n}. \qquad [\because \rho = 1/\kappa]$$

Differentiating this with respect to s, we get

$$\frac{d\mathbf{r}_1}{ds_1}\frac{ds_1}{ds} = \mathbf{t} + \frac{1}{\kappa}(\tau\mathbf{b} - \kappa\mathbf{t}) - \frac{\kappa'}{\kappa^2}\mathbf{n}$$

or $$\mathbf{t}_1\frac{ds_1}{ds} = -\frac{\kappa'}{\kappa^2}\mathbf{n} + \frac{\tau}{\kappa}\mathbf{b}.$$

Squaring, we get

$$\left(\frac{ds_1}{ds}\right)^2 = \frac{\kappa'^2}{\kappa^4} + \frac{\tau^2}{\kappa^2} = \frac{\kappa'^2 + \kappa^2\tau^2}{\kappa^4}. \qquad [\because \mathbf{t}_1^{\,2} = \mathbf{n}^2 = \mathbf{b}^2 = 1]$$

$$\therefore \quad \frac{ds_1}{ds} = \frac{\sqrt{(\kappa^2\tau^2 + \kappa'^2)}}{\kappa^2}. \tag{1}$$

Substituting $\tau = \dfrac{1}{\sigma}$, $\kappa = \dfrac{1}{\rho}$, $\kappa' = -\dfrac{\rho'}{\rho^2}$ in (1), we get

$$\frac{ds_1}{ds} = \frac{\sqrt{\left(\dfrac{1}{\rho^2\sigma^2} + \dfrac{\rho'^2}{\rho^4}\right)}}{\dfrac{1}{\rho^2}}$$

$$= \rho^2\sqrt{\left(\frac{\rho^2 + \rho'^2\sigma^2}{\rho^4\sigma^2}\right)}$$

$$= \sqrt{\left(\frac{\rho^2 + \rho'^2\sigma^2}{\sigma^2}\right)}$$

i.e., $$\frac{ds_1}{ds} = \sqrt{\left[\left(\frac{\rho}{\sigma}\right)^2 + \rho'^2\right]}. \tag{2}$$

From (1) and (2), we get

$$\frac{ds_1}{ds} = \frac{\sqrt{(\kappa^2\,\tau^2 + \kappa'^2)}}{\kappa^2} = \sqrt{\left[\left(\frac{\rho}{\sigma}\right)^2 + \rho'^2\right]}.$$

Example 6. If ψ is such that $d\psi = \tau\, ds$, show that

(i) $\rho_1 = \dfrac{1}{\kappa_1} = \rho + \dfrac{d^2\rho}{d\psi^2}$

(ii) $\sigma_1 = \dfrac{1}{\tau_1} = \dfrac{\rho}{\sigma}\left(\rho + \dfrac{d^2\rho}{d\psi^2}\right) = \dfrac{R\rho}{\sigma}\dfrac{dR}{d\rho}$

(iii) $R^2 = \rho^2 + \left(\dfrac{d\rho}{d\psi}\right)^2$

(iv) $R\dfrac{dR}{d\rho} = \rho + \dfrac{d^2\rho}{d\psi^2} = \rho_1$

(v) $R\dfrac{dR}{ds} = \sqrt{(R^2 - \rho^2)}\dfrac{ds_1}{ds}$,

where ρ_1 and σ_1 are the radii of curvature and torsion of the locus of the centre of spherical curvature and R_1 is the radius of the osculating sphere.

Solution. Given that

$$\frac{d\psi}{ds} = \tau = \frac{1}{\sigma}. \tag{1}$$

(i) We know that

$$\rho_1 = \frac{1}{\kappa_1} = \rho + \sigma\frac{d}{ds}(\rho'\sigma)$$

$$= \rho + \sigma\frac{d}{d\psi}(\rho'\sigma)\frac{d\psi}{ds}$$

$$= \rho + \sigma\frac{d}{d\psi}\left(\frac{d\rho}{ds}\frac{ds}{d\psi}\right)\frac{1}{\sigma} \qquad \text{[using (1)]}$$

i.e.,

$$\rho_1 = \frac{1}{\kappa_1} = \rho + \frac{d^2\rho}{d\psi^2}. \tag{2}$$

(ii) We know that

$$\sigma_1 = \frac{1}{\tau_1} = \frac{\rho^2}{\sigma} + \rho\frac{d}{ds}(\rho'\sigma)$$

$$= \frac{\rho^2}{\sigma} + \rho\frac{d}{d\psi}(\rho'\sigma)\frac{d\psi}{ds}$$

$$= \frac{\rho^2}{\sigma} + \rho\frac{d}{d\psi}\left(\frac{d\rho}{ds}\frac{ds}{d\psi}\right)\frac{1}{\sigma} \qquad \text{[using (1)]}$$

$$= \frac{\rho^2}{\sigma} + \frac{\rho}{\sigma}\frac{d^2\rho}{d\psi^2}$$

i.e.,
$$\sigma_1 = \frac{1}{\tau_1} = \frac{\rho}{\sigma}\left(\rho + \frac{d^2\rho}{d\psi^2}\right). \qquad (3)$$

Again, we know that

$$R^2 = \rho^2 + \rho'^2\sigma^2$$

or
$$R^2 = \rho^2 + \left(\frac{d\rho}{ds}\frac{ds}{d\psi}\right)^2 \qquad \text{[using (1)]}$$

or
$$R^2 = \rho^2 + \left(\frac{d\rho}{d\psi}\right)^2. \qquad (4)$$

Differentiating (4) with respect to ρ, we get

$$2R\frac{dR}{d\rho} = 2\rho + 2\frac{d\rho}{d\psi}\frac{d^2\rho}{d\psi^2}\frac{d\psi}{d\rho}$$

$$\Rightarrow \quad R\frac{dR}{d\rho} = \rho + \frac{d^2\rho}{d\psi^2} \qquad (5)$$

$$\Rightarrow \quad \frac{R\rho}{\sigma}\frac{dR}{d\rho} = \frac{\rho}{\sigma}\left(\rho + \frac{d^2\rho}{d\psi^2}\right). \qquad (6)$$

From (3) and (6), we get

$$\sigma_1 = \frac{1}{\tau_1} = \frac{\rho}{\sigma}\left(\rho + \frac{d^2\rho}{d\psi^2}\right) = \frac{R\rho}{\sigma}\frac{dR}{d\rho}.$$

(iii) It follows from equation (4) above.

(iv) It follows from equations (2) and (5) above.

(v) We know that if $\mathbf{r}_1$ denotes the position vector of the centre of spherical curvature, then

$$\mathbf{r}_1 = \mathbf{r} + \rho\mathbf{n} + \rho'\sigma\mathbf{b}.$$

Differentiating this with respect to s, we get

$$\frac{d\mathbf{r}_1}{ds_1}\frac{ds_1}{ds} = \mathbf{t} + \rho'\mathbf{n} + \rho(\tau\mathbf{b} - \kappa\mathbf{t}) + \mathbf{b}\frac{d}{ds}(\sigma\rho') + (\sigma\rho')(-\tau\mathbf{n})$$

or
$$\mathbf{t}_1\frac{ds_1}{ds} = \left[\frac{\rho}{\sigma} + \frac{d}{ds}(\sigma\rho')\right]\mathbf{b}. \qquad [\because \rho = 1/\kappa \text{ and } \sigma = 1/\tau]$$

Squaring both sides of this equation, we get

$$\left(\frac{ds_1}{ds}\right)^2=\left[\frac{\rho}{\sigma}+\frac{d}{ds}(\sigma\rho')\right]^2 \qquad [\because \mathbf{t}_1^{\ 2}=\mathbf{b}^{\ 2}=1]$$

$$\Rightarrow \qquad \frac{ds_1}{ds}=\frac{\rho}{\sigma}+\frac{d}{ds}(\sigma\rho')$$

$$=\rho\frac{d\psi}{ds}+\frac{d}{d\psi}\left(\frac{ds}{d\psi}\frac{d\rho}{ds}\right)\frac{d\psi}{ds} \qquad \text{[using (1)]}$$

$$=\left(\rho+\frac{d^2\rho}{d\psi^2}\right)\frac{d\psi}{ds}$$

$$=R\frac{dR}{d\rho}\frac{d\psi}{ds} \qquad \text{[using (5)]}$$

$$\Rightarrow \qquad R\frac{dR}{ds}=\frac{d\rho}{d\psi}\frac{ds_1}{ds}$$

or $$R\frac{dR}{ds}=\sqrt{(R^2-\rho^2)}\,\frac{ds_1}{ds}. \qquad \text{[using (4)]}$$

Example 7. If a curve lies on a sphere, show that ρ and σ are connected by

$$\frac{\rho}{\sigma}+\frac{d}{ds}(\sigma\rho')=0.$$

Solution. If a curve lies on a sphere, then that sphere is the osculating sphere for every point and consequently, the centre of spherical curvature for every point of the curve is a fixed point.

Therefore if $\mathbf{r}_1$ denotes the position vector of the centre of spherical curvature for the point $\mathbf{r}(s)$ of the given curve, then $\mathbf{r}_1$ is a constant vector, i.e.,

$$\frac{d\mathbf{r}_1}{ds}=\mathbf{0}. \qquad (1)$$

But, we know that

$$\mathbf{r}_1=\mathbf{r}+\rho\mathbf{n}+\rho'\sigma\mathbf{b}.$$

Differentiating this with respect to s, we get

$$\frac{d\mathbf{r}_1}{ds}=\mathbf{t}+\rho'\mathbf{n}+\rho(\tau\mathbf{b}-\kappa\mathbf{t})+\mathbf{b}\frac{d}{ds}(\sigma\rho')+(\sigma\rho')(-\tau\mathbf{n})$$

$$\Rightarrow \qquad \mathbf{t}+\rho'\mathbf{n}+\rho(\tau\mathbf{b}-\kappa\mathbf{t})+\mathbf{b}\frac{d}{ds}(\sigma\rho')+(\sigma\rho')(-\tau\mathbf{n})=\mathbf{0} \qquad \text{[using (1)]}$$

$$\Rightarrow \quad \left[\frac{\rho}{\sigma}+\frac{d}{ds}(\sigma\rho')\right]\mathbf{b}=0 \qquad [\because \rho = 1/\kappa \text{ and } \sigma = 1/\tau]$$

$$\Rightarrow \quad \frac{\rho}{\sigma}+\frac{d}{ds}(\sigma\rho')=0.$$

Example 8. If the radius of spherical curvature of a curve is constant, show that the curve either lies on a sphere or has constant curvature.

Solution. If the radius R of spherical curvature of a curve is constant, then

$$\frac{dR}{ds}=0. \tag{1}$$

But, we know that

$$R^2 = \rho^2 + \rho'^2\sigma^2. \tag{2}$$

Differentiating this with respect to s, we get

$$2R\frac{dR}{ds}=2\rho\rho' + 2\sigma\rho'\frac{d}{ds}(\sigma\rho')$$

$$\Rightarrow \quad \rho\rho' + \sigma\rho'\frac{d}{ds}(\sigma\rho')=0 \qquad \text{[using (1)]}$$

$$\Rightarrow \quad \sigma\rho'\left[\frac{\rho}{\sigma}+\frac{d}{ds}(\sigma\rho')\right]=0.$$

Hence,

either $\quad \rho'=0 \qquad (3)$

or $\quad \dfrac{\rho}{\sigma}+\dfrac{d}{ds}(\sigma\rho')=0. \qquad (4)$

If $\rho'=0$, then $\rho = \text{constant}$.

We now show that the condition $\frac{\rho}{\sigma}+\frac{d}{ds}(\sigma\rho')=0$ implies that the curve lies on a sphere.

If (4) is true, then

$$2\rho\rho' + 2\sigma\rho'\frac{d}{ds}(\sigma\rho')=0.$$

Integrating this, we get

$$\rho^2 + (\sigma\rho')^2 = \text{constant}$$

$$\Rightarrow \quad R^2 = \text{constant} \qquad \text{[using (2)]}$$

$$\Rightarrow \quad R = \text{constant}.$$

This shows that the radius of the osculating sphere is independent of the position of a point on the curve.

Again, we know that the position vector $\mathbf{r}_1$ of the centre of spherical curvature is given by

$$\mathbf{r}_1 = \mathbf{r} + \rho\mathbf{n} + \rho'\sigma\,\mathbf{b}.$$

Differentiating this with respect to s, we get

$$\frac{d\mathbf{r}_1}{ds} = \mathbf{t} + \rho'\mathbf{n} + \rho\,(\tau\mathbf{b} - \kappa\mathbf{t}) + \mathbf{b}\frac{d}{ds}(\sigma\rho') + (\sigma\rho')\,(-\tau\mathbf{n})$$

or $$\frac{d\mathbf{r}_1}{ds} = \left[\frac{\rho}{\sigma} + \frac{d}{ds}(\sigma\rho')\right]\mathbf{b} \qquad [\because \rho = 1/\kappa \text{ and } \sigma = 1/\tau]$$

$$= \mathbf{0}. \qquad \text{[using (4)]}$$

This shows that the centre of the osculating sphere is independent of the position of a point on the curve.

Hence condition (4) implies that the curve lies on a sphere.

Example 9. Prove that the shortest distance between the principal normals at consecutive points distance s apart is $s\rho/\sqrt{(\rho^2 + \sigma^2)}$ and that it divides the radius of curvature in the ratio $\rho^2 : \sigma^2$.

Solution. Let Γ be a space curve and let P and Q be any two consecutive points on Γ, whose position vectors are $\mathbf{r}$ and $\mathbf{r} + d\mathbf{r}$ respectively.
Again, let arc $AP = s$ and arc $AQ = s + \delta s$. Then arc $PQ = \delta s$.

Let $\mathbf{n}$ and $\mathbf{n} + d\mathbf{n}$ denote the unit principal normals to Γ at P and Q respectively. Again, let C and H be the centres of curvature of Γ at P and Q respectively.

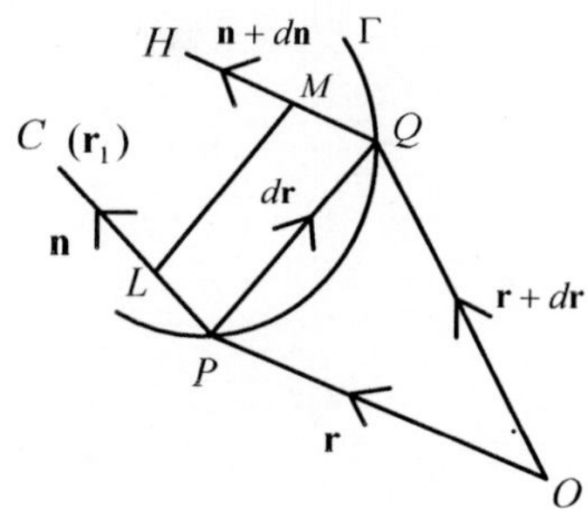

If LM is the shortest distance between PC and QH and $\mathbf{r}_1$ is the position vector of C, then

$$LM = \text{Projection of } \overrightarrow{PQ} \text{ on } \overrightarrow{LM}$$

$$= \text{Projection of } d\mathbf{r} \text{ on } \overrightarrow{LM}$$

$$= d\mathbf{r}\,.\ \text{unit vector along } \overrightarrow{LM}$$

$$= d\mathbf{r} \,.\, \text{unit vector perpendicular to each of } \mathbf{n} \text{ and } \mathbf{n} + d\mathbf{n}$$

$$= d\mathbf{r} \,.\, \text{unit vector parallel to } \mathbf{n} \times (\mathbf{n} + d\mathbf{n})$$

$$= d\mathbf{r} . \frac{\mathbf{n} \times (\mathbf{n} + d\mathbf{n})}{|\mathbf{n} \times (\mathbf{n} + d\mathbf{n})|}$$

$$= d\mathbf{r} \,.\, \frac{\mathbf{n} \times d\mathbf{n}}{|\mathbf{n} \times d\mathbf{n}|}$$

$$= \frac{\dfrac{d\mathbf{r}}{ds} . \left(\mathbf{n} \times \dfrac{d\mathbf{n}}{ds}\right) ds}{\left|\mathbf{n} \times \dfrac{d\mathbf{n}}{ds}\right|}$$

$$= \frac{\mathbf{t} \,.\, [\mathbf{n} \times (\tau \mathbf{b} - \kappa \mathbf{t})]\, ds}{|\mathbf{n} \times (\tau \mathbf{b} - \kappa \mathbf{t})|}$$

$$= \frac{\mathbf{t} \,.\, (\tau \mathbf{t} + \kappa \mathbf{b})\, ds}{|\tau \mathbf{t} + \kappa \mathbf{b}|}$$

$$= \frac{\tau\, ds}{\sqrt{(\tau^2 + \kappa^2)}}$$

$$= \frac{\dfrac{1}{\sigma} ds}{\sqrt{\left(\dfrac{1}{\sigma^2} + \dfrac{1}{\rho^2}\right)}} \qquad [\because \rho = 1/\kappa \text{ and } \sigma = 1/\tau]$$

$$= \frac{\rho\, ds}{\sqrt{(\rho^2 + \sigma^2)}}$$

i.e., $$LM = \frac{\rho\, ds}{\sqrt{(\rho^2 + \sigma^2)}}. \tag{1}$$

Since, in our case, the actual distance between P and Q is given to be s, the required shortest distance is obtained by replacing ds by s in (1), i.e., we have

$$LM = \frac{\rho s}{\sqrt{(\rho^2 + \sigma^2)}}.$$

Let L divide CP in the ratio $\lambda : 1$.

We have to show that $\lambda : 1 = \rho^2 : \sigma^2$.

Clearly, the position vector of $L = \dfrac{\lambda \mathbf{r} + \mathbf{r}_1}{\lambda + 1}$

$$= \frac{\lambda \mathbf{r} + (\mathbf{r} + \rho \mathbf{n})}{\lambda + 1} \qquad [\because \mathbf{r}_1 = \mathbf{r} + \rho \mathbf{n}]$$

$$= \frac{(\lambda + 1)\,\mathbf{r} + \rho \mathbf{n}}{\lambda + 1}$$

i.e., the position vector of $L = \mathbf{r} + \dfrac{\rho}{\lambda + 1}\mathbf{n}$. (2)

Let us call the plane LMQ as the α-plane. Then, clearly,

$(\mathbf{n} + d\mathbf{n})$, $\overrightarrow{LM}$ and $\overrightarrow{QL}$ lie in α-plane.

$\Rightarrow$ $(\mathbf{n} + d\mathbf{n})$, $\mathbf{n} \times (\mathbf{n} + d\mathbf{n})$ and $\left(\mathbf{r} + \dfrac{\rho}{\lambda + 1}\mathbf{n}\right) - (\mathbf{r} + d\mathbf{r})$ are coplanar.

$\Rightarrow$ $(\mathbf{n} + d\mathbf{n})$, $\mathbf{n} \times (\mathbf{n} + d\mathbf{n})$ and $\left(\dfrac{\rho}{\lambda + 1}\mathbf{n} - d\mathbf{r}\right)$ are coplanar.

$\Rightarrow$ $(\mathbf{n} + d\mathbf{n})$, $\mathbf{n} \times d\mathbf{n}$ and $\left(\dfrac{\rho}{\lambda + 1}\mathbf{n} - d\mathbf{r}\right)$ are coplanar.

$$\Rightarrow \quad (\mathbf{n} + d\mathbf{n}) \times (\mathbf{n} \times d\mathbf{n}) \,.\left(\frac{\rho}{\lambda + 1}\mathbf{n} - d\mathbf{r}\right) = 0$$

$$\Rightarrow \quad [\mathbf{n} \times (\mathbf{n} \times d\mathbf{n}) + d\mathbf{n} \times (\mathbf{n} \times d\mathbf{n})] \,.\left(\frac{\rho}{\lambda + 1}\mathbf{n} - d\mathbf{n}\right) = 0$$

$$\Rightarrow \quad [(\mathbf{n}\,.\,d\mathbf{n})\,\mathbf{n} - (\mathbf{n}\,.\,\mathbf{n})\,d\mathbf{n} + (d\mathbf{n}\,.\,d\mathbf{n})\,\mathbf{n} - (d\mathbf{n}\,.\,\mathbf{n})\,d\mathbf{n}] \,.\left(\frac{\rho}{\lambda + 1}\mathbf{n} - d\mathbf{r}\right) = 0$$

$$\Rightarrow \quad \frac{\rho}{\lambda + 1}\left[(\mathbf{n}\,.\,d\mathbf{n}) - (d\mathbf{n}\,.\,\mathbf{n}) + (d\mathbf{n}\,.\,d\mathbf{n}) - (d\mathbf{n}\,.\,\mathbf{n})^2\right]$$
$$- \left[(\mathbf{n}\,.\,d\mathbf{n})(\mathbf{n}\,.\,d\mathbf{r}) - (d\mathbf{n}\,.\,d\mathbf{r}) + (d\mathbf{n}\,.\,d\mathbf{n})(\mathbf{n}\,.\,d\mathbf{r})\right.$$
$$\left. - (d\mathbf{n}\,.\,\mathbf{n})\,(d\mathbf{n}\,.\,d\mathbf{r})\right] = 0$$

$$\Rightarrow \quad \frac{\rho}{\lambda + 1}\left[(d\mathbf{n}\,.\,d\mathbf{n}) - (d\mathbf{n}\,.\,\mathbf{n})^2\right] - \left[(\mathbf{n}\,.\,d\mathbf{n})(\mathbf{n}\,.\,d\mathbf{r})\right.$$
$$\left. - (d\mathbf{n}\,.\,d\mathbf{r}) + (d\mathbf{n}\,.\,d\mathbf{n})(\mathbf{n}\,.\,d\mathbf{r}) - (d\mathbf{n}\,.\,\mathbf{n})(d\mathbf{n}\,.\,d\mathbf{r})\right] = 0$$

$$\Rightarrow \quad \frac{\rho}{\lambda+1}\left[\left(\frac{d\mathbf{n}}{ds}\cdot\frac{d\mathbf{n}}{ds}\right)ds^2-\left(\frac{d\mathbf{n}}{ds}\cdot\mathbf{n}\right)^2 ds^2\right]$$

$$-\left[\left(\mathbf{n}\cdot\frac{d\mathbf{n}}{ds}\right)\left(\mathbf{n}\cdot\frac{d\mathbf{r}}{ds}\right)ds^2-\left(\frac{d\mathbf{n}}{ds}\cdot\frac{d\mathbf{r}}{ds}\right)ds^2\right.$$

$$\left.+\left(\frac{d\mathbf{n}}{ds}\cdot\frac{d\mathbf{n}}{ds}\right)\left(\mathbf{n}\cdot\frac{d\mathbf{r}}{ds}\right)ds^3-\left(\frac{d\mathbf{n}}{ds}\cdot\mathbf{n}\right)\left(\frac{d\mathbf{n}}{ds}\cdot\frac{d\mathbf{r}}{ds}\right)ds^3\right]=0$$

$$\Rightarrow \quad \frac{\rho}{\lambda+1}[(\tau\mathbf{b}-\kappa\mathbf{t})\cdot(\tau\mathbf{b}-\kappa\mathbf{t})-\{(\tau\mathbf{b}-\kappa\mathbf{t})\cdot\mathbf{n}\}^2]\,ds^2$$

$$-[\{\mathbf{n}\cdot(\tau\mathbf{b}-\kappa\mathbf{t})\}(\mathbf{n}\cdot\mathbf{t})-(\tau\mathbf{b}-\kappa\mathbf{t})\cdot\mathbf{t}]\,ds^2$$

$$+(\tau\mathbf{b}-\kappa\mathbf{t})\cdot(\tau\mathbf{b}-\kappa\mathbf{t})(\mathbf{n}\cdot\mathbf{t})\,ds^3$$

$$-\{(\tau\mathbf{b}-\kappa\mathbf{t})\cdot\mathbf{n}\}\{(\tau\mathbf{b}-\kappa\mathbf{t})\cdot\mathbf{t}\}ds^3=0$$

$$\Rightarrow \quad \frac{\rho}{\lambda+1}(\tau^2+\kappa^2)\,ds^2-\kappa\,ds^2=0$$

$$\Rightarrow \quad \frac{\rho}{\lambda+1}(\tau^2+\kappa^2)\,ds^2=\kappa\,ds^2$$

$$\Rightarrow \quad \frac{\rho}{\lambda+1}(\tau^2+\kappa^2)=\kappa$$

$$\Rightarrow \quad \lambda+1=\frac{\rho(\tau^2+\kappa^2)}{\kappa}=\frac{\tau^2+\kappa^2}{\kappa^2}$$

$$\Rightarrow \quad \lambda+1=\frac{\tau^2}{\kappa^2}+1$$

$$\Rightarrow \quad \lambda=\frac{\tau^2}{\kappa^2}=\frac{\rho^2}{\sigma^2}$$

$$\Rightarrow \quad \lambda:1=\rho^2:\sigma^2.$$

Example 10. Prove that the shortest distance between consecutive radii of spherical curvature divides the radius in the ratio

$$\sigma^2:\rho^2\left(\frac{dR}{d\rho}\right)^2.$$

Solution. Let Γ be a space curve and let P and Q be any two consecutive points on Γ whose position vectors are $\mathbf{r}$ and $\mathbf{r}+\delta\mathbf{r}$ respectively.

Again, let arc $AP=s$ and arc $AQ=s+\delta s$. Then arc $PQ=\delta s$.

Let $\mathbf{n}$ and $\mathbf{n}+d\mathbf{n}$ denote the unit principal normals to Γ at P and Q respectively. Again, let C and H be the centres of spherical curvature of Γ at P and Q respectively.

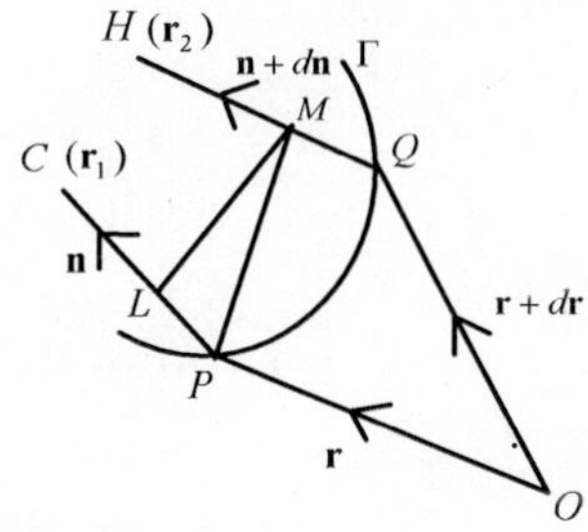

Let LM be the shortest distance between PC and QH.

Let $\mathbf{r}_1$ and $\mathbf{r}_2$ be the position vectors of C and H respectively, and let M divide QH in the ratio $\lambda : 1$.

We have to show that

$$\lambda:1 = \sigma^2 : \rho^2\left(\frac{dR}{d\rho}\right)^2.$$

We know that

$$\mathbf{r}_1 = \mathbf{r} + \rho\mathbf{n} + \sigma\rho'\mathbf{b}$$

or $\quad \mathbf{r}_1 = \mathbf{r} + \mathbf{w}$, where $\mathbf{w} = \rho\mathbf{n} + \sigma\rho'\mathbf{b}$

and

$$\mathbf{r}_2 = (\mathbf{r} + d\mathbf{r}) + (\mathbf{w} + d\mathbf{w}).$$

Now,

$$\begin{aligned}
\text{the position vector of } M &= \frac{\lambda\mathbf{r}_2 + (\mathbf{r}+d\mathbf{r})}{\lambda+1} \\
&= \frac{\lambda[(\mathbf{r}+d\mathbf{r}) + (\mathbf{w}+d\mathbf{w})] + (\mathbf{r}+d\mathbf{r})}{\lambda+1} \\
&= \frac{(\lambda+1)(\mathbf{r}+d\mathbf{r}) + \lambda(\mathbf{w}+d\mathbf{w})}{\lambda+1} \\
&= (\mathbf{r}+d\mathbf{r}) + \frac{\lambda}{\lambda+1}(\mathbf{w}+d\mathbf{w}).
\end{aligned}$$

Let us call the plane PLM as the α-plane. Then, clearly,

$\overrightarrow{PC}$, $\overrightarrow{LM}$ and $\overrightarrow{PM}$ lie in α-plane.

$\Rightarrow \quad \overrightarrow{PC}$, $\overrightarrow{LM}$ and $\overrightarrow{PM}$ are coplanar.

$$\Rightarrow \quad \overrightarrow{PC} \times \overrightarrow{LM} \cdot \overrightarrow{PM} = 0. \tag{1}$$

Now,

$$\overrightarrow{PC} = \mathbf{r}_1 - \mathbf{r} = \mathbf{w}, \tag{2}$$

$$\overrightarrow{QH} = \mathbf{r}_2 - (\mathbf{r} + d\mathbf{r}) = w + d\mathbf{w},$$

$$\overrightarrow{LM} = \text{ a vector parallel to } \overrightarrow{PC} \times \overrightarrow{QH}$$

$$\Rightarrow \quad \overrightarrow{LM} = \text{a vector parallel to } \mathbf{w} \times (\mathbf{w} + d\mathbf{w})$$

$$\Rightarrow \quad \overrightarrow{LM} = \text{ a vector parallel to } \mathbf{w} \times d\mathbf{w}. \qquad (3)$$

Again,

$$\overrightarrow{PM} = \left[(\mathbf{r} + d\mathbf{r}) + \frac{\lambda}{\lambda + 1}(\mathbf{w} + d\mathbf{w}) \right] - \mathbf{r}$$

$$\Rightarrow \quad \overrightarrow{PM} = d\mathbf{r} + \frac{\lambda}{\lambda + 1}(\mathbf{w} + d\mathbf{w}). \qquad (4)$$

Using (2), (3) and (4), (1) gives

$$\mathbf{w} \times (\mathbf{w} \times d\mathbf{w}) \,.\left[d\mathbf{r} + \frac{\lambda}{\lambda + 1}(\mathbf{w} + d\mathbf{w}) \right] = 0$$

$$\Rightarrow \quad \mathbf{w} \times \left(\mathbf{w} \times \frac{d\mathbf{w}}{ds} ds \right) . \left[\frac{d\mathbf{r}}{ds} ds + \frac{\lambda}{\lambda + 1} \left(\mathbf{w} + \frac{d\mathbf{w}}{ds} ds \right) \right] = 0$$

$$\Rightarrow \quad \mathbf{w} \times \left(\mathbf{w} \times \mathbf{w}'ds \right) . \left[\mathbf{t}\, ds + \frac{\lambda}{\lambda + 1} \left(\mathbf{w} + \mathbf{w}'ds \right) \right] = 0$$

$$\Rightarrow \quad \mathbf{w} \times (\mathbf{w} \times \mathbf{w}') \,.\, \mathbf{t}\, ds^2 + \mathbf{w} \times (\mathbf{w} \times \mathbf{w}') \,.\, \mathbf{w} \frac{\lambda}{\lambda + 1} ds$$

$$+ \mathbf{w} \times (\mathbf{w} \times \mathbf{w}') \,.\, \mathbf{w}' \frac{\lambda}{\lambda + 1} ds^2 = 0$$

$$\Rightarrow \quad \mathbf{w} \times (\mathbf{w} \times \mathbf{w}') \,.\, \mathbf{t}\, ds^2 + \mathbf{w} \times (\mathbf{w} \times \mathbf{w}') \,.\, \mathbf{w}' \frac{\lambda}{\lambda + 1} ds^2 = 0$$

$$\Rightarrow \quad \mathbf{w} \times (\mathbf{w} \times \mathbf{w}') \,.\, \mathbf{t} + \mathbf{w} \times (\mathbf{w} \times \mathbf{w}') \,.\, \mathbf{w}' \frac{\lambda}{\lambda + 1} = 0. \qquad (5)$$

Now,

$$\mathbf{w} = \rho \mathbf{n} + \sigma \rho' \mathbf{b}$$

$$\Rightarrow \quad \mathbf{w}' = \rho' \mathbf{n} + \rho\, (\tau \mathbf{b} - \kappa \mathbf{t}) + \mathbf{b} \frac{d}{ds}(\sigma \rho') + \sigma \rho' (-\tau \mathbf{n})$$

$$= \left[\frac{\rho}{\sigma} + \frac{d}{ds}(\sigma \rho') \right] \mathbf{b} - \mathbf{t} \qquad [\because \rho = 1/\kappa \text{ and } \sigma = 1/\tau]$$

$$\Rightarrow \qquad \mathbf{w}\times\mathbf{w}' = \rho\left[\frac{\rho}{\sigma}+\frac{d}{ds}(\sigma\rho')\right]\mathbf{t}-(\sigma\rho')\,\mathbf{n}+\rho\mathbf{b}$$

$$\Rightarrow \qquad \mathbf{w}\times(\mathbf{w}\times\mathbf{w}') = \left[\rho^2+(\sigma\rho')^2\right]\mathbf{t}+\sigma\rho\rho'\left[\frac{\rho}{\sigma}+\frac{d}{ds}(\sigma\rho')\right]\mathbf{n} - \rho^2\left[\frac{\rho}{\sigma}+\frac{d}{ds}(\sigma\rho')\right]\mathbf{b}$$

$$\Rightarrow \qquad \mathbf{w}\times(\mathbf{w}\times\mathbf{w}')\,.\,\mathbf{t} = \rho^2+(\sigma\rho')^2 \tag{6}$$

and

$$\mathbf{w}\times(\mathbf{w}\times\mathbf{w}')\,.\,\mathbf{w}' = -\left[\rho^2+(\sigma\rho')^2\right]-\rho^2\left[\frac{\rho}{\sigma}+\frac{d}{ds}(\sigma\rho')\right]^2. \tag{7}$$

Substituting the values of $\mathbf{w}\times(\mathbf{w}\times\mathbf{w}')\,.\,\mathbf{t}$ and $\mathbf{w}\times(\mathbf{w}\times\mathbf{w}')\,.\,\mathbf{w}'$ from (6) and (7) respectively in (5), we get

$$\rho^2+(\sigma\rho')^2-\frac{\lambda}{\lambda+1}\left[\rho^2+(\sigma\rho')^2\right]-\frac{\lambda}{\lambda+1}\left[\rho^2\left\{\frac{\rho}{\sigma}+\frac{d}{ds}(\sigma\rho')\right\}^2\right]=0$$

or $$\frac{1}{\lambda+1}\left[\rho^2+(\sigma\rho')^2\right]-\frac{\lambda}{\lambda+1}\left[\rho^2\left\{\frac{\rho}{\sigma}+\frac{d}{ds}(\sigma\rho')\right\}^2\right]=0$$

or $$\left[\rho^2+(\sigma\rho')^2\right]-\lambda\rho^2\left[\frac{\rho}{\sigma}+\frac{d}{ds}(\sigma\rho')\right]^2=0.$$

$$\therefore \qquad \lambda = \frac{\rho^2+(\sigma\rho')^2}{\rho^2\left[\frac{\rho}{\sigma}+\frac{d}{ds}(\sigma\rho')\right]^2}. \tag{8}$$

Again, we know that

$$R^2 = \rho^2+(\sigma\rho')^2. \tag{9}$$

Differentiating (9) with respect to ρ, we get

$$2R\frac{dR}{d\rho} = 2\rho+2\,(\sigma\rho')\left[\frac{d}{ds}(\sigma\rho')\right]\frac{1}{\rho'}$$

or $$R\frac{dR}{d\rho} = \rho+\sigma\frac{d}{ds}(\sigma\rho')$$

or $$\frac{R}{\sigma}\frac{dR}{d\rho} = \frac{\rho}{\sigma}+\frac{d}{ds}(\sigma\rho'). \tag{10}$$

Using (9) and (10), (8) reduces to

$$\lambda = \frac{R^2}{\rho^2\left[\dfrac{R}{\sigma}\dfrac{dR}{d\rho}\right]^2} = \frac{\sigma^2}{\rho^2\left(\dfrac{dR}{d\rho}\right)^2}$$

i.e., $\lambda : 1 = \sigma^2 : \rho^2\left(\dfrac{dR}{d\rho}\right)^2.$

Example 11. Prove that the normal plane at any point to the locus of centre of circular curvature of any curve bisects the radius of spherical curvature at the corresponding points of the given curve.

Solution. Let Γ be a space curve and let Γ_1 be the locus of the centre of curvature of Γ.

Let the quantities of Γ_1 be distinguished from those of Γ by using the suffix unity.

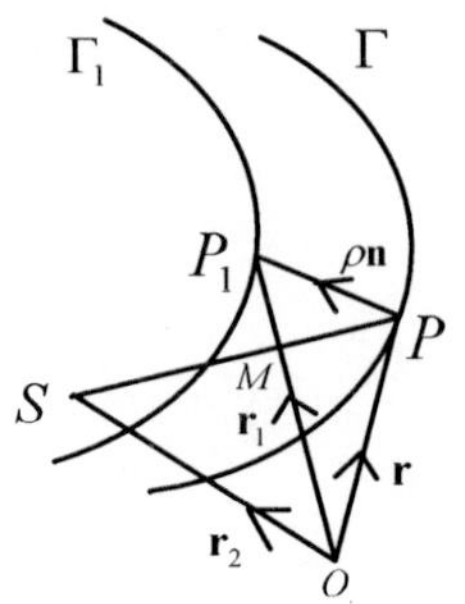

Take any pair of corresponding points P and P_1 on Γ and Γ_1 respectively.

The equation of the normal plane at P_1 is given by

$$(\mathbf{R} - \mathbf{r}_1)\,.\,\mathbf{t}_1 = 0. \qquad (1)$$

Again, we have

$$\mathbf{r}_1 = \mathbf{r} + \rho\mathbf{n}. \qquad (2)$$

Differentiating this with respect to s, we get

$$\frac{d\mathbf{r}_1}{ds_1}\frac{ds_1}{ds} = \mathbf{t} + \rho'\mathbf{n} + \rho\,(\tau\mathbf{b} - \kappa\mathbf{t})$$

$$\Rightarrow \quad \mathbf{t}_1\frac{ds_1}{ds} = \rho'\mathbf{n} + \frac{\rho}{\sigma}\,\mathbf{b} \qquad [\because \rho = 1/\kappa \text{ and } \sigma = 1/\tau]$$

$$\Rightarrow \quad \mathbf{t}_1 = \left(\rho'\mathbf{n} + \frac{\rho}{\sigma}\mathbf{b}\right)\frac{ds}{ds_1}. \qquad (3)$$

Substituting the values of $\mathbf{r}_1$ and $\mathbf{t}_1$ from (2) and (3) in (1), we get

$$[\mathbf{R} - (\mathbf{r} + \rho\mathbf{n})]\,.\left(\rho'\mathbf{n} + \frac{\rho}{\sigma}\mathbf{b}\right)\frac{ds}{ds_1} = 0$$

$$\Rightarrow \quad [(\mathbf{R} - \mathbf{r}) - \rho\mathbf{n}]\,.\left(\rho'\mathbf{n} + \frac{\rho}{\sigma}\mathbf{b}\right) = 0$$

$$\Rightarrow \qquad (\mathbf{R}-\mathbf{r})\cdot\left(\rho'\mathbf{n}+\frac{\rho}{\sigma}\mathbf{b}\right)-\rho\rho'=0. \qquad (4)$$

Let $\mathbf{r}_2$ be the position vector of the centre of spherical curvature S of Γ at P and let M be the middle point of SP.

$$\therefore \qquad \text{The position vector of } M=\frac{\mathbf{r}+\mathbf{r}_2}{2}=\frac{\mathbf{r}+(\mathbf{r}+\rho\mathbf{n}+\sigma\rho'\mathbf{b})}{2}$$

$$\text{i.e.,} \qquad \text{The position vector of } M=\mathbf{r}+\frac{1}{2}(\rho\mathbf{n}+\sigma\rho'\mathbf{b}). \qquad (5)$$

If the normal plane at P_1 bisects the radius of spherical curvature at M, then the position vector of M given by (5) must satisfy (4).

Substituting for $\mathbf{R}$ the RHS of (5) in (4), we get

$$\left[\left\{\mathbf{r}+\frac{1}{2}(\rho\mathbf{n}+\sigma\rho'\mathbf{b})\right\}-\mathbf{r}\right]\cdot\left(\rho'\mathbf{n}+\frac{\rho}{\sigma}\mathbf{b}\right)-\rho\rho'=0$$

$$\text{or} \qquad \frac{1}{2}(\rho\mathbf{n}+\sigma\rho'\mathbf{b})\cdot\left(\rho'\mathbf{n}+\frac{\rho}{\sigma}\mathbf{b}\right)-\rho\rho'=0$$

$$\text{or} \qquad \frac{1}{2}(\rho\rho'+\rho\rho')-\rho\rho'=0$$

$$\text{or} \qquad \rho\rho'-\rho\rho'=0,$$

which is true.

Therefore the position vector of middle point of the radius of spherical curvature at P satisfies the equation of the normal plane at P_1.

Example 12. Show that the radius of curvature of the locus of the centre of curvature of a curve is given by

$$\left[\left\{\frac{\rho^2\sigma}{R^3}\frac{d}{ds}\left(\frac{\sigma\rho'}{\rho}\right)-\frac{1}{R}\right\}^2+\frac{\rho'^2\sigma^4}{\rho^2R^4}\right]^{-\frac{1}{2}}.$$

Solution. Let Γ be a space curve and let Γ_1 be the locus of the centre of curvature of Γ.

Let the quantities of Γ_1 be distinguished from those of Γ by using the suffix unity.

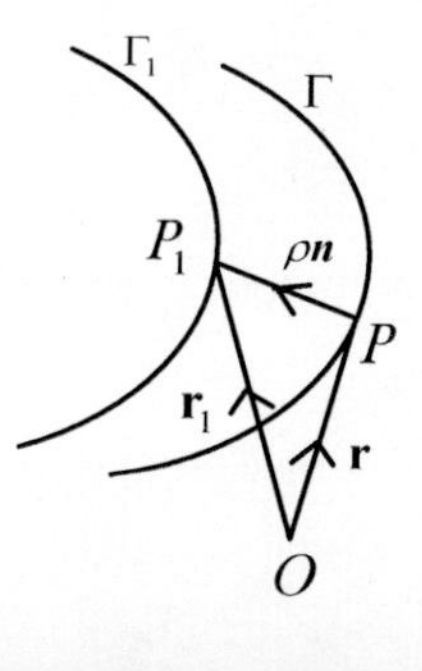

Take any pair of corresponding points P and P_1 on Γ and Γ_1 respectively.

We have

$$\mathbf{r}_1 = \mathbf{r} + \rho\mathbf{n}. \tag{1}$$

Differentiating this with respect to s, we get

$$\frac{d\mathbf{r}_1}{ds_1}\frac{ds_1}{ds} = \mathbf{t} + \rho'\mathbf{n} + \rho(\tau\mathbf{b} - \kappa\mathbf{t})$$

or $$\mathbf{t}_1 s_1' = \rho'\mathbf{n} + \frac{\rho}{\sigma}\mathbf{b}. \tag{2}$$

Squaring both sides of (2), we get

$$s_1'^2 = \rho'^2 + \left(\frac{\rho}{\sigma}\right)^2 = \frac{\rho'^2\sigma^2 + \rho^2}{\sigma^2}$$

or $$s_1'^2 = \frac{R^2}{\sigma^2},$$ where R is the radius of spherical curvature of Γ at P

or $$s_1' = \frac{R}{\sigma}. \tag{3}$$

Again, from (2), we have

$$\frac{\sigma}{\rho}s_1'\mathbf{t}_1 = \frac{\sigma\rho'}{\rho}\mathbf{n} + \mathbf{b}. \tag{4}$$

Differentiating this with respect to s, we get

$$\left[\frac{d}{ds}\left(\frac{\sigma}{\rho}s_1'\right)\right]\mathbf{t}_1 + \left(\frac{\sigma}{\rho}s_1'\right)\frac{d\mathbf{t}_1}{ds_1}s_1' = \left[\frac{d}{ds}\left(\frac{\sigma\rho'}{\rho}\right)\right]\mathbf{n} + \frac{\sigma\rho'}{\rho}(\tau\mathbf{b} - \kappa\mathbf{t}) - \tau\mathbf{n}$$

or $$\left[\frac{d}{ds}\left(\frac{\sigma}{\rho}s_1'\right)\right]\mathbf{t}_1 + \left(\frac{\sigma}{\rho}s_1'^2\kappa_1\right)\mathbf{n}_1 = -\frac{\sigma\rho'}{\rho^2}\mathbf{t} + \left[\frac{d}{ds}\left(\frac{\sigma\rho'}{\rho}\right) - \frac{1}{\sigma}\right]\mathbf{n} + \frac{\rho'}{\rho}\mathbf{b}. \tag{5}$$

Taking the cross product of (4) and (5), we get

$$\frac{\sigma^2}{\rho^2}s_1'^3\kappa_1\mathbf{b}_1 = \left[\frac{\sigma\rho'^2}{\rho^2} - \frac{d}{ds}\left(\frac{\sigma\rho'}{\rho}\right) + \frac{1}{\sigma}\right]\mathbf{t} - \frac{\sigma\rho'}{\rho^2}\mathbf{n} + \frac{\sigma^2\rho'^2}{\rho^3}\mathbf{b}$$

or $$\frac{\sigma^2}{\rho^2}s_1'^3\kappa_1\mathbf{b}_1 = \left[\frac{\sigma^2\rho'^2 + \rho^2}{\rho^2\sigma} - \frac{d}{ds}\left(\frac{\sigma\rho'}{\rho}\right)\right]\mathbf{t} - \frac{\sigma\rho'}{\rho^2}\mathbf{n} + \frac{\sigma^2\rho'^2}{\rho^3}\mathbf{b}$$

or $$\frac{\sigma^2}{\rho^2} s_1'^3 \kappa_1 \mathbf{b}_1 = \left[\frac{R^2}{\rho^2\sigma} - \frac{d}{ds}\left(\frac{\sigma\rho'}{\rho}\right)\right]\mathbf{t} - \frac{\sigma\rho'}{\rho^2}\mathbf{n} + \frac{\sigma^2\rho'^2}{\rho^3}\mathbf{b}.$$

Squaring both sides, we get

$$\frac{\sigma^4}{\rho^4} s_1'^6 \kappa_1^2 = \left[\frac{R^2}{\rho^2\sigma} - \frac{d}{ds}\left(\frac{\sigma\rho'}{\rho}\right)\right]^2 + \frac{\sigma^2\rho'^2}{\rho^4} + \frac{\sigma^4\rho'^4}{\rho^6}$$

$$\Rightarrow \quad \frac{\sigma^4}{\rho^4}\frac{R^6}{\sigma^6}\kappa_1^2 = \left[\frac{R^2}{\rho^2\sigma} - \frac{d}{ds}\left(\frac{\sigma\rho'}{\rho}\right)\right]^2 + \frac{\sigma^2\rho'^2}{\rho^6}(\rho^2 + \sigma^2\rho'^2) \qquad \text{[using (3)]}$$

$$\Rightarrow \quad \frac{R^6}{\rho^4\sigma^2}\kappa_1^2 = \left[\frac{d}{ds}\left(\frac{\sigma\rho'}{\rho}\right) - \frac{R^2}{\rho^2\sigma}\right]^2 + \frac{\sigma^2\rho'^2}{\rho^6}R^2$$

$$\Rightarrow \quad \kappa_1^2 = \left[\frac{\rho^2\sigma}{R^3}\frac{d}{ds}\left(\frac{\sigma\rho'}{\rho}\right) - \frac{1}{R}\right]^2 + \frac{\rho'^2\sigma^4}{\rho^2 R^4}$$

$$\Rightarrow \quad \kappa_1 = \left[\left\{\frac{\rho^2\sigma}{R^3}\frac{d}{ds}\left(\frac{\sigma\rho'}{\rho}\right) - \frac{1}{R}\right\}^2 + \frac{\rho'^2\sigma^4}{\rho^2 R^4}\right]^{1/2}$$

$$\Rightarrow \quad \rho_1 = \left[\left\{\frac{\rho^2\sigma}{R^3}\frac{d}{ds}\left(\frac{\sigma\rho'}{\rho}\right) - \frac{1}{R}\right\}^2 + \frac{\rho'^2\sigma^4}{\rho^2 R^4}\right]^{-\frac{1}{2}}.$$

Exercises

1. Find the equation of the osculating sphere and osculating circle at (1, 2, 3) on the curve $\mathbf{r} = (2t+1,\ 3t^2+2,\ 4t^3+3)$.

$$\left[\begin{array}{l}\textbf{Ans. } 3x^2+3y^2+3z^2-6x-16y-18z+50=0; \\ 3x^2+3y^2+3z^2-6x-16y-18z+50=0,\ z-3=0\end{array}\right]$$

2. Prove that the curve $\mathbf{r} = (a\sin^2\theta,\ a\sin\theta\cos\theta,\ a\cos\theta)$ lies on a sphere.

3. Show that the radius of spherical curvature of a circular helix is equal to the radius of circular curvature.

4. For a curve drawn on a sphere, show that the centre of curvature at any point is the foot of the perpendicular from the centre of the sphere upon the osculating plane at that point.

5. Show that the tangent to the locus of the centre of the osculating sphere passes through the centre of the osculating circle.

1.18 SPHERICAL INDICATRIX

(i) The Spherical Indicatrix of the Tangent

Definition. The locus of a point whose position vector is equal to the unit tangent **t** at any point of a given curve is called the spherical indicatrix of the tangent.

Curvature and Torsion. Let the suffix unity be used to distinguish the quantities belonging to this spherical indicatrix.

Then, by definition,

$$\mathbf{r}_1 = \mathbf{t}. \tag{1}$$

Differentiating this relation with respect to s, we get

$$\frac{d\mathbf{r}_1}{ds_1}\frac{ds_1}{ds} = \kappa\mathbf{n}$$

$$\Rightarrow \quad \mathbf{t}_1\frac{ds_1}{ds} = \kappa\mathbf{n} \tag{2}$$

$\Rightarrow$ The tangent to the spherical indicatrix is parallel to the principal normal of the given curve.

We may measure s_1 so that

$$\mathbf{t}_1 = \mathbf{n}. \tag{3}$$

From (2) and (3), we get

$$\frac{ds_1}{ds} = \kappa. \tag{4}$$

Differentiating (3) with respect to s, we get

$$\frac{d\mathbf{t}_1}{ds_1}\frac{ds_1}{ds} = \tau\mathbf{b} - \kappa\mathbf{t}$$

or

$$\kappa_1\kappa\mathbf{n}_1 = \tau\mathbf{b} - \kappa\mathbf{t}. \tag{5}$$

[using (4)]

Squaring both sides of (5), we get

$$\kappa_1^{\,2}\kappa^2 = \tau^2 + \kappa^2$$

$$\Rightarrow \quad \kappa_1 = \frac{\sqrt{(\tau^2 + \kappa^2)}}{\kappa} \tag{6}$$

$\Rightarrow$ The curvature of the spherical indicatrix is the ratio of the screw curvature to the circular curvature of the given curve.

Again,

$$\mathbf{b}_1 = \mathbf{t}_1 \times \mathbf{n}_1$$

$$= \mathbf{n} \times \frac{1}{\kappa_1 \kappa}(\tau \mathbf{b} - \kappa \mathbf{t}) \qquad \text{[using (3) and (5)]}$$

$$\Rightarrow \qquad \mathbf{b}_1 = \frac{\tau \mathbf{t} + \kappa \mathbf{b}}{\kappa_1 \kappa}$$

$$\Rightarrow \qquad \kappa_1 \kappa \mathbf{b}_1 = \tau \mathbf{t} + \kappa \mathbf{b}. \tag{7}$$

Differentiating (7) with respect to s, we get

$$\left[\frac{d}{ds}(\kappa_1 \kappa)\right]\mathbf{b}_1 + \kappa_1 \kappa \frac{d\mathbf{b}_1}{ds_1}\frac{ds_1}{ds} = \tau' \mathbf{t} + \tau(\kappa \mathbf{n}) + \kappa' \mathbf{b} + \kappa(-\tau \mathbf{n})$$

or

$$-\kappa_1 \kappa^2 \tau_1 \mathbf{n}_1 + \mathbf{b}_1 \frac{d}{ds}(\kappa_1 \kappa) = \tau' \mathbf{t} + \kappa' \mathbf{b}. \tag{8}$$

Taking the dot product of (5) and (8), we get

$$-\kappa_1^2 \kappa^3 \tau_1 = \tau \kappa' - \kappa \tau'$$

or

$$\tau_1 = \frac{\kappa \tau' - \tau \kappa'}{\kappa_1^2 \kappa^3}$$

or

$$\tau_1 = \frac{\kappa \tau' - \tau \kappa'}{\kappa(\kappa^2 + \tau^2)}. \tag{9}$$

[using (6)]

(ii) The Spherical Indicatrix of the Principal Normal

Definition. The locus of a point whose position vector is equal to the unit principal normal $\mathbf{n}$ at any point of a given curve is called the spherical indicatrix of the principal normal.

Curvature and Torsion. Let the suffix unity be used to distinguish the quantities belonging to this spherical indicatrix.

Then, by definition,

$$\mathbf{r}_1 = \mathbf{n}. \tag{10}$$

Differentiating this relation with respect to s, we get

$$\frac{d\mathbf{r}_1}{ds_1}\frac{ds_1}{ds} = \tau \mathbf{b} - \kappa \mathbf{t}$$

or

$$\mathbf{t}_1 \frac{ds_1}{ds} = \tau \mathbf{b} - \kappa \mathbf{t}. \tag{11}$$

Squaring both sides of (11), we get

$$\left(\frac{ds_1}{ds}\right)^2 = \tau^2 + \kappa^2. \tag{12}$$

Differentiating (11) with respect to s, we get

$$\left(\frac{d\mathbf{t}_1}{ds_1}\frac{ds_1}{ds}\right)\frac{ds_1}{ds}+\mathbf{t}_1\frac{d^2s_1}{ds^2}=\tau'\mathbf{b}+\tau(-\tau\mathbf{n})-\kappa'\mathbf{t}-\kappa(\kappa\mathbf{n})$$

or
$$\kappa_1\mathbf{n}_1\left(\frac{ds_1}{ds}\right)^2+\mathbf{t}_1\frac{d^2s_1}{ds^2}=-\kappa'\mathbf{t}-(\kappa^2+\tau^2)\mathbf{n}+\tau'\mathbf{b}. \tag{13}$$

Now, taking the cross product of (11) and (13), we get

$$\kappa_1\left(\frac{ds_1}{ds}\right)^3\mathbf{b}_1=\tau(\kappa^2+\tau^2)\mathbf{t}+(\kappa\tau'-\kappa'\tau)\mathbf{n}+\kappa(\kappa^2+\tau^2)\mathbf{b}. \tag{14}$$

Squaring (14), we get

$$\kappa_1^2\left(\frac{ds_1}{ds}\right)^6=\tau^2(\kappa^2+\tau^2)^2+(\kappa\tau'-\kappa'\tau)^2+\kappa^2(\kappa^2+\tau^2)^2$$

or
$$\kappa_1^2\left(\frac{ds_1}{ds}\right)^6=(\kappa^2+\tau^2)^3+(\kappa\tau'-\kappa'\tau)^2 \tag{15}$$

or
$$\kappa_1^2(\kappa^2+\tau^2)^3=(\kappa^2+\tau^2)^3+(\kappa\tau'-\kappa'\tau)^2 \qquad \text{[using (12)]}$$

or
$$\kappa_1^2=1+\frac{(\kappa\tau'-\kappa'\tau)^2}{(\kappa^2+\tau^2)^3}. \tag{16}$$

Differentiating (14) with respect to s, we get

$$\left[\frac{d}{ds}\left(\kappa_1\left(\frac{ds_1}{ds}\right)^3\right)\right]\mathbf{b}_1+\kappa_1\left(\frac{ds_1}{ds}\right)^3\left(\frac{d\mathbf{b}_1}{ds_1}\frac{ds_1}{ds}\right)$$
$$=[\tau'(\kappa^2+\tau^2)+\tau(2\kappa\kappa'+2\tau\tau')]\mathbf{t}+\tau(\kappa^2+\tau^2)\kappa\mathbf{n}$$
$$+[\kappa'\tau'+\kappa\tau''-(\kappa''\tau+\kappa'\tau')]\mathbf{n}+(\kappa\tau'-\kappa'\tau)(\tau\mathbf{b}-\kappa\mathbf{t})$$
$$+[\kappa'(\kappa^2+\tau^2)+\kappa(2\kappa\kappa'+2\tau\tau')]\mathbf{b}+\kappa(\kappa^2+\tau^2)(-\tau\mathbf{n})$$

or
$$\left[\frac{d}{ds}\left(\kappa_1\left(\frac{ds_1}{ds}\right)^3\right)\right]\mathbf{b}_1-\kappa_1\tau_1\left(\frac{ds_1}{ds}\right)^4\mathbf{n}_1$$
$$=3\tau(\kappa\kappa'+\tau\tau')\mathbf{t}+(\kappa\tau''-\tau\kappa'')\mathbf{n}+3\kappa(\kappa\kappa'+\tau\tau')\mathbf{b}. \tag{17}$$

Taking the dot product of (13) and (17), we get

$$-\kappa_1^2\tau_1\left(\frac{ds_1}{ds}\right)^6=-3\kappa'\tau(\kappa\kappa'+\tau\tau')-(\kappa^2+\tau^2)(\kappa\tau''-\tau\kappa'')$$
$$+3\kappa\tau'(\kappa\kappa'+\tau\tau')$$

or
$$\kappa_1^2\tau_1\left(\frac{ds_1}{ds}\right)^6=3(\kappa'\tau-\kappa\tau')(\kappa\kappa'+\tau\tau')+(\kappa^2+\tau^2)(\kappa\tau''-\tau\kappa'')$$

or $\quad \tau_1[(\kappa^2+\tau^2)^3+(\kappa\tau'-\kappa'\tau)^2]$

$$=3(\kappa'\tau-\kappa\tau')(\kappa\kappa'+\tau\tau')+(\kappa^2+\tau^2)(\kappa\tau''-\tau\kappa'').$$

[using (15)]

$$\therefore \quad \tau_1=\frac{3(\kappa'\tau-\kappa\tau')(\kappa\kappa'+\tau\tau')+(\kappa^2+\tau^2)(\kappa\tau''-\tau\kappa'')}{(\kappa^2+\tau^2)^3+(\kappa\tau'-\kappa'\tau)^2}. \tag{18}$$

(iii) The Spherical Indicatrix of the Binormal

Definition. The locus of a point whose position vector is equal to the unit binormal **b** at any point of a given curve is called the spherical indicatrix of the binormal.

Curvature and Torsion. Let the suffix unity be used to distinguish the quantities belonging to this spherical indicatrix.

Then, by definition,

$$\mathbf{r}_1=\mathbf{b}. \tag{19}$$

Differentiating this relation with respect to s, we get

$$\frac{d\mathbf{r}_1}{ds_1}\frac{ds_1}{ds}=-\tau\mathbf{n}$$

or $$\mathbf{t}_1\frac{ds_1}{ds}=-\tau\mathbf{n}. \tag{20}$$

We may measure s_1 so that

$$\mathbf{t}_1=-\mathbf{n}. \tag{21}$$

From (20) and (21), we get

$$\frac{ds_1}{ds}=\tau. \tag{22}$$

Differentiating (21) with respect to s, we get

$$\frac{d\mathbf{t}_1}{ds_1}\frac{ds_1}{ds}=-(\tau\mathbf{b}-\kappa\mathbf{t})$$

or $$\kappa_1\tau\mathbf{n}_1=-(\tau\mathbf{b}-\kappa\mathbf{t}). \tag{23}$$

[using (22)]

Squaring both sides of (5), we get

$$\kappa_1^2\tau^2=\tau^2+\kappa^2$$

$$\Rightarrow \quad \kappa_1=\frac{\sqrt{(\tau^2+\kappa^2)}}{\tau} \tag{24}$$

⇒ The curvature of the spherical indicatrix is the ratio of the screw curvature to the torsion of the given curve.

Again,

$$\mathbf{b}_1 = \mathbf{t}_1 \times \mathbf{n}_1$$

$$\Rightarrow \quad \mathbf{b}_1 = (-\mathbf{n}) \times \frac{1}{\kappa_1 \tau}[-(\tau \mathbf{b} - \kappa \mathbf{t})] \qquad \text{[using (21) and (23)]}$$

$$\Rightarrow \quad \mathbf{b}_1 = \frac{\tau \mathbf{t} + \kappa \mathbf{b}}{\kappa_1 \tau}$$

$$\Rightarrow \quad \kappa_1 \tau \mathbf{b}_1 = \tau \mathbf{t} + \kappa \mathbf{b}. \tag{25}$$

Differentiating (25) with respect to s, we get

$$\left[\frac{d}{ds}(\kappa_1 \tau)\right]\mathbf{b}_1 + \kappa_1 \tau \frac{d\mathbf{b}_1}{ds_1}\frac{ds_1}{ds} = \tau' \mathbf{t} + \tau(\kappa \mathbf{n}) + \kappa' \mathbf{b} + \kappa(-\tau \mathbf{n})$$

or
$$\left[\frac{d}{ds}(\kappa_1 \tau)\right]\mathbf{b}_1 + \kappa_1 \tau \; (-\tau_1 \mathbf{n}_1)\,\tau = \tau' \mathbf{t} + \kappa' \mathbf{b}. \tag{26}$$

Taking the cross product of (25) and (26), we get

$$\kappa_1^2 \tau^3 \tau_1 \mathbf{t}_1 = (\kappa \tau' - \tau \kappa')\,\mathbf{n}$$

or
$$\tau_1 = \frac{\kappa \tau' - \tau \kappa'}{\kappa_1^2 \tau^3} \qquad \text{[using (21)]}$$

or
$$\tau_1 = \frac{\kappa \tau' - \tau \kappa'}{\tau(\kappa^2 + \tau^2)}. \tag{27}$$

[using (24)]

1.19 INVOLUTES AND EVOLUTES

Definition. If the tangents to a curve C are normals to another curve C_1, then C_1 is called an involute of C and C is called an evolute of C_1.

An involute may be generated in the following manner.

Let one end of an inextensible string be tied to a fixed point, say A, of the curve C, and let the string be wrapped on C on its convex side, keeping it taut all the time.

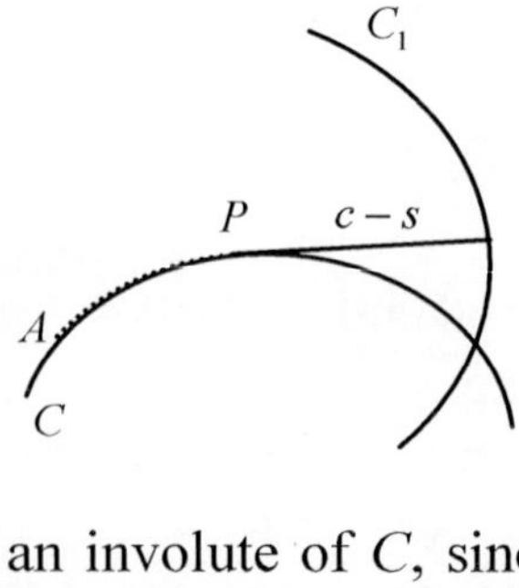

Then, any particle of the string describes an involute of C, since at each instant the free part of the string is a tangent to the curve C, whereas the direction of motion of the particle is at right angles to this tangent.

Equation of the Involute. Let $\mathbf{r} = \mathbf{r}(s)$ be the equation of the given curve C and let C_1 denote the involute of C.

Let the quantities of C_1 be distinguished from those of C by using the suffix unity.

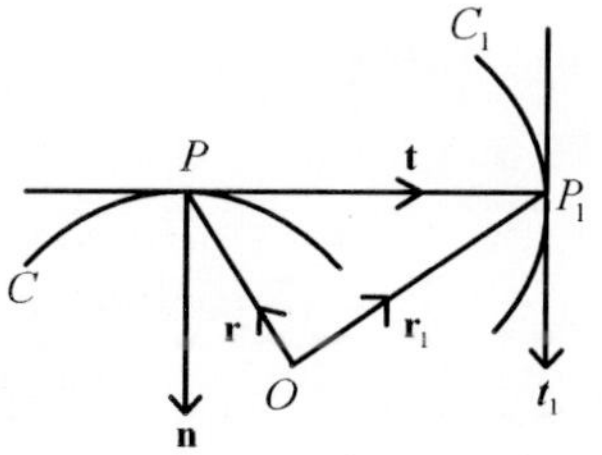

Then, the position vector $\mathbf{r}_1$ of any point P_1 on C_1 is given by

$$\mathbf{r}_1 = \mathbf{r} + \lambda \mathbf{t}, \tag{1}$$

where λ is to be determined.

Differentiating (1) with respect to s, we get

$$\frac{d\mathbf{r}_1}{ds_1}\frac{ds_1}{ds} = \mathbf{t} + \lambda \frac{d\mathbf{t}}{ds} + \lambda' \mathbf{t}$$

or
$$\mathbf{t}_1 \frac{ds_1}{ds} = \mathbf{t} + \lambda(\kappa \mathbf{n}) + \lambda' \mathbf{t}. \tag{2}$$

Taking the dot product of both sides of (2) with $\mathbf{t}$ and using $\mathbf{t} . \mathbf{t}_1 = 0$, we get

$$0 = 1 + \lambda'$$

$$\Rightarrow \quad \frac{d\lambda}{ds} = -1.$$

Integrating, we get

$$\lambda = -s + c,$$

where c is the constant of integration.

Substituting the value of λ in equation (1), we get

$$\mathbf{r}_1 = \mathbf{r} + (c - s)\mathbf{t}. \tag{3}$$

Equation (3) is the required equation of the involute C_1 of the curve C.

Curvature and Torsion of the Involute.

Differentiating (3) with respect to s, we get

$$\frac{d\mathbf{r}_1}{ds_1}\frac{ds_1}{ds} = \mathbf{t} + (0 - 1)\mathbf{t} + (c - s)\kappa \mathbf{n}$$

$$\Rightarrow \quad \mathbf{t}_1 \frac{ds_1}{ds} = (c - s)\kappa \mathbf{n} \tag{4}$$

$\Rightarrow$ The tangent to the involute is parallel to the principal normal to the given curve C.

We choose the positive direction along the involute such that

$$\mathbf{t}_1 = \mathbf{n}. \tag{5}$$

From (4) and (5), we get

$$\frac{ds_1}{ds} = (c - s)\,\kappa. \tag{6}$$

Differentiating (5) with respect to s, we get

$$\frac{d\mathbf{t}_1}{ds_1}\frac{ds_1}{ds} = \tau\,\mathbf{b} - \kappa\,\mathbf{t}$$

or $$\kappa_1\kappa\,(c - s)\,\mathbf{n}_1 = \tau\,\mathbf{b} - \kappa\,\mathbf{t}. \tag{7}$$

[using (6)]

Squaring both sides of (7), we get

$$\kappa_1^{\,2}\kappa^2(c - s)^2 = \tau^2 + \kappa^2$$

$$\Rightarrow \quad \kappa_1^{\,2} = \frac{\tau^2 + \kappa^2}{\kappa^2(c - s)^2}$$

$$\Rightarrow \quad \kappa_1 = \frac{\sqrt{(\tau^2 + \kappa^2)}}{\kappa(c - s)}. \tag{8}$$

Again,

$$\mathbf{b}_1 = \mathbf{t}_1 \times \mathbf{n}_1$$

$$= \mathbf{n} \times \frac{1}{\kappa_1\kappa\,(c - s)}(\tau\,\mathbf{b} - \kappa\,\mathbf{t}) \qquad \text{[using (5) and (7)]}$$

i.e., $$\mathbf{b}_1 = \frac{\tau\,\mathbf{t} + \kappa\,\mathbf{b}}{\kappa_1\kappa\,(c - s)}$$

or $$\kappa_1\kappa\,(c - s)\,\mathbf{b}_1 = \tau\,\mathbf{t} + \kappa\,\mathbf{b}. \tag{9}$$

Differentiating (9) with respect to s, we get

$$\left[\frac{d}{ds}\left(\kappa_1\kappa\,(c - s)\right)\right]\mathbf{b}_1 + \kappa_1\kappa\,(c - s)\frac{d\mathbf{b}_1}{ds_1}\frac{ds_1}{ds} = \tau'\,\mathbf{t} + \tau\,(\kappa\,\mathbf{n})$$

$$+ \kappa'\,\mathbf{b} + \kappa\,(-\tau\,\mathbf{n})$$

or $$\left[\frac{d}{ds}\left(\kappa_1\kappa\,(c - s)\right)\right]\mathbf{b}_1 - \kappa_1\kappa^2(c - s)^2\,\tau_1\mathbf{n}_1 = \tau'\,\mathbf{t} + \kappa'\,\mathbf{b}. \tag{10}$$

Taking the dot product of (7) and (10), we get

$$-\kappa_1^{\,2}\kappa^3(c - s)^3\,\tau_1 = \tau\kappa' - \kappa\tau'$$

or $$\tau_1 = \frac{\kappa\tau' - \tau\kappa'}{\kappa_1^{\,2}\kappa^3(c - s)^3}$$

or $$\tau_1 = \frac{\kappa\tau' - \tau\kappa'}{\kappa(c-s)(\kappa^2+\tau^2)}. \tag{11}$$

[using (8)]

Equation of the Evolute. Let $\mathbf{r} = \mathbf{r}(s)$ be the equation of the given curve C and let C_1 denote the evolute of C.

Let the quantities of C_1 be distinguished from those of C by using the suffix unity.

By definition, C is the involute of C_1 and therefore, the tangents to C_1 are normals to C. This implies that the vector $\overrightarrow{PP_1}$ lies in the normal plane to C at P. Hence, we have

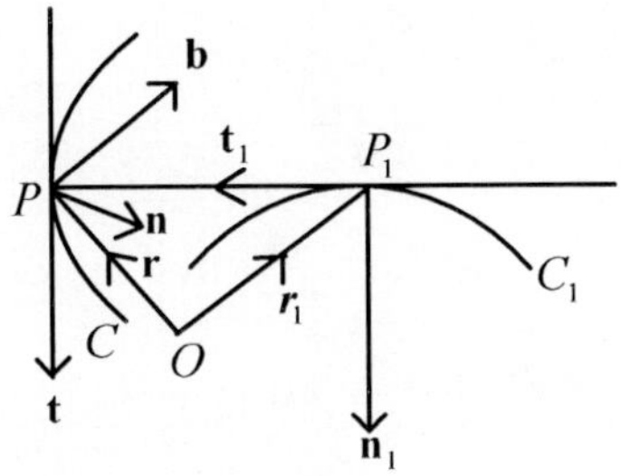

$$\overrightarrow{PP_1} = \lambda\mathbf{n} + \mu\mathbf{b}$$

or $$\mathbf{r}_1 - \mathbf{r} = \lambda\mathbf{n} + \mu\mathbf{b}$$

or $$\mathbf{r}_1 = \mathbf{r} + \lambda\mathbf{n} + \mu\mathbf{b}, \tag{12}$$

where λ and μ are some functions of s to be determined.

Differentiating (12) with respect to s, we get

$$\frac{d\mathbf{r}_1}{ds_1}\frac{ds_1}{ds} = \mathbf{t} + \lambda'\mathbf{n} + \lambda(\tau\mathbf{b} - \kappa\mathbf{t}) + \mu'\mathbf{b} + \mu(-\tau\mathbf{n})$$

or $$\mathbf{t}_1\frac{ds_1}{ds} = (1-\lambda\kappa)\mathbf{t} + (\lambda' - \mu\tau)\mathbf{n} + (\mu' + \lambda\tau)\mathbf{b}. \tag{13}$$

As $\mathbf{t}_1$ lies in the normal plane of C at P and is parallel to $\overrightarrow{PP_1}$, we have

$$\mathbf{t}_1 = u(\lambda\mathbf{n} + \mu\mathbf{b}), \tag{14}$$

where u is any constant.

From (13) and (14), we get

$$u(\lambda\mathbf{n} + \mu\mathbf{b})\frac{ds_1}{ds} = (1-\lambda\kappa)\mathbf{t} + (\lambda' - \mu\tau)\mathbf{n} + (\mu' + \lambda\tau)\mathbf{b}.$$

Equating the coefficients of **t**, **n**, **b** on both sides, we get

$$1 - \lambda\kappa = 0,\ \lambda' - \mu\tau = u\lambda\frac{ds_1}{ds},\ \mu' + \lambda\tau = u\mu\frac{ds_1}{ds}.$$

$\therefore$ $$\lambda = \frac{1}{\kappa} = \rho \tag{15}$$

and $$\frac{\lambda' - \mu\tau}{\mu' + \lambda\tau} = \frac{\lambda}{\mu}. \tag{16}$$

From (16), we have

$$\mu\ (\lambda' - \mu\tau) = \lambda\,(\mu' + \lambda\tau)$$

or $$\tau\,(\lambda^2 + \mu^2) = \lambda'\mu - \lambda\mu'$$

or $$\tau = \frac{\lambda'\mu - \lambda\mu'}{\lambda^2 + \mu^2}$$

or $$\tau = \frac{d}{ds}\left[\tan^{-1}\left(-\frac{\mu}{\lambda}\right)\right].$$

Integrating with respect to s, we get

$$\int \tau\, ds + c = \tan^{-1}\left(-\frac{\mu}{\lambda}\right).$$

We write

$$\int \tau\, ds = \psi \text{ so that } \psi' = \tau. \tag{17}$$

Then, we have

$$\psi + c = \tan^{-1}\left(-\frac{\mu}{\lambda}\right)$$

$$\Rightarrow \quad \tan\,(\psi + c) = -\frac{\mu}{\lambda}$$

$$\Rightarrow \quad \mu = -\rho \tan\,(\psi + c). \tag{18}$$

[using (15)]

Substituting the values of λ and μ from (15) and (18) in (12), we get

$$\mathbf{r}_1 = \mathbf{r} + \rho\mathbf{n} - \rho \tan\,(\psi + c)\,\mathbf{b}. \tag{19}$$

Equation (18) is the equation of the evolute C_1 of the given curve C.

Curvature and Torsion of the Evolute.

Differentiating (19) with respect to s, we get

$$\frac{d\mathbf{r}_1}{ds_1}\frac{ds_1}{ds} = \mathbf{t} + \rho'\mathbf{n} + \rho\,(\tau\mathbf{b} - \kappa\mathbf{t})$$
$$- [\rho' \tan\,(\psi + c) + \rho\,\psi' \sec^2(\psi + c)]\,\mathbf{b}$$
$$- [\rho \tan\,(\psi + c)](-\tau\mathbf{n})$$

or $$\mathbf{t}_1\frac{ds_1}{ds} = [\rho' + \rho\tau \tan\,(\psi + c)]\,\mathbf{n}$$
$$-[\rho' \tan\,(\psi + c) + \rho\tau \tan^2(\psi + c)]\,\mathbf{b} \qquad [\because \psi' = \tau]$$

or $$\mathbf{t}_1 \frac{ds_1}{ds} = [\rho' + \rho\tau \tan(\psi + c)][\mathbf{n} - \tan(\psi + c)\,\mathbf{b}]. \tag{20}$$

Squaring both sides of (20), we get

$$\left(\frac{ds_1}{ds}\right)^2 = [\rho' + \rho\tau \tan\ (\psi + c)]^2\,[1 + \tan^2(\psi + c)].$$

$$\therefore \quad \frac{ds_1}{ds} = [\rho' + \rho\tau \tan(\psi + c)]\sec(\psi + c). \tag{21}$$

From (20) and (21), we get

$$\mathbf{t}_1 = \cos(\psi + c)\,\mathbf{n} - \sin(\psi + c)\,\mathbf{b}. \tag{22}$$

Differentiating (22) with respect to s, we get

$$\frac{d\mathbf{t}_1}{ds_1}\frac{ds_1}{ds} = -\psi'\sin(\psi + c)\,\mathbf{n} + \cos(\psi + c)(\tau\mathbf{b} - \kappa\mathbf{t})$$

$$-\psi'\cos(\psi + c)\,\mathbf{b} - \sin(\psi + c)(-\tau\mathbf{n})$$

$$\Rightarrow \quad \kappa_1\mathbf{n}_1\frac{ds_1}{ds} = -\kappa\cos(\psi + c)\,\mathbf{t}. \tag{23}$$

$\Rightarrow$ The principal normal to the evolute is parallel to the tangent to the given curve C.

We choose the direction such that

$$\mathbf{n}_1 = -\mathbf{t}. \tag{24}$$

From (23) and (24), we get

$$\kappa_1\frac{ds_1}{ds} = \kappa\cos(\psi + c)$$

$$\Rightarrow \quad \kappa_1[\rho' + \rho\tau\tan(\psi + c)]\sec(\psi + c) = \kappa\cos(\psi + c) \qquad \text{[using (21)]}$$

$$\Rightarrow \quad \kappa_1 = \frac{\kappa\cos^2(\psi + c)}{\rho' + \rho\tau\tan(\psi + c)}$$

$$= \frac{\kappa^3\cos^2(\psi + c)}{\kappa\tau\tan(\psi + c) - \kappa'} \qquad \left[\because \rho = \frac{1}{\kappa} \text{ and } \rho' = -\frac{\kappa'}{\kappa^2}\right]$$

i.e., $$\kappa_1 = \frac{\kappa^3\cos^3(\psi + c)}{\kappa\tau\sin(\psi + c) - \kappa'\cos(\psi + c)}. \tag{25}$$

Again,

$$\mathbf{b}_1 = \mathbf{t}_1 \times \mathbf{n}_1$$

$$= [\cos(\psi + c)\,\mathbf{n} - \sin(\psi + c)\,\mathbf{b}] \times (-\mathbf{t}) \qquad \text{[using (22) and (24)]}$$

i.e., $$\mathbf{b}_1 = \cos\ (\psi + c)\,\mathbf{b} + \sin(\psi + c)\,\mathbf{n}. \tag{26}$$

Differentiating (26) with respect to s, we get

$$\frac{d\mathbf{b}_1}{ds_1}\frac{ds_1}{ds} = -\psi'\sin(\psi+c)\,\mathbf{b} + \cos(\psi+c)(-\tau\mathbf{n})$$

$$+\psi'\cos(\psi+c)\,\mathbf{n} + \sin(\psi+c)(\tau\mathbf{b}-\kappa\mathbf{t})$$

$$\Rightarrow \quad -\tau_1\mathbf{n}_1\frac{ds_1}{ds} = [-\kappa\sin(\psi+c)]\,\mathbf{t} \qquad [\because \psi' = \tau]$$

$$\Rightarrow \quad \tau_1\frac{ds_1}{ds} = -\kappa\sin(\psi+c) \qquad \text{[using (24)]}$$

$$\Rightarrow \quad \tau_1[\rho' + \rho\tau\tan(\psi+c)]\sec(\psi+c) = -\kappa\sin(\psi+c) \qquad \text{[using (21)]}$$

$$\Rightarrow \quad \tau_1 = -\frac{\kappa\sin(\psi+c)}{[\rho'+\rho\tau\tan(\psi+c)]\sec(\psi+c)}$$

$$= -\frac{\kappa^3\sin(\psi+c)}{[-\kappa'+\kappa\tau\tan(\psi+c)]\sec(\psi+c)} \qquad \left[\because \rho = \frac{1}{\kappa},\ \rho' = -\frac{\kappa'}{\kappa^2}\right]$$

i.e.,
$$\tau_1 = -\frac{\kappa^3\sin(\psi+c)\cos^2(\psi+c)}{\kappa\tau\sin(\psi+c)-\kappa'\cos(\psi+c)}. \qquad (27)$$

1.20 BERTRAND CURVES

Definition. Any two curves having their principal normals in common are said to be conjugate or associate Bertrand curves.

Properties of Bertrand Curves.

(i) The distance between corresponding points of a pair of conjugate or associate Bertrand curves is constant.

Proof. Let C and C_1 be any two conjugate or associate Bertrand curves and let the quantities of C_1 be distinguished from those of C by using the suffix unity.

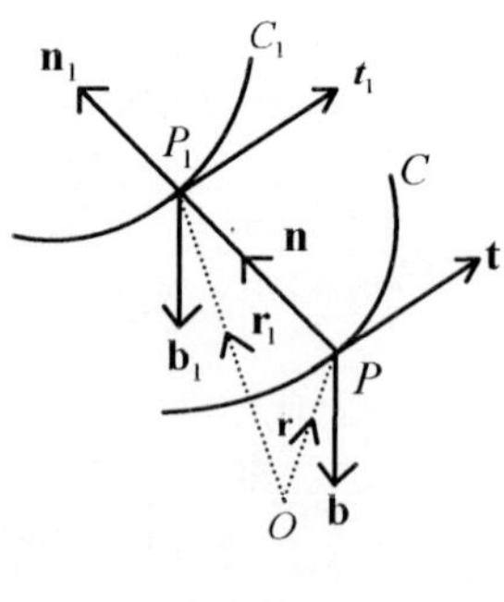

Let P and P_1 denote the corresponding points of C and C_1 respectively. Then, by definition,

$$\mathbf{n}_1 = \mathbf{n}. \qquad (1)$$

If $\mathbf{r}$ and $\mathbf{r}_1$ be the position vector of the points P and P_1 respectively, then

$$\mathbf{r}_1 = \mathbf{r} + a\mathbf{n}, \qquad (2)$$

where a is a suitable function of s and denotes the distance between the corresponding points of the two curves.

Differentiating (2) with respect to s, we get

$$\frac{d\mathbf{r}_1}{ds_1}\frac{ds_1}{ds} = \mathbf{t} + a'\mathbf{n} + a(\tau\mathbf{b} - \kappa\mathbf{t})$$

or $$\mathbf{t}_1 \frac{ds_1}{ds} = (1 - a\kappa)\,\mathbf{t} + a'\mathbf{n} + a\tau\mathbf{b}. \tag{3}$$

Taking the dot product of (1) and (3), we get

$$\mathbf{n}_1 . \mathbf{t}_1 \frac{ds_1}{ds} = \mathbf{n} . [(1 - a\kappa)\,\mathbf{t} + a'\mathbf{n} + a\tau\mathbf{b}]$$

$\Rightarrow \quad 0 = a'$ $\left[\because \mathbf{t}, \mathbf{n}, \mathbf{b} \text{ are mutually perpendicular and also, } \mathbf{t}_1, \mathbf{n}_1, \mathbf{b}_1 \text{ are mutually perpendicular.}\right]$

$\Rightarrow \quad a = \text{constant.}$

(ii) The tangents and the binormals at the corresponding points of the two curves are inclined at a constant angle α.

Proof. We have

$$\frac{d}{ds}(\mathbf{t} . \mathbf{t}_1) = \frac{d\mathbf{t}}{ds} . \mathbf{t}_1 + \mathbf{t} . \left(\frac{d\mathbf{t}_1}{ds_1}\frac{ds_1}{ds}\right)$$

$$= (\kappa\mathbf{n}) . \mathbf{t}_1 + \mathbf{t} . \left(\kappa_1 \mathbf{n}_1 \frac{ds_1}{ds}\right)$$

$$= \kappa\,(\mathbf{n}_1 . \mathbf{t}_1) + (\mathbf{t} . \mathbf{n})\,\kappa_1 \frac{ds_1}{ds} \qquad \text{[using (1)]}$$

$\Rightarrow \quad \dfrac{d}{ds}(\mathbf{t} . \mathbf{t}_1) = 0$

$\Rightarrow \quad \mathbf{t} . \mathbf{t}_1 = \text{constant}$

$\Rightarrow \quad \cos\alpha = \text{constant},$

where α is the angle between $\mathbf{t}$ and $\mathbf{t}_1$.

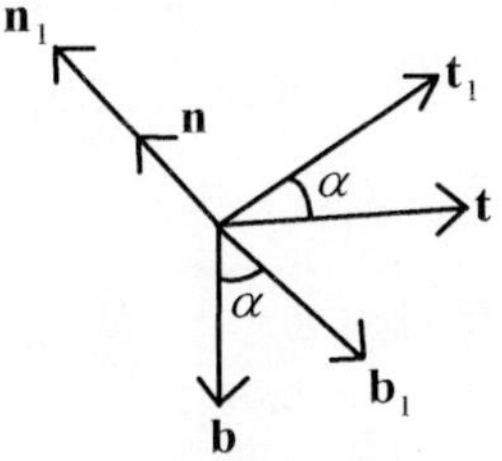

Since the principal normals of C and C_1 at the corresponding points are coincident, their rectifying planes at the corresponding points are parallel. Thus it follows that since $\mathbf{t}$ and $\mathbf{t}_1$ are inclined at a constant angle α, therefore $\mathbf{b}$ and $\mathbf{b}_1$ are also inclined at the same constant angle α.

(iii) Curvature and torsion of either curve are connected by a linear relation.

Proof. Since $a' = 0$, therefore equation (3) reduces to

$$\mathbf{t}_1 \frac{ds_1}{ds} = (1 - a\kappa)\,\mathbf{t} + a\tau\mathbf{b}. \tag{4}$$

Taking the dot product of both sides of (4) with $\mathbf{b}_1$, we get

$$\mathbf{b}_1 . \mathbf{t}_1 \frac{ds_1}{ds} = (1 - a\kappa)\,\mathbf{b}_1 . \mathbf{t} + a\tau\mathbf{b}_1 . \mathbf{b}$$

or $$0 = (1 - a\kappa)\,\mathbf{b}_1 . \mathbf{t} + a\tau\mathbf{b}_1 . \mathbf{b}$$

or $$0 = (1 - a\kappa)\cos\left(\frac{\pi}{2} - \alpha\right) + a\tau\cos\alpha$$

or $$(1 - a\kappa)\sin\alpha + a\tau\cos\alpha = 0, \tag{5}$$

which is a linear relation with constant coefficients between the curvature and torsion of C.

The relation (5) may be put in the form

$$\tau = \left(\kappa - \frac{1}{a}\right)\tan\alpha. \tag{6}$$

Now, the relation between the curves C and C_1 is a reciprocal one. The point $\mathbf{r}$ is at a distance $-a$ along the principal normal at $\mathbf{r}_1$ and $\mathbf{t}$ is inclined at angle $-\alpha$ with $\mathbf{t}_1$. Hence corresponding to (6), we shall have

$$\tau_1 = -\left(\kappa_1 + \frac{1}{a}\right)\tan\alpha. \tag{7}$$

(iv) The torsions of the two associated Bertrand curves have the same sign and their product is constant.

Proof. The vectors $\mathbf{t}, \mathbf{b}, \mathbf{t}_1, \mathbf{b}_1$ are coplanar, since each of them is perpendicular to the same vector $\mathbf{n}$. Hence

$$\mathbf{t}_1 = l\mathbf{t} + m\mathbf{b} \tag{8}$$

$$\Rightarrow \quad \left.\begin{array}{l} \mathbf{t}_1 . \mathbf{t} = l, \\ \mathbf{t}_1 . \mathbf{b} = m. \end{array}\right\}$$

$$\Rightarrow \quad \left.\begin{array}{r} \cos\alpha = l, \\ -\sin\alpha = m. \end{array}\right\} \tag{9}$$

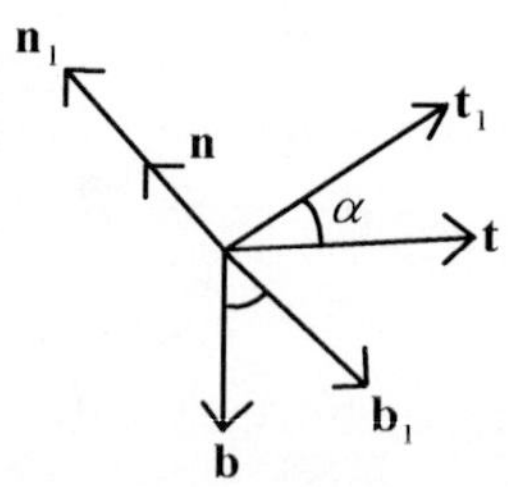

Substituting the values of l and m from (9) in (8), we get

$$\mathbf{t}_1 = \mathbf{t}\cos\alpha - \mathbf{b}\sin\alpha. \tag{10}$$

Comparing (4) and (10), we get

$$\frac{\frac{ds_1}{ds}}{1} = \frac{1 - a\kappa}{\cos\alpha} = \frac{a\tau}{-\sin\alpha}$$

$$\Rightarrow \qquad 1 - a\kappa = \cos\alpha \frac{ds_1}{ds} \tag{11}$$

and $$a\tau = -\sin\alpha \frac{ds_1}{ds}. \tag{12}$$

Since the relation between the curves C and C_1 is reciprocal, the reciprocal results corresponding to (11) and (12) are given by

$$1 + a\kappa_1 = \cos\alpha \frac{ds}{ds_1} \tag{13}$$

and $$-a\tau_1 = \sin\alpha \frac{ds}{ds_1}. \tag{14}$$

Multiplying (11) by (13) and (12) by (14), we get

$$(1 - a\kappa)(1 + a\kappa_1) = \cos^2\alpha \tag{15}$$

and $$\tau\tau_1 = \frac{\sin^2\alpha}{a^2}. \tag{16}$$

It follows from relation (16) that torsions of the two curves have the same sign and their product is constant.

(v) If P, P_1 are the corresponding points on two conjugate Bertrand curves, and O, O_1 are their centres of curvature, then the cross ratio of the range (POP_1O_1) is constant and equal to $\sec^2\alpha$.

Proof. First, we define the cross ratio.

If A, B, C, D are four points lying on a straight line, then the ratio of AB to BC divided by the ratio of AD to DC is called the cross ratio of the four points A, B, C, D and is denoted by $(ABCD)$, i.e.,

$$(ABCD) = \frac{AB}{BC} \Big/ \frac{AD}{DC} = \frac{AB.CD}{BC.DA}.$$

Such a set of points A, B, C, D lying on a straight line is called range.

Now, we shall prove the main theorem.

Here, the four points are P, O, P_1, O_1. Hence their cross ratio is given by

$$(POP_1O_1) = \frac{PO.P_1O_1}{OP_1.O_1P}. \tag{17}$$

We know that if the position vectors of the points P and P_1 are $\mathbf{r}$ and $\mathbf{r}_1$ respectively, then the position vectors of their centres of curvature O and O_1 are $\mathbf{r} + \rho\mathbf{n}$ and $\mathbf{r}_1 + \rho_1\mathbf{n}_1$ respectively.

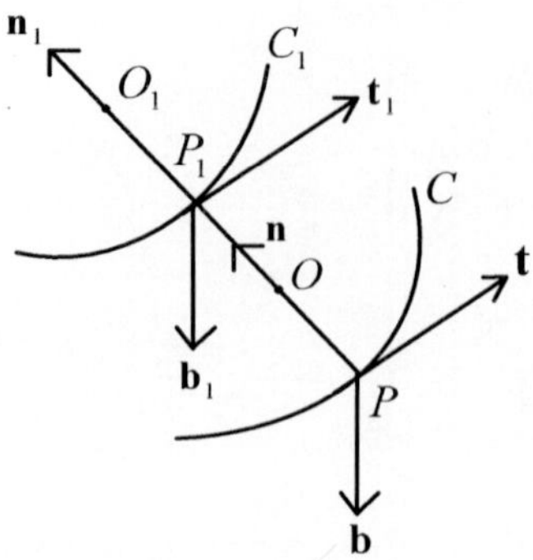

Therefore

$$PO = |(\mathbf{r} + \rho\mathbf{n}) - \mathbf{r}| = |\rho\mathbf{n}| = \rho,$$

$$P_1O_1 = |(\mathbf{r}_1 + \rho_1\mathbf{n}_1) - \mathbf{r}_1| = |\rho_1\mathbf{n}_1| = \rho_1,$$

$$\begin{aligned} OP_1 &= |\mathbf{r}_1 - (\mathbf{r} + \rho\mathbf{n})| \\ &= |(\mathbf{r} + a\mathbf{n}) - (\mathbf{r} + \rho\mathbf{n})| \qquad \text{[using (2)]} \\ &= |(a - \rho)\,\mathbf{n}| = |a - \rho| \end{aligned}$$

and

$$\begin{aligned} O_1P &= |\mathbf{r} - (\mathbf{r}_1 + \rho_1\mathbf{n}_1)| \\ &= |\mathbf{r} - [(\mathbf{r} + a\mathbf{n}) + \rho_1\mathbf{n}]| \qquad \text{[using (1) and (2)]} \\ &= |-(a + \rho_1)\,\mathbf{n}| = |a + \rho_1|. \end{aligned}$$

Substituting these values in (17), we get

$$\begin{aligned} (POP_1O_1) &= \frac{\rho\rho_1}{|a-\rho|\,|a+\rho_1|} \\ &= \frac{\dfrac{1}{\kappa\kappa_1}}{\left|a - \dfrac{1}{\kappa}\right| \left|a + \dfrac{1}{\kappa_1}\right|} \\ &= \frac{1}{|a\kappa - 1|\,|a\kappa_1 + 1|} \\ &= \frac{1}{|1 - a\kappa|\,|1 + a\kappa_1|} \end{aligned}$$

$$= \frac{1}{|(1-a\kappa)(1+a\kappa_1)|}$$

$$= \sec^2 \alpha. \qquad \text{[using (15)]}$$

Solved Examples

Example 1. Prove that the involutes of a circular helix are plane curves.

Solution. We know that the torsion τ_1 of an involute for the curve $\mathbf{r} = \mathbf{r}(s)$ is given by

$$\tau_1 = \frac{\kappa\tau' - \tau\kappa'}{\kappa(c-s)(\kappa^2 + \tau^2)}. \qquad (1)$$

For a circular helix,

$$\frac{\kappa}{\tau} = a \,(\text{constant}).$$

Therefore

$$\kappa = a\tau, \ \kappa' = a\tau'. \qquad (2)$$

Substituting the values of κ and κ_1 from (2) in (1), we get

$$\tau_1 = 0$$

$\Rightarrow$ The torsion of the involute is zero.

$\Rightarrow$ The involute is a plane curve.

Example 2. Show that the distance between corresponding points of two involutes is constant.

Solution. Let $\mathbf{r}_1$ and $\mathbf{r}_2$ denote the position vectors of the corresponding points on the two involutes determined by $c = c_1$ and $c = c_2$ respectively. Then, we know that

$$\mathbf{r}_1 = \mathbf{r} + (c_1 - s)\,\mathbf{t} \qquad (1)$$

and

$$\mathbf{r}_2 = \mathbf{r} + (c_2 - s)\,\mathbf{t}. \qquad (2)$$

From (1) and (2), we get

$$\mathbf{r}_1 - \mathbf{r}_2 = (c_1 - c_2)\,\mathbf{t}$$

$$\Rightarrow \quad |\mathbf{r}_1 - \mathbf{r}_2| = |c_1 - c_2| = \text{constant}.$$

Example 3. Show that the locus of the centre of curvature is an evolute only when the curve is plane.

Solution. Let the locus of the centre of curvature be an evolute.
We have to prove that the curve is plane.

We know that the locus of the centre of curvature is given by

$$\mathbf{r}_1 = \mathbf{r} + \rho \mathbf{n}. \tag{1}$$

Also, the equation of an evolute is given by

$$\mathbf{r}_1 = \mathbf{r} + \rho \mathbf{n} - \rho \tan(\psi + c)\,\mathbf{b}. \tag{2}$$

Comparing (1) and (2), we get

$$\rho \tan(\psi + c) = 0$$

$$\Rightarrow \quad \tan(\psi + c) = 0$$

$$\Rightarrow \quad \psi + c = n\pi$$

$$\Rightarrow \quad \int \tau\, ds = n\pi - c. \qquad \left[\because \psi = \int \tau\, ds\right]$$

Differentiating both sides with respect to s, we get

$$\tau = 0.$$

Hence the curve is a plane curve.

Example 4. Find the involutes and evolutes of the circular helix

$$x = a\cos\theta,\ y = a\sin\theta,\ z = a\,\theta\tan\alpha.$$

Solution. We have

$$\mathbf{r} = a(\cos\theta, \sin\theta,\ \theta\tan\alpha). \tag{1}$$

Differentiating both sides with respect to s, we get

$$\frac{d\mathbf{r}}{ds} = a(-\sin\theta, \cos\theta,\ \tan\alpha)\frac{d\theta}{ds}$$

or

$$\mathbf{t} = a(-\sin\theta, \cos\theta,\ \tan\alpha)\frac{d\theta}{ds}. \tag{2}$$

Squaring both sides of (2), we get

$$1 = a^2\left(\frac{d\theta}{ds}\right)^2(\sin^2\theta + \cos^2\theta + \tan^2\alpha)$$

$$= a^2\left(\frac{d\theta}{ds}\right)^2(1 + \tan^2\alpha)$$

$$\Rightarrow \quad 1 = a^2\left(\frac{d\theta}{ds}\right)^2\sec^2\alpha$$

$$\Rightarrow \quad \frac{ds}{d\theta} = a\sec\alpha. \tag{3}$$

From (2) and (3), we get

$$\mathbf{t} = \cos\alpha(-\sin\theta, \cos\theta,\ \tan\alpha). \tag{4}$$

Now, integrating (3), we get

$$s = \int_0^\theta a\sec\alpha \, d\theta = a\theta\sec\alpha. \tag{5}$$

We know that the equation of an involute of the space curve $\mathbf{r} = \mathbf{r}(s)$ is given by

$$\mathbf{r}_1 = \mathbf{r} + (c - s)\,\mathbf{t}. \tag{6}$$

Substituting the values of $\mathbf{r}$, $\mathbf{t}$ and s from (1), (4) and (5) respectively in (6), we get

$$\mathbf{r}_1 = a(\cos\theta, \sin\theta, \theta\tan\alpha) + (c - a\theta\sec\alpha)\cos\alpha\,(-\sin\theta, \cos\theta, \tan\alpha).$$

Therefore, if $\mathbf{r}_1 = (x, y, z)$, then the equations of the involutes are given by

$$\left.\begin{aligned} x &= a\cos\theta - \cos\alpha\sin\theta(c - a\theta\sec\alpha), \\ y &= a\sin\theta + \cos\alpha\cos\theta(c - a\theta\sec\alpha), \\ z &= a\theta\tan\alpha + \sin\alpha(c - a\theta\sec\alpha). \end{aligned}\right\} \tag{7}$$

Again, we know that the equation of an evolute of the space curve $\mathbf{r} = \mathbf{r}(s)$ is given by

$$\mathbf{r}_1 = \mathbf{r} + \rho\mathbf{n} - \rho\tan(\psi + c)\,\mathbf{b}. \tag{8}$$

Differentiating both sides of (4) with respect to s, we get

$$\frac{d\mathbf{t}}{ds} = \cos\alpha\,(-\cos\theta, -\sin\theta, 0)\frac{d\theta}{ds}$$

or $$\kappa\mathbf{n} = (-\cos\theta, -\sin\theta, 0)\frac{\cos^2\alpha}{a}. \tag{9}$$

[using (3)]

Squaring both sides of (9), we get

$$\kappa^2 = (\cos^2\theta + \sin^2\theta)\frac{\cos^4\alpha}{a^2}$$

$$\Rightarrow \quad \kappa = \frac{\cos^2\alpha}{a}$$

$$\Rightarrow \quad \rho = a\sec^2\alpha. \tag{10}$$

Substituting the value of κ in (9), we get

$$\mathbf{n} = (-\cos\theta, -\sin\theta, 0). \tag{11}$$

Taking the cross product of (4) and (11), we get

$$\mathbf{t}\times\mathbf{n} = \cos\alpha\,(-\sin\theta, \cos\theta, \tan\alpha)\times(-\cos\theta, -\sin\theta, 0)$$

or $$\mathbf{b} = \cos\alpha\,(\sin\theta\tan\alpha, -\cos\theta\tan\alpha, 1). \tag{12}$$

Differentiating both sides of (12) with respect to s, we get

$$\frac{d\mathbf{b}}{ds} = \cos\alpha\,(\cos\theta\tan\alpha, \sin\theta\tan\alpha,\ 0)\frac{d\theta}{ds}$$

or $$-\tau\mathbf{n} = \frac{\cos^2\alpha}{a}\tan\alpha\,(\cos\theta, \sin\theta,\ 0). \tag{13}$$

Squaring both sides of (13), we get

$$\tau^2 = \left(\frac{\cos^2\alpha}{a}\right)^2 \tan^2\alpha\,(\cos^2\theta + \sin^2\theta)$$

$$\Rightarrow \quad \tau = \frac{\cos^2\alpha}{a}\tan\alpha = \frac{\cos\alpha\sin\alpha}{a}$$

$$\Rightarrow \quad \psi = \int \tau\, ds = \int \frac{\cos\alpha\sin\alpha}{a} ds = \frac{s\cos\alpha\sin\alpha}{a}$$

$$\Rightarrow \quad \psi = \theta\sin\alpha. \tag{14}$$

[using (5)]

Substituting the values of $\mathbf{r}$, ρ, $\mathbf{n}$, $\mathbf{b}$ and ψ from (1), (10), (11), (12) and (14) respectively in (8), we get

$$\mathbf{r}_1 = a\,(\cos\theta, \sin\theta,\ \theta\tan\alpha) + a\sec^2\alpha\,(-\cos\theta, -\sin\theta,\ 0)$$
$$-\,a\sec\alpha\tan(c + \theta\sin\alpha)(\sin\theta\tan\alpha, -\cos\theta\tan\alpha, 1).$$

Therefore, if $\mathbf{r}_1 = (x, y, z)$, then the equations of the evolutes are given by

$$\left.\begin{aligned} x &= -a\cos\theta\tan^2\alpha - a\sec\alpha\tan\alpha\sin\theta\tan(c+\theta\sin\alpha), \\ y &= -a\sin\theta\tan^2\alpha + a\sec\alpha\tan\alpha\cos\theta\tan(c+\theta\sin\alpha), \\ z &= a\theta\tan\alpha - a\sec\alpha\tan(c+\theta\sin\alpha). \end{aligned}\right\} \tag{15}$$

Equations (7) and (15) give the required involutes and evolutes respectively.

2

Concept of a Surface and Envelopes and Developable

2.1 INTRODUCTION

The locus of a point whose position vector is a function of two independent variable parameters is known as a surface.

Let the two independent variable parameters be denoted by u and v. Then the parametric equation of the surface in vector form is given by

$$\mathbf{r} = \mathbf{r}\,(u, v). \tag{1}$$

If (x, y, z) are the coordinates of the point whose position vector is $\mathbf{r}$ and if $x\,(u, v), y\,(u, v), z\,(u, v)$ are the components of the vector function $\mathbf{r}\,(u, v)$, then equation (1) can be written as follows.

$$(x, y, z) = (x\,(u, v), y\,(u, v), z\,(u, v)).$$

Equating the corresponding components on both sides, we get

$$x = x\,(u, v), \;\; y = y\,(u, v), \;\; z = z\,(u, v). \tag{2}$$

Equations (2) are the parametric equations of the surface in Cartesian form.

If we eliminate u and v between the equations (2), then we get an equation of the form

$$F(x, y, z) = 0. \tag{3}$$

This form of equation of a surface is called the implicit or the constraint equation of the surface.

If the equation (3) of the surface is reduced to the form

$$z = f\,(x, y), \tag{4}$$

then it is called the equation of a surface in Monge's form.

Some Useful Notations. Let the parametric equation of a surface in vector form be given by

$$\mathbf{r} = \mathbf{r}\,(u, v).$$

The independent parameters u and v are known as the first and the second parameters. The partial differentiations with respect to u and v are denoted by using the suffixes 1 and 2 respectively. Thus

$$\mathbf{r}_1 = \frac{\partial \mathbf{r}}{\partial u}, \; \mathbf{r}_2 = \frac{\partial \mathbf{r}}{\partial v}, \; \mathbf{r}_{11} = \frac{\partial^2 \mathbf{r}}{\partial u^2},$$

$$\mathbf{r}_{22} = \frac{\partial^2 \mathbf{r}}{\partial v^2}, \; \mathbf{r}_{12} = \frac{\partial^2 \mathbf{r}}{\partial u\,\partial v}, \; \mathbf{r}_{21} = \frac{\partial^2 \mathbf{r}}{\partial v\,\partial u}.$$

Obviously,

$$\mathbf{r}_{12} = \mathbf{r}_{21}.$$

If $\mathbf{r} = (x(u, v), y(u, v), z(u, v))$ is the parametric equation of a surface, then

$$\mathbf{r}_1 = \left(\frac{\partial x}{\partial u}, \frac{\partial y}{\partial u}, \frac{\partial z}{\partial u}\right)$$

and $$\mathbf{r}_2 = \left(\frac{\partial x}{\partial v}, \frac{\partial y}{\partial v}, \frac{\partial z}{\partial v}\right).$$

For example, consider the parametric equations of a surface given by

$$x = u\cos v, \ y = u\sin v, \ z = av.$$

In vector form, these equations can be written as

$$\mathbf{r} = (u\cos v, \ u\sin v, \ av).$$

$$\therefore \quad \mathbf{r}_1 = (\cos v, \ \sin v, \ 0),$$

$$\mathbf{r}_2 = (-u\sin v, \ u\cos v, \ a),$$

$$\mathbf{r}_{11} = (0, \ 0, \ 0),$$

$$\mathbf{r}_{22} = (-u\cos v, -u\sin v, \ 0),$$

$$\mathbf{r}_{12} = (-\sin v, \ \cos v, \ 0),$$

$$\mathbf{r}_{21} = (-\sin v, \ \cos v, \ 0).$$

2.2 ORDINARY POINTS AND SINGULARITIES

Let the parametric equation of a surface in vector form be given by

$$\mathbf{r} = \mathbf{r}(u, v). \tag{1}$$

A point on the surface (1) is said to be an ordinary point or a regular point if $\mathbf{r}_1 \times \mathbf{r}_2 \neq \mathbf{0}$ at that point.

If, however, $\mathbf{r}_1 \times \mathbf{r}_2 = \mathbf{0}$ at a point, then we say that the point is a singularity of the surface.

The singularities are classified into the following two classes.

(i) Essential Singularities. These singularities are due to the nature of the surface and are independent of the choice of the parametric representation of the surface, e.g., the vertex of a cone is an essential singularity.

(ii) Artificial Singularities. These singularities are due to the choice of a particular parametric representation of the surface. For example, consider a plane surface. If we take a point O on this plane as origin and z-axis perpendicular to this plane, then the equation of the plane can be written as

$$\mathbf{r} = (r\cos\theta,\ r\sin\theta,\ 0),$$

where (r,θ) are the polar coordinates of any point on the plane.

$\therefore \qquad \mathbf{r}_1 = (\cos\theta,\ \sin\theta,\ 0)$,

$\qquad \mathbf{r}_2 = (-r\sin\theta,\ r\cos\theta,\ 0)$,

$\Rightarrow \qquad \mathbf{r}_1 \times \mathbf{r}_2 = (0, 0, r)$

$\Rightarrow \qquad \mathbf{r}_1 \times \mathbf{r}_2 = r\mathbf{k}$

$\Rightarrow \qquad [\mathbf{r}_1 \times \mathbf{r}_2]_{\text{at the pole}} = 0$

$\Rightarrow$ The pole is a singularity of the plane.

Since every point of the plane can be made a singularity by taking the point as the pole, this is an artificial singularity.

2.3 CURVILINEAR EQUATIONS OF THE CURVE ON THE SURFACE

Let the parametric equation of a surface in vector form be given by

$$\mathbf{r} = \mathbf{r}(u, v). \tag{1}$$

If the parameters u and v are functions of a single parameter t, then the position vector $\mathbf{r}$ becomes a function of only one parameter t and hence its locus is a curve lying on the surface (1).

Thus, if $u = u(t), v = v(t)$, then $\mathbf{r} = \mathbf{r}(u(t), v(t))$ is the equation of a curve lying on the surface (1).

The equations $u = u(t), v = v(t)$ are called the curvilinear equations of the curve.

2.4 PARAMETRIC CURVES

Definition. Let the parametric equation of a surface in vector form be given by

$$\mathbf{r} = \mathbf{r}(u, v). \tag{1}$$

If $u =$ constant, then the position vector $\mathbf{r}$ given by (1) is a function of a single variable parameter v, and therefore the locus of such points is a curve on the surface (1).

Similarly, if $v =$ constant, then the position vector $\mathbf{r}$ given by (1) is a function of a single variable parameter u, and therefore the locus of such points is a curve on the surface (1).

These curves are known as the parametric curves on the surface (1).

Vectors Tangential to Parametric Curves. Let the parametric equation of a surface in vector form be given by

$$\mathbf{r} = \mathbf{r}(u, v).$$

Then the equation of the parametric curve $u = a$ (constant) is

$$\mathbf{r} = \mathbf{r}(a, v)$$

$$\Rightarrow \quad \frac{\partial \mathbf{r}}{\partial v} = \mathbf{r}_2$$

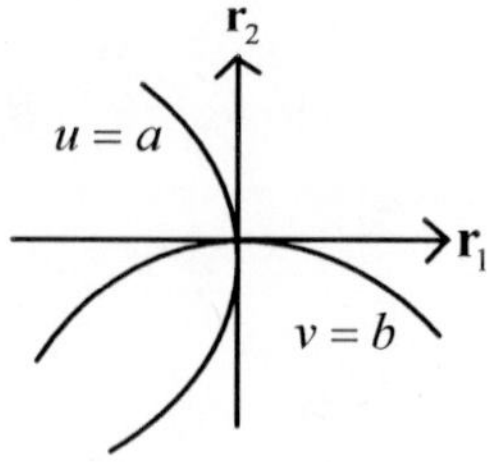

$\Rightarrow$ $\mathbf{r}_2$ = a vector along the tangent to the curve $u = a$.

Similarly, the equation of the parametric curve $v = b$ (const.) is

$$\mathbf{r} = \mathbf{r}(u, b)$$

$$\Rightarrow \quad \frac{\partial \mathbf{r}}{\partial u} = \mathbf{r}_1$$

$\Rightarrow$ $\mathbf{r}_1$ = a vector along the tangent to the curve $v = b$.

Thus $\mathbf{r}_1$ and $\mathbf{r}_2$ are vectors along the tangent to the parametric curves $v = b$ and $u = a$ respectively.

2.5 TANGENT PLANE AND NORMAL

Tangent Plane. Let the parametric equation of a surface in vector form be given by

$$\mathbf{r} = \mathbf{r}(u, v). \tag{1}$$

Let $u = u(t), v = v(t)$ be the equations of any curve drawn on the surface (1) through the point P. Then a vector along the tangent to this curve is

$$\frac{d\mathbf{r}}{dt} = \frac{\partial \mathbf{r}}{\partial u}\frac{du}{dt} + \frac{\partial \mathbf{r}}{\partial v}\frac{dv}{dt} = \mathbf{r}_1\frac{du}{dt} + \mathbf{r}_2\frac{dv}{dt}. \tag{2}$$

Since the vectors $\mathbf{r}_1$ and $\mathbf{r}_2$ are non-zero and independent, the plane passing through P and containing the vectors $\mathbf{r}_1$ and $\mathbf{r}_2$ is a fixed plane at P and it contains all the tangent lines to the surface at P. This plane is called the tangent plane to the surface at the point P.

Normal. The normal to a surface at a point P is a line passing through P and perpendicular to the tangent plane to the surface at P.

Since the tangent plane to the surface (1) at P contains the vectors $\mathbf{r}_1$ and $\mathbf{r}_2$, the normal to the surface (1) at P is a line perpendicular to both $\mathbf{r}_1$ and $\mathbf{r}_2$. But we know that the vector $\mathbf{r}_1 \times \mathbf{r}_2$ is perpendicular to both $\mathbf{r}_1$ and $\mathbf{r}_2$. Hence it follows that the normal to the surface (1) at P is parallel to the vector $\mathbf{r}_1 \times \mathbf{r}_2$.

Let $|\mathbf{r}_1 \times \mathbf{r}_2| = H$. Then

$$\frac{|\mathbf{r}_1 \times \mathbf{r}_2|}{H} = \text{a unit vector along the normal to the surface (1) at } P.$$

We write

$$\frac{\mathbf{r}_1 \times \mathbf{r}_2}{H} = \mathbf{N}.$$

$$\therefore \qquad \mathbf{r}_1 \times \mathbf{r}_2 = H\mathbf{N}.$$

Equation of the Tangent Plane in Vector Form.

Let the parametric equation of a surface in vector form be given by

$$\mathbf{r} = \mathbf{r}\,(u, v). \tag{1}$$

Let P be any point on this surface corresponding to the parametric coordinates (u, v) and let the position vector of P be $\mathbf{r}$.

The tangent plane to this surface at P passes through the point $\mathbf{r}$ and contains the vectors $\mathbf{r}_1$ and $\mathbf{r}_2$. Therefore, if $\mathbf{R}$ is the position vector of any current point on the tangent plane at P, then the vectors $\mathbf{R} - \mathbf{r}$, $\mathbf{r}_1$ and $\mathbf{r}_2$ are coplanar. Hence, we have

$$\mathbf{R} - \mathbf{r} = a\mathbf{r}_1 + b\mathbf{r}_2, \text{ where } a \text{ and } b \text{ are parameters.}$$

$$\therefore \qquad \mathbf{R} = \mathbf{r} + a\mathbf{r}_1 + b\mathbf{r}_2. \tag{2}$$

This is the equation of the tangent plane to the surface (1) at the point P.

Also, the tangent plane at P is perpendicular to the vector $\mathbf{r}_1 \times \mathbf{r}_2$. Therefore the vector $\mathbf{r}_1 \times \mathbf{r}_2$ is perpendicular to the vector $\mathbf{R} - \mathbf{r}$ which lies in the tangent plane. Hence, we have

$$(\mathbf{R} - \mathbf{r}) \cdot (\mathbf{r}_1 \times \mathbf{r}_2) = 0. \tag{3}$$

This is another form of the equation of the tangent plane to the surface (1) at P.

Equation of the Normal in Vector Form.

The normal to the surface (1) at P is parallel to the vector $\mathbf{r}_1 \times \mathbf{r}_2$. Therefore, if $\mathbf{R}$ is the position vector of any current point on the normal at P, then the vectors $\mathbf{R} - \mathbf{r}$ and $\mathbf{r}_1 \times \mathbf{r}_2$ are parallel. Hence, we have

$$\mathbf{R} - \mathbf{r} = a\,(\mathbf{r}_1 \times \mathbf{r}_2), \text{ where } a \text{ is a parameter.}$$

$$\therefore \qquad \mathbf{R} = \mathbf{r} + a\,(\mathbf{r}_1 \times \mathbf{r}_2). \tag{4}$$

This is the equation of normal to the surface (1) at the point P.

Equation of the Tangent Plane in Cartesian Form.

Let the equation of a surface be given by

$$F(x, y, z) = 0. \tag{1}$$

Consider any curve drawn on this surface. Let s be the arc length of the curve measured from a fixed point to the current point $P(x, y, z)$. Since the function F has the same value at all points of the surface, it remains constant along the curve as s varies. Hence

$$\frac{dF}{ds} = 0$$

$\Rightarrow$ $$\frac{\partial F}{\partial x}\frac{dx}{ds} + \frac{\partial F}{\partial y}\frac{dy}{ds} + \frac{\partial F}{\partial z}\frac{dz}{ds} = 0$$

$\Rightarrow$ The vector $\left(\frac{dx}{ds}, \frac{dy}{ds}, \frac{dz}{ds}\right)$ is perpendicular to the vector $\left(\frac{\partial F}{\partial x}, \frac{\partial F}{\partial y}, \frac{\partial F}{\partial z}\right)$.

$\Rightarrow$ All the tangent lines to the surface (1) are perpendicular to the vector $\left(\frac{\partial F}{\partial x}, \frac{\partial F}{\partial y}, \frac{\partial F}{\partial z}\right)$.

$$\left[\because \left(\frac{dx}{ds}, \frac{dy}{ds}, \frac{dz}{ds}\right) \text{ is the unit tangent to the curve at } P(x, y, z)\right]$$

Therefore, if (X, Y, Z) are the coordinates of any current point Q on the tangent plane to the surface (1) at $P(x, y, z)$, then the line PQ is perpendicular to the vector $\left(\frac{\partial F}{\partial x}, \frac{\partial F}{\partial y}, \frac{\partial F}{\partial z}\right)$. Hence

$$(X - x)\frac{\partial F}{\partial x} + (Y - y)\frac{\partial F}{\partial y} + (Z - z)\frac{\partial F}{\partial z} = 0. \tag{2}$$

This is the equation of the tangent plane to the surface (1) at the point P.

Equation of the Normal in Cartesian Form.

If (X, Y, Z) are the coordinates of any current point Q on the normal to the surface (1) at $P(x, y, z)$, then the line PQ is parallel to the vector $\left(\frac{\partial F}{\partial x}, \frac{\partial F}{\partial y}, \frac{\partial F}{\partial z}\right)$. Hence

$$\frac{X-x}{\frac{\partial F}{\partial x}}=\frac{Y-y}{\frac{\partial F}{\partial y}}=\frac{Z-z}{\frac{\partial F}{\partial z}}. \quad (3)$$

This is the equation of normal to the surface (1) at the point P.

Solved Examples

Example 1. Find the equations of the tangent plane and the normal on the surface $xyz=4$ at the point (1, 2, 2).

Solution. We know that the equations of the tangent plane and the normal to the surface

$$F(x, y, z)=0 \quad (1)$$

at the point $P(x, y, z)$ are

$$(X-x)\frac{\partial F}{\partial x}+(Y-y)\frac{\partial F}{\partial y}+(Z-z)\frac{\partial F}{\partial z}=0 \quad (2)$$

and
$$\frac{X-x}{\frac{\partial F}{\partial x}}=\frac{Y-y}{\frac{\partial F}{\partial y}}=\frac{Z-z}{\frac{\partial F}{\partial z}} \quad (3)$$

respectively.

The equation of the given surface is

$$F(x, y, z) \equiv xyz-4=0.$$

$$\therefore \quad \frac{\partial F}{\partial x}=yz=4 \text{ at } (1, 2, 2),$$

$$\frac{\partial F}{\partial y}=xz=2 \text{ at } (1, 2, 2),$$

$$\frac{\partial F}{\partial z}=xy=2 \text{ at } (1, 2, 2).$$

Hence, from (2), the equation of the tangent plane at (1, 2, 2) is

$$(X-1)(4)+(Y-2)(2)+(Z-2)(2)=0$$

or $\quad 4X+2Y+2Z=12$

or $\quad 2X+Y+Z=6.$

Again, from (3), the equation of the normal at (1, 2, 2) is

$$\frac{X-1}{4}=\frac{Y-2}{2}=\frac{Z-2}{2}$$

or
$$\frac{X-1}{2}=\frac{Y-2}{1}=\frac{Z-2}{1}.$$

Example 2. Show that the sum of the squares of the intercepts on the coordinate axes by the tangent plane to the surface

$$x^{2/3} + y^{2/3} + z^{2/3} = a^{2/3}$$

is constant.

Solution. We know that the equation of the tangent plane to the surface

$$F(x, y, z) = 0 \tag{1}$$

at the point $P(x, y, z)$ is

$$(X - x)\frac{\partial F}{\partial x} + (Y - y)\frac{\partial F}{\partial y} + (Z - z)\frac{\partial F}{\partial z} = 0. \tag{2}$$

The equation of the given surface is

$$F(x, y, z) \equiv x^{2/3} + y^{2/3} + z^{2/3} - a^{2/3} = 0. \tag{3}$$

$$\therefore \quad \frac{\partial F}{\partial x} = \frac{2}{3}x^{-1/3}, \quad \frac{\partial F}{\partial y} = \frac{2}{3}y^{-1/3}, \quad \frac{\partial F}{\partial z} = \frac{2}{3}z^{-1/3}.$$

Hence, from (2), the equation of the tangent plane at (x, y, z) is

$$(X - x)\left(\frac{2}{3}x^{-1/3}\right) + (Y - y)\left(\frac{2}{3}y^{-1/3}\right) + (Z - z)\left(\frac{2}{3}z^{-1/3}\right) = 0$$

or

$$Xx^{-1/3} + Yx^{-1/3} + Xz^{-1/3} = x^{2/3} + y^{2/3} + z^{2/3}$$

$$= a^{2/3} \qquad \text{[using (3)]}$$

or

$$\frac{X}{a^{2/3}x^{1/3}} + \frac{Y}{a^{2/3}y^{1/3}} + \frac{Z}{a^{2/3}z^{1/3}} = 1.$$

Therefore, if x_1, y_1, z_1 are the intercepts on the coordinate axes by the tangent plane to the surface (3), then

$$x_1 = a^{2/3}x^{1/3}, y_1 = a^{2/3}y^{1/3}, z_1 = a^{2/3}z^{1/3}$$

$\Rightarrow$ Sum of the squares of the intercepts on the coordinate axes

$$= x_1^{\ 2} + y_1^{\ 2} + z_1^{\ 2}$$

$$= (a^{2/3}x^{1/3})^2 + (a^{2/3}y^{1/3})^2 + (a^{2/3}z^{1/3})^2$$

$$= a^{4/3}(x^{2/3} + y^{2/3} + z^{2/3})$$

$$= a^{4/3}(a^{2/3}) \qquad \text{[using (3)]}$$

$$= a^2$$

$$= \text{constant.}$$

Example 3. Prove that at points common to the surface

$$a(yz + zx + xy) = xyz$$

and a sphere whose centre is the origin, the tangent plane to the surface makes intercepts on the axes whose sum is constant.

Solution. We know that the equation of the tangent plane to the surface

$$F(x, y, z) = 0 \tag{1}$$

at the point $P(x, y, z)$ is

$$(X - x)\frac{\partial F}{\partial x} + (Y - y)\frac{\partial F}{\partial y} + (Z - z)\frac{\partial F}{\partial z} = 0. \tag{2}$$

The equation of the given surface is

$$F(x, y, z) \equiv a(yz + zx + xy) - xyz = 0. \tag{3}$$

$$\therefore \quad \frac{\partial F}{\partial x} = a(z + y) - yz,$$

$$\frac{\partial F}{\partial y} = a(z + x) - xz,$$

$$\frac{\partial F}{\partial z} = a(y + x) - xy.$$

Equation of the sphere with centre as origin and radius r is

$$x^2 + y^2 + z^2 = r^2. \tag{4}$$

Hence, from (2), the equation of the tangent plane at a point $P(x, y, z)$ common to surface (3) and sphere (4) is

$$(X - x)[a(z + y) - yz] + (Y - y)[a(z + x) - xz] + (Z - z)[a(y + x) - xy] = 0$$

or
$$X[a(z + y) - yz] + Y[a(z + x) - xz] + Z[a(y + x) - xy]$$
$$= x[a(z + y) - yz] + y[a(z + x) - xz] + z[a(y + x) - xy]$$
$$= 2a(yz + zx + xy) - 3xyz$$
$$= 2xyz - 3xyz \quad \text{[using (3)]}$$
$$= -xyz$$

or
$$\frac{X}{x}[a(zx + xy) - xyz] + \frac{Y}{y}[a(yz + xy) - xyz] + \frac{Z}{z}[a(yz + zx) - xyz] = -xyz$$

or
$$\frac{X}{x}(-ayz) + \frac{Y}{y}(-azx) + \frac{Z}{z}(-axy) = -xyz \quad \text{[using (3)]}$$

or
$$\frac{X}{x^2/a} + \frac{Y}{y^2/a} + \frac{Z}{z^2/a} = 1.$$

Therefore, if x_1, y_1, z_1 are the intercepts on the coordinate axes, then

$$x_1 = x^2/a, \; y_1 = y^2/a, \; z_1 = z^2/a$$

$\Rightarrow$ Sum of the intercepts on the coordinate axes

$$= x_1 + y_1 + z_1$$

$$= \frac{x^2}{a} + \frac{y^2}{a} + \frac{z^2}{a}$$

$$= \frac{r^2}{a}, \text{ which is constant.} \qquad \text{[using (4)]}$$

Example 4. Prove that any tangent plane to the surface

$$a(x^2 + y^2) + xyz = 0$$

meets it again in a conic whose projection on xy-plane is a rectangular hyperbola.

Solution. We know that the equation of the tangent plane to the surface

$$F(x, y, z) = 0 \tag{1}$$

at the point (x, y, z) is

$$(X - x)\frac{\partial F}{\partial x} + (Y - y)\frac{\partial F}{\partial y} + (Z - z)\frac{\partial F}{\partial z} = 0. \tag{2}$$

The equation of the given surface is

$$F(x, y, z) \equiv a(x^2 + y^2) + xyz = 0. \tag{3}$$

$$\therefore \quad \frac{\partial F}{\partial x} = 2ax + yz = 2ax_1 + y_1 z_1 \text{ at } P(x_1, y_1, z_1),$$

$$\frac{\partial F}{\partial y} = 2ay + xz = 2ay_1 + x_1 z_1 \text{ at } P(x_1, y_1, z_1),$$

$$\frac{\partial F}{\partial z} = xy = x_1 y_1 \text{ at } P(x_1, y_1, z_1).$$

Hence, from (2), the equation of the tangent plane at $P(x_1, y_1, z_1)$ is

$$(x - x_1)(2ax_1 + y_1 z_1) + (y - y_1)(2ay_1 + x_1 z_1) + (z - z_1)x_1 y_1 = 0. \tag{4}$$

Since the point $P(x_1, y_1, z_1)$ lies on the surface (3), we have

$$a(x_1^2 + y_1^2) + x_1 y_1 z_1 = 0. \tag{5}$$

To find the projection of the curve of intersection of (3) and (4) on xy-plane, we eliminate z between (3) and (4). Then, we get

$$(x - x_1)(2ax_1 + y_1 z_1) + (y - y_1)(2ay_1 + x_1 z_1) - \left[\frac{a(x^2 + y^2)}{xy} + z_1\right]x_1 y_1 = 0$$

or $$x(2ax_1 + y_1z_1) + y(2ay_1 + x_1z_1) - x_1(2ax_1 + y_1z_1) - y_1(2ay_1 + x_1z_1) = \frac{a(x^2+y^2)}{xy}x_1y_1 + x_1y_1z_1$$

or $$x(2ax_1 + y_1z_1) + y(2ay_1 + x_1z_1) - [2a(x_1^2 + y_1^2) + 3x_1y_1z_1] = \frac{a(x^2+y^2)}{xy}x_1y_1$$

or $$x(2ax_1 + y_1z_1) + y(2ay_1 + x_1z_1) - x_1y_1z_1 = \frac{a(x^2+y^2)}{xy}x_1y_1$$

[using (5)]

or $$xy[x(2ax_1 + y_1z_1) + y(2ay_1 + x_1z_1) - x_1y_1z_1] = a(x^2+y^2)x_1y_1$$

or $$xy\left[\frac{x}{x_1}(2ax_1^2 + x_1y_1z_1) + \frac{y}{y_1}(2ay_1^2 + x_1y_1z_1) - x_1y_1z_1\right] = a(x^2+y^2)x_1y_1$$

or $$xy\left[\frac{x}{x_1}(ax_1^2 - ay_1^2) + \frac{y}{y_1}(ay_1^2 - ax_1^2) - x_1y_1z_1\right] = a(x^2+y^2)x_1y_1$$

[using (5)]

or $$xy\left[\frac{x}{x_1}(ax_1^2 - ay_1^2) + \frac{y}{y_1}(ay_1^2 - ax_1^2)\right] = a(x^2+y^2)x_1y_1 + x_1y_1z_1xy$$

or $$xy\left[\frac{x}{x_1}(ax_1^2 - ay_1^2) + \frac{y}{y_1}(ay_1^2 - ax_1^2)\right] = a(x^2+y^2)x_1y_1 - a(x_1^2+y_1^2)xy$$

[using (5)]

or $$xy(x_1^2 - y_1^2)\left(\frac{x}{x_1} - \frac{y}{y_1}\right) = xx_1^2y_1\left(\frac{x}{x_1} - \frac{y}{y_1}\right) - yx_1y_1^2\left(\frac{x}{x_1} - \frac{y}{y_1}\right)$$

or $$xy(x_1^2 - y_1^2) = xx_1^2y_1 - yx_1y_1^2. \tag{6}$$

Equation (6) is the equation of a cylinder passing through the curve of intersection of (3) and (4). The projection of this curve on the plane $z = 0$ is given by

$$z = 0,\ xy(x_1^2 - y_1^2) = x(x_1^2y_1) - y(x_1y_1^2), \tag{7}$$

which is a rectangular hyperbola.

Exercises

1. Find the equation of the tangent plane and the normal to the surface $z = x^2 + y^2$ at the point (1, – 1, 2).

$$\left[\text{Ans. } 2x - 2y - z = 2; \frac{x-1}{-2} = \frac{y+1}{2} = \frac{z-2}{1}\right]$$

2. Find a unit normal to the surface $x^2 y + 2xz = 4$ at the point $(2, -2, 3)$.

$$\left[\text{Ans.} \left(-\frac{1}{3}, \frac{2}{3}, \frac{2}{3}\right)\right]$$

3. Prove that the tangent plane to the surface $xyz = a^2$ and the coordinate planes bound a tetrahedron of constant volume.

4. The normal at a point of the ellipsoid $\frac{x^2}{a^2} + \frac{y^2}{b^2} + \frac{z^2}{c^2} = 1$ meets the coordinate planes in G_1, G_2, G_3. Prove that the ratios $PG_1 : PG_2 : PG_3$ are constant.

2.6 CHARACTERISTICS AND ENVELOPE OF ONE PARAMETER FAMILY OF SURFACES

Characteristic. The equation

$$F(a) \equiv F(x, y, z, a) = 0 \tag{1}$$

represents a surface if a is a constant. However, if a be a variable parameter, then (1) represents a family of surfaces.

The curve of intersection of two members of the family (1) for the parametric values a and $a + \delta a$ is given by

$$F(a) = 0, \quad F(a + \delta a) = 0. \tag{2}$$

The curve of intersection given by (2) is also given by

$$F(a) = 0, \quad \frac{F(a + \delta a) - F(a)}{\delta a} = 0. \tag{3}$$

Now, make $\delta a \to 0$. Then the limiting case of the curve of intersection given by (3) is given by

$$F(a) = 0, \quad \frac{\partial}{\partial a} F(a) = 0. \tag{4}$$

The curve given by (4) is known as the characteristic of the member of the family (1) for the parametric value a.

Envelope. Equation (4) gives a number of characteristics for different members of the family (1) for different values of the parameter a.

Thus varying a, we get a number of characteristics by (4). The locus of all these characteristics is known as the envelope of the family of surfaces (1). The equation of the envelope of the family (1) is obtained by eliminating the parameter a from equations (4).

Theorem. The envelope of a family of surfaces touches each member of the family at all points of its characteristics.

Proof. Let a family of surfaces be given by

$$F(a) \equiv F(x, y, z, a) = 0, \tag{1}$$

where a is a variable parameter which varies from member to member of the family (1).

The characteristic of a member of family (1) is given by

$$F(a) = 0, \ \frac{\partial}{\partial a} F(a) = 0. \tag{2}$$

The locus of the characteristics given by (2) is known as the envelope of family (1).

Thus the equation of the envelope of the family of surfaces (1) is obtained by eliminating the parameter a from equations (2).

Take any member

$$F(a) = 0 \tag{3}$$

of the family (1).

We have to prove that the envelope of family (1) touches member (3) of (1) along the characteristic of (3) given by (2).

The normal to (3) at $P(x, y, z)$ is parallel to the vector

$$\left(\frac{\partial F}{\partial x}, \frac{\partial F}{\partial y}, \frac{\partial F}{\partial z}\right). \tag{4}$$

The envelope of (1) is given by $F(a) = 0$, provided a is given by

$$\frac{\partial}{\partial a} F(a) = 0. \tag{5}$$

Therefore, the normal to the envelope at $P(x, y, z)$ is parallel to the vector

$$\left(\frac{\partial F}{\partial x} + \frac{\partial F}{\partial a}\frac{\partial a}{\partial x}, \frac{\partial F}{\partial y} + \frac{\partial F}{\partial a}\frac{\partial a}{\partial y}, \frac{\partial F}{\partial z} + \frac{\partial F}{\partial a}\frac{\partial a}{\partial z}\right)$$

i.e.,
$$\left(\frac{\partial F}{\partial x}, \frac{\partial F}{\partial y}, \frac{\partial F}{\partial z}\right). \tag{6}$$

[using (5)]

Clearly, (4) and (6) represent the same vector. Hence it follows that the surface (3) and the envelope have the same normal, and therefore the same tangent plane at $P\ (x, y, z)$. This implies that the envelope of the family (1) touches the member (3) at $P(x, y, z)$.

But P is an arbitrary point on the characteristic (2) of (3). Therefore the envelope of the family (1) touches the member (3) along its characteristic given by (2).

2.7 EDGE OF REGRESSION

Definition. The edge of regression of a one parameter family of surfaces is the locus of the ultimate intersection of two consecutive characteristics.

Since each characteristic lies on the envelope, it follows that the edge of regression is a curve lying on the envelope.

Equation of the Edge of Regression.

Let a family of surfaces be given by

$$F(a) \equiv F\ (x, y, z, a) = 0, \tag{1}$$

where a is a variable parameter which varies from member to member of the family (1).

The characteristic of a member of the family (1) for the parametric value a is given by

$$F(a) = 0, \quad \frac{\partial}{\partial a} F(a) = 0. \tag{2}$$

The characteristic of a member of the family (1) for the parametric value $a + da$ is given by

$$F(a + da) = 0, \quad \frac{\partial}{\partial a} F(a + da) = 0$$

i.e., $$F(a) + \frac{\partial F(a)}{\partial a} da = 0, \quad \frac{\partial F(a)}{\partial a} + \frac{\partial^2 F(a)}{\partial a^2} da = 0. \tag{3}$$

[by Taylor's expansion, neglecting higher powers of da]

The edge of regression of the family (1) is obtained by eliminating a from (2) and (3), i.e., the edge of regression of the family (1) is a-eliminant of the equations

$$F(a) = 0, \quad \frac{\partial}{\partial a} F(a) = 0, \quad \frac{\partial^2}{\partial a^2} F(a) = 0. \tag{4}$$

Thus (4) are the equations of the edge of regression if a is regarded as the parameter.

Theorem. Each characteristic of a one parameter family of surfaces touches the edge of regression.

Proof. Let a one parameter family of surfaces be given by

$$F(a) \equiv F(x, y, z, a) = 0. \tag{1}$$

The characteristic of a member of the family (1) for the parametric value a is given by

$$F(a) = 0, \quad \frac{\partial}{\partial a} F(a) = 0. \tag{2}$$

The edge of regression of the family (1) is obtained by eliminating a from the equations

$$F(a) = 0, \quad \frac{\partial}{\partial a} F(a) = 0, \quad \frac{\partial^2}{\partial a^2} F(a) = 0. \tag{3}$$

We have to show that (2) touches (3).

Since the edge of regression is the locus of the ultimate intersection of two consecutive characteristics, (3) intersects (2) at some point, say $P(x, y, z)$.

Now, the characteristic given by (2) lies on each of the two surfaces

$$F(a) = 0, \quad \frac{\partial}{\partial a} F(a) = 0.$$

$\Rightarrow$ The tangent to the characteristic given by (2) is also tangential to each of the two surfaces

$$F(a) = 0, \quad \frac{\partial}{\partial a} F(a) = 0.$$

$\Rightarrow$ The tangent to the characteristic given by (2) is perpendicular to the normals to each of the two surfaces

$$F(a) = 0, \quad \frac{\partial}{\partial a} F(a) = 0.$$

$\Rightarrow$ The tangent to the characteristic given by (2) is perpendicular to each of the two vectors

$$\left(\frac{\partial F}{\partial x}, \frac{\partial F}{\partial y}, \frac{\partial F}{\partial z}\right) \text{ and } \left(\frac{\partial^2 F}{\partial x\, \partial a}, \frac{\partial^2 F}{\partial y\, \partial a}, \frac{\partial^2 F}{\partial z\, \partial a}\right).$$

$\Rightarrow$ The tangent to the characteristic given by (2) is parallel to

$$\left(\frac{\partial F}{\partial x}, \frac{\partial F}{\partial y}, \frac{\partial F}{\partial z}\right) \times \left(\frac{\partial^2 F}{\partial x\, \partial a}, \frac{\partial^2 F}{\partial y\, \partial a}, \frac{\partial^2 F}{\partial z\, \partial a}\right). \tag{4}$$

Again, the edge of regression given by (3) may be regarded as the curve of intersection of the surfaces (2), provided a is a function of x, y, z given by

$$\frac{\partial^2}{\partial a^2} F(a) = 0. \tag{5}$$

$\Rightarrow$ The tangent to the edge of regression given by (3) is perpendicular to the normals to each of the two surfaces given by (2), provided (5) is satisfied.

$\Rightarrow$ The tangent to the edge of regression given by (3) is perpendicular to each of the two vectors

$$\left(\frac{\partial F}{\partial x} + \frac{\partial F}{\partial a}\frac{\partial a}{\partial x}, \frac{\partial F}{\partial y} + \frac{\partial F}{\partial a}\frac{\partial a}{\partial y}, \frac{\partial F}{\partial z} + \frac{\partial F}{\partial a}\frac{\partial a}{\partial z}\right)$$

and

$$\left(\frac{\partial^2 F}{\partial x\,\partial a} + \frac{\partial^2 F}{\partial a^2}\frac{\partial a}{\partial x}, \frac{\partial^2 F}{\partial y\,\partial a} + \frac{\partial^2 F}{\partial a^2}\frac{\partial a}{\partial y}, \frac{\partial^2 F}{\partial z\,\partial a} + \frac{\partial^2 F}{\partial a^2}\frac{\partial a}{\partial z}\right).$$

$\Rightarrow$ The tangent to the edge of regression given by (3) is parallel to

$$\left(\frac{\partial F}{\partial x} + \frac{\partial F}{\partial a}\frac{\partial a}{\partial x}, \frac{\partial F}{\partial y} + \frac{\partial F}{\partial a}\frac{\partial a}{\partial y}, \frac{\partial F}{\partial z} + \frac{\partial F}{\partial a}\frac{\partial a}{\partial z}\right)$$
$$\times\left(\frac{\partial^2 F}{\partial x\,\partial a} + \frac{\partial^2 F}{\partial a^2}\frac{\partial a}{\partial x}, \frac{\partial^2 F}{\partial y\,\partial a} + \frac{\partial^2 F}{\partial a^2}\frac{\partial a}{\partial y}, \frac{\partial^2 F}{\partial z\,\partial a} + \frac{\partial^2 F}{\partial a^2}\frac{\partial a}{\partial z}\right).$$

$\Rightarrow$ The tangent to the edge of regression given by (3) is parallel to

$$\left(\frac{\partial F}{\partial x}, \frac{\partial F}{\partial y}, \frac{\partial F}{\partial z}\right) \times \left(\frac{\partial^2 F}{\partial x\,\partial a}, \frac{\partial^2 F}{\partial y\,\partial a}, \frac{\partial^2 F}{\partial z\,\partial a}\right). \tag{6}$$

[using (2) and (5)]

It follows from (4) and (6) that the tangent to the characteristic given by (2) is parallel to the tangent to the edge of regression given by (3). Therefore the two curves touch each other at their common point.

Solved Examples

Example 1. Prove that the envelope of the family of paraboloids $x^2 + y^2 = 4a(z - a)$ is the circular cone $x^2 + y^2 = z^2$.

Solution. Equation of the given family of paraboloids is

$$F(a) \equiv x^2 + y^2 - 4a(z - a) = 0. \tag{1}$$

Differentiating (1) partially with respect to a, we get

$$\frac{\partial}{\partial a} F(a) \equiv -4z + 8a = 0. \tag{2}$$

The required envelope is obtained by eliminating a from (1) and (2) as under.

From (2), we get

$$2a = z. \tag{3}$$

Substituting the value of a from (3) in (1), we get

$$x^2 + y^2 - 2z\left(z - \frac{z}{2}\right) = 0$$

or $\quad x^2 + y^2 = z^2.$

This is the required envelope of (1).

Example 2. Spheres of constant radius b have their centres on the fixed circle $x^2 + y^2 = a^2, z = 0$. Prove that their envelope is the surface

$$(x^2 + y^2 + z^2 + a^2 - b^2)^2 = 4a^2(x^2 + y^2).$$

Solution. The fixed circle is

$$x^2 + y^2 = a^2, z = 0. \tag{1}$$

Any point on (1) is given by $P(a\cos\theta, a\sin\theta, 0)$.

The equation of the sphere whose radius is b and whose centre is $P(a\cos\theta, a\sin\theta, 0)$ is given by

$$(x - a\cos\theta)^2 + (y - a\sin\theta)^2 + (z - 0)^2 = b^2$$

i.e., $\quad F(\theta) \equiv x^2 + y^2 + z^2 - 2ax\cos\theta - 2ay\sin\theta + (a^2 - b^2) = 0. \qquad (2)$

Clearly, (2) represents a family of spheres with θ as the parameter of the family.

We have to find the envelope of the family (2).

Differentiating (2) partially with respect to θ, we get

$$\frac{\partial}{\partial\theta}F(\theta) \equiv 2ax\sin\theta - 2ay\cos\theta = 0. \tag{3}$$

The required envelope is obtained by eliminating a from (2) and (3) as under.

From (2), we get

$$2ax\cos\theta + 2ay\sin\theta = x^2 + y^2 + z^2 + a^2 - b^2. \tag{4}$$

Squaring and adding (3) and (4), we get

$$4a^2(x^2 + y^2) = (x^2 + y^2 + z^2 + a^2 - b^2)^2.$$

This is the required envelope of (2).

Example 3. Show that the envelope of the family of surfaces $F(x, y, z, a, b) = 0$, in which the parameters a, b are connected by the equation $f(a, b) = 0$, is found by eliminating a and b from the equations

$$F = 0, \quad f = 0, \quad \frac{F_a}{f_a} = \frac{F_b}{f_b}.$$

Solution. The equation of the given family of surfaces is

$$F(x, y, z, a, b) = 0, \tag{1}$$

where a and b are the parameters of the family which are connected by the relation

$$f(a, b) = 0. \tag{2}$$

It follows from (2) that either a can be expressed in terms of b or vice-versa, and consequently, (1) represents a one parameter family of surfaces. For definiteness, let b be expressed in terms of a. Then (1) represents a one parameter family of surfaces with a as the parameter of the family. Therefore, from (1), we have

$$F(a) \equiv F(x, y, z, a, b) = 0. \tag{3}$$

Differentiating (3) with respect to a, we get

$$\frac{dF}{da} \equiv \frac{\partial F}{\partial a} + \frac{\partial F}{\partial b}\frac{db}{da} = 0. \tag{4}$$

Now, from (2), we have

$$df = 0$$

$$\Rightarrow \quad \frac{\partial f}{\partial a}da + \frac{\partial f}{\partial b}db = 0$$

$$\Rightarrow \quad f_a da + f_b db = 0$$

$$\Rightarrow \quad \frac{db}{da} = -\frac{f_a}{f_b}. \tag{5}$$

Substituting the value of $\frac{db}{da}$ from (5) in (4), we get

$$\frac{dF}{da} \equiv F_a + F_b\left(-\frac{f_a}{f_b}\right) = 0$$

$$\Rightarrow \quad \frac{F_a}{f_a} = \frac{F_b}{f_b}. \tag{6}$$

The required envelope is obtained by eliminating a and b from (2), (3) and (6).

Example 4. Find the envelope of the family of planes $3a^2x - 3ay + z = a^3$ and show that the edge of regression is the curve of intersection of the surfaces $xz = y^2, xy = z$.

Solution. The equation of the given family of planes is

$$F(a) \equiv 3a^2x - 3ay + z - a^3 = 0, \tag{1}$$

where a is the parameter of the family.

Differentiating (1) partially with respect to a, we get

$$\frac{\partial}{\partial a} F(a) \equiv 6ax - 3y - 3a^2 = 0. \tag{2}$$

Again, differentiating (2) partially with respect to a, we get

$$\frac{\partial^2}{\partial a^2} F(a) \equiv 6x - 6a = 0. \tag{3}$$

The equation of the envelope is obtained by eliminating a from (1) and (2) as under.

Multiplying (1) by 3 and (2) by a, and subtracting, we get

$$3a^2x - 6ay + 3z = 0$$

or $$a^2x - 2ay + z = 0. \tag{4}$$

Also, from (2), we have

$$a^2 - 2ax + y = 0. \tag{5}$$

Solving (4) and (5), we get

$$\frac{a^2}{-2y^2 + 2zx} = \frac{a}{z - xy} = \frac{1}{-2x^2 + 2y}$$

$$\Rightarrow \quad a^2 = \frac{xz - y^2}{y - x^2} \text{ and } a = \frac{z - xy}{2(y - x^2)}$$

$$\Rightarrow \quad 4(xz - y^2)(y - x^2) = (z - xy)^2$$

$$\Rightarrow \quad (xy - z)^2 = 4(xz - y^2)(y - x^2).$$

This is the required envelope of the family (1).

The edge of regression is obtained by eliminating a from (1), (2) and (3) as under.

From (3), we get

$$a = x. \tag{6}$$

Eliminating a from (2) and (6), we get

$$6x^2 - 3y - 3x^2 = 0$$

or $$x^2 = y. \tag{7}$$

Eliminating a from (1) and (6), we get

$$3x^3 - 3xy + z - x^3 = 0$$

or $$2x^3 - 3xy + z = 0$$

or $$2xy - 3xy + z = 0 \qquad \text{[using (7)]}$$

or $$-xy + z = 0$$

or $\quad xy = z.$ (8)

Now, squaring (7), we get

$$\begin{aligned} y^2 &= x^4 \\ &= x\,(x^3) \\ &= x\,(xy) && \text{[using (7)]} \\ &= xz && \text{[using (8)]} \end{aligned}$$

i.e., $\quad y^2 = xz.$ (9)

The edge of regression is the curve of intersection of (8) and (9).

Exercises

1. Find the edge of regression of the envelope of the family of planes $x\sin\theta - y\cos\theta + z = a\theta$, θ being the parameter.

   [**Ans.** $z = a\tan^{-1}(y/x), x^2 + y^2 = a^2$]

2. Find the envelope of the family of cones $(ax + x + y + z - 1)(ay + z) = ax(x + y + z - 1)$, a being the parameter.

   $$\left[\textbf{Ans.} \left(\frac{x^2 - y^2 - yz - x + y}{2y} + x + y + z - 1\right)\left(\frac{x^2 - y^2 - yz - x + y}{2x} + z\right) - \left(\frac{x^2 - y^2 - yz - x + y}{2y}\right)(x + y + z - 1) = 0\right]$$

3. Prove that the characteristics of the family of osculating spheres of a twisted curve are its circles of curvature, and the edge of regression is the curve itself.

4. Find the envelope of the plane

   $$\frac{x}{a+u} + \frac{y}{b+u} + \frac{z}{c+u},$$

   where u is the parameter, and determine the edge of regression.

   $$\left[\begin{array}{l} \textbf{Ans.}\ (9v - \mu\lambda)^2 = 4(\mu^2 - 3\lambda v)(\lambda^2 - 3\mu), \text{ where} \\ \lambda = a + b + c - (x + y + z), \\ \mu = ab + bc + ca - x(b + c) - y(c + a) - z(a + b), \\ v = abc - (bcx + cay + abz); \\ x = \dfrac{(a+u)^3}{(c-a)(b-a)}, y = \dfrac{(b+u)^3}{(c-b)(a-b)}, z = \dfrac{(c+u)^3}{(a-c)(b-c)}. \end{array}\right]$$

2.8 DEVELOPABLE SURFACES OR DEVELOPABLES

Definition. The envelope of a single parameter family of planes is called a developable surface, or briefly a developable. The characteristics in this case are known as the generators of the developable surface.

The name 'developable' is due to the fact that the developable surface has the tendency to unroll or unfold or develop into a plane without stretching or tearing it, since the consecutive generators of the developable are coplanar.

Some Important Properties.

(i) The tangents to the edge of regression are the generators of the developable surface. Also, the osculating plane of the edge of regression at any point is the tangent plane to the developable at that point.

Proof. Let the developable surface be the envelope of the one parameter family of planes whose equation is

$$\mathbf{r}.\mathbf{a} = q, \tag{1}$$

where $\mathbf{r}$ is the position vector of a point in the plane, $\mathbf{a}$ is a vector perpendicular to the plane, $\dfrac{q}{|\mathbf{a}|}$ is the perpendicular length from the vector origin to the plane and u is the parameter of the family of planes (1).

Clearly, $\mathbf{a}$ and q are functions of u, whereas $\mathbf{r}$ is not a function of u.

Differentiating (1) with respect to u, we get

$$\mathbf{r}.\dot{\mathbf{a}} = \dot{q}, \tag{2}$$

where dot denotes the differentiation with respect to u.

Again, differentiating (2) with respect to u, we get

$$\mathbf{r}.\ddot{\mathbf{a}} = \ddot{q}. \tag{3}$$

The equation of the edge of regression is obtained by eliminating u from (1), (2) and (3). Therefore the edge of regression is given by (1), (2) and (3) if $\mathbf{r}$ is regarded as a function of u. Also, for different values of u, the lines of intersection of (1) and (2) give different generators of the developable.

Differentiating (1) with respect to u, regarding $\mathbf{r}$ as a function of u, we get

$$\left(\frac{d\mathbf{r}}{ds}\frac{ds}{du}\right).\mathbf{a} + \mathbf{r}.\frac{d\mathbf{a}}{du} = \frac{dq}{du}$$

or $(\mathbf{t}\dot{s}).\mathbf{a}+\mathbf{r}.\dot{\mathbf{a}}=\dot{q}$, where $\mathbf{t}$ denotes the unit tangent to the edge of regression

or $(\mathbf{t}\dot{s}).\mathbf{a}=0$ [using (2)]

or $\mathbf{t}.\mathbf{a}=0.$ (4)

Again, differentiating (2) with respect to u, regarding $\mathbf{r}$ as a function of u, we get

$$\left(\frac{d\mathbf{r}}{ds}\frac{ds}{du}\right).\dot{\mathbf{a}}+\mathbf{r}.\frac{d\dot{\mathbf{a}}}{du}=\frac{d\dot{q}}{du}$$

or $(\mathbf{t}\dot{s}).\dot{\mathbf{a}}+\mathbf{r}.\ddot{\mathbf{a}}=\ddot{q}$

or $(\mathbf{t}\dot{s}).\dot{\mathbf{a}}=0$ [using (3)]

or $\mathbf{t}.\dot{\mathbf{a}}=0.$ (5)

From (4) and (5), we get

$\mathbf{t}.\mathbf{a}=0$ and $\mathbf{t}.\dot{\mathbf{a}}=0$

$\Rightarrow$ $\mathbf{t}$ is perpendicular to each of the vectors $\mathbf{a}$ and $\dot{\mathbf{a}}$.

$\Rightarrow$ $\mathbf{t}$ is parallel to the vector $\mathbf{a}\times\dot{\mathbf{a}}$. (6)

Again, since the line of intersection of the planes (1) and (2) lies in each of the planes, it is perpendicular to each of the vectors $\mathbf{a}$ and $\dot{\mathbf{a}}$.

$\therefore$ The line of intersection of (1) and (2) is parallel to $\mathbf{a}\times\dot{\mathbf{a}}$. (7)

From (6) and (7), it follows that $\mathbf{t}$ is parallel to the line of intersection of (1) and (2), i.e., the tangent $\mathbf{t}$ to the edge of regression is parallel to generator of the developable. Hence the tangents to the edge of regression are the generators of the developable.

Now, differentiating (4) with respect to u, regarding $\mathbf{r}$ as a function of u, we get

$$\left(\frac{d\mathbf{t}}{ds}\frac{ds}{du}\right).\mathbf{a}+\mathbf{t}.\frac{d\mathbf{a}}{du}=0$$

or $(\kappa\mathbf{n}\,\dot{s}).\mathbf{a}+\mathbf{t}.\dot{\mathbf{a}}=0$, where $\mathbf{n}$ denotes the unit normal to the edge of regression and κ is its curvature

or $(\kappa\mathbf{n}\,\dot{s}).\mathbf{a}=0$ [using (5)]

or $\mathbf{n}.\mathbf{a}=0.$ (8)

From (4) and (8), we get

$\mathbf{t}.\mathbf{a}=0$ and $\mathbf{n}.\mathbf{a}=0$

$\Rightarrow$ $\mathbf{a}$ is perpendicular to each of the vectors $\mathbf{t}$ and $\mathbf{n}$.

$\Rightarrow$ $\mathbf{a}$ is perpendicular to the osculating plane of the edge of regression.

$\Rightarrow$ The osculating plane of the edge of regression is coincident with the plane (1). [$\because$ $\mathbf{a}$ is perpendicular to the plane (1)]

⇒ The osculating plane of the edge of regression at any point is the tangent plane to the developable at that point.

$$\left[\because \begin{array}{l}\text{Each plane belonging to the family (1)} \\ \text{is tangent plane of the developable}\end{array}\right]$$

(ii) The tangent plane is the same at all points of the generators of a developable surface.

Proof. Let the equation of the edge of regression of the developable be given by

$$\mathbf{r} = \mathbf{r}(s). \tag{1}$$

Since the tangents to the edge of regression are the generators of the developable, the equation of the developable can be taken as

$$\mathbf{R} = \mathbf{r}(s) + \lambda \mathbf{t}(s), \tag{2}$$

where $\mathbf{R}$ is the position vector of any current point on the developable, $\mathbf{t}(s)$ is the unit tangent vector at any point $P(\mathbf{r}(s))$ on the edge of regression (1) and λ, s are the two independent parameters of the developable.

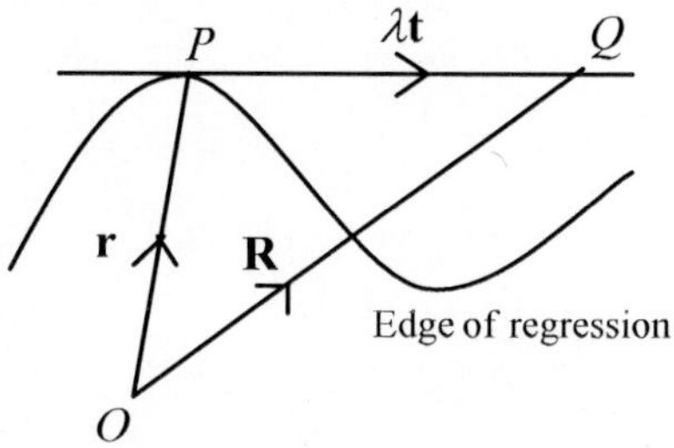

Differentiating (1) partially with respect to λ and s, we get

$$\left.\begin{array}{l}\dfrac{\partial \mathbf{R}}{\partial \lambda} = \mathbf{R}_1 = \mathbf{t}, \\ \dfrac{\partial \mathbf{R}}{\partial s} = \mathbf{R}_2 = \mathbf{t} + \lambda\kappa\mathbf{n},\end{array}\right\} \tag{3}$$

where $\mathbf{n}$ is the unit normal vector at any point $P(\mathbf{r}(s))$ on the edge of regression (1) and κ is its curvature.

If $\mathbf{N}$ is the unit normal vector to the developable, then

$$\mathbf{N} = \frac{\mathbf{R}_1 \times \mathbf{R}_2}{H}$$

⇒ $$\mathbf{N} = \frac{\lambda\kappa\mathbf{b}}{H}$$ [using (3)]

⇒ The unit normal vector $\mathbf{N}$ to the developable is parallel to the unit binormal vector $\mathbf{b}$ to the edge of regression.

⇒ We can take

$$\mathbf{N} = \mathbf{b}.$$

⇒ $\mathbf{N}$ is a function of s only. [$\because$ $\mathbf{b}$ is a function of s only]

$\Rightarrow$ **N** is independent of λ.

$\Rightarrow$ The tangent plane is the same at all points of the generators of a developable surface.

Characterization of a Developable Surface

Theorem. A necessary and sufficient condition for the surface $z = f(x, y)$ to be a developable surface is that $rt - s^2 = 0$, where

$$r = \frac{\partial^2 f}{\partial x^2}, \quad s = \frac{\partial^2 f}{\partial x\, \partial y}, \ t = \frac{\partial^2 f}{\partial y^2}.$$

Proof. The equation of the given surface is

$$z = f(x, y). \tag{1}$$

We know that the equation of the tangent plane to the surface (1) at the point $P\,(x, y, z)$ is

$$(X - x)\frac{\partial f}{\partial x} + (Y - y)\frac{\partial f}{\partial y} - (Z - z) = 0$$

or $$p\,(X - x) + q\,(Y - y) - (Z - z) = 0, \tag{2}$$

where $p = \dfrac{\partial f}{\partial x}, \ q = \dfrac{\partial f}{\partial y}.$

Necessary condition. Let the given surface (1) be a developable surface.

To prove : $rt - s^2 = 0$

Since the surface (1) is developable, the equation of the tangent plane to it given by (2) must be expressible in terms of a single parameter. Hence there must be some relation between p and q, which we may write as

$$p = \phi\,(q). \tag{3}$$

Differentiating (3) partially with respect to x and y, we get

$$\frac{\partial p}{\partial x} = \phi'(q)\frac{\partial q}{\partial x},$$

$$\frac{\partial p}{\partial y} = \phi'(q)\frac{\partial q}{\partial y}.$$

Eliminating $\phi'(q)$ from the above two equations, we get

$$\frac{\partial p}{\partial x}\frac{\partial q}{\partial y} = \frac{\partial p}{\partial y}\frac{\partial q}{\partial x}$$

or $$rt = s^2$$

or $$rt - s^2 = 0,$$

where

$$r = \frac{\partial p}{\partial x} = \frac{\partial^2 f}{\partial x^2}, \quad s = \frac{\partial p}{\partial y} = \frac{\partial q}{\partial x} = \frac{\partial^2 f}{\partial x \partial y}, \quad t = \frac{\partial q}{\partial y} = \frac{\partial^2 f}{\partial y^2}.$$

Sufficient condition. Now, we suppose that

$$rt - s^2 = 0. \tag{4}$$

To prove : The surface (1) is developable.

Substituting $r = \frac{\partial p}{\partial x}$, $s = \frac{\partial p}{\partial y} = \frac{\partial q}{\partial x}$, $t = \frac{\partial q}{\partial y}$ in (4), we get

$$\frac{\partial p}{\partial x}\frac{\partial q}{\partial y} = \frac{\partial p}{\partial y}\frac{\partial q}{\partial x}$$

$$\Rightarrow \quad \begin{vmatrix} \frac{\partial p}{\partial x} & \frac{\partial p}{\partial y} \\ \frac{\partial q}{\partial x} & \frac{\partial q}{\partial y} \end{vmatrix} = 0$$

$$\Rightarrow \quad \frac{\partial (p, q)}{\partial (x, y)} = 0$$

$\Rightarrow$ The Jacobian of the functions p and q vanishes.

$\Rightarrow$ The functions p and q depend on a single parameter.

$\Rightarrow$ The equation of the tangent plane given by (2) depends on a single parameter.

$\Rightarrow$ The surface (1) is developable.

2.9 DEVELOPABLES ASSOCIATED WITH A SPACE CURVE

Osculating Developable

Definition. The envelope of the family of osculating planes of a twisted curve is called its osculating developable.

Theorem. The generators of the osculating developable of a twisted curve are the tangents to the curve and the edge of regression of the osculating developable of the twisted curve is the curve itself.

Proof. Let the equation of the space curve be given by

$$\mathbf{r} = \mathbf{r}(s), \tag{1}$$

where s denotes the arc length measured along the space curve.

The equation of the osculating plane of (1) at a point $P(\mathbf{r}(s))$ is given by

$$(\mathbf{R} - \mathbf{r}) . \mathbf{b} = 0, \tag{2}$$

where $\mathbf{R}$ is the position vector of any current point on the osculating plane, and $\mathbf{r}$ and $\mathbf{b}$ are functions of s.

Clearly, (2) represents a one parameter family of planes, the parameter being s. The envelope of (2) is the osculating developable of (1).

Differentiating (2) with respect to s, we get

$$-\mathbf{t}\,.\,\mathbf{b}+(\mathbf{R}-\mathbf{r})\,.\,(-\tau\mathbf{n})=0$$

i.e., $$(\mathbf{R}-\mathbf{r})\,.\,\mathbf{n}=0. \qquad (3)$$

$[\because \mathbf{t}\,.\,\mathbf{b}=0 \text{ and } \tau\neq 0]$

The generator of the osculating developable is the line of intersection of the planes (2) and (3). Since (2) and (3) are the osculating and rectifying planes of (1) at the point $P(\mathbf{r}(s))$, it follows that the line of intersection of the planes (2) and (3) is the tangent to the curve (1) at the point $P(\mathbf{r}(s))$. Hence the generators of the osculating developable of the curve (1) are the tangents to the curve.

To find the edge of regression, we differentiate (3) with respect to s. Then we get

$$-\mathbf{t}\,.\,\mathbf{n}+(\mathbf{R}-\mathbf{r})\,.\,(\tau\mathbf{b}-\kappa\mathbf{t})=0$$

or $$(\mathbf{R}-\mathbf{r})\,.\,(\tau\mathbf{b}-\kappa\mathbf{t})=0 \qquad [\because \mathbf{t}\,.\,\mathbf{n}=0]$$

or $$\tau(\mathbf{R}-\mathbf{r})\,.\,\mathbf{b}-\kappa(\mathbf{R}-\mathbf{r})\,.\,\mathbf{t}=0$$

or $$\kappa(\mathbf{R}-\mathbf{r})\,.\,\mathbf{t}=0 \qquad [\text{using } (2)]$$

or $$(\mathbf{R}-\mathbf{r})\,.\,\mathbf{t}=0 \qquad (4)$$

$[\because \kappa\neq 0]$

The point of intersection of the planes (2), (3) and (4) gives the characteristic point. Since (2), (3) and (4) are the osculating, rectifying and normal planes of (1) at the point $P(\mathbf{r}(s))$, the point of intersection of the planes (2), (3) and (4) is the point $P(\mathbf{r}(s))$.

Thus the characteristic point is the point $P(\mathbf{r}(s))$ itself. Since the edge of regression is the locus of characteristic points, it is the curve (1) itself.

Polar Developable

Definition. The polar developable of a space curve is the envelope of the family of normal planes of the space curve. The generators of the polar developable are known as the polar lines.

Theorem. The polar line is the axis of the circle of curvature and the edge of regression of the polar developable is the locus of the centre of spherical curvature.

Proof. Let the equation of the space curve be given by

$$\mathbf{r}=\mathbf{r}(s), \qquad (1)$$

where s denotes the arc length measured along the space curve.

The equation of the normal plane of (1) at a point $P(\mathbf{r}(s))$ is given by

$$(\mathbf{R}-\mathbf{r})\,.\,\mathbf{t}=0, \qquad (2)$$

where $\mathbf{R}$ is the position vector of any current point on the normal plane and $\mathbf{r}$ and $\mathbf{t}$ are functions of s.

Clearly, (2) represents a one parameter family of planes, the parameter being s. The envelope of (2) is the polar developable of (1).

Differentiating (2) with respect to s, we get

$$-\mathbf{t}\,.\,\mathbf{t}+(\mathbf{R}-\mathbf{r})\,.\,(\kappa\mathbf{n})=0$$

or $$-1+(\mathbf{R}-\mathbf{r})\,.\,(\kappa\mathbf{n})=0 \qquad [\because \mathbf{t}\,.\,\mathbf{t}=1]$$

or $$(\mathbf{R}-\mathbf{r})\,.\,\mathbf{n}=\frac{1}{\kappa}$$

or $$(\mathbf{R}-\mathbf{r})\,.\,\mathbf{n}=\rho \qquad (3)$$

Since $\mathbf{n}\,.\,\mathbf{n}=1$, therefore (3) can also be written as

$$(\mathbf{R}-\mathbf{r})\,.\,\mathbf{n}=\rho\mathbf{n}\,.\,\mathbf{n}$$

or $$(\mathbf{R}-\mathbf{r}-\rho\mathbf{n})\,.\,\mathbf{n}=0$$

or $$[\mathbf{R}-(\mathbf{r}+\rho\mathbf{n})\,]\,.\,\mathbf{n}=0. \qquad (3')$$

The generator of the polar developable, which is called the polar line, is the line of intersection of the planes (2) and (3) [or (3')]. Since (2) is the equation of the normal plane of (1) at a point $P(\mathbf{r}(s))$ and (3') represents a plane through the centre of curvature perpendicular to the principal normal, it follows that the intersection of the planes (2) and (3') is a straight line passing through the centre of curvature and parallel to the binormal $\mathbf{b}$. Thus the polar line is the axis of the circle of curvature.

To find the edge of regression, we differentiate (3) with respect to s. Then we get

$$-\mathbf{t}\,.\,\mathbf{n}+(\mathbf{R}-\mathbf{r})\,.\,(\tau\mathbf{b}-\kappa\mathbf{t})=\rho'$$

or $$(\mathbf{R}-\mathbf{r})\,.\,(\tau\mathbf{b}-\kappa\mathbf{t})=\rho' \qquad [\because \mathbf{t}\,.\,\mathbf{n}=0]$$

or $$\tau\,(\mathbf{R}-\mathbf{r})\,.\,\mathbf{b}-\kappa\,(\mathbf{R}-\mathbf{r})\,.\,\mathbf{t}=\rho'$$

or $$\tau\,(\mathbf{R}-\mathbf{r})\,.\,\mathbf{b}=\rho' \qquad \text{[using (2)]}$$

or $$(\mathbf{R}-\mathbf{r})\,.\,\mathbf{b}=\frac{\rho'}{\tau}$$

or $$(\mathbf{R}-\mathbf{r})\,.\,\mathbf{b}=\rho'\sigma. \qquad (4)$$

The point of intersection of the planes (2), (3) and (4) gives the characteristic point and the locus of the characteristic point is the edge of regression of the polar developable of the curve (1).

From (2), we get

$$(\mathbf{R}-\mathbf{r})\,.\,\mathbf{t}=0$$

$\Rightarrow$ $\mathbf{R}-\mathbf{r}$ is perpendicular to $\mathbf{t}$.

$\Rightarrow$ $\mathbf{R}-\mathbf{r}$ lies in the plane of $\mathbf{n}$ and $\mathbf{b}$.

$\Rightarrow$ There exist scalars λ and μ such that

$$\mathbf{R}-\mathbf{r}=\lambda\mathbf{n}+\mu\mathbf{b}. \tag{5}$$

Substituting the value of $\mathbf{R}-\mathbf{r}$ from (5) in (3) and (4), we get

$$(\lambda\mathbf{n}+\mu\mathbf{b})\,.\,\mathbf{n}=\rho,$$
$$(\lambda\mathbf{n}+\mu\mathbf{b})\,.\,\mathbf{b}=\rho'\sigma$$

$$\Rightarrow \quad \left.\begin{array}{l}\lambda=\rho,\\ \mu=\rho'\sigma.\end{array}\right\} \tag{6}$$

Now, substituting the values of λ and μ from (6) in (5), we get

$$\mathbf{R}-\mathbf{r}=\rho\mathbf{n}+\rho'\sigma\mathbf{b}$$

or $$\mathbf{R}=\mathbf{r}+\rho\mathbf{n}+\rho'\sigma\mathbf{b}. \tag{7}$$

The right hand side of (7) is the position vector of the characteristic point, which is also the position vector of the centre of osculating sphere (or spherical curvature) of the curve (1) at the point $P(\mathbf{r}(s))$.

Hence it follows that the edge of regression of the polar developable of a space curve is the locus of the centre of spherical curvature.

Rectifying Developable

Definition. The rectifying developable of a space curve is the envelope of the family of rectifying planes of the space curve. The generators of the rectifying developable are known as the rectifying lines.

Theorem. The generator of the rectifying developable of a space curve is inclined to the tangent to the curve at an angle ϕ, where

$$\tan\phi=\frac{\kappa}{\tau}$$

and the edge of regression of the rectifying developable of a space curve has the equation

$$\mathbf{R}=\mathbf{r}+\frac{\kappa\,(\tau\mathbf{t}+\kappa\mathbf{b})}{\kappa'\tau-\kappa\tau'}.$$

Proof. Let the equation of the space curve be given by

$$\mathbf{r}=\mathbf{r}\,(s), \tag{1}$$

where s denotes the arc length measured along the space curve.

The equation of the rectifying plane of (1) at a point $P\,(\mathbf{r}(s))$ is given by

$$(\mathbf{R}-\mathbf{r})\,.\,\mathbf{n}=0, \tag{2}$$

where $\mathbf{R}$ is the position vector of any current point on the rectifying plane, and $\mathbf{r}$ and $\mathbf{n}$ are functions of s.

Clearly, (2) represents a one parameter family of planes, the parameter being s. The envelope of (2) is the rectifying developable of (1).

Differentiating (2) with respect to s, we get

$$-\mathbf{t}\,.\,\mathbf{n}+(\mathbf{R}-\mathbf{r})\,.\,(\tau\mathbf{b}-\kappa\mathbf{t})=0$$

or $$(\mathbf{R}-\mathbf{r})\,.\,(\tau\mathbf{b}-\kappa\mathbf{t})=0. \tag{3}$$

$[\because \mathbf{t}\,.\,\mathbf{n}=0]$

The locus of the point of intersection of (2) and (3) is the rectifying developable of (1).

To find the edge of regression, we differentiate (3) with respect to s. Then we get

$$-\mathbf{t}\,.\,(\tau\mathbf{b}-\kappa\mathbf{t})+(\mathbf{R}-\mathbf{r})\,.\,(\tau'\mathbf{b}+\tau\mathbf{b}'-\kappa'\mathbf{t}-\kappa\mathbf{t}')=0$$

or $$-\mathbf{t}\,.\,(\tau\mathbf{b}-\kappa\mathbf{t})+(\mathbf{R}-\mathbf{r})\,.\,(\tau'\mathbf{b}-\tau^2\mathbf{n}-\kappa'\mathbf{t}-\kappa^2\mathbf{n})=0$$

or $$-\mathbf{t}\,.\,(\tau\mathbf{b}-\kappa\mathbf{t})+(\mathbf{R}-\mathbf{r})\,.\,[\tau'\mathbf{b}-\kappa'\mathbf{t}-(\tau^2+\kappa^2)\,\mathbf{n}]=0$$

or $$-\mathbf{t}\,.\,(\tau\mathbf{b}-\kappa\mathbf{t})+(\mathbf{R}-\mathbf{r})\,.\,(\tau'\mathbf{b}-\kappa'\mathbf{t})=0 \qquad \text{[using (2)]}$$

or $$\kappa+(\mathbf{R}-\mathbf{r})\,.\,(\tau'\mathbf{b}-\kappa'\mathbf{t})=0$$

or $$(\mathbf{R}-\mathbf{r})\,.\,(\tau'\mathbf{b}-\kappa'\mathbf{t})+\kappa=0. \tag{4}$$

The point of intersection of the planes (2), (3) and (4) gives the characteristic point and the locus of the characteristic points is the edge of regression of the rectifying developable of the curve (1).

From (2) and (3), we get

$$(\mathbf{R}-\mathbf{r})\,.\,\mathbf{n}=0 \text{ and } (\mathbf{R}-\mathbf{r})\,.\,(\tau\mathbf{b}-\kappa\mathbf{t})=0$$

$\Rightarrow$ $\mathbf{R}-\mathbf{r}$ is perpendicular to each of $\mathbf{n}$ and $\tau\mathbf{b}-\kappa\mathbf{t}$.

$\Rightarrow$ $\mathbf{R}-\mathbf{r}$ is parallel to $\mathbf{n}\times(\tau\mathbf{b}-\kappa\mathbf{t})$.

$\Rightarrow$ $\mathbf{R}-\mathbf{r}$ is parallel to $\tau\mathbf{t}+\kappa\mathbf{b}$.

$\Rightarrow$ There exists some scalar λ such that

$$\mathbf{R}-\mathbf{r}=\lambda\,(\tau\mathbf{t}+\kappa\mathbf{b}). \tag{5}$$

It follows from (5) that

$$(\mathbf{R}-\mathbf{r})\,.\,\mathbf{t}=\lambda\tau,$$

$$(\mathbf{R}-\mathbf{r})\,.\,\mathbf{b}=\lambda\kappa.$$

Therefore, if ϕ is the angle between $\mathbf{R}-\mathbf{r}$ and $\mathbf{t}$, we have

$$|\mathbf{R}-\mathbf{r}|\,|\mathbf{t}|\cos\phi=\lambda\tau,$$

$$|\mathbf{R}-\mathbf{r}|\,|\mathbf{b}|\cos\left(\frac{\pi}{2}-\phi\right)=\lambda\kappa$$

or $$|\mathbf{R}-\mathbf{r}|\,|\mathbf{t}|\cos\phi=\lambda\tau,$$

$$|\mathbf{R}-\mathbf{r}|\,|\mathbf{b}|\sin\phi=\lambda\kappa,$$

or $$\tan\phi=\frac{\kappa}{\tau}.$$

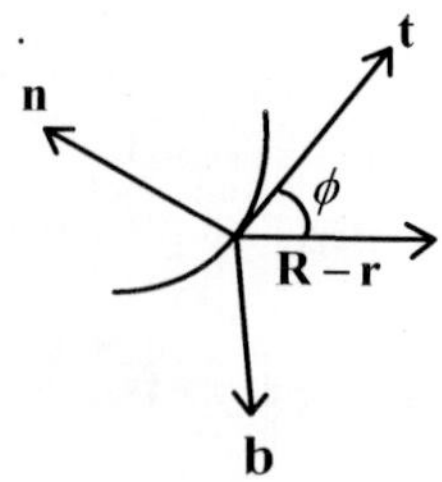

Substituting the value of $\mathbf{R}-\mathbf{r}$ from (5) in (4), we get

$$\lambda\,(\tau\mathbf{t}+\kappa\mathbf{b})\,.\,(\tau'\mathbf{b}-\kappa'\mathbf{t})+\kappa=0$$

or $$\lambda\,(\kappa\tau'-\tau\kappa')+\kappa=0$$

or $$\lambda=\frac{\kappa}{\kappa'\tau-\kappa\tau'}. \tag{6}$$

Now, substituting the value of λ from (6) in (5), we get

$$\mathbf{R}-\mathbf{r}=\frac{\kappa}{\kappa'\tau-\kappa\tau'}(\tau\mathbf{t}+\kappa\mathbf{b})$$

or $$\mathbf{R}=\mathbf{r}+\frac{\kappa\,(\tau\mathbf{t}+\kappa\mathbf{b})}{\kappa'\tau-\kappa\tau'}. \tag{7}$$

This is the required equation of the edge of regression of the rectifying developable of the space curve (1).

Solved Examples

Example 1. Find the equation to the developable surface which has the curve $x=6t,\ y=3t^2,\ z=2t^3$ for its edge of regression.

Solution. The position vector of a point on the given edge of regression is

$$\mathbf{r}=(x,y,z)$$

or $$\mathbf{r}=(6t,3t^2,2t^3). \tag{1}$$

Differentiating (1) with respect to t, we get

$$\frac{d\mathbf{r}}{dt}=(6,6t,6t^2). \tag{2}$$

We know that $\dfrac{d\mathbf{r}}{dt}$ is a vector along the tangent to the edge of regression (1).

Also, we know that the tangents to the edge of regression are the generators of the developable.

Therefore if **R** is the position vector of any current point (X, Y, Z) on the developable, then equation of the developable can be written as

$$\mathbf{R} = \mathbf{r} + \lambda \frac{d\mathbf{r}}{dt}$$

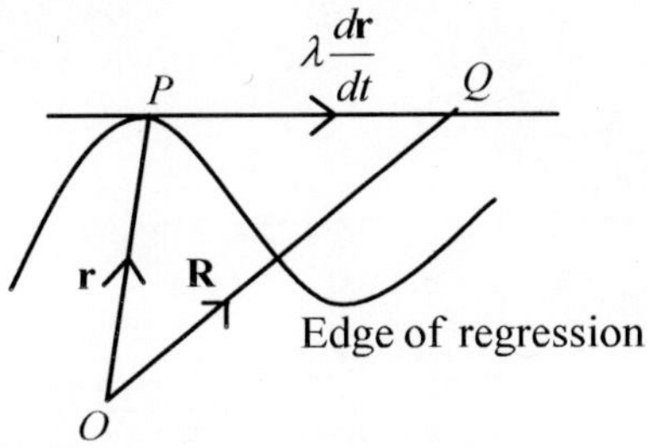

or $\quad (X, Y, Z) = (6t, 3t^2, 2t^3) + \lambda (6, 6t, 6t^2)$ [using (1) and (2)]

or $\quad (X - 6t, Y - 3t^2, Z - 2t^3) = \lambda (6, 6t, 6t^2)$

or $\quad X - 6t = 6\lambda,\ Y - 3t^2 = 6\lambda t,\ Z - 2t^3 = 6\lambda t^2.$ (3)

From first and second relations of (3), we have

$$\frac{X - 6t}{Y - 3t^2} = \frac{1}{t}$$

or $\quad t(X - 6t) = Y - 3t^2$

or $\quad 3t^2 = Xt - Y.$

or $\quad 3t^2 - Xt + Y = 0.$ (4)

From second and third relations of (3), we have

$$\frac{Y - 3t^2}{Z - 2t^3} = \frac{1}{t}$$

or $\quad t(Y - 3t^2) = Z - 2t^3$

or $\quad t^3 = Yt - Z$

or $\quad 3t^3 = 3Yt - 3Z$

or $\quad t(Xt - Y) = 3Yt - 3Z$ [using (4)]

or $\quad Xt^2 - 4Yt + 3Z = 0.$ (5)

Now, from (4) and (5), we get

$$\frac{t^2}{-3XZ + 4Y^2} = \frac{t}{XY - 9Z} = \frac{1}{-12Y + X^2}$$

$\Rightarrow \quad t = \dfrac{XY - 9Z}{X^2 - 12Y}$ and $t^2 = \dfrac{4Y^2 - 3XZ}{X^2 - 12Y}$

$\Rightarrow \quad \left(\dfrac{XY - 9Z}{X^2 - 12Y}\right)^2 = \dfrac{4Y^2 - 3XZ}{X^2 - 12Y}$

$\Rightarrow \quad (XY - 9Z)^2 = (X^2 - 12Y)(4Y^2 - 3XZ).$

This is the required equation of the developable surface.

Example 2. Find the equation to the developable surface which has the helix $x = a\cos t, y = a\sin t, z = ct$ for its edge of regression.

Solution. The position vector of a point on the given edge of regression is

$$\mathbf{r} = (x, y, z)$$

or $$\mathbf{r} = (a\cos t, a\sin t, ct). \quad (1)$$

Differentiating (1) with respect to t, we get

$$\frac{d\mathbf{r}}{dt} = (-a\sin t, a\cos t, c). \quad (2)$$

We know that $\frac{d\mathbf{r}}{dt}$ is a vector along the tangent to the edge of regression (1).

Also, we know that the tangents to the edge of regression are the generators of the developable. Therefore if $\mathbf{R}$ is the position vector of any current point (X, Y, Z) on the developable, then equation of the developable can be written as

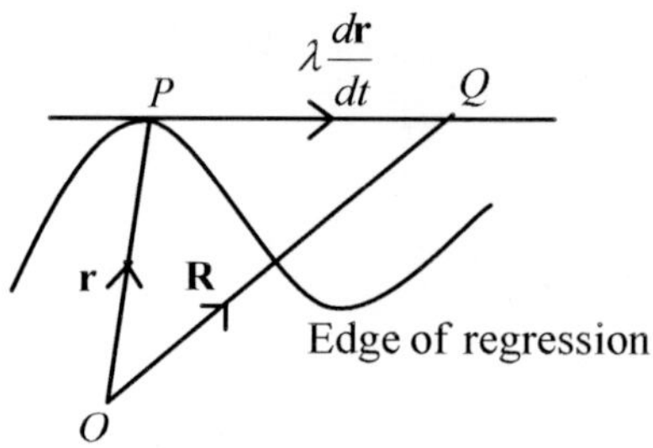

$$\mathbf{R} = \mathbf{r} + \lambda\frac{d\mathbf{r}}{dt}$$

or $$(X, Y, Z) = (a\cos t, a\sin t, ct) + \lambda(-a\sin t, a\cos t, c)$$

[using (1) and (2)]

or $$X = a(\cos t - \lambda\sin t),\ Y = a(\sin t + \lambda\cos t),\ Z = c(t + \lambda).$$

This is the required equation of the developable surface, where λ and t are parameters.

Example 3. Prove that the surface $xy = (z-c)^2$ is developable.

Solution. The equation of the given surface is

$$xy = (z-c)^2$$

or $$z - c = \sqrt{xy}$$

or $$z = c + \sqrt{xy}$$

or $$z = f(x, y), \text{ where } f(x, y) = c + \sqrt{xy}.$$

Thus the equation of the given surface is in Monge's form and we have

$$p = \frac{\partial f}{\partial x} = \frac{\partial z}{\partial x} = \frac{1}{2\sqrt{xy}} y = \frac{1}{2}\sqrt{\frac{y}{x}}$$

$$q = \frac{\partial f}{\partial y} = \frac{\partial z}{\partial y} = \frac{1}{2\sqrt{xy}} x = \frac{1}{2}\sqrt{\frac{x}{y}}$$

$$r = \frac{\partial^2 f}{\partial x^2} = \frac{\partial^2 z}{\partial x^2} = \frac{\partial}{\partial x}\left(\frac{1}{2}\sqrt{\frac{y}{x}}\right) = \frac{1}{4\sqrt{y/x}}\left(-\frac{y}{x^2}\right) = -\frac{1}{4}\frac{\sqrt{y}}{x\sqrt{x}},$$

$$s = \frac{\partial^2 f}{\partial x\,\partial y} = \frac{\partial^2 z}{\partial x\,\partial y} = \frac{\partial}{\partial x}\left(\frac{1}{2}\sqrt{\frac{x}{y}}\right) = \frac{1}{4\sqrt{x/y}}\left(\frac{1}{y}\right) = \frac{1}{4\sqrt{xy}},$$

$$t = \frac{\partial^2 f}{\partial y^2} = \frac{\partial^2 z}{\partial y^2} = \frac{\partial}{\partial y}\left(\frac{1}{2}\sqrt{\frac{x}{y}}\right) = \frac{1}{4\sqrt{x/y}}\left(-\frac{x}{y^2}\right) = -\frac{1}{4}\frac{\sqrt{x}}{y\sqrt{y}}.$$

$$\therefore \qquad rt - s^2 = \left(-\frac{1}{4}\frac{\sqrt{y}}{x\sqrt{x}}\right)\left(-\frac{1}{4}\frac{\sqrt{x}}{y\sqrt{y}}\right) - \left(\frac{1}{4\sqrt{xy}}\right)^2$$

$$= \frac{1}{16xy} - \frac{1}{16xy}$$

$$= 0.$$

Hence the given surface is developable.

Example 4. By considering the value of $rt - s^2$, determine if the surface $xyz = a^3$ is developable.

Solution. The equation of the given surface is

$$xyz = a^3$$

or $$z = \frac{a^3}{xy}$$

or $$z = f(x, y), \text{where } f(x, y) = \frac{a^3}{xy}.$$

Thus the equation of the given surface is in Monge's form and we have

$$p = \frac{\partial f}{\partial x} = \frac{\partial z}{\partial x} = -\frac{a^3}{x^2 y},$$

$$q = \frac{\partial f}{\partial y} = \frac{\partial z}{\partial y} = -\frac{a^3}{xy^2},$$

$$r = \frac{\partial^2 f}{\partial x^2} = \frac{\partial^2 z}{\partial x^2} = \frac{\partial}{\partial x}\left(-\frac{a^3}{x^2 y}\right) = \frac{2a^3}{x^3 y},$$

$$s = \frac{\partial^2 f}{\partial x\, \partial y} = \frac{\partial^2 z}{\partial x\, \partial y} = \frac{\partial}{\partial x}\left(-\frac{a^3}{xy^2}\right) = \frac{a^3}{x^2 y^2},$$

$$t = \frac{\partial^2 f}{\partial y^2} = \frac{\partial^2 z}{\partial y^2} = \frac{\partial}{\partial y}\left(-\frac{a^3}{xy^2}\right) = \frac{2a^3}{xy^3}.$$

$$\therefore \qquad rt - s^2 = \left(\frac{2a^3}{x^3 y}\right)\left(\frac{2a^3}{xy^3}\right) - \left(\frac{a^3}{x^2 y^2}\right)^2$$

$$= \frac{4a^6}{x^4 y^4} - \frac{a^6}{x^4 y^4}$$

$$= \frac{3a^6}{x^4 y^4}$$

$$\neq 0.$$

Hence the given surface is not developable.

Example 5. Find the equation of the developable surface which passes through the curves $z = 0, y^2 = 4ax$; $x = 0, y^2 = 4bz$.

Solution. Equation of any line tangent to the first parabola $z = 0, y^2 = 4ax$ is

$$z = 0, y = mx + \frac{a}{m}.$$

Therefore equation of any plane which touches the first parabola $z = 0, y^2 = 4ax$ is

$$y - \left(mx + \frac{a}{m}\right) + \lambda z = 0$$

or $$y = mx + \frac{a}{m} - \lambda z. \tag{1}$$

If the plane (1) touches the second parabola $x = 0, y^2 = 4bz,$ then the straight line

$$y = -\lambda z + \frac{a}{m} \tag{2}$$

touches the parabola

$$y^2 = 4bz. \tag{3}$$

$$\therefore \qquad \frac{a}{m} = -\frac{b}{\lambda}$$

$$\left[\begin{array}{l}\because \text{The condition for the straight line } Y = MZ + C \\ \text{to touch the parabola } Y^2 = 4AZ \text{ is } C = A/M.\end{array}\right]$$

or $$\lambda = -\frac{bm}{a}. \tag{4}$$

Substituting the value of λ from (4) in (1), we get

$$y = mx + \frac{a}{m} + \frac{bm}{a} z$$

or $$F(m) \equiv y - m\left(x + \frac{b}{a} z\right) - \frac{a}{m} = 0. \tag{5}$$

The equation of the required developable is given by

$$F(m) = 0, F'(m) = 0.$$

Now, $F'(m) = 0$ gives

$$-\left(x + \frac{b}{a} z\right) + \frac{a}{m^2} = 0$$

or $$x + \frac{b}{a} z = \frac{a}{m^2}. \tag{6}$$

Using (6), (5) gives

$$y - m\left(\frac{a}{m^2}\right) - \frac{a}{m} = 0$$

or $$y - \frac{2a}{m} = 0$$

or $$m = \frac{2a}{y}. \tag{7}$$

Substituting the above value of m in (6), we get

$$x + \frac{b}{a} z = \frac{a}{(2a/y)^2}$$

or $$\frac{ax + bz}{a} = \frac{y^2}{4a}$$

or $$y^2 = 4ax + 4bz.$$

This is the required equation of the developable surface.

Example 6. Show that the edge of regression of the developable that passes through the parabolas $x=0, z^2=4ay\,; x=a, y^2=4az$ is given by

$$\frac{3x}{y}=\frac{y}{z}=\frac{z}{3(a-x)}.$$

Solution. Equation of any line tangent to the first parabola $x=0, z^2=4ay$ is

$$x=0, z=my+\frac{a}{m}.$$

Therefore equation of any plane which touches the first parabola $x=0, z^2=4ay$ is

$$z-\left(my+\frac{a}{m}\right)+\lambda x=0$$

or $$y=\frac{z}{m}-\frac{a}{m}\left(\frac{1}{m}-\frac{\lambda x}{a}\right). \tag{1}$$

If the plane (1) touches the second parabola $x=a, y^2=4az,$ then the straight line

$$y=\frac{z}{m}-\frac{a}{m}\left(\frac{1}{m}-\lambda\right) \tag{2}$$

touches the parabola

$$y^2=4az. \tag{3}$$

Therefore

$$-\frac{a}{m}\left(\frac{1}{m}-\lambda\right)=\frac{a}{1/m}$$

$\left[\because \text{The condition for the straight line } Y=MZ+C \text{ to touch the parabola } Y^2=4AZ \text{ is } C=A/M.\right]$

or $$-\frac{1}{m}\left(\frac{1}{m}-\lambda\right)=m$$

or $$-\frac{1}{m}\left(\frac{1-m\lambda}{m}\right)=m$$

or $$1-m\lambda=-m^3$$

or $$\lambda=\frac{1+m^3}{m}. \tag{4}$$

Substituting the value of λ from (4) in (1), we get

$$y = \frac{z}{m} - \frac{a}{m}\left(\frac{1}{m} - \frac{1+m^3}{m}\frac{x}{a}\right)$$

or $$y = \frac{z}{m} - \frac{a}{m^2} + \frac{1+m^3}{m^2}x$$

or $$F(m) \equiv m^3x - m^2y + mz + (x-a) = 0. \tag{5}$$

The equation of the required edge of regression is given by

$$F(m) = 0,\ F'(m) = 0,\ F''(m) = 0.$$

Now, $F'(m) = 0$ gives

$$3m^2x - 2my + z = 0 \tag{6}$$

and $F''(m) = 0$ gives

$$6mx - 2y = 0$$

i.e., $$m = \frac{y}{3x}. \tag{7}$$

Dividing (6) by $3x$, we get

$$m^2 - \frac{2my}{3x} + \frac{z}{3x} = 0$$

or $$m^2 - 2m^2 + \frac{z}{3x} = 0 \qquad \text{[using (7)]}$$

or $$m^2 = \frac{z}{3x}. \tag{8}$$

Again, dividing (5) by x, we get

$$m^3 - \frac{m^2y}{x} + \frac{mz}{x} + 1 - \frac{a}{x} = 0$$

or $$m^3 - 3m^3 + 3m^3 + 1 - \frac{a}{x} = 0 \qquad \text{[using (7) and (8)]}$$

or $$m^3 = \frac{a-x}{x}. \tag{9}$$

Now, $m^3 . m = (m^2)^2$ gives

$$\frac{a-x}{x} \cdot \frac{y}{3x} = \left(\frac{z}{3x}\right)^2 \qquad \text{[using (7), (8) and (9)]}$$

or $$\frac{y}{z} = \frac{z}{3(a-x)}. \tag{10}$$

Again, $m^2 . m = m^3$ gives

$$\frac{z}{3x}\cdot\frac{y}{3x}=\frac{a-x}{x}$$ [using (7), (8) and (9)]

or $$\frac{3x}{y}=\frac{z}{3(a-x)}. \qquad (11)$$

From (10) and (11), we get

$$\frac{3x}{y}=\frac{y}{z}=\frac{z}{3(a-x)}.$$

This is the required equation of the edge of regression.

Example 7. Prove that the edge of regression of the developable that passes through the parabolas $z=0, y^2=4ax\,; x=0, (y-a)^2=4az$ is the curve of intersection of the surfaces

$$(a+y)^2=3a(x+y+z),\ (a+y)^3=27a^2x.$$

Solution. Equation of any line tangent to the first parabola $z=0, y^2=4ax$ is

$$z=0, y=mx+\frac{a}{m}.$$

Therefore equation of any plane which touches the first parabola $z=0, y^2=4ax$ is

$$y-\left(mx+\frac{a}{m}\right)+\lambda z=0$$

or $$y=mx+\frac{a}{m}-\lambda z. \qquad (1)$$

If the plane (1) touches the second parabola $x=0, (y-a)^2=4az$, then the straight line

$$y=-\lambda z+\frac{a}{m}$$

or $$y-a=-\lambda z+a\left(\frac{1}{m}-1\right) \qquad (2)$$

touches the parabola

$$(y-a)^2=4az. \qquad (3)$$

Therefore

$$a\left(\frac{1}{m}-1\right)=-\frac{a}{\lambda}$$

$\left[\because\right.$ The condition for the straight line $Y=MZ+C$ to touch the parabola $Y^2=4AZ$ is $\left.C=A/M.\right]$

or $$\frac{1-m}{m}=-\frac{1}{\lambda}$$

or $$\lambda=\frac{m}{m-1}. \tag{4}$$

Substituting the value of λ from (4) in (1), we get

$$y=mx+\frac{a}{m}-\frac{m}{m-1}z$$

or $$y=\frac{m^2(m-1)\,x+a\,(m-1)-m^2z}{m\,(m-1)}$$

or $$m^2(m-1)\,x+a\,(m-1)-m^2z=m\,(m-1)\,y$$

or $$F(m)\equiv m^3x-m^2(x+y+z)+m\,(a+y)-a=0. \tag{5}$$

The equation of the required edge of regression is given by

$$F(m)=0,\;\; F'(m)=0,\;\; F''(m)=0.$$

Now, $F'(m)=0$ gives

$$3m^2x-2m\,(x+y+z)+(a+y)=0 \tag{6}$$

and $F''(m)=0$ gives

$$6mx-2\,(x+y+z)=0$$

i.e., $$m=\frac{x+y+z}{3x}. \tag{7}$$

Dividing (6) by $3x$, we get

$$m^2-\frac{2m\,(x+y+z)}{3x}+\frac{a+y}{3x}=0$$

or $$m^2-2m^2+\frac{a+y}{3x}=0 \qquad \text{[using (7)]}$$

or $$m^2=\frac{a+y}{3x}. \tag{8}$$

Again, dividing (5) by x, we get

$$m^3-\frac{m^2(x+y+z)}{x}+\frac{m\,(a+y)}{x}-\frac{a}{x}=0$$

or $$m^3-3m^3+3m^3-\frac{a}{x}=0 \qquad \text{[using (7) and (8)]}$$

or $$m^3=\frac{a}{x}. \tag{9}$$

Now, $m^3.m=(m^2)^2$ gives

$$\frac{a}{x} \cdot \frac{x+y+z}{3x} = \left(\frac{a+y}{3x}\right)^2 \qquad \text{[using (7), (8) and (9)]}$$

or $$(a+y)^2 = 3a(x+y+z). \qquad (10)$$

Again, $(m^2)^3 = (m^3)^2$ gives

$$\left(\frac{a+y}{3x}\right)^3 = \left(\frac{a}{x}\right)^2 \qquad \text{[using (8) and (9)]}$$

or $$(a+y)^3 = 27a^2 x. \qquad (11)$$

The curve of intersection of (10) and (11) gives the required edge of regression.

Example 8. An ellipsoid $\frac{x^2}{a^2} + \frac{y^2}{b^2} + \frac{z^2}{c^2} = 1$ is surrounded by a luminous ring $x = 0, y^2 + z^2 = a^2$. Show that the boundary of the shadow cast on the plane $z = 0$ is given by

$$\frac{x^2}{a^2} + \frac{y^2}{b^2 - c^2} = \frac{a^2}{a^2 - c^2}.$$

Solution. We have to find the equation of the developable which touches the given ellipsoid and the luminous ring.

Let the equation of plane which touches the ellipsoid and the ring both be

$$lx + my + nz = 1. \qquad (1)$$

Then, we have

$$a^2 l^2 + b^2 m^2 + c^2 n^2 = 1 \qquad (2)$$

and $$a^2(m^2 + n^2) = 1. \qquad (3)$$

Eliminating n^2 between (2) and (3), we get

$$a^4 l^2 + a^2(b^2 - c^2)\, m^2 = a^2 - c^2. \qquad (4)$$

Again, plane (1) intersects $z = 0$ in

$$lx + my = 1. \qquad (5)$$

Therefore the required developable is the envelope of (5) subject to the condition (4).

Eliminating m between (4) and (5), we get

$$a^4 l^2 + a^2(b^2 - c^2)\left(\frac{1 - lx}{y}\right)^2 = a^2 - c^2$$

or $$a^4 l^2 y^2 + a^2(b^2 - c^2)(1 - lx)^2 = (a^2 - c^2) y^2$$

or $\quad a^4l^2y^2 + a^2(b^2-c^2)(1-2lx+l^2x^2) = (a^2-c^2)y^2$

or $\quad F(l) \equiv a^2l^2[a^2y^2+(b^2-c^2)x^2] - 2lxa^2(b^2-c^2)$

$$+a^2(b^2-c^2)-(a^2-c^2)y^2 = 0. \qquad (6)$$

The equation of the required developable is given by

$$F(l)=0, F'(l)=0.$$

Now, $F'(l)=0$ gives

$$2a^2l[a^2y^2+(b^2-c^2)x^2]-2xa^2(b^2-c^2)=0$$

or $\quad l = \dfrac{x(b^2-c^2)}{a^2y^2+(b^2-c^2)x^2} \qquad (7)$

or $\quad l[a^2y^2+(b^2-c^2)x^2] = x(b^2-c^2). \qquad (8)$

Using (8), (6) gives

$$lxa^2(b^2-c^2)-2lxa^2(b^2-c^2)+a^2(b^2-c^2)-(a^2-c^2)y^2=0$$

or $\quad -lxa^2(b^2-c^2)+a^2(b^2-c^2)-(a^2-c^2)y^2=0$

or $\quad lx = \dfrac{a^2(b^2-c^2)-(a^2-c^2)y^2}{a^2(b^2-c^2)}. \qquad (9)$

Multiplying (7) by x, we get

$$lx = \frac{x^2(b^2-c^2)}{a^2y^2+(b^2-c^2)x^2}. \qquad (10)$$

From (9) and (10), we get

$$\frac{a^2(b^2-c^2)-(a^2-c^2)y^2}{a^2(b^2-c^2)} = \frac{x^2(b^2-c^2)}{a^2y^2+(b^2-c^2)x^2}$$

or $\quad 1-\dfrac{(a^2-c^2)y^2}{a^2(b^2-c^2)} = \dfrac{x^2(b^2-c^2)}{a^2y^2+(b^2-c^2)x^2}$

or $\quad \dfrac{(a^2-c^2)y^2}{a^2(b^2-c^2)} = 1-\dfrac{x^2(b^2-c^2)}{a^2y^2+(b^2-c^2)x^2}$

or $\quad \dfrac{(a^2-c^2)y^2}{a^2(b^2-c^2)} = \dfrac{a^2y^2}{a^2y^2+(b^2-c^2)x^2}$

or $\quad \dfrac{(a^2-c^2)}{a^2(b^2-c^2)} = \dfrac{a^2}{a^2y^2+(b^2-c^2)x^2}$

or $\quad a^2y^2(a^2-c^2)+x^2(b^2-c^2)(a^2-c^2) = a^4(b^2-c^2)$

or $\quad \dfrac{y^2}{b^2-c^2}+\dfrac{x^2}{a^2} = \dfrac{a^2}{a^2-c^2}$

or $$\frac{x^2}{a^2}+\frac{y^2}{b^2-c^2}=\frac{a^2}{a^2-c^2}.$$

Example 9. A developable surface is drawn through the curves $\lambda^2x^2+y^2=\lambda^2, z=c\,; x^2+y^2=1, z=-c$. Show that its section by the plane $z=0$ is given by $2x=\sin\alpha+\sin\beta,\ 2y=\cos\alpha+\lambda\cos\beta$, where $\tan\alpha=\lambda\tan\beta$.

Solution. Equation of any line tangent to the ellipse $\lambda^2x^2+y^2=\lambda^2, z=c$ is

$$z=c,\ y=mx+\sqrt{(m^2+\lambda^2)}.$$

Therefore equation of any plane which touches the ellipse $\lambda^2x^2+y^2=\lambda^2, z=c$ is

$$y-\left[mx+\sqrt{(m^2+\lambda^2)}\right]+\mu\,(z-c)=0$$

or $$y=mx+\sqrt{(m^2+\lambda^2)}-\mu\,(z-c). \tag{1}$$

If the plane (1) touches the circle $x^2+y^2=1, z=-c$, then the straight line

$$y=mx+\sqrt{(m^2+\lambda^2)}+2\mu c \tag{2}$$

touches the circle

$$x^2+y^2=1. \tag{3}$$

Therefore

$$\frac{\sqrt{(m^2+\lambda^2)}+2\mu c}{\sqrt{(1+m^2)}}=1$$

$$\left[\because \text{The condition for the straight line } Y=MZ+C \text{ to touch the circle } X^2+Y^2=r^2 \text{ is } \frac{C}{\sqrt{(1+m^2)}}=r\right]$$

or $$2\mu c=\sqrt{(1+m^2)}-\sqrt{(m^2+\lambda^2)}$$

or $$\mu=\frac{\sqrt{(1+m^2)}-\sqrt{(m^2+\lambda^2)}}{2c}. \tag{4}$$

Substituting the value of μ from (4) in (1), we get

$$y=mx+\sqrt{(m^2+\lambda^2)}+\frac{\left[\sqrt{(m^2+\lambda^2)}-\sqrt{(1+m^2)}\right](z-c)}{2c}. \tag{5}$$

Since (5) contains only one parameter m, envelope of (5) is a developable.

If $z = 0$, then it reduces to

$$F(m) \equiv y - mx - \frac{\left[\sqrt{(m^2+\lambda^2)} + \sqrt{(1+m^2)}\right]}{2} = 0. \tag{6}$$

The equation of the required envelope is given by

$$F(m) = 0, \; F'(m) = 0.$$

Now, $F'(m) = 0$ gives

$$-x - \frac{m}{2\sqrt{(m^2+\lambda^2)}} - \frac{m}{2\sqrt{(1+m^2)}} = 0$$

or $$x + \frac{m}{2\sqrt{(m^2+\lambda^2)}} + \frac{m}{2\sqrt{(1+m^2)}} = 0. \tag{7}$$

Now, we put

$$\tan\alpha = \lambda \tan\beta = -m. \tag{8}$$

Then

$$\sin\alpha = -\frac{m}{\sqrt{(1+m^2)}}, \; \sin\beta = -\frac{m}{\sqrt{(m^2+\lambda^2)}} \tag{9}$$

and $$\sec\alpha = \sqrt{(1+m^2)}, \; \sec\beta = \frac{\sqrt{(m^2+\lambda^2)}}{\lambda}. \tag{10}$$

Using (9), (7) gives

$$x - \frac{1}{2}\sin\beta - \frac{1}{2}\sin\alpha = 0$$

or $$2x = \sin\alpha + \sin\beta. \tag{11}$$

Using (8), (10) and (11), (6) gives

$$y + \tan\alpha\left(\frac{\sin\alpha + \sin\beta}{2}\right) - \frac{1}{2}\lambda\sec\beta - \frac{1}{2}\sec\alpha = 0$$

or
$$\begin{aligned}
2y &= -\tan\alpha\,(\sin\alpha + \sin\beta) + (\lambda\sec\beta + \sec\alpha) \\
&= (\sec\alpha - \tan\alpha\sin\alpha) + (\lambda\sec\beta - \tan\alpha\sin\beta) \\
&= (\sec\alpha - \tan\alpha\sin\alpha) + (\lambda\sec\beta - \lambda\tan\beta\sin\beta) \qquad \text{[using (8)]} \\
&= \left(\frac{1}{\cos\alpha} - \frac{\sin^2\alpha}{\cos\alpha}\right) + \lambda\left(\frac{1}{\cos\beta} - \frac{\sin^2\beta}{\cos\beta}\right) \\
&= \left(\frac{1-\sin^2\alpha}{\cos\alpha}\right) + \lambda\left(\frac{1-\sin^2\beta}{\cos\beta}\right) \\
&= \left(\frac{\cos^2\alpha}{\cos\alpha}\right) + \lambda\left(\frac{\cos^2\beta}{\cos\beta}\right)
\end{aligned}$$

i.e., $2y = \cos\alpha + \lambda\cos\beta.$ (12)

Equations (11) and (12) give the required solution.

Example 10. Find the osculating developable of the circular helix.

Solution. Let the equation of the circular helix be

$$\mathbf{r} = (a\cos\theta, a\sin\theta, c\theta). \quad (1)$$

Differentiating (1) with respect to θ, we get

$$\frac{d\mathbf{r}}{d\theta} = (-a\sin\theta, a\cos\theta, c). \quad (2)$$

We know that $\frac{d\mathbf{r}}{d\theta}$ is a vector along the tangent to the curve (1).

Also, we know that the generators of the osculating developable of a space curve are the tangents to the curve.

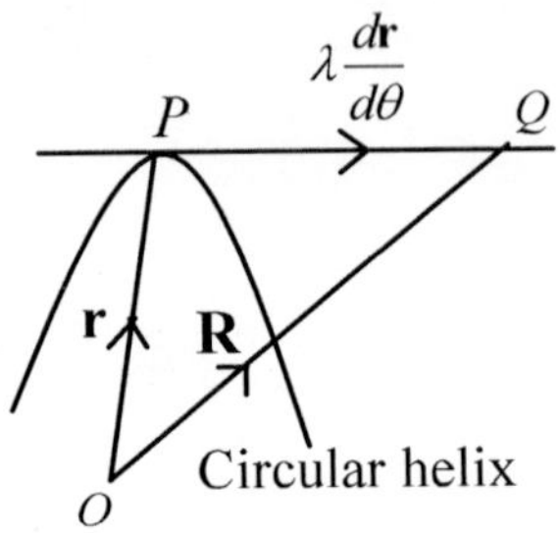

Therefore if $\mathbf{R}$ is the position vector of any current point on the developable, then equation of the developable can be written as

$$\mathbf{R} = \mathbf{r} + \lambda\frac{d\mathbf{r}}{d\theta}$$

or $$(X, Y, Z) = (a\cos\theta, a\sin\theta, c\theta) + \lambda(-a\sin\theta, a\cos\theta, c)$$

[using (1) and (2)]

or $$X = a(\cos\theta - \lambda\sin\theta), Y = a(\sin\theta + \lambda\cos\theta), Z = c(\theta + \lambda).$$

This is the required equation of the developable surface, where λ and θ are parameters.

Example 11. Prove that the rectifying developable of a curve is the polar developable of its involutes and conversely.

Solution. Let $\mathbf{r} = \mathbf{r}(s)$ be the equation of the given curve C and let C_1 denote the involute of C.

Let the quantities of C_1 be distinguished from those of C by using the suffix unity. Then, the position vector $\mathbf{r}_1$ of any point P_1 on C_1 is given by

$$\mathbf{r}_1 = \mathbf{r} + (c - s)\,\mathbf{t}, \quad (1)$$

where c is a scalar constant and s is the arc length.

The rectifying developable of C is the envelope of rectifying planes of C given by

$$(\mathbf{R}-\mathbf{r})\,.\,\mathbf{n}=0, \tag{2}$$

where $\mathbf{R}$ is the position vector of any current point on the rectifying plane of C.

The polar developable of C_1 is the envelope of normal planes of C_1 given by

$$(\mathbf{R}-\mathbf{r}_1)\,.\,\mathbf{t}_1=0, \tag{3}$$

where $\mathbf{R}$ is the position vector of any current point on the normal plane of C_1.

We have to prove that (2) and (3) are the same.

Differentiating (1) with respect to s, we get

$$\mathbf{t}_1\frac{ds_1}{ds}=\mathbf{t}+(-1)\,\mathbf{t}+(c-s)\,\kappa\mathbf{n}$$

or $$\mathbf{t}_1\frac{ds_1}{ds}=(c-s)\,\kappa\mathbf{n}. \tag{4}$$

Squaring (4), we get

$$\left(\frac{ds_1}{ds}\right)^2=(c-s)^2\,\kappa^2.$$

$$\therefore \quad \frac{ds_1}{ds}=\kappa\,(c-s). \tag{5}$$

Substituting the value of $\frac{ds_1}{ds}$ from (5) in (4), we get

$$\mathbf{t}_1=\mathbf{n}. \tag{6}$$

Substituting the values of $\mathbf{r}$ and $\mathbf{n}$ from (1) and (6) in (2), we get

$$[\mathbf{R}-(\mathbf{r}_1-(c-s)\,\mathbf{t})]\,.\,\mathbf{t}_1=0$$

or $$[(\mathbf{R}-\mathbf{r}_1)-(c-s)\,\mathbf{t})]\,.\,\mathbf{t}_1=0$$

or $$(\mathbf{R}-\mathbf{r}_1)\,.\,\mathbf{t}_1-(c-s)\,\mathbf{t}\,.\,\mathbf{t}_1=0$$

or $$(\mathbf{R}-\mathbf{r}_1)\,.\,\mathbf{t}_1=0, \qquad [\because\ \mathbf{t}\,.\,\mathbf{t}_1=\mathbf{t}\,.\,\mathbf{n}=0]$$

which is the same as (3).

Therefore the rectifying developable of C is the polar developable of C_1.

Conversely, substituting the values of $\mathbf{r}_1$ and $\mathbf{t}_1$ from (1) and (6) in (3), we get

$$[\mathbf{R}-(\mathbf{r}+(c-s)\,\mathbf{t})]\,.\,\mathbf{n}=0$$

or $$[(\mathbf{R}-\mathbf{r})-(c-s)\,\mathbf{t})]\,.\,\mathbf{n}=0$$

or $$(\mathbf{R}-\mathbf{r})\,.\,\mathbf{n}-(c-s)\,\mathbf{t}\,.\,\mathbf{n}=0$$

or $$(\mathbf{R}-\mathbf{r})\,.\,\mathbf{n}=0,$$

which is the same as (2).

Therefore the polar developable of C_1 is the rectifying developable of C.

Example 12. Show that the radius of curvature of the edge of regression of rectifying developable is equal to

$$\operatorname{cosec}\phi \frac{d}{d\phi}\left(\sin^2\phi \frac{ds}{d\phi}\right),$$

where $\tan\phi = \dfrac{\sigma}{\rho}$, and that the radius of torsion is equal to

$$-\rho \frac{d}{ds}\left(\sin^2\phi \frac{ds}{d\phi}\right).$$

Solution. Let the equation of the space curve be given by

$$\mathbf{r} = \mathbf{r}(s), \tag{1}$$

where s denotes the arc length measured along the space curve.

The equation of the rectifying plane of (1) at a point $P(\mathbf{r}(s))$ is given by

$$(\mathbf{R} - \mathbf{r}) \cdot \mathbf{n} = 0, \tag{2}$$

where $\mathbf{R}$ is the position vector of any current point on the rectifying plane, and $\mathbf{r}$ and $\mathbf{n}$ are functions of s.

Clearly, (2) represents a one parameter family of planes, the parameter being s. The envelope of (2) is the rectifying developable of (1).

Differentiating (2) with respect to s, we get

$$-\mathbf{t} \cdot \mathbf{n} + (\mathbf{R} - \mathbf{r}) \cdot (\tau \mathbf{b} - \kappa \mathbf{t}) = 0$$

or $\quad (\mathbf{R} - \mathbf{r}) \cdot (\tau \mathbf{b} - \kappa \mathbf{t}) = 0 \qquad [\because \mathbf{t} \cdot \mathbf{n} = 0]$

or $\quad (\mathbf{R} - \mathbf{r}) \cdot [(\kappa \cot\phi)\, \mathbf{b} - \kappa \mathbf{t})] = 0 \qquad \left[\because \tan\phi = \dfrac{\sigma}{\rho} = \dfrac{\kappa}{\tau} \text{ given}\right]$

or $\quad (\mathbf{R} - \mathbf{r}) \cdot [(\cot\phi)\, \mathbf{b} - \mathbf{t})] = 0. \qquad (3)$

Again, differentiating (3) with respect to s, we get

$$-\mathbf{t} \cdot [(\cot\phi)\, \mathbf{b} - \mathbf{t})] +$$

$$(\mathbf{R} - \mathbf{r}) \cdot \left[-\operatorname{cosec}^2\phi \frac{d\phi}{ds} \mathbf{b} + \cot\phi\,(-\tau \mathbf{n}) - \kappa \mathbf{n}\right] = 0 \tag{4}$$

or $\quad 1 + (\mathbf{R} - \mathbf{r}) \cdot \left(-\operatorname{cosec}^2\phi \dfrac{d\phi}{ds} \mathbf{b}\right) = 0 \quad$ [using (2) and $\mathbf{t} \cdot \mathbf{b} = 0, \mathbf{t} \cdot \mathbf{t} = 1$]

or $$(\mathbf{R}-\mathbf{r}).\ \mathbf{b}=\sin^2\phi\frac{ds}{d\phi}=\lambda\,(\text{say}), \tag{5}$$

where
$$\frac{1}{\lambda}=\operatorname{cosec}^2\phi\frac{d\phi}{ds}=-\frac{d}{ds}(\cot\phi). \tag{6}$$

The edge of regression is given by (2), (3) and (4) [or (5)].

From (2) and (3), we get

$$(\mathbf{R}-\mathbf{r}).\mathbf{n}=0 \text{ and } (\mathbf{R}-\mathbf{r}).[(\cot\phi)\,\mathbf{b}-\ \mathbf{t})]=0$$

$\Rightarrow$ $\mathbf{R}-\mathbf{r}$ is perpendicular to each of $\mathbf{n}$ and $(\cot\phi)\,\mathbf{b}-\ \mathbf{t}$.

$\Rightarrow$ $\mathbf{R}-\mathbf{r}$ is parallel to $\mathbf{n}\times[(\cot\phi)\,\mathbf{b}-\ \mathbf{t})]$.

$\Rightarrow$ $\mathbf{R}-\mathbf{r}$ is parallel to $(\cot\phi)\,\mathbf{t}+\ \mathbf{b}$.

$\Rightarrow$ There exists some scalar μ such that
$$\mathbf{R}-\mathbf{r}=\mu\,[(\cot\phi)\,\mathbf{t}+\mathbf{b}]. \tag{7}$$

Substituting the value of $\mathbf{R}-\mathbf{r}$ from (7) in (5), we get

$$\mu\,[(\cot\phi)\,\mathbf{t}+\mathbf{b}].\mathbf{b}=\lambda$$

$$\Rightarrow\quad \mu=\lambda. \tag{8}$$

Substituting the value of μ from (8) in (7), we get

$$\mathbf{R}-\mathbf{r}=\lambda\,[(\cot\phi)\,\mathbf{t}+\mathbf{b}]$$

$$\Rightarrow\quad \mathbf{R}=\mathbf{r}+\lambda\,[(\cot\phi)\,\mathbf{t}+\mathbf{b}], \tag{9}$$

which gives the position vector of any current point on the edge of regression.

Now, we have to find the curvature and torsion of the locus of **R**. Let the quantities belonging to the locus of **R** be distinguished from those of the original curve by using the suffix unity.

Differentiating (9) with respect to s, we get

$$\mathbf{t}_1\frac{ds_1}{ds}=\mathbf{t}+\frac{d\lambda}{ds}[(\cot\phi)\,\mathbf{t}+\mathbf{b}]+\lambda\left[-\operatorname{cosec}^2\phi\frac{d\phi}{ds}\mathbf{t}+(\cot\phi)(\kappa\mathbf{n})-\tau\mathbf{n}\right]$$

or $$\mathbf{t}_1\frac{ds_1}{ds}=\mathbf{t}+\frac{d\lambda}{ds}[(\cot\phi)\,\mathbf{t}+\mathbf{b}]+\lambda\left[-\frac{1}{\lambda}\mathbf{t}+(\cot\phi)(\kappa\mathbf{n})-\tau\mathbf{n}\right]$$

[using (6)]

or $$\mathbf{t}_1\frac{ds_1}{ds}=[(\cot\phi)\,\mathbf{t}+\mathbf{b}]\frac{d\lambda}{ds}. \tag{10}$$

$$\left[\because \cot\phi=\frac{\rho}{\sigma}=\frac{\tau}{\kappa}\text{ given }\right]$$

Squaring (10), we get

$$\left(\frac{ds_1}{ds}\right)^2 = (\cot^2\phi + 1)\left(\frac{d\lambda}{ds}\right)^2.$$

$$\therefore \quad \frac{ds_1}{ds} = \operatorname{cosec}\phi\left(\frac{d\lambda}{ds}\right). \tag{11}$$

Using (11), (10) gives

$$\mathbf{t}_1 = (\cos\phi)\,\mathbf{t} + (\sin\phi)\,\mathbf{b}. \tag{12}$$

Differentiating (12) with respect to s, we get

$$\kappa_1\mathbf{n}_1\frac{ds_1}{ds} = \left(-\sin\phi\frac{d\phi}{ds}\right)\mathbf{t} + (\cos\phi)\,\kappa\mathbf{n} + \left(\cos\phi\frac{d\phi}{ds}\right)\mathbf{b} + (\sin\phi)(-\tau\mathbf{n})$$

or $$\kappa_1\mathbf{n}_1\frac{ds_1}{ds} = [(-\sin\phi)\,\mathbf{t} + (\cos\phi)\,\mathbf{b}]\frac{d\phi}{ds} + (\kappa\cos\phi - \tau\sin\phi)\,\mathbf{n}$$

or $$\kappa_1\mathbf{n}_1\frac{ds_1}{ds} = [(-\sin\phi)\,\mathbf{t} + (\cos\phi)\,\mathbf{b}]\frac{d\phi}{ds}. \tag{13}$$

$[\because \cot\phi = \rho/\sigma = \tau/\kappa \text{ given}]$

Squaring (13), we get

$$\left(\kappa_1\frac{ds_1}{ds}\right)^2 = \left(\frac{d\phi}{ds}\right)^2.$$

$$\therefore \quad \kappa_1\frac{ds_1}{ds} = \frac{d\phi}{ds} \tag{14}$$

Using (14), (13) reduces to

$$\mathbf{n}_1 = (-\sin\phi)\,\mathbf{t} + (\cos\phi)\,\mathbf{b}. \tag{15}$$

Now, $$\rho_1 = \frac{1}{\kappa_1}$$

$$= \frac{ds_1}{ds}\frac{ds}{d\phi} \qquad \text{[using (14)]}$$

$$= \left[\operatorname{cosec}\phi\left(\frac{d\lambda}{ds}\right)\right]\frac{ds}{d\phi} \qquad \text{[using (11)]}$$

$$= \left[\operatorname{cosec}\phi\frac{d}{ds}\left(\sin^2\phi\frac{ds}{d\phi}\right)\right]\frac{ds}{d\phi} \qquad \text{[using (5)]}$$

i.e., $$\rho_1 = \operatorname{cosec}\phi\frac{d}{d\phi}\left(\sin^2\phi\frac{ds}{d\phi}\right). \tag{16}$$

This gives the radius of curvature of the edge of regression of the rectifying developable.

Taking the cross product of (12) and (15), we get

$$\mathbf{t}_1 \times \mathbf{n}_1 = [(\cos\phi)\,\mathbf{t} + (\sin\phi)\,\mathbf{b}] \times [(-\sin\phi)\,\mathbf{t} + (\cos\phi)\,\mathbf{b}]$$

or $$\mathbf{b}_1 = -\mathbf{n}. \tag{17}$$

Differentiating (17) with respect to s, we get

$$-\tau_1 \mathbf{n}_1 \frac{ds_1}{ds} = -(\tau\mathbf{b} - \kappa\mathbf{t})$$

or $$\tau_1 \mathbf{n}_1 \frac{ds_1}{ds} = \tau\mathbf{b} - \kappa\mathbf{t}$$

or $$\tau_1 \mathbf{n}_1 \frac{ds_1}{ds} = (\kappa\cot\phi)\,\mathbf{b} - \kappa\mathbf{t} \qquad \left[\because \cot\phi = \frac{\rho}{\sigma} = \frac{\tau}{\kappa} \text{ given}\right]$$

or $$\tau_1 \mathbf{n}_1 \frac{ds_1}{ds} = \kappa[(\cot\phi)\,\mathbf{b} - \mathbf{t}]. \tag{18}$$

Squaring (18), we get

$$\left(\tau_1 \frac{ds_1}{ds}\right)^2 = \kappa^2 (\cot^2\phi + 1).$$

$\therefore$ $$\tau_1 \frac{ds_1}{ds} = \kappa \operatorname{cosec}\phi$$

or $$\frac{1}{\tau_1}\frac{ds}{ds_1} = \frac{1}{\kappa}\sin\phi$$

or $$\sigma_1 \frac{ds}{ds_1} = \rho\sin\phi$$

or $$\sigma_1 = \rho\sin\phi \frac{ds_1}{ds}$$

$$= \rho\sin\phi\left[\operatorname{cosec}\phi\left(\frac{d\lambda}{ds}\right)\right] \qquad \text{[using (11)]}$$

$$= \rho\frac{d\lambda}{ds}$$

$$= \rho\frac{d}{ds}\left(\sin^2\phi\frac{ds}{d\phi}\right). \qquad \text{[using (6)]}$$

Therefore we can choose the direction of $\mathbf{n}_1$ such that

$$\sigma_1 = -\rho\frac{d}{ds}\left(\sin^2\phi\frac{ds}{d\phi}\right) \tag{19}$$

This gives the radius of torsion of the edge of regression of the rectifying developable.

Exercises

1. Find the equations to the edge of regression of the developable $y = tx - t^3, z = t^3 y - t^6$. [**Ans.** $x = 3t^2, y = 2t^3, z = t^6$]

2. Find the equation of the developable surface whose generating lines pass through the curves $y^2 = 4ax, z = 0; x^2 = 4ay, z = c$ and show that its edge of regression is given by $cx^2 - 3ayz = 0 = cy^2 - 3ax(c - z)$.

$$\left[\begin{array}{l}\textbf{Ans.}\ \text{The developable is } m \text{ - eliminant of the equations} \\ (am^3 + a)\frac{z}{c} + my - m^2 x - a = 0, (3am^2)\frac{z}{c} + y - 2mx = 0\end{array}\right]$$

3. Show that two cones pass through the curves $x^2 + y^2 = 4a^2, z = 0; x = 0, y^2 = 4a(z + a)$ and their vertices are the points $(2a, 0, -2a), (-2a, 0, -2a)$.

4. Prove that the edge of regression of the developable that passes through the circles $z = 0, x^2 + y^2 = a^2; x = 0, y^2 + z^2 = b^2$ lies on the cylinder $\left(\frac{x}{a^2}\right)^{2/3} - \left(\frac{z}{b^2}\right)^{2/3} = \left(\frac{1}{a^2} - \frac{1}{b^2}\right)^{1/3}$.

5. Prove that the developable surface that envelopes the sphere $x^2 + y^2 + z^2 = c^2$ and the hyperboloid $\frac{x^2}{a^2} + \frac{y^2}{b^2} - \frac{z^2}{c^2} = 1$ meets the plane $y = 0$ in the conic

$$\frac{x^2}{b^2 - a^2} + \frac{z^2}{b^2 + c^2} = \frac{c^2}{b^2 - c^2}.$$

6. Prove that the section by the xy-plane of the developable generated by the tangents to the curve $x^2 + y^2 + z^2 = r^2, \frac{x^2}{a^2} + \frac{y^2}{b^2} = \frac{z^2}{c^2}$ is given by

$$z = 0, \frac{a^2(a^2 + c^2)}{x^2} + \frac{b^2(b^2 + c^2)}{y^2} = \frac{(a^2 - b^2)^2}{r^2}.$$

2.10 ENVELOPE OF TWO PARAMETER FAMILY OF SURFACES

Definition. The equation of a two parameter family of surfaces is given by

$$F(a,b) \equiv F(x,y,z,a,b) = 0, \tag{1}$$

where a and b are two independent parameters.

The curve of intersection of two members of the family (1) for the parametric values a, b and $a+da, b+db$ is given by

$$F(a,b) = 0, \quad F(a+da, b+db) = 0. \tag{2}$$

But, by Taylor's expansion of two variables,

$$F(a+da, b+db) = 0 \text{ gives}$$

$$F(a,b) + \left(\frac{\partial F}{\partial a}da + \frac{\partial F}{\partial b}db\right)$$

$$+\frac{1}{2}\left(\frac{\partial^2 F}{\partial a^2}da^2 + 2\frac{\partial^2 F}{\partial a\,\partial b}da\,db + \frac{\partial^2 F}{\partial b^2}db^2\right) + \ldots = 0$$

or $$F(a,b) + \left(\frac{\partial F}{\partial a}da + \frac{\partial F}{\partial b}db\right) = 0.$$

[neglecting higher powers of da, db]

Therefore, the curve of intersection given by (2) is also given by the equations

$$F(a,b) = 0, \quad F(a,b) + \left(\frac{\partial F}{\partial a}da + \frac{\partial F}{\partial b}db\right) = 0,$$

or $$F(a,b) = 0, \quad \frac{\partial F}{\partial a}da + \frac{\partial F}{\partial b}db = 0. \tag{3}$$

Since da, db are independent, the curve of intersection given by (3) is also given by the equations

$$F(a,b) = 0, \quad \frac{\partial}{\partial a}F(a,b) = 0, \quad \frac{\partial}{\partial b}F(a,b) = 0. \tag{4}$$

The points of intersection given by (4) are known as the characteristic points of the family (1) and the locus of the characteristic points is known as the envelope of the family of surfaces (1).

The equation of the envelope of the family (1) is obtained by eliminating the parameters a and b from equations (4).

Theorem. The envelope of a two parameter family of surfaces touches each surface at the characteristic points.

Proof. Let a two parameter family of surfaces be given by

$$F(a,b) \equiv F(x,y,z,a,b) = 0, \tag{1}$$

where a and b are two independent parameters.

The characteristic points of this family (1) are given by

$$F(a,b) = 0, \quad \frac{\partial}{\partial a}F(a,b) = 0, \quad \frac{\partial}{\partial b}F(a,b) = 0. \tag{2}$$

The locus of the characteristic points given by (2) is known as the envelope of the family (1).

Thus, the equation of the envelope of the family of surfaces (1) is obtained by eliminating the parameters a and b from equations (2).

Take any member

$$F(a,b) = 0 \tag{3}$$

of the family (1).

We have to prove that the envelope of the family (1) touches the member (3) of (1) at the characteristic points of (3) given by (2).

The normal to (3) at P (x, y, z) is parallel to the vector

$$\left(\frac{\partial F}{\partial x}, \frac{\partial F}{\partial y}, \frac{\partial F}{\partial z}\right). \tag{4}$$

The envelope of (1) is given by $F(a,b) = 0$, provided a, b are functions of x, y, z given by

$$\frac{\partial}{\partial a}F(a,b) = 0, \quad \frac{\partial}{\partial b}F(a,b) = 0. \tag{5}$$

Therefore, the normal to the envelope at P (x, y, z) is parallel to the vector

$$\left(\frac{\partial F}{\partial x} + \frac{\partial F}{\partial a}\frac{\partial a}{\partial x} + \frac{\partial F}{\partial b}\frac{\partial b}{\partial x}, \quad \frac{\partial F}{\partial y} + \frac{\partial F}{\partial a}\frac{\partial a}{\partial y} + \frac{\partial F}{\partial b}\frac{\partial b}{\partial y}, \quad \frac{\partial F}{\partial z} + \frac{\partial F}{\partial a}\frac{\partial a}{\partial z} + \frac{\partial F}{\partial b}\frac{\partial b}{\partial z}\right)$$

i.e., $$\left(\frac{\partial F}{\partial x}, \frac{\partial F}{\partial y}, \frac{\partial F}{\partial z}\right). \tag{6}$$

[using (5)]

Clearly, (4) and (6) represent the same vector. Hence it follows that the surface (3) and the envelope have the same normal, and therefore the same tangent plane at $P(x,y,z)$. This implies that the envelope of the family (1) touches the surface (3) at $P(x,y,z)$.

But, $P(x,y,z)$ is an arbitrary characteristic point. Therefore the envelope of the family (1) touches the surface (3) at the characteristic points.

Solved Examples

Example 1. Show that the envelope of the plane

$$\frac{x}{a}\cos\theta\sin\phi+\frac{y}{b}\sin\theta\sin\phi+\frac{z}{c}\cos\phi=1,$$

where θ, ϕ are independent parameters, is the ellipsoid

$$\frac{x^2}{a^2}+\frac{y^2}{b^2}+\frac{z^2}{c^2}=1.$$

Solution. The equation of the given family of planes is

$$F(\theta,\phi)\equiv\frac{x}{a}\cos\theta\sin\phi+\frac{y}{b}\sin\theta\sin\phi+\frac{z}{c}\cos\phi-1=0, \tag{1}$$

where θ, ϕ are independent parameters.
Differentiating (1) partially with respect to θ, we get

$$\frac{\partial F}{\partial\theta}\equiv-\frac{x}{a}\sin\theta\sin\phi+\frac{y}{b}\cos\theta\sin\phi=0. \tag{2}$$

Again, differentiating (1) partially with respect to ϕ, we get

$$\frac{\partial F}{\partial\phi}\equiv\frac{x}{a}\cos\theta\cos\phi+\frac{y}{b}\sin\theta\cos\phi-\frac{z}{c}\sin\phi=0. \tag{3}$$

The equation of the envelope is obtained by eliminating θ and ϕ from (1), (2) and (3) as under.
From (1), we have

$$\left(\frac{x}{a}\cos\theta+\frac{y}{b}\sin\theta\right)\sin\phi=1-\frac{z}{c}\cos\phi. \tag{4}$$

From (3), we have

$$\left(\frac{x}{a}\cos\theta+\frac{y}{b}\sin\theta\right)\cos\phi=\frac{z}{c}\sin\phi. \tag{5}$$

From (4) and (5), we get

$$\frac{\sin\phi}{\cos\phi}=\frac{1-\frac{z}{c}\cos\phi}{\frac{z}{c}\sin\phi}$$

or $$\cos\phi-\frac{z}{c}\cos^2\phi=\frac{z}{c}\sin^2\phi$$

or $$\cos\phi=\frac{z}{c}(\cos^2\phi+\sin^2\phi)$$

or $$\cos\phi=\frac{z}{c},\ \sin\phi=\frac{\sqrt{c^2-z^2}}{c}. \tag{6}$$

Eliminating ϕ from (5) and (6), we get

$$\frac{x}{a}\cos\theta+\frac{y}{b}\sin\theta=\frac{\sqrt{c^2-z^2}}{c}. \tag{7}$$

From (2), we have

$$\frac{x}{a}\sin\theta-\frac{y}{b}\cos\theta=0. \tag{8}$$

Squaring and adding (7) and (8), we get

$$\frac{x^2}{a^2}+\frac{y^2}{b^2}=\frac{c^2-z^2}{c^2}$$

or $$\frac{x^2}{a^2}+\frac{y^2}{b^2}+\frac{z^2}{c^2}=1.$$

This is the required envelope of the family (1).

Example 2. Prove that the envelope of the plane

$$(u-v)\,bcx+(1+uv)\,cay+(1-uv)\,abz=(u+v)\,abc,$$

where u, v are parameters, is the hyperboloid

$$\frac{x^2}{a^2}+\frac{y^2}{b^2}-\frac{z^2}{c^2}=1.$$

Solution. The equation of the given plane is

$$F(u,v)\equiv(u-v)\,bcx+(1+uv)\,cay+(1-uv)\,abz$$
$$-(u+v)\,abc=0, \tag{1}$$

where u, v are parameters.

Differentiating (1) partially with respect to u, we get

$$\frac{\partial F}{\partial u}\equiv bcx+vcay-vabz-abc=0. \tag{2}$$

Again, differentiating (1) partially with respect to v, we get

$$\frac{\partial F}{\partial v}\equiv -bcx+ucay-uabz-abc=0. \tag{3}$$

The equation of the envelope is obtained by eliminating u and v from (1), (2) and (3) as under.

From (2), we have

$$bc\,(x-a)+va\,(cy-bz)=0$$

or $$v=-\frac{bc\,(x-a)}{a\,(cy-bz)}. \tag{4}$$

From (3), we have

$$-bc\,(x+a)+ua\,(cy-bz)=0$$

or $$u = \frac{bc\,(x+a)}{a\,(cy-bz)}. \tag{5}$$

From (4) and (5), we get

$$u - v = \frac{2bcx}{a\,(cy-bz)}, \tag{6}$$

$$u + v = \frac{2bc}{cy-bz}, \tag{7}$$

$$1 - uv = 1 + \frac{b^2c^2(x^2-a^2)}{a^2(cy-bz)^2}, \tag{8}$$

$$1 + uv = 1 - \frac{b^2c^2(x^2-a^2)}{a^2(cy-bz)^2}. \tag{9}$$

Using (6), (7), (8) and (9), (1) gives

$$\frac{2b^2c^2x^2}{a\,(cy-bz)} + \left[1 - \frac{b^2c^2(x^2-a^2)}{a^2(cy-bz)^2}\right] cay + \left[1 + \frac{b^2c^2(x^2-a^2)}{a^2(cy-bz)^2}\right] abz - \frac{2ab^2c^2}{cy-bz} = 0$$

or $$\frac{2b^2c^2x^2}{a\,(cy-bz)} + a\,(cy+bz) - \frac{b^2c^2(x^2-a^2)}{a\,(cy-bz)^2}(cy-bz) - \frac{2ab^2c^2}{cy-bz} = 0$$

or $$\frac{2b^2c^2x^2}{a\,(cy-bz)} + a\,(cy+bz) - \frac{b^2c^2(x^2-a^2)}{a\,(cy-bz)} - \frac{2ab^2c^2}{cy-bz} = 0$$

or $$2b^2c^2x^2 + a^2(c^2y^2 - b^2z^2) - b^2c^2(x^2-a^2) - 2a^2b^2c^2 = 0$$

or $$b^2c^2x^2 + a^2c^2y^2 - a^2b^2z^2 = a^2b^2c^2$$

or $$\frac{x^2}{a^2} + \frac{y^2}{b^2} - \frac{z^2}{c^2} = 1.$$

Example 3. Prove that the envelope of a plane which forms with the coordinate planes a tetrahedron of constant volume is a surface

$$xyz = \text{constant}.$$

Solution. Let the equation of the plane be

$$F(a, b) \equiv \frac{x}{a} + \frac{y}{b} + \frac{z}{c} - 1 = 0, \tag{1}$$

where a, b are independent parameters.

The equations of three coordinate planes are

$$x = 0, \tag{2}$$

$$y = 0, \tag{3}$$

$$z = 0. \tag{4}$$

The vertices of the tetrahedron are obtained by solving equations (1), (2), (3) and (4), taking three at a time.

Therefore, the vertices of the tetrahedron are the points O (0, 0, 0), A (a, 0, 0), B (0, b, 0) and C (0, 0, c).

$\therefore$ Volume of the tetrahedron $OABC$ is given by

$$\frac{1}{6}\begin{vmatrix} 0 & 0 & 0 & 1 \\ a & 0 & 0 & 1 \\ 0 & b & 0 & 1 \\ 0 & 0 & c & 1 \end{vmatrix} = -\frac{1}{6}abc.$$

But, by supposition, the volume of the tetrahedron $OABC$ is constant. Therefore,

$$-\frac{1}{6}abc = \text{constant}$$

or $\quad abc = K$, where K is a constant

or $\quad f(a,b) \equiv abc - K = 0. \qquad (5)$

We have to find the envelope of the family of planes (1), where a, b, c are connected by (5).

Differentiating (1) with respect to a and b, we get

$$\frac{dF}{da} \equiv -\frac{x}{a^2} - \frac{z}{c^2}\frac{dc}{da} = 0 \qquad (6)$$

and $$\frac{dF}{db} \equiv -\frac{y}{b^2} - \frac{z}{c^2}\frac{dc}{db} = 0. \qquad (7)$$

Differentiating (5) with respect to a and b, we get

$$\frac{df}{da} \equiv bc + ab\frac{dc}{da} = 0 \qquad (8)$$

and $$\frac{df}{db} \equiv ac + ab\frac{dc}{db} = 0. \qquad (9)$$

The equation of the envelope is obtained by eliminating a, b, c from (1), (5), (6), (7), (8) and (9) as under.

From (6) and (8), we have

$$-\frac{xc^2}{za^2} = -\frac{c}{a}$$

or $$\frac{x}{a} = \frac{z}{c}. \qquad (10)$$

From (7) and (9), we have

$$-\frac{yc^2}{zb^2} = -\frac{c}{b}$$

or $$\frac{y}{b}=\frac{z}{c}. \tag{11}$$

Now, from (10) and (11), we have

$$\frac{x}{a}=\frac{y}{b}=\frac{z}{c}. \tag{12}$$

Using (12), (1) gives

$$\frac{x}{a}=\frac{y}{b}=\frac{z}{c}=\frac{1}{3}$$

$$\Rightarrow \quad a=3x, b=3y, c=3z. \tag{13}$$

Substituting he values of a, b, c from (13) in (5), we get

$$27xyz=K$$

or $$xyz=\frac{K}{27}, \text{ which is constant.}$$

Example 4. Prove that the envelope of a plane, the sum of the squares of whose intercepts on the axes is constant, is a surface

$$x^{2/3}+y^{2/3}+z^{2/3}=\text{constant.}$$

Solution. Let the equation of the plane be

$$F(a,b)\equiv\frac{x}{a}+\frac{y}{b}+\frac{z}{c}-1=0, \tag{1}$$

where a, b are independent parameters.

By supposition,

$$a^2+b^2+c^2=K, \text{ where } K \text{ is a constant}$$

$$\Rightarrow \quad f(a,b)\equiv a^2+b^2+c^2-K=0. \tag{2}$$

We have to find the envelope of the family of planes (1), where a, b, c are connected by (2).

Differentiating (1) with respect to a and b, we get

$$\frac{dF}{da}\equiv-\frac{x}{a^2}-\frac{z}{c^2}\frac{dc}{da}=0 \tag{3}$$

and $$\frac{dF}{db}\equiv-\frac{y}{b^2}-\frac{z}{c^2}\frac{dc}{db}=0. \tag{4}$$

Differentiating (2) with respect to a and b, we get

$$\frac{df}{da}\equiv 2a+2c\frac{dc}{da}=0 \tag{5}$$

and $$\frac{df}{db}\equiv 2b+2c\frac{dc}{db}=0. \tag{6}$$

The equation of the envelope is obtained by eliminating a, b, c from (1), (2), (3), (4), (5) and (6) as under.
From (3) and (5), we have

$$-\frac{xc^2}{za^2} = -\frac{a}{c}$$

or $$\frac{x}{a^3} = \frac{z}{c^3}. \tag{7}$$

From (4) and (6), we have

$$-\frac{yc^2}{zb^2} = -\frac{b}{c}$$

or $$\frac{y}{b^3} = \frac{z}{c^3}. \tag{8}$$

Now, from (7) and (8), we have

$$\frac{x}{a^3} = \frac{y}{b^3} = \frac{z}{c^3} = \lambda \text{ (let)}$$

$$\Rightarrow \quad x = \lambda a^3, y = \lambda b^3, z = \lambda c^3. \tag{9}$$

Substituting the values of x, y, z from (9) in (1), we get

$$\lambda (a^2 + b^2 + c^2) = 1$$

$$\Rightarrow \quad \lambda K = 1 \qquad \text{[using (2)]}$$

$$\Rightarrow \quad \lambda = \frac{1}{K}. \tag{10}$$

Substituting the value of λ from (10) in (9), we get

$$x = \frac{a^3}{K}, y = \frac{b^3}{K}, z = \frac{c^3}{K}.$$

$$\therefore \quad a = (Kx)^{1/3}, b = (Ky)^{1/3}, c = (Kz)^{1/3}. \tag{11}$$

Now, substituting the values of a, b, c from (11) in (2), we get

$$K^{2/3}(x^{2/3} + y^{2/3} + z^{2/3}) = K$$

$$\Rightarrow \quad x^{2/3} + y^{2/3} + z^{2/3} = K^{1/3}, \text{ which is constant.}$$

Example 5. Prove that the envelope of the surface $F(x, y, z, a, b, c) = 0$, where a, b, c are parameters connected by the equation $f(a, b, c) = 0$, is obtained by eliminating a, b, c from the equations

$$F = 0, \quad f = 0, \quad \frac{F_a}{f_a} = \frac{F_b}{f_b} = \frac{F_c}{f_c}.$$

Solution. The equation of the given family of surfaces is

$$F(x, y, z, a, b, c) = 0, \tag{1}$$

where a, b, c are parameters connected by the equation

$$f(a,b,c)=0. \tag{2}$$

It follows from (2) that any one of the variables a, b, c can be expressed in terms of the remaining two, and consequently, (1) represents a two parameter family of surfaces.

For definiteness, let c be expressed in terms of a, b. Then (1) represents a two parameter family of surfaces with a, b as the parameters of the family. Therefore, from (1), we have

$$F(a,b) \equiv F(x,y,z,a,b,c)=0. \tag{3}$$

Differentiating (3) with respect to a and b, we get

$$\frac{dF}{da} \equiv \frac{\partial F}{\partial a}+\frac{\partial F}{\partial c}\frac{dc}{da}=0 \tag{4}$$

and
$$\frac{dF}{db} \equiv \frac{\partial F}{\partial b}+\frac{\partial F}{\partial c}\frac{dc}{db}=0. \tag{5}$$

Differentiating (2) with respect to a and b, we get

$$\frac{df}{da} \equiv \frac{\partial f}{\partial a}+\frac{\partial f}{\partial c}\frac{dc}{da}=0 \tag{6}$$

and
$$\frac{df}{db} \equiv \frac{\partial f}{\partial b}+\frac{\partial f}{\partial c}\frac{dc}{db}=0. \tag{7}$$

From (4) and (6), we have

$$-\frac{\dfrac{\partial F}{\partial a}}{\dfrac{\partial F}{\partial c}}=-\frac{\dfrac{\partial f}{\partial a}}{\dfrac{\partial f}{\partial c}}$$

or
$$\frac{F_a}{F_c}=\frac{f_a}{f_c}$$

or
$$\frac{F_a}{f_a}=\frac{F_c}{f_c}. \tag{8}$$

From (5) and (7), we have

$$-\frac{\dfrac{\partial F}{\partial b}}{\dfrac{\partial F}{\partial c}}=-\frac{\dfrac{\partial f}{\partial b}}{\dfrac{\partial f}{\partial c}}$$

or
$$\frac{F_b}{F_c}=\frac{f_b}{f_c}$$

or
$$\frac{F_b}{f_b}=\frac{F_c}{f_c}. \tag{9}$$

Now, from (8) and (9), we have

$$\frac{F_a}{f_a} = \frac{F_b}{f_b} = \frac{F_c}{f_c}. \tag{10}$$

The required envelope is obtained by eliminating a, b, c from (2), (3) and (10).

Example 6. Prove that the envelope of the plane $lx + my + nz = p$, where $p^2 = a^2l^2 + b^2m^2 + c^2n^2$ is an ellipsoid.

Solution. Let us take

$$\frac{l}{p} = \lambda, \frac{m}{p} = \mu, \frac{n}{p} = \nu.$$

Then the equation of the given plane can be written as

$$F(\lambda, \mu, \nu) \equiv \lambda x + \mu y + \nu z - 1 = 0, \tag{1}$$

where the parameters λ, μ, ν are connected by the equation

$$f(\lambda, \mu, \nu) \equiv a^2\lambda^2 + b^2\mu^2 + c^2\nu^2 - 1 = 0. \tag{2}$$

For the envelope, we have

$$\frac{F_\lambda}{f_\lambda} = \frac{F_\mu}{f_\mu} = \frac{F_\nu}{f_\nu}$$

or $$\frac{x}{2a^2\lambda} = \frac{y}{2b^2\mu} = \frac{z}{2c^2\nu}$$

or $$\frac{x}{a^2\lambda} = \frac{y}{b^2\mu} = \frac{z}{c^2\nu} = K \text{ (let)}. \tag{3}$$

The equation of the envelope is obtained by eliminating λ, μ, ν from (1), (2) and (3) as under.

From (3), we have

$$x = Ka^2\lambda, y = Kb^2\mu, z = Kc^2\nu. \tag{4}$$

Substituting the values of x, y, z from (4) in (1), we get

$$K(a^2\lambda^2 + b^2\mu^2 + c^2\nu^2) = 1.$$

$$\therefore \quad K = 1. \tag{5}$$

[using (2)]

Substituting the values of K from (5) in (4), we get

$$x = a^2\lambda, y = b^2\mu, z = c^2\nu$$

or $$\lambda = \frac{x}{a^2}, \ \mu = \frac{y}{b^2}, \ \nu = \frac{z}{c^2}. \tag{6}$$

Now, substituting the values of λ, μ, ν from (6) in (2), we get

$$\frac{x^2}{a^2}+\frac{y^2}{b^2}+\frac{z^2}{c^2}=1.$$

This is the required envelope of the family (1).

Exercises

1. A plane makes intercepts a, b, c on the coordinate axes such that

$$\frac{1}{a^2}+\frac{1}{b^2}+\frac{1}{c^2}=\frac{1}{K^2}.$$

Prove that its envelope is a conicoid with equi-conjugate diameters along the axes.

2. From a point P on $a^2x^2+b^2y^2+c^2z^2=1$, perpendiculars PL, PM, PN are drawn to the coordinate planes. Find the envelope of the plane LNM. [**Ans.** $(ax)^{2/3}+(by)^{2/3}+(cz)^{2/3}=2^{2/3}$]

3. A tangent plane to the ellipsoid $\frac{x^2}{a^2}+\frac{y^2}{b^2}+\frac{z^2}{c^2}=1$ meets the axes in A, B, C. Show that the envelope of the sphere $OABC$ is

$$(ax)^{2/3}+(by)^{2/3}+(cz)^{2/3}=(x^2+y^2+z^2)^{2/3}.$$

4. Prove that the envelope of the surface $F(x,y,z,a,b,c,d)=0$, where a, b, c, d are parameters connected by the equation $f(a,b,c,d)=0$, is obtained by eliminating a, b, c, d from the equations

$$F=0,\ f=0,\ \frac{F_a}{f_a}=\frac{F_b}{f_b}=\frac{F_c}{f_c}=\frac{F_d}{f_d}.$$

3

Fundamental Forms and Directions on a Surface

3.1 FIRST FUNDAMENTAL FORM

Definition. Let the parametric equation of a surface in vector form be given by

$$\mathbf{r} = \mathbf{r}\,(u, v). \qquad (1)$$

We introduce

$$E = \mathbf{r}_1 . \mathbf{r}_1 = \mathbf{r}_1^{\,2}, \; F = \mathbf{r}_1 . \mathbf{r}_2, \; G = \mathbf{r}_2 . \mathbf{r}_2 = \mathbf{r}_2^{\,2}.$$

The quadratic differential form

$$E\,du^2 + 2F\,du\,dv + G\,dv^2$$

is known as the first fundamental form or the metric of the surface. The quantities E, F, G are known as the first fundamental coefficients or the fundamental magnitudes of the first order of the surface.

Geometrical Interpretation of the First Fundamental Form. Let the parametric equation of a surface in vector form be given by

$$\mathbf{r} = \mathbf{r}\,(u, v). \qquad (1)$$

Take two points P and Q on the surface corresponding to the parametric coordinates (u, v) and $(u + du, v + dv)$ respectively. Let $\mathbf{r}$ and $\mathbf{r} + d\mathbf{r}$ be the position vectors of P and Q respectively.

Let ds denote an elementary arc length on the surface joining P and Q. Since P and Q are adjacent points on the surface, we have

$$\text{arc } PQ = \text{chord } PQ$$

or $$ds = |\,d\mathbf{r}|$$

or $$ds^2 = d\mathbf{r}^2$$

or $$ds^2 = \left(\frac{\partial \mathbf{r}}{\partial u} du + \frac{\partial \mathbf{r}}{\partial v} dv \right)^2 = (\mathbf{r}_1 du + \mathbf{r}_2 dv)^2$$

or $$ds^2 = \mathbf{r}_1^{\,2} du^2 + 2\mathbf{r}_1 . \mathbf{r}_2\, du\, dv + \mathbf{r}_2^{\,2} dv^2$$

or $$ds^2 = E\,du^2 + 2F\,du\,dv + G\,dv^2, \qquad (2)$$

which gives the required geometrical interpretation.

Since the first fundamental form is mainly used for the calculation of lengths on the surface, it is called the metric of the surface.

If $u = u(t), v = v(t)$ are the equations of a curve lying on the surface, then the arc length s of the curve is related to the parameter t by

$$\left(\frac{ds}{dt}\right)^2 = E\left(\frac{du}{dt}\right)^2 + 2F\frac{du}{dt}\frac{dv}{dt} + G\left(\frac{dv}{dt}\right)^2.$$

On the parametric curve $u = \text{constant}$, we have $du = 0$ and hence the metric (2) reduces to $ds^2 = G\,dv^2$.

Similarly, on the parametric curve $v = \text{constant}$, we have $dv = 0$ and hence the metric (2) reduces to $ds^2 = E\,du^2$.

Relation between the Coefficients *E, F, G* and *H.*

If ω is the angle between the vectors $\mathbf{r}_1$ and $\mathbf{r}_2$, then, from the definition of the cross product of two vectors, we have

$$\mathbf{r}_1 \times \mathbf{r}_2 = |\mathbf{r}_1||\mathbf{r}_2|(\sin\omega)\,\mathbf{N},$$

where $\mathbf{N}$ is the unit normal vector to the surface at the point $\mathbf{r}$.

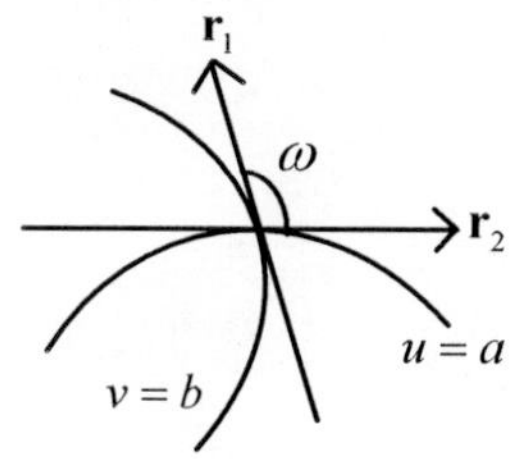

$$\therefore \quad (\mathbf{r}_1 \times \mathbf{r}_2)^2 = |\mathbf{r}_1|^2\,|\mathbf{r}_2|^2 \sin^2\omega \qquad [\because \mathbf{N}^2 = 1]$$

or $$|\mathbf{r}_1 \times \mathbf{r}_2|^2 = |\mathbf{r}_1|^2\,|\mathbf{r}_2|^2(1 - \cos^2\omega)$$

or $$H^2 = EG\left[1 - \left(\frac{\mathbf{r}_1 \cdot \mathbf{r}_2}{|\mathbf{r}_1||\mathbf{r}_2|}\right)^2\right] = EG\left(1 - \frac{F^2}{EG}\right)$$

or $$H^2 = EG - F^2.$$

Since H is a positive quantity, it follows that $EG - F^2$ is always a positive quantity and H is equal to the positive square root of $EG - F^2$.

Two Important Properties of the Metric.

Property 1. The metric or the first fundamental form is a positive definite quadratic form in *du, dv*.

Proof. We have

$$E\,du^2 + 2F\,du\,dv + G\,dv^2 = (\mathbf{r}_1 du + \mathbf{r}_2 dv)^2$$
$$= |\,\mathbf{r}_1 du + \mathbf{r}_2 dv\,|^2$$
$$\geq 0 \text{ for all real values of } du,\ dv.$$

Also,

$$E\,du^2 + 2F\,du\,dv + G\,dv^2 = 0 \Rightarrow |\,\mathbf{r}_1 du + \mathbf{r}_2 dv\,|^2 = 0$$
$$\Rightarrow du = 0,\ dv = 0.$$

But both *du* and *dv* cannot vanish together.

Hence $E\,du^2 + 2F\,du\,dv + G\,dv^2$ is a positive definite quadratic form in *du*, *dv*.

Property 2. Invariance Property. The metric is invariant under a transformation of parameters.

Proof. Let the parametric equation of a surface in vector form be given by

$$\mathbf{r} = \mathbf{r}\,(u, v) \tag{1}$$

and let the parameters u, v be transformed to the parameters u', v' by the relations

$$u' = \phi\,(u, v),\ v' = \psi\,(u, v).$$

Then

$$\mathbf{r}_1' = \frac{\partial \mathbf{r}}{\partial u'} = \frac{\partial \mathbf{r}}{\partial u}\frac{\partial u}{\partial u'} + \frac{\partial \mathbf{r}}{\partial v}\frac{\partial v}{\partial u'} = \mathbf{r}_1 \frac{\partial u}{\partial u'} + \mathbf{r}_2 \frac{\partial v}{\partial u'} \tag{2}$$

and
$$\mathbf{r}_2' = \frac{\partial \mathbf{r}}{\partial v'} = \frac{\partial \mathbf{r}}{\partial u}\frac{\partial u}{\partial v'} + \frac{\partial \mathbf{r}}{\partial v}\frac{\partial v}{\partial v'} = \mathbf{r}_1 \frac{\partial u}{\partial v'} + \mathbf{r}_2 \frac{\partial v}{\partial v'}. \tag{3}$$

Also,

$$du = \frac{\partial u}{\partial u'}du' + \frac{\partial u}{\partial v'}dv' \tag{4}$$

and
$$dv = \frac{\partial v}{\partial u'}du' + \frac{\partial v}{\partial v'}dv'. \tag{5}$$

Now,

$$E'\,du'^2 + 2F'\,du'\,dv' + G'\,dv'^2$$
$$= \mathbf{r}_1'^2\,du'^2 + 2\,\mathbf{r}_1'.\mathbf{r}_2'\,du'\,dv' + \mathbf{r}_2'^2\,dv'^2$$
$$= (\mathbf{r}_1'\,du' + \mathbf{r}_2'\,dv')^2$$
$$= \left[\left(\mathbf{r}_1 \frac{\partial u}{\partial u'} + \mathbf{r}_2 \frac{\partial v}{\partial u'}\right)du' + \left(\mathbf{r}_1 \frac{\partial u}{\partial v'} + \mathbf{r}_2 \frac{\partial v}{\partial v'}\right)dv'\right]^2 \quad \text{[using (2) and (3)]}$$

$$= \left[\mathbf{r}_1 \left(\frac{\partial u}{\partial u'} du' + \frac{\partial u}{\partial v'} dv' \right) + \mathbf{r}_2 \left(\frac{\partial v}{\partial u'} du' + \frac{\partial v}{\partial v'} dv' \right) \right]^2$$

$$= (\mathbf{r}_1 du + \mathbf{r}_2 dv)^2 \qquad \text{[using (4) and (5)]}$$

$$= \mathbf{r}_1{}^2 du^2 + 2\,\mathbf{r}_1 . \mathbf{r}_2\, du\, dv + \mathbf{r}_2{}^2 dv^2$$

$$= E\, du^2 + 2F\, du\, dv + G\, dv^2.$$

Hence the metric is invariant.

Angle between the Parametric Curves.

Let P be the point of intersection of the parametric curves $u = \text{constant}$ and $v = \text{constant}$ and let $\mathbf{r}$ be the position vector of P. Then the vectors $\mathbf{r}_1$ and $\mathbf{r}_2$ are along the tangents to the curves $v = \text{constant}$ and $u = \text{constant}$ respectively.

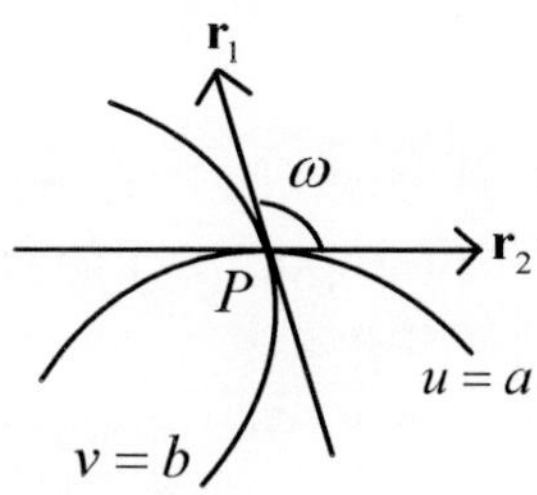

Therefore, if ω is the angle between the parametric curves, then

$$\cos\omega = \frac{\mathbf{r}_1 . \mathbf{r}_2}{|\mathbf{r}_1||\mathbf{r}_2|} = \frac{F}{\sqrt{EG}},$$

$$\sin\omega = \frac{|\mathbf{r}_1 \times \mathbf{r}_2|}{|\mathbf{r}_1||\mathbf{r}_2|} = \frac{H}{\sqrt{EG}} = \frac{\sqrt{EG - F^2}}{\sqrt{EG}},$$

$$\tan\omega = \frac{H}{F}.$$

Orthogonal System of Parametric Curves.

Definition. The parametric curves are said to form an orthogonal system if they cut at right angles at all points of the surface.

Theorem. The necessary and sufficient condition for the parametric curves to form an orthogonal system is that $F = 0$ at each point of the surface.

Proof. If P is the point of intersection of the parametric curves $u = \text{const.}$ and $v = \text{const.}$, and ω is the angle between them at P, then

$$\cos\omega = \frac{\mathbf{r}_1 . \mathbf{r}_2}{|\mathbf{r}_1||\mathbf{r}_2|} = \frac{F}{\sqrt{EG}}. \qquad (1)$$

If these parametric curves cut orthogonally at each point of the surface, then

$$\omega = \frac{\pi}{2}$$

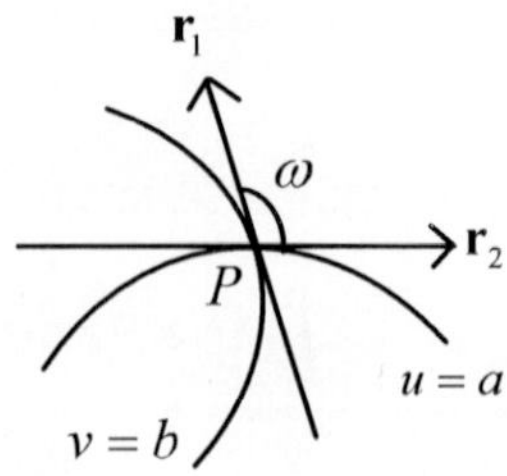

$\Rightarrow$ $\cos\omega = 0$

$\Rightarrow$ $F = 0$ [using (1)]

$\Rightarrow$ The condition is necessary.

Conversely,

$F = 0$

$\Rightarrow$ $\cos\omega = 0$ [by (1)]

$\Rightarrow$ $\omega = \dfrac{\pi}{2}$

$\Rightarrow$ The parametric curves cut orthogonally at each point of the surface.

$\Rightarrow$ The condition is sufficient.

Element of Area.

Consider the figure with vertices $P(u, v)$, $Q(u + du, v)$, $R(u + du, v + dv)$ and $S(u, v + dv)$ joined by parametric curves. If du and dv are small and positive, then this figure is approximately a parallelogram.

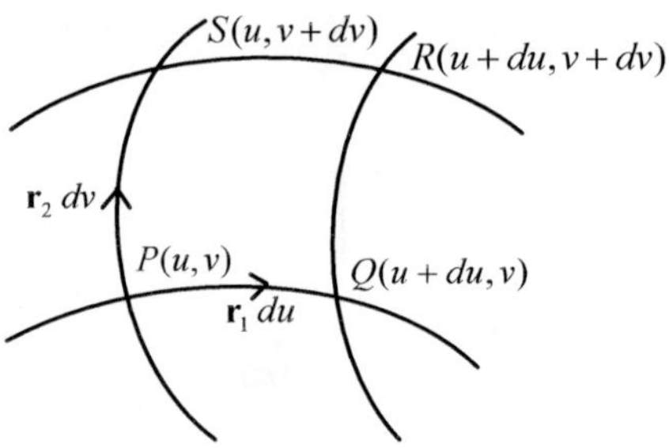

We have

$$\begin{aligned}
\overrightarrow{PQ} &= \text{Position vector of } Q - \text{Position vector of } P \\
&= \mathbf{r}(u + du, v) - \mathbf{r}(u, v) \\
&= \left[\mathbf{r}(u, v) + \frac{\partial \mathbf{r}}{\partial u} du\right] - \mathbf{r}(u, v) \\
&= \frac{\partial \mathbf{r}}{\partial u} du \\
&= \mathbf{r}_1\, du.
\end{aligned}$$

Similarly,

$$\overrightarrow{PS} = \mathbf{r}_2\, dv.$$

Therefore, if dS is the area of the parallelogram, then

$$dS = |\mathbf{r}_1\, du \times \mathbf{r}_2\, dv| = |\mathbf{r}_1 \times \mathbf{r}_2|\, du\, dv = H\, du\, dv.$$

Thus $H\, du\, dv$ is the expression for an elementary area on the surface at the point (u, v).

3.2 SECOND FUNDAMENTAL FORM

Definition. Let the parametric equation of a surface in vector form be given by

$$\mathbf{r} = \mathbf{r}\,(u,v) \tag{1}$$

and let $\mathbf{N}$ be the unit normal vector to this surface at the point $\mathbf{r}(u,v)$. Then

$$\mathbf{N} = \frac{\mathbf{r}_1 \times \mathbf{r}_2}{|\mathbf{r}_1 \times \mathbf{r}_2|} = \frac{\mathbf{r}_1 \times \mathbf{r}_2}{H}.$$

We introduce

$$L = \mathbf{r}_{11} \,.\, \mathbf{N},\ M = \mathbf{r}_{12} \,.\, \mathbf{N} = \mathbf{r}_{21} .\, \mathbf{N},\ N = \mathbf{r}_{22} \,.\, \mathbf{N}.$$

The quadratic differential form

$$L\, du^2 + 2M\, du\, dv + N\, dv^2$$

is known as the second fundamental form of a surface. The quantities L, M, N are known as the second fundamental coefficients or the fundamental magnitudes of the second order of the surface.

We write

$$LN - M^2 = T^2.$$

It should be noted that $LN - M^2 = T^2$ is not necessarily positive, just as $EG - F^2 = H^2$ is always positive.

Expressions for *L, M, N* in terms of Scalar Triple Products.

We may express L, M, N in terms of scalar triple products of vectors as follows.

$$[\mathbf{r}_{11},\ \mathbf{r}_1,\ \mathbf{r}_2] = HL,$$
$$[\mathbf{r}_{12},\ \mathbf{r}_1,\ \mathbf{r}_2] = HM,$$
$$[\mathbf{r}_{22},\ \mathbf{r}_1,\ \mathbf{r}_2] = HN.$$

Proof. We have

$$[\mathbf{r}_{11},\ \mathbf{r}_1,\ \mathbf{r}_2] = \mathbf{r}_{11} .\, \mathbf{r}_1 \times \mathbf{r}_2 = \mathbf{r}_{11} .\, H\mathbf{N} = H(\mathbf{r}_{11} .\, \mathbf{N}) = HL.$$

Similarly,

$$[\mathbf{r}_{12},\ \mathbf{r}_1,\ \mathbf{r}_2] = \mathbf{r}_{12} .\, \mathbf{r}_1 \times \mathbf{r}_2 = \mathbf{r}_{12} .\, H\mathbf{N} = H(\mathbf{r}_{12} .\, \mathbf{N}) = HM$$

and $\quad [\mathbf{r}_{22}, \mathbf{r}_1, \mathbf{r}_2] = \mathbf{r}_{22} \cdot \mathbf{r}_1 \times \mathbf{r}_2 = \mathbf{r}_{22} \cdot H\mathbf{N} = H(\mathbf{r}_{22} \cdot \mathbf{N}) = HN.$

Alternative Expressions for *L, M, N.*

We may express L, M, N in terms of the derivatives of the unit normal vector $\mathbf{N}$ as follows.

$$L = -\mathbf{r}_1 \cdot \mathbf{N}_1, \quad M = -\mathbf{r}_1 \cdot \mathbf{N}_2 = -\mathbf{r}_2 \cdot \mathbf{N}_1, \quad N = -\mathbf{r}_2 \cdot \mathbf{N}_2 .$$

Proof. Since the vectors $\mathbf{r}_1$ and $\mathbf{N}$ are tangential and normal to the surface at the same point, we have

$$\mathbf{r}_1 \cdot \mathbf{N} = 0$$

$$\Rightarrow \quad \frac{\partial}{\partial u}(\mathbf{r}_1 \cdot \mathbf{N}) = 0, \quad \frac{\partial}{\partial v}(\mathbf{r}_1 \cdot \mathbf{N}) = 0$$

$$\Rightarrow \quad \mathbf{r}_{11} \cdot \mathbf{N} + \mathbf{r}_1 \cdot \mathbf{N}_1 = 0, \quad \mathbf{r}_{12} \cdot \mathbf{N} + \mathbf{r}_1 \cdot \mathbf{N}_2 = 0$$

$$\Rightarrow \quad L + \mathbf{r}_1 \cdot \mathbf{N}_1 = 0, \quad M + \mathbf{r}_1 \cdot \mathbf{N}_2 = 0$$

$$\Rightarrow \quad L = -\mathbf{r}_1 \cdot \mathbf{N}_1, \quad M = -\mathbf{r}_1 \cdot \mathbf{N}_2. \tag{1}$$

Similarly, since the vectors $\mathbf{r}_2$ and $\mathbf{N}$ are tangential and normal to the surface at the same point, we have

$$\mathbf{r}_2 \cdot \mathbf{N} = 0$$

$$\Rightarrow \quad \frac{\partial}{\partial u}(\mathbf{r}_2 \cdot \mathbf{N}) = 0, \quad \frac{\partial}{\partial v}(\mathbf{r}_2 \cdot \mathbf{N}) = 0$$

$$\Rightarrow \quad \mathbf{r}_{21} \cdot \mathbf{N} + \mathbf{r}_2 \cdot \mathbf{N}_1 = 0, \quad \mathbf{r}_{22} \cdot \mathbf{N} + \mathbf{r}_2 \cdot \mathbf{N}_2 = 0$$

$$\Rightarrow \quad M + \mathbf{r}_2 \cdot \mathbf{N}_1 = 0, \quad N + \mathbf{r}_2 \cdot \mathbf{N}_2 = 0$$

$$\Rightarrow \quad M = -\mathbf{r}_2 \cdot \mathbf{N}_1, \quad N = -\mathbf{r}_2 \cdot \mathbf{N}_2. \tag{2}$$

From (1) and (2), we get

$$L = -\mathbf{r}_1 \cdot \mathbf{N}_1, \quad M = -\mathbf{r}_1 \cdot \mathbf{N}_2 = -\mathbf{r}_2 \cdot \mathbf{N}_1, \quad N = -\mathbf{r}_2 \cdot \mathbf{N}_2 .$$

Geometrical Interpretation of the Second Fundamental Form.

Let the parametric equation of a surface in vector form be given by

$$\mathbf{r} = \mathbf{r}\,(u, v). \tag{1}$$

Take two points P and Q on the surface corresponding to the parametric coordinates (u, v) and $(u + du, v + dv)$ respectively. Let $\mathbf{r}$ and $\mathbf{r} + d\mathbf{r}$ be the position vectors of P and Q respectively.

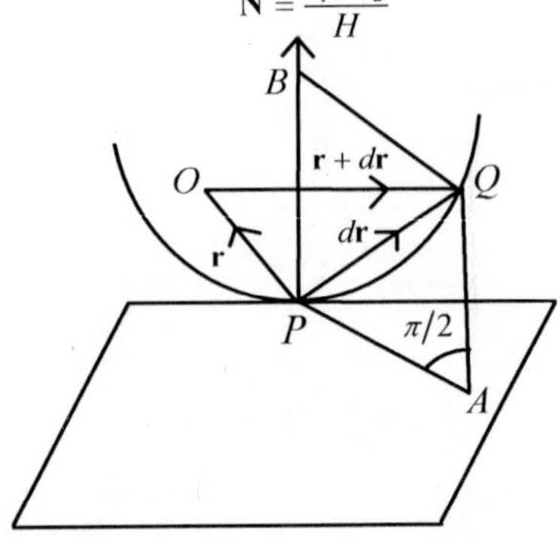

QA is perpendicular from Q to the tangent plane of the surface at P.
QB is perpendicular from Q to the normal to the surface at P.
We have

$$\mathbf{r} + d\mathbf{r} = \mathbf{r}\,(u + du, v + dv)$$

$$= \mathbf{r}\,(u,v) + [\mathbf{r}_1\,du + \mathbf{r}_2\,dv] + \frac{1}{2}[\mathbf{r}_{11}du^2 + 2\mathbf{r}_{12}du\,dv + \mathbf{r}_{22}dv^2]$$

$$\left[\begin{array}{l}\text{by Taylor's expansion, neglecting terms}\\ \text{of order higher than two in } du, dv\end{array}\right]$$

$$\Rightarrow \quad d\mathbf{r} = [\mathbf{r}_1\,du + \mathbf{r}_2\,dv] + \frac{1}{2}[\mathbf{r}_{11}du^2 + 2\mathbf{r}_{12}du\,dv + \mathbf{r}_{22}dv^2]$$

$$\Rightarrow \quad d\mathbf{r}\,.\,\mathbf{N} = [\mathbf{r}_1\,du + \mathbf{r}_2\,dv]\,.\,\mathbf{N} + \frac{1}{2}[\mathbf{r}_{11}du^2 + 2\mathbf{r}_{12}du\,dv + \mathbf{r}_{22}dv^2]\,.\,\mathbf{N}$$

$$\Rightarrow \quad PB = \frac{1}{2}[L\,du^2 + 2M\,du\,dv + N\,dv^2] \qquad [\because \mathbf{r}_1\,.\,\mathbf{N} = 0, \mathbf{r}_2\,.\,\mathbf{N} = 0]$$

$$\Rightarrow \quad QA = \frac{1}{2}[L\,du^2 + 2M\,du\,dv + N\,dv^2]$$

$$\Rightarrow \quad 2\,QA = L\,du^2 + 2M\,du\,dv + N\,dv^2$$

$\Rightarrow$ The second fundamental form of' a surface at a point P is twice the perpendicular length drawn from a neighbouring point Q on the surface to the tangent plane to the surface at the point P.

3.3 DERIVATIVES OF N (WEINGARTEN EQUATIONS)

In terms of E, F, G, L, M, N, the Weingarten equations are

$$H^2\mathbf{N}_1 = (FM - GL)\,\mathbf{r}_1 + (FL - EM)\,\mathbf{r}_2,$$

$$H^2\mathbf{N}_2 = (FN - GM)\,\mathbf{r}_1 + (FM - EN)\,\mathbf{r}_2$$

and hence

$$H\mathbf{N}_1 \times \mathbf{N}_2 = (LN - M^2)\,\mathbf{N}.$$

Proof. We know that

$$\mathbf{N}\,.\,\mathbf{N} = 1$$

$$\Rightarrow \quad \frac{\partial}{\partial u}(\mathbf{N}\,.\,\mathbf{N}) = 0, \quad \frac{\partial}{\partial v}(\mathbf{N}\,.\,\mathbf{N}) = 0$$

$$\Rightarrow \quad \mathbf{N}_1\,.\,\mathbf{N} + \mathbf{N}\,.\,\mathbf{N}_1 = 0, \quad \mathbf{N}_2\,.\,\mathbf{N} + \mathbf{N}\,.\,\mathbf{N}_2 = 0$$

$$\Rightarrow \quad 2\mathbf{N}_1\,.\,\mathbf{N} = 0, \quad 2\mathbf{N}_2\,.\,\mathbf{N} = 0$$

$$\Rightarrow \quad \mathbf{N}_1\,.\,\mathbf{N} = 0, \quad \mathbf{N}_2\,.\,\mathbf{N} = 0$$

$\Rightarrow$ Either $\mathbf{N}_1 = \mathbf{0}$ or $\mathbf{N}_1$ is perpendicular to $\mathbf{N}$,
Either $\mathbf{N}_2 = \mathbf{0}$ or $\mathbf{N}_2$ is perpendicular to $\mathbf{N}$.

$\Rightarrow$ Either **N** is independent of u or $\mathbf{N}_1$ is perpendicular to **N**,
Either **N** is independent of v or $\mathbf{N}_2$ is perpendicular to **N**.

$\Rightarrow$ Either **N** is independent of u and v or $\mathbf{N}_1$ and $\mathbf{N}_2$ are each perpendicular to **N**.

If **N** is independent of u and v, then the surface is a plane surface.

If $\mathbf{N}_1$ and $\mathbf{N}_2$ are each perpendicular to **N**, then the surface is not necessarily a plane surface, and $\mathbf{N}_1$ and $\mathbf{N}_2$ both lie in the plane of $\mathbf{r}_1$ and $\mathbf{r}_2$. Hence there exist scalars a, b, c, d such that

$$\mathbf{N}_1 = a\mathbf{r}_1 + b\mathbf{r}_2, \tag{1}$$

$$\mathbf{N}_2 = c\mathbf{r}_1 + d\mathbf{r}_2. \tag{2}$$

Taking dot product of both sides of (1) with $\mathbf{r}_1$ and then with $\mathbf{r}_2$, we have

$$\mathbf{N}_1 \cdot \mathbf{r}_1 = a\mathbf{r}_1 \cdot \mathbf{r}_1 + b\mathbf{r}_2 \cdot \mathbf{r}_1,$$

$$\mathbf{N}_1 \cdot \mathbf{r}_2 = a\mathbf{r}_1 \cdot \mathbf{r}_2 + b\mathbf{r}_2 \cdot \mathbf{r}_2$$

$$\Rightarrow \quad -L = aE + bF,$$

$$-M = aF + bG$$

$$\Rightarrow \quad aE + bF + L = 0, \tag{3}$$

$$aF + bG + M = 0. \tag{4}$$

Solving (3) and (4), we get

$$\frac{a}{FM - GL} = \frac{b}{FL - EM} = \frac{1}{EG - F^2}$$

or $$a = \frac{FM - GL}{H^2}, \quad b = \frac{FL - EM}{H^2}. \qquad [\because EG - F^2 = H^2]$$

Substituting the values of a and b in (1), we have

$$H^2\mathbf{N}_1 = (FM - GL)\,\mathbf{r}_1 + (FL - EM)\,\mathbf{r}_2. \tag{5}$$

Similarly, taking the dot product of both sides of (2) with $\mathbf{r}_1$ and then with $\mathbf{r}_2$, we have

$$\mathbf{N}_2 \cdot \mathbf{r}_1 = c\mathbf{r}_1 \cdot \mathbf{r}_1 + d\mathbf{r}_2 \cdot \mathbf{r}_1,$$

$$\mathbf{N}_2 \cdot \mathbf{r}_2 = c\mathbf{r}_1 \cdot \mathbf{r}_2 + d\mathbf{r}_2 \cdot \mathbf{r}_2$$

$$\Rightarrow \quad -M = cE + dF,$$

$$-N = cF + dG$$

$$\Rightarrow \quad cE + dF + M = 0, \tag{6}$$

$$cF + dG + N = 0. \tag{7}$$

Solving (6) and (7), we get

$$\frac{c}{FN-GM}=\frac{d}{FM-EN}=\frac{1}{EG-F^2}$$

or $\quad c=\dfrac{FN-GM}{H^2},\ d=\dfrac{FM-EN}{H^2}.\qquad [\because EG-F^2=H^2]$

Substituting the values of c and d in (2), we have

$$H^2\mathbf{N}_2=(FN-GM)\,\mathbf{r}_1+(FM-EN)\,\mathbf{r}_2. \tag{8}$$

Now, taking the cross product of (5) and (8), we have

$$H^4\mathbf{N}_1\times\mathbf{N}_2=[(FM-GL)(FM-EN)]\,H\mathbf{N}$$
$$-[(FL-EM)(FN-GM)]\,H\mathbf{N}$$

or $\quad H^4\mathbf{N}_1\times\mathbf{N}_2=[(LN-M^2)(EG-F^2)]\,H\mathbf{N}$

or $\quad H^4\mathbf{N}_1\times\mathbf{N}_2=(LN-M^2)\,H^3\mathbf{N}$

or $\quad H\mathbf{N}_1\times\mathbf{N}_2=(LN-M^2)\,\mathbf{N}.$

3.4 FUNDAMENTAL MAGNITUDES FOR SOME IMPORTANT SURFACES

1. Surface in Monge's Form. If the equation of the surface is in the form

$$z=f(x,y), \tag{1}$$

then it is called the equation of a surface in Monge's form.

It follows from (1) that x, y, z are not all independent variables. Therefore we can regard x and y as the independent variables.

Thus if $\mathbf{r}$ is the position vector of any current point on the surface (1), then

$$\mathbf{r}=(x,y,z), \tag{2}$$

where z is a function of x and y.

We write

$$\frac{\partial z}{\partial x}=p,\ \frac{\partial z}{\partial y}=q,\ \frac{\partial^2 z}{\partial x^2}=r,\ \frac{\partial^2 z}{\partial x\,\partial y}=s,\ \frac{\partial^2 z}{\partial y^2}=t.$$

Let x and y be taken as the first and the second parameters. Then, from (2), we have

$$\mathbf{r}_1=(1,0,p),\quad \mathbf{r}_2=(0,1,q),$$
$$\mathbf{r}_{11}=(0,0,r),\ \mathbf{r}_{12}=(0,0,s),\ \mathbf{r}_{22}=(0,0,t).$$

Therefore the fundamental magnitudes of the first order are

$$E=\mathbf{r}_1^{\ 2}=1+p^2,\ F=\mathbf{r}_1\cdot\mathbf{r}_2=pq,\ G=\mathbf{r}_2^{\ 2}=1+q^2.$$

Again, we have

$$H^2=EG-F^2=1+p^2+q^2$$

$\Rightarrow \quad H=\sqrt{1+p^2+q^2},$

$$\mathbf{r}_1 \times \mathbf{r}_2 = (-p, -q, 1)$$

and $$\mathbf{N} = \frac{\mathbf{r}_1 \times \mathbf{r}_2}{H} = \frac{1}{H}(-p, -q, 1).$$

Therefore the fundamental magnitudes of the second order are

$$L = \mathbf{r}_{11} \,.\, \mathbf{N} = \frac{r}{H}, \; M = \mathbf{r}_{12} \,.\, \mathbf{N} = \frac{s}{H}, \; N = \mathbf{r}_{22} \,.\, \mathbf{N} = \frac{t}{H}.$$

2. Surface of Revolution. A surface generated by the rotation of a plane curve about an axis in its plane is called the surface of revolution.

Let us take the z-axis as the axis of revolution.

If u is the perpendicular distance of a point P (x, y, z) on the surface of revolution from z-axis and ϕ is the angle through which the plane curve is rotated about z-axis, i.e., the angle between the planes ZOX and ZOP, then

$$x = u\cos\phi, \; y = u\sin\phi, \; z = f(u).$$

The angle ϕ is called the longitude of the point P.

Therefore the position vector $\mathbf{r}$ of any current point on the surface of revolution is given by

$$\mathbf{r} = (u\cos\phi, u\sin\phi, f(u)).$$

Taking u and ϕ as the first and the second parameters, we have

$$\mathbf{r}_1 = (\cos\phi, \sin\phi, f_1),$$
$$\mathbf{r}_2 = (-u\sin\phi, u\cos\phi, 0),$$
$$\mathbf{r}_{11} = (0, 0, f_{11}),$$
$$\mathbf{r}_{12} = (-\sin\phi, \cos\phi, 0),$$
$$\mathbf{r}_{22} = (-u\cos\phi, -u\sin\phi, 0).$$

Therefore the fundamental magnitudes of the first order are

$$E = \mathbf{r}_1^{\;2} = \cos^2\phi + \sin^2\phi + f_1^{\;2} = 1 + f_1^{\;2},$$
$$F = \mathbf{r}_1 \,.\, \mathbf{r}_2 = -u\cos\phi\sin\phi + u\sin\phi\cos\phi + 0 = 0,$$
$$G = \mathbf{r}_2^{\;2} = u^2\sin^2\phi + u^2\cos^2\phi + 0 = u^2.$$

Again, we have

$$H^2 = EG - F^2 = u^2(1 + f_1^{\;2})$$

$\Rightarrow$ $$H = u\sqrt{1 + f_1^{\;2}},$$

$$\mathbf{r}_1 \times \mathbf{r}_2 = (-f_1 u\cos\phi, -f_1 u\sin\phi, u)$$

and $$\mathbf{N} = \frac{\mathbf{r}_1 \times \mathbf{r}_2}{H} = \frac{1}{H}(-f_1 u\cos\phi, -f_1 u\sin\phi, u).$$

Therefore the fundamental magnitudes of the second order are

$$L = \mathbf{r}_{11} \cdot \mathbf{N} = \frac{uf_{11}}{H}, \; M = \mathbf{r}_{12} \cdot \mathbf{N} = 0, \; N = \mathbf{r}_{22} \cdot \mathbf{N} = \frac{u^2 f_1}{H}.$$

3. Helicoids. A helicoid is the surface generated by a curve which is simultaneously rotated about a fixed axis and translated in the direction of the axis with a velocity proportional to the angular velocity of rotation. The plane sections through the axis are called meridians.

There is no loss of generality if we assume the generating curve to be a plane curve. Let the surface be generated by the helicoidal motion of a meridian.

Let us take the z-axis as the axis of revolution.

If u is the perpendicular distance of a point $P\ (x, y, z)$ on the surface from z-axis and ϕ is the inclination of the meridian plane through the point P to ZOX-plane, then

$$x = u\cos\phi,\ y = u\sin\phi,\ z = f(u) + c\phi,$$

where c is a constant and $2\pi c$ is called the pitch of the helicoidal motion.

Therefore the position vector $\mathbf{r}$ of any current point on the helicoid is given by

$$\mathbf{r} = (u\cos\phi, u\sin\phi, f(u) + c\phi).$$

Taking u and ϕ as the first and the second parameters, we have

$$\mathbf{r}_1 = (\cos\phi, \sin\phi, f_1),$$

$$\mathbf{r}_2 = (-u\sin\phi, u\cos\phi, c),$$

$$\mathbf{r}_{11} = (0, 0, f_{11}),$$

$$\mathbf{r}_{12} = (-\sin\phi, \cos\phi, 0),$$

$$\mathbf{r}_{22} = (-u\cos\phi, -u\sin\phi, 0).$$

Therefore the fundamental magnitudes of the first order are

$$E = \mathbf{r}_1^{\,2} = \cos^2\phi + \sin^2\phi + f_1^{\,2} = 1 + f_1^{\,2},$$

$$F = \mathbf{r}_1 \cdot \mathbf{r}_2 = -u\cos\phi\sin\phi + u\sin\phi\cos\phi + cf_1 = cf_1,$$

$$G = \mathbf{r}_2^{\,2} = u^2\sin^2\phi + u^2\cos^2\phi + c^2 = u^2 + c^2.$$

Again, we have

$$H^2 = EG - F^2 = c^2 + u^2(1 + f_1^{\,2})$$

$$\Rightarrow \qquad H = u\sqrt{c^2 + u^2(1 + f_1^{\,2})},$$

$$\mathbf{r}_1 \times \mathbf{r}_2 = (c\sin\phi - f_1 u\cos\phi, -c\cos\phi - f_1 u\sin\phi, u)$$

and $\quad \mathbf{N} = \dfrac{\mathbf{r}_1 \times \mathbf{r}_2}{H}$

$$= \frac{1}{H}(c\sin\phi - f_1 u\cos\phi, -c\cos\phi - f_1 u\sin\phi, u).$$

Therefore the fundamental magnitudes of the second order are

$$L = \mathbf{r}_{11}\,.\,\mathbf{N} = \frac{uf_{11}}{H},\ M = \mathbf{r}_{12}\,.\,\mathbf{N} = -\frac{c}{H},\ N = \mathbf{r}_{22}\,.\,\mathbf{N} = \frac{u^2 f_1}{H}.$$

4. Right Helicoid. A right helicoid is a surface generated by the helicoidal motion of a straight line meeting the axis at right angles.

If we take the axis to be the z-axis, then the position vector $\mathbf{r}$ of any current point on the right helicoid is given by

$$\mathbf{r} = (u\cos\phi, u\sin\phi, c\phi).$$

Taking u and ϕ as the first and the second parameters, we have

$$\mathbf{r}_1 = (\cos\phi, \sin\phi, 0),$$
$$\mathbf{r}_2 = (-u\sin\phi, u\cos\phi, c),$$
$$\mathbf{r}_{11} = (0, 0, 0),$$
$$\mathbf{r}_{12} = (-\sin\phi, \cos\phi, 0),$$
$$\mathbf{r}_{22} = (-u\cos\phi, -u\sin\phi, 0).$$

Therefore the fundamental magnitudes of the first order are

$$E = \mathbf{r}_1^{\,2} = \cos^2\phi + \sin^2\phi + 0 = 1,$$
$$F = \mathbf{r}_1 . \mathbf{r}_2 = -u\cos\phi\sin\phi + u\sin\phi\cos\phi + 0 = 0,$$
$$G = \mathbf{r}_2^{\,2} = u^2\sin^2\phi + u^2\cos^2\phi + c^2 = u^2 + c^2.$$

Since $F = 0$, the parametric curves are orthogonal.

Again, we have

$$H^2 = EG - F^2 = 1(u^2 + c^2) - 0 = u^2 + c^2$$

$\Rightarrow \qquad H = \sqrt{EG - F^2},$

$$\mathbf{r}_1 \times \mathbf{r}_2 = (c\sin\phi, -c\cos\phi, u)$$

and $\quad \mathbf{N} = \dfrac{\mathbf{r}_1 \times \mathbf{r}_2}{H} = \left(\dfrac{c\sin\phi}{\sqrt{u^2 + c^2}}, -\dfrac{c\cos\phi}{\sqrt{u^2 + c^2}}, \dfrac{u}{\sqrt{u^2 + c^2}}\right).$

Therefore the fundamental magnitudes of the second order are

$$L = \mathbf{r}_{11}\,.\,\mathbf{N}$$
$$= (0, 0, 0)\,.\left(\frac{c\sin\phi}{\sqrt{u^2 + c^2}}, -\frac{c\cos\phi}{\sqrt{u^2 + c^2}}, \frac{u}{\sqrt{u^2 + c^2}}\right)$$
$$= 0,$$

$$M = \mathbf{r}_{12} \cdot \mathbf{N}$$

$$= (-\sin\phi, \cos\phi, 0) \cdot \left(\frac{c \sin\phi}{\sqrt{u^2+c^2}}, -\frac{c\cos\phi}{\sqrt{u^2+c^2}}, \frac{u}{\sqrt{u^2+c^2}} \right)$$

$$= -\frac{c\sin^2\phi}{\sqrt{u^2+c^2}} - \frac{c\cos^2\phi}{\sqrt{u^2+c^2}} + 0$$

$$= -\frac{c}{\sqrt{u^2+c^2}},$$

$$N = \mathbf{r}_{22} \cdot \mathbf{N}$$

$$= (-u\cos\phi, -u\sin\phi, 0) \cdot \left(\frac{c \sin\phi}{\sqrt{u^2+c^2}}, -\frac{c\cos\phi}{\sqrt{u^2+c^2}}, \frac{u}{\sqrt{u^2+c^2}} \right)$$

$$= -\frac{uc\cos\phi\sin\phi}{\sqrt{u^2+c^2}} + \frac{uc\sin\phi\cos\phi}{\sqrt{u^2+c^2}} + 0$$

$$= 0.$$

Solved Examples

Example 1. Calculate the fundamental magnitudes for the right helicoid given by $x = u\cos\phi, y = u\sin\phi, z = c\phi$, the parameters being u and ϕ.

Solution. See Section 3.4.

Example 2. Show that on the surface given by

$$x = a(u+v), y = b(u-v), z = uv,$$

the parametric curves are straight lines. Also, calculate the fundamental magnitudes for this surface.

Solution. The equation of the given surface is

$$x = a(u+v), y = b(u-v), z = uv. \quad (1)$$

Therefore, the parametric curve $u = c$ (constant) is given by

$$x = a(c+v), y = b(c-v), z = cv, \quad (2)$$

where the parameter is v.

Eliminating the parameter v, we get

$$x = a\left(c + \frac{z}{c}\right), y = b\left(c - \frac{z}{c}\right),$$

which are the equations of two planes whose intersection is a straight line. Therefore the parametric curves $u = c$ (constant) are straight lines.

Similarly, we can show that the parametric curves $v =$ constant are straight lines.

Now, from (1), we have

$$\mathbf{r} = (a\,(u+v), b\,(u-v), uv).$$

$$\therefore \qquad \mathbf{r}_1 = (a, b, v),\ \mathbf{r}_2 = (a, -b, u),$$

$$\mathbf{r}_{11} = (0,0,0),\ \mathbf{r}_{12} = (0,0,1),\ \mathbf{r}_{22} = (0,0,0).$$

Therefore the fundamental magnitudes of the first order are

$$E = \mathbf{r}_1^{\,2} = a^2 + b^2 + v^2,$$

$$F = \mathbf{r}_1 \cdot \mathbf{r}_2 = a^2 - b^2 + uv,$$

$$G = \mathbf{r}_2^{\,2} = a^2 + b^2 + u^2.$$

Again, we have

$$\begin{aligned} H^2 &= EG - F^2 \\ &= (a^2+b^2+v^2)(a^2+b^2+u^2) - (a^2-b^2+uv)^2 \\ &= 4a^2b^2 + a^2(u-v)^2 + b^2(u+v)^2 \end{aligned}$$

$$\Rightarrow \qquad H = \sqrt{4a^2b^2 + a^2(u-v)^2 + b^2(u+v)^2},$$

$$\mathbf{r}_1 \times \mathbf{r}_2 = (b\,(u+v), a\,(v-u), -2ab)$$

and $$\mathbf{N} = \frac{\mathbf{r}_1 \times \mathbf{r}_2}{H} = \left(\frac{b\,(u+v)}{H}, \frac{a\,(v-u)}{H}, -\frac{2ab}{H}\right).$$

Therefore the fundamental magnitudes of the second order are

$$\begin{aligned} L &= \mathbf{r}_{11} \cdot \mathbf{N} \\ &= (0,0,0) \cdot \left(\frac{b\,(u+v)}{H}, \frac{a\,(v-u)}{H}, -\frac{2ab}{H}\right) \\ &= 0, \end{aligned}$$

$$\begin{aligned} M &= \mathbf{r}_{12} \cdot \mathbf{N} \\ &= (0,0,1) \cdot \left(\frac{b\,(u+v)}{H}, \frac{a\,(v-u)}{H}, -\frac{2ab}{H}\right) \\ &= -\frac{2ab}{H}, \end{aligned}$$

$$\begin{aligned} N &= \mathbf{r}_{22} \cdot \mathbf{N} \\ &= (0,0,0) \cdot \left(\frac{b\,(u+v)}{H}, \frac{a\,(v-u)}{H}, -\frac{2ab}{H}\right) \\ &= 0. \end{aligned}$$

Example 3. Obtain Weingarten equations and deduce the formulae

$$H[\mathbf{N},\ \mathbf{N}_1,\ \mathbf{r}_1] = EM - FL,$$
$$H[\mathbf{N},\ \mathbf{N}_1,\ \mathbf{r}_2] = FM - GL,$$
$$H[\mathbf{N},\ \mathbf{N}_2,\ \mathbf{r}_1] = EN - FM,$$
$$H[\mathbf{N},\ \mathbf{N}_2,\ \mathbf{r}_2] = FN - GM.$$

Solution. For Weingarten equations, see Section 3.3. These are

$$H^2\mathbf{N}_1 = (FM - GL)\,\mathbf{r}_1 + (FL - EM)\,\mathbf{r}_2, \tag{1}$$
$$H^2\mathbf{N}_2 = (FN - GM)\,\mathbf{r}_1 + (FM - EN)\,\mathbf{r}_2. \tag{2}$$

Now,

$$H[\mathbf{N},\ \mathbf{N}_1,\ \mathbf{r}_1] = H\mathbf{N}\,.\,\mathbf{N}_1 \times \mathbf{r}_1$$
$$= H\mathbf{N}\,.\left[\left(\frac{FM - GL}{H^2}\right)\mathbf{r}_1 + \left(\frac{FL - EM}{H^2}\right)\mathbf{r}_2\right] \times \mathbf{r}_1$$

[using (1)]

$$= H\mathbf{N}\,.\left[\left(\frac{FL - EM}{H^2}\right)\mathbf{r}_2 \times \mathbf{r}_1\right] \qquad [\because\ \mathbf{r}_2 \times \mathbf{r}_2 = \mathbf{0}]$$
$$= H\mathbf{N}\,.\left[\left(\frac{FL - EM}{H^2}\right)(-H\mathbf{N})\right]$$
$$= EM - FL.$$

Similarly,

$$H[\mathbf{N},\ \mathbf{N}_1,\ \mathbf{r}_2] = H\mathbf{N}\,.\,\mathbf{N}_1 \times \mathbf{r}_2$$
$$= H\mathbf{N}\,.\left[\left(\frac{FM - GL}{H^2}\right)\mathbf{r}_1 + \left(\frac{FL - EM}{H^2}\right)\mathbf{r}_2\right] \times \mathbf{r}_2$$

[using (1)]

$$= H\mathbf{N}\,.\left[\left(\frac{FM - GL}{H^2}\right)\mathbf{r}_1 \times \mathbf{r}_2\right] \qquad [\because\ \mathbf{r}_2 \times \mathbf{r}_2 = \mathbf{0}]$$
$$= H\mathbf{N}\,.\left[\left(\frac{FM - GL}{H^2}\right)(H\mathbf{N})\right]$$
$$= FM - GL.$$

Again,

$$H[\mathbf{N},\ \mathbf{N}_2,\ \mathbf{r}_1] = H\mathbf{N}\,.\,\mathbf{N}_2 \times \mathbf{r}_1$$
$$= H\mathbf{N}\,.\left[\left(\frac{FN - GM}{H^2}\right)\mathbf{r}_1 + \left(\frac{FM - EN}{H^2}\right)\mathbf{r}_2\right] \times \mathbf{r}_1$$

[using (2)]

$$= H\mathbf{N} \,.\left[\left(\frac{FM - EN}{H^2}\right)\mathbf{r}_2 \times \mathbf{r}_1\right] \qquad [\because \mathbf{r}_1 \times \mathbf{r}_1 = \mathbf{0}]$$

$$= H\mathbf{N} \,.\left[\left(\frac{FM - EN}{H^2}\right)(-H\mathbf{N})\right]$$

$$= EN - FM.$$

Similarly,

$$H\,[\mathbf{N},\ \mathbf{N}_2,\ \mathbf{r}_2] = H\mathbf{N} \,. \mathbf{N}_2 \times \mathbf{r}_2$$

$$= H\mathbf{N} \,.\left[\left(\frac{FN - GM}{H^2}\right)\mathbf{r}_1 + \left(\frac{FM - EN}{H^2}\right)\mathbf{r}_2\right] \times \mathbf{r}_2$$

[using (2)]

$$= H\mathbf{N} \,.\left[\left(\frac{FN - GM}{H^2}\right)\mathbf{r}_1 \times \mathbf{r}_2\right] \qquad [\because \mathbf{r}_1 \times \mathbf{r}_1 = \mathbf{0}]$$

$$= H\mathbf{N} \,.\left[\left(\frac{FN - GM}{H^2}\right)(H\mathbf{N})\right]$$

$$= FN - GM.$$

Example 4. Prove the formulae

$$H\mathbf{N} \times \mathbf{N}_1 = M\mathbf{r}_1 - L\mathbf{r}_2$$

and $\quad H\mathbf{N} \times \mathbf{N}_2 = N\mathbf{r}_1 - M\mathbf{r}_2.$

Solution. We have

$$H\mathbf{N} = \mathbf{r}_1 \times \mathbf{r}_2. \tag{1}$$

Differentiating (1) with respect to u, we get

$$H_1\mathbf{N} + H\mathbf{N}_1 = \mathbf{r}_{11} \times \mathbf{r}_2 + \mathbf{r}_1 \times \mathbf{r}_{21}. \tag{2}$$

Taking the cross product of both sides of (2) with $\mathbf{N}$, we get

$$H\mathbf{N} \times \mathbf{N}_1 = \mathbf{N} \times (\mathbf{r}_{11} \times \mathbf{r}_2) + \mathbf{N} \times (\mathbf{r}_1 \times \mathbf{r}_{21})$$

$$= [(\mathbf{N} \,. \mathbf{r}_2)\,\mathbf{r}_{11} - (\mathbf{N} \,. \mathbf{r}_{11})\,\mathbf{r}_2] + [(\mathbf{N} \,. \mathbf{r}_{21})\,\mathbf{r}_1 - (\mathbf{N} \,. \mathbf{r}_1)\,\mathbf{r}_{21}]$$

$$[\because \mathbf{a} \times (\mathbf{b} \times \mathbf{c}) = (\mathbf{a} \,. \mathbf{c})\,\mathbf{b} - (\mathbf{a} \,. \mathbf{b})\,\mathbf{c}]$$

$$= M\mathbf{r}_1 - L\mathbf{r}_2.$$

$$[\because \mathbf{N} \,. \mathbf{r}_1 = 0, \mathbf{N} \,. \mathbf{r}_2 = 0, \mathbf{N} \,. \mathbf{r}_{11} = L,\ \mathbf{N} \,. \mathbf{r}_{21} = M]$$

Now, differentiating (1) with respect to v, we get

$$H_2\mathbf{N} + H\mathbf{N}_2 = \mathbf{r}_{12} \times \mathbf{r}_2 + \mathbf{r}_1 \times \mathbf{r}_{22}. \tag{3}$$

Taking the cross product of both sides of (3) with $\mathbf{N}$, we get

$$\begin{aligned} H\mathbf{N}\times\mathbf{N}_2 &= \mathbf{N}\times(\mathbf{r}_{12}\times\mathbf{r}_2)+\mathbf{N}\times(\mathbf{r}_1\times\mathbf{r}_{22}) \\ &= [(\mathbf{N}\,.\,\mathbf{r}_2)\,\mathbf{r}_{12}-(\mathbf{N}\,.\,\mathbf{r}_{12})\,\mathbf{r}_2]+[(\mathbf{N}\,.\,\mathbf{r}_{22})\,\mathbf{r}_1-(\mathbf{N}\,.\,\mathbf{r}_1)\,\mathbf{r}_{22}] \\ &= N\mathbf{r}_1 - M\mathbf{r}_2. \end{aligned}$$

$$[\because \mathbf{N}\,.\,\mathbf{r}_1 = 0, \boldsymbol{N}\,.\,\mathbf{r}_2 = 0, \mathbf{N}\,.\,\mathbf{r}_{12} = M,\ \mathbf{N}\,.\,\mathbf{r}_{22} = N]$$

Example 5. Show that if L, M, N vanish everywhere on a surface, then the surface is a part of a plane.

Solution. We know that

$$L = -\mathbf{N}_1.\mathbf{r}_1,\ M = -\mathbf{N}_1.\mathbf{r}_2 = -\mathbf{N}_2.\mathbf{r}_1,\ N = -\mathbf{N}_2.\mathbf{r}_2. \tag{1}$$

Since $L = 0$, we have

$-\mathbf{N}_1.\mathbf{r}_1 = 0$ [using (1)]

$\Rightarrow$ $\mathbf{N}_1.\mathbf{r}_1 = 0$

$\Rightarrow$ $\mathbf{N}_1 = \mathbf{0}$ or $\mathbf{r}_1 = \mathbf{0}$ or $\mathbf{N}_1$ is perpendicular to $\mathbf{r}_1$.

$\Rightarrow$ $\mathbf{N}_1 = \mathbf{0}$ or $\mathbf{N}_1$ is perpendicular to $\mathbf{r}_1$. (2)

$[\because \mathbf{r}_1 \neq 0]$

Similarly, since $M = 0$, we have

$-\mathbf{N}_1.\mathbf{r}_2 = 0,\ -\mathbf{N}_2.\mathbf{r}_1 = 0$ [using (1)]

$\Rightarrow$ $\mathbf{N}_1.\mathbf{r}_2 = 0,\ \mathbf{N}_2.\mathbf{r}_1 = 0$

$\Rightarrow$ $\mathbf{N}_1 = \mathbf{0}$ or $\mathbf{r}_2 = \mathbf{0}$ or $\mathbf{N}_1$ is perpendicular to $\mathbf{r}_2$,

$\mathbf{N}_2 = \mathbf{0}$ or $\mathbf{r}_1 = \mathbf{0}$ or $\mathbf{N}_2$ is perpendicular to $\mathbf{r}_1$.

$\Rightarrow$ $\mathbf{N}_1 = \mathbf{0}$ or $\mathbf{N}_1$ is perpendicular to $\mathbf{r}_2$, (3)

$[\because \mathbf{r}_2 \neq 0]$

$\mathbf{N}_2 = \mathbf{0}$ or $\mathbf{N}_2$ is perpendicular to $\mathbf{r}_1$. (4)

$[\because \mathbf{r}_1 \neq 0]$

Again, since $N = 0$, we have

$-\mathbf{N}_2.\mathbf{r}_2 = 0$ [using (1)]

$\Rightarrow$ $\mathbf{N}_2.\mathbf{r}_2 = 0$

$\Rightarrow$ $\mathbf{N}_2 = \mathbf{0}$ or $\mathbf{r}_2 = \mathbf{0}$ or $\mathbf{N}_2$ is perpendicular to $\mathbf{r}_2$.

$\Rightarrow$ $\mathbf{N}_2 = \mathbf{0}$ or $\mathbf{N}_2$ is perpendicular to $\mathbf{r}_2$. (5)

$[\because \mathbf{r}_2 \neq 0]$

Now, suppose, if possible, that $\mathbf{N}_1 \neq \mathbf{0}$. Then

$\mathbf{N}_1$ is perpendicular to $\mathbf{r}_1$, [using (2)]

$\mathbf{N}_1$ is perpendicular to $\mathbf{r}_2$. [using (3)]

$\Rightarrow$ $\mathbf{N}_1$ is parallel to $\mathbf{r}_1 \times \mathbf{r}_2$.

$\Rightarrow$ $\mathbf{N}_1$ is parallel to $\mathbf{N}$. $[\because \mathbf{r}_1 \times \mathbf{r}_2 = H\mathbf{N}]$

But, this is absurd, since

$$\mathbf{N}.\mathbf{N} = 1$$

$$\Rightarrow \quad \frac{\partial}{\partial u}(\mathbf{N}.\mathbf{N}) = 0$$

$$\Rightarrow \quad \mathbf{N}_1.\mathbf{N} + \mathbf{N}.\mathbf{N}_1 = 0$$

$$\Rightarrow \quad \mathbf{N}_1.\mathbf{N} + \mathbf{N}_1.\mathbf{N} = 0$$

$$\Rightarrow \quad 2\mathbf{N}_1.\mathbf{N} = 0$$

$$\Rightarrow \quad \mathbf{N}_1.\mathbf{N} = 0$$

$\Rightarrow$ $\mathbf{N}_1$ is perpendicular to $\mathbf{N}$.

Hence our supposition that $\mathbf{N}_1 \neq \mathbf{0}$ is false, and therefore

$$\mathbf{N}_1 = \mathbf{0}$$

$$\Rightarrow \quad \frac{\partial \mathbf{N}}{\partial u} = 0$$

$\Rightarrow$ $\mathbf{N}$ is independent of u. (6)

Similarly, it can be proved that

$\mathbf{N}$ is independent of v. (7)

It follows from (6) and (7) that $\mathbf{N}$ is a constant vector. Therefore the surface is a part of a plane.

Example 6. Prove that a real surface for which the equations

$$\frac{E}{L} = \frac{F}{M} = \frac{G}{N}$$

hold is either spherical or plane.

Solution. Let

$$\frac{L}{E} = \frac{M}{F} = \frac{N}{G} = \lambda.$$

Then $L = E\lambda,\ M = F\lambda,\ N = G\lambda.$ (1)

We know that

$$L = -\mathbf{N}_1.\mathbf{r}_1$$

$$\Rightarrow \quad \mathbf{N}_1.\mathbf{r}_1 + L = 0$$

$$\Rightarrow \quad \mathbf{N}_1.\mathbf{r}_1 + E\lambda = 0 \qquad \text{[using (1)]}$$

$$\Rightarrow \quad \mathbf{N}_1.\mathbf{r}_1 + \lambda\mathbf{r}_1.\mathbf{r}_1 = 0$$

$$\Rightarrow \quad (\mathbf{N}_1 + \lambda\mathbf{r}_1).\mathbf{r}_1 = 0. \qquad (2)$$

Similarly,

$$M = -\mathbf{N}_1.\mathbf{r}_2 = -\mathbf{N}_2.\mathbf{r}_1$$

$$\Rightarrow \quad \mathbf{N}_1.\mathbf{r}_2 + M = 0,\ \mathbf{N}_2.\mathbf{r}_1 + M = 0$$

$\Rightarrow \quad \mathbf{N}_1 . \mathbf{r}_2 + F\lambda = 0,\ \mathbf{N}_2 . \mathbf{r}_1 + F\lambda = 0$ [using (1)]

$\Rightarrow \quad \mathbf{N}_1 . \mathbf{r}_2 + \lambda \mathbf{r}_1 . \mathbf{r}_2 = 0,\ \mathbf{N}_2 . \mathbf{r}_1 + \lambda \mathbf{r}_2 . \mathbf{r}_1 = 0$

$\Rightarrow \quad (\mathbf{N}_1 + \lambda \mathbf{r}_1) . \mathbf{r}_2 = 0,\ (\mathbf{N}_2 + \lambda \mathbf{r}_2) . \mathbf{r}_1 = 0.$ (3)

Again,

$$N = -\mathbf{N}_2 . \mathbf{r}_2$$

$\Rightarrow \quad \mathbf{N}_2 . \mathbf{r}_2 + N = 0$

$\Rightarrow \quad \mathbf{N}_2 . \mathbf{r}_2 + G\lambda = 0$ [using (1)]

$\Rightarrow \quad \mathbf{N}_2 . \mathbf{r}_2 + \lambda \mathbf{r}_2 . \mathbf{r}_2 = 0$

$\Rightarrow \quad (\mathbf{N}_2 + \lambda \mathbf{r}_2) . \mathbf{r}_2 = 0.$ (4)

Since $\mathbf{N}_1$ and $\mathbf{N}_2$ are both perpendicular to $\mathbf{N}$, each of them lies in the plane of $\mathbf{r}_1$ and $\mathbf{r}_2$. Hence the vectors $\mathbf{N}_1 + \lambda \mathbf{r}_1$ and $\mathbf{N}_2 + \lambda \mathbf{r}_2$ lie in the plane of $\mathbf{r}_1$ and $\mathbf{r}_2$.

Now, from (2) and (3), we have

$$(\mathbf{N}_1 + \lambda \mathbf{r}_1) . \mathbf{r}_1 = 0,\ (\mathbf{N}_1 + \lambda \mathbf{r}_1) . \mathbf{r}_2 = 0.$$

Since $\mathbf{r}_1 \neq 0$, $\mathbf{r}_2 \neq 0$ and the vector $\mathbf{N}_1 + \lambda \mathbf{r}_1$ cannot be perpendicular to both $\mathbf{r}_1$ and $\mathbf{r}_2$, the above equations can hold only if

$$\mathbf{N}_1 + \lambda \mathbf{r}_1 = 0$$

i.e., $\quad \mathbf{N}_1 = -\lambda \mathbf{r}_1.$ (5)

Similarly, from (3) and (4), we get

$$\mathbf{N}_2 = -\lambda \mathbf{r}_2. \tag{6}$$

Differentiating (5) with respect to v, we get

$$\mathbf{N}_{12} = -\lambda_2 \mathbf{r}_1 - \lambda \mathbf{r}_{12}. \tag{7}$$

Differentiating (6) with respect to u, we get

$$\mathbf{N}_{21} = -\lambda_1 \mathbf{r}_2 - \lambda \mathbf{r}_{21}. \tag{8}$$

From (7) and (8), we have

$\lambda_2 \mathbf{r}_1 = \lambda_1 \mathbf{r}_2$ $[\because \mathbf{N}_{12} = \mathbf{N}_{21},\ \mathbf{r}_{12} = \mathbf{r}_{21}]$

$\Rightarrow \quad \lambda_1 = 0,\ \lambda_2 = 0$ $[\because \mathbf{r}_1$ is not parallel to $\mathbf{r}_2]$

$\Rightarrow \quad \lambda$ is independent of both u and v.

$\Rightarrow \quad \lambda$ is constant.

Now, if $\lambda \neq 0$, then, from (5), we have

$$\mathbf{r}_1 = -\frac{1}{\lambda} \mathbf{N}_1.$$

Integrating, we get

$\mathbf{r} = -\dfrac{1}{\lambda} \mathbf{N} + \mathbf{r}_0$, where $\mathbf{r}_0$ is constant of integration.

$$\Rightarrow \quad (\mathbf{r}-\mathbf{r}_0)^2 = \frac{1}{\lambda^2}\mathbf{N}^2$$

$$\Rightarrow \quad (\mathbf{r}-\mathbf{r}_0)^2 = \frac{1}{\lambda^2} \qquad [\because \mathbf{N}^2 = 1]$$

$\Rightarrow$ Locus of **r** is a sphere.

Again, if $\lambda = 0$, then, from (5) and (6), we have

$$\mathbf{N}_1 = 0,\ \mathbf{N}_2 = 0$$

$\Rightarrow$ **N** is independent of both u and v.

$\Rightarrow$ **N** is a constant vector.

$\Rightarrow$ The surface is a plane surface.

Exercises

1. Find the fundamental magnitudes and the unit normal for the surface of revolution $x = u\cos v,\ y = u\sin v,\ z = f(u)$.

$$\left[\begin{array}{l}\textbf{Ans. } E = 1+f'^2,\ F = 0,\ G = u^2, H = u\sqrt{1+f'^2}, \\ \mathbf{N} = \dfrac{u}{H}(-f'\cos v, f'\sin v, 1), L = \dfrac{uf''}{H}, M = 0, N = \dfrac{u^2 f'}{H}.\end{array}\right]$$

2. Find the metric for the paraboloid $\mathbf{r} = (u, v, u^2 - v^2)$

[**Ans.** $(1+4u^2)\,du^2 - 8uv\,du\,dv + (1+4v^2)\,dv^2$.]

3. On the right helicoid given by $x = u\cos v, y = u\sin v, z = cv$, show that the parametric curves are circular helices and straight lines.

4. Prove the formulae

$$T^2\mathbf{r}_1 = (FM - EN)\,\mathbf{N}_1 + (EM - FL)\,\mathbf{N}_2,$$

$$T^2\mathbf{r}_2 = (GM - FN)\mathbf{N}_1 + (FM - GL)\mathbf{N}_2.$$

3.5 DIRECTIONS ON A SURFACE

Any direction on a surface from a given point (u, v) is determined by the increments du, dv in the parameters for a small displacement in that direction.

Let ds be the length of the displacement $d\mathbf{r}$ corresponding to the increments du, dv. Let δs be the length of another displacement $\delta\mathbf{r}$ corresponding to the increments δu, δv. Then

$$d\mathbf{r} = \mathbf{r}_1 du + \mathbf{r}_2 dv,$$

$$\delta\mathbf{r} = \mathbf{r}_1 \delta u + \mathbf{r}_2 \delta v.$$

Therefore if ψ is the angle between these two directions and if ψ be measured in the sense of rotation which carries the direction of $\mathbf{r}_1$ to that of $\mathbf{r}_2$ through an angle between 0 and π, then

$$d\mathbf{r}.\delta\mathbf{r} = (\mathbf{r}_1 du + \mathbf{r}_2 dv).(\mathbf{r}_1\delta u + \mathbf{r}_2\delta v)$$

$$\Rightarrow \quad |d\mathbf{r}||\delta\mathbf{r}|\cos\psi = E\,du\,\delta u + F\,(du\,\delta v + dv\,\delta u) + G\,dv\,\delta v$$

$$\Rightarrow \quad ds\,\delta s\cos\psi = E\,du\,\delta u + F(du\,\delta v + dv\,\delta u) + G\,dv\,\delta v$$

$$\Rightarrow \quad \cos\psi = E\frac{du}{ds}\frac{\delta u}{\delta s} + F\left(\frac{du}{ds}\frac{\delta v}{\delta s} + \frac{dv}{ds}\frac{\delta u}{\delta s}\right) + G\frac{dv}{ds}\frac{\delta v}{\delta s}. \qquad (1)$$

Similarly,

$$d\mathbf{r}\times\delta\mathbf{r} = (\mathbf{r}_1 du + \mathbf{r}_2 dv)\times(\mathbf{r}_1\delta u + \mathbf{r}_2\delta v)$$

$$\Rightarrow \quad |d\mathbf{r}||\delta\mathbf{r}|\sin\psi\,\mathbf{N} = (du\,\delta v - dv\,\delta u)(\mathbf{r}_1\times\mathbf{r}_2)$$

$$\Rightarrow \quad ds\,\delta s\sin\psi\,\mathbf{N} = (du\,\delta v - dv\,\delta u)\,H\mathbf{N} \qquad [\because \mathbf{r}_1\times\mathbf{r}_2 = H\mathbf{N}]$$

$$\Rightarrow \quad ds\,\delta s\sin\psi = H(du\,\delta v - dv\,\delta u)$$

$$\Rightarrow \quad \sin\psi = H\left(\frac{du}{ds}\frac{\delta v}{\delta s} - \frac{dv}{ds}\frac{\delta u}{\delta s}\right). \qquad (2)$$

$$\therefore \quad \tan\psi = \frac{H(du\,\delta v - dv\,\delta u)}{E\,du\,\delta u + F(du\,\delta v + dv\,\delta u) + G\,dv\,\delta v}.$$

Note. If ψ is the angle between the directions *du*, *dv* and δu, δv such that $0 \le \psi \le \pi$, then

$$d\mathbf{r}.\delta\mathbf{r} = (\mathbf{r}_1 du + \mathbf{r}_2 dv).(\mathbf{r}_1\delta u + \mathbf{r}_2\delta v)$$

$$\Rightarrow \quad ds\,\delta s\cos\psi = E\,du\,\delta u + F\,(du\,\delta v + dv\,\delta u) + G\,dv\,\delta v$$

$$\Rightarrow \quad \cos\psi = E\frac{du}{ds}\frac{\delta u}{\delta s} + F\left(\frac{du}{ds}\frac{\delta v}{\delta s} + \frac{dv}{ds}\frac{\delta u}{\delta s}\right) + G\frac{dv}{ds}\frac{\delta v}{\delta s}.$$

Similarly,

$$d\mathbf{r}\times\delta\mathbf{r} = (\mathbf{r}_1 du + \mathbf{r}_2 dv)\times(\mathbf{r}_1\delta u + \mathbf{r}_2\delta v)$$

$$\Rightarrow \quad |d\mathbf{r}\times\delta\mathbf{r}| = |(\mathbf{r}_{11}\,du + \mathbf{r}_2\,dv)\times(\mathbf{r}_{11}\,\delta u + \mathbf{r}_2\,\delta v)|$$

$$\Rightarrow \quad ds\,\delta s\sin\psi = H\,|du\,\delta v - dv\,\delta u|.$$

$$\Rightarrow \quad \sin\psi = H\left|\frac{du}{ds}\frac{\delta v}{\delta s} - \frac{dv}{ds}\frac{\delta u}{\delta s}\right|.$$

$$\therefore \quad \tan\psi = \frac{H\,|du\,\delta v - dv\,\delta u|}{E\,du\,\delta u + F\,(du\,\delta v + dv\,\delta u) + G\,dv\,\delta v}.$$

Corollary 1. If the directions *du*, *dv* and δu, δv are orthogonal, then $\cos\psi = 0$. Therefore, from (1), we have

$$E\frac{du}{ds}\frac{\delta u}{\delta s}+F\left(\frac{du}{ds}\frac{\delta v}{\delta s}+\frac{dv}{ds}\frac{\delta u}{\delta s}\right)+G\frac{dv}{ds}\frac{\delta v}{\delta s}=0.$$

Corollary 2. If $\theta\,(0\le\theta\le\pi)$ is the angle between the direction *du*, *dv* and the parametric curve $v=$ constant, then $\delta v=0$ and $\delta s=\sqrt{E}\,\delta u$. Therefore

$$\cos\theta=\frac{1}{\sqrt{E}}\left(E\frac{du}{ds}+F\frac{dv}{ds}\right),$$

$$\sin\theta=\frac{H}{\sqrt{E}}\left|\frac{dv}{ds}\right|.$$

Corollary 3. If $\phi\,(0\le\phi\le\pi)$ is the angle between the direction *du*, *dv* and the parametric curve $u=$ constant, then $\delta u=0$ and $\delta s=\sqrt{G}\,\delta v$. Therefore

$$\cos\phi=\frac{1}{\sqrt{G}}\left(F\frac{du}{ds}+G\frac{dv}{ds}\right),$$

$$\sin\phi=\frac{H}{\sqrt{G}}\left|\frac{du}{ds}\right|.$$

3.6 DIRECTION COEFFICIENTS

Definition. At a point $P(\mathbf{r}(u,v))$ on a surface $\mathbf{r}=\mathbf{r}(u,v)$, there are three independent vectors, $\mathbf{N}$ (the unit normal to the surface) and $\mathbf{r}_1,\mathbf{r}_2$ (the tangential vectors). Therefore any arbitrary vector $\mathbf{a}$ at *P* can be expressed in the form

$$\mathbf{a}=a_n\mathbf{N}+\lambda\mathbf{r}_1+\mu\mathbf{r}_2, \tag{1}$$

where $\lambda\mathbf{r}_1+\mu\mathbf{r}_2$ is the tangential part of $\mathbf{a}$ and a_n, λ, μ are the scalars defined uniquely by (1).

The tangential part of $\mathbf{a}$, i.e., $\lambda\mathbf{r}_1+\mu\mathbf{r}_2$ is called the tangential vector, written as (λ,μ) and is denoted by $\mathbf{T}$, i.e.,

$$\mathbf{T}=(\lambda,\mu)=\lambda\mathbf{r}_1+\mu\mathbf{r}_2. \tag{2}$$

Therefore

$$|\mathbf{T}|=|\lambda\mathbf{r}_1+\mu\mathbf{r}_2|=(E\lambda^2+2F\lambda\mu+G\mu^2)^{1/2}. \tag{3}$$

The scalar a_n is called the normal component of $\mathbf{a}$, and is given by

$$a_n=\mathbf{a}\,.\,\mathbf{N}. \tag{4}$$

To describe the direction in the tangent plane at *P*, we use the component of a unit vector $\mathbf{e}$ in that direction. These components are written as (l, m) and called direction coefficients.

Therefore the direction coefficients of a direction on a surface can be defined as follows.

Let $\mathbf{T}$ be a vector tangential to the surface at the point P and let $\mathbf{e}$ be the unit vector in the direction of $\mathbf{T}$. If (l, m) are the components of $\mathbf{e}$, i.e., if $\mathbf{e} = l\mathbf{r}_1 + m\mathbf{r}_2$, then (l, m) are called the direction coefficients of the direction represented by the vector $\mathbf{T}$.

Since $\mathbf{e} = l\mathbf{r}_1 + m\mathbf{r}_2$, we have

$$\mathbf{e}^2 = |\, l\mathbf{r}_1 + m\mathbf{r}_2 |^2 = El^2 + 2Flm + Gm^2,$$

i.e., $$El^2 + 2Flm + Gm^2 = 1. \qquad (5)$$

Angle between Two Directions on a Surface at a Point whose Direction Coefficients are known.

Let $P\,((u, v))$ be any point on the surface $\mathbf{r} = \mathbf{r}\,(u, v)$.

Let (l, m) and (l', m') be direction coefficients of two directions on the surface at P, and let $\mathbf{e}$ and $\mathbf{e}'$ be unit vectors in the directions (l, m) and (l', m') respectively. Then

$$\mathbf{e} = l\mathbf{r}_1 + m\mathbf{r}_2$$

and $$\mathbf{e}' = l'\,\mathbf{r}_1 + m'\,\mathbf{r}_2.$$

Therefore if θ is the angle between these two directions, then

$$\mathbf{e}\,.\,\mathbf{e}' = (l\mathbf{r}_1 + m\mathbf{r}_2)\,.\,(l'\,\mathbf{r}_1 + m'\,\mathbf{r}_2)$$

$$\Rightarrow \quad |\,\mathbf{e}\,|\,|\,\mathbf{e}'\,|\cos\theta = ll'\,\mathbf{r}_1\,.\,\mathbf{r}_1 + (lm' + l'm)\,\mathbf{r}_1\,.\,\mathbf{r}_2 + mm'\,\mathbf{r}_2\,.\,\mathbf{r}_2$$

$$\Rightarrow \quad \cos\theta = Ell' + F(lm' + l'm) + Gmm'. \qquad (1)$$

Similarly,

$$\mathbf{e} \times \mathbf{e}' = (l\mathbf{r}_1 + m\mathbf{r}_2) \times (l'\,\mathbf{r}_1 + m'\,\mathbf{r}_2)$$

$$\Rightarrow \quad |\,\mathbf{e}\,|\,|\,\mathbf{e}'\,|\sin\theta\,\mathbf{N} = (lm' - l'm)\;(\mathbf{r}_1 \times \mathbf{r}_2)$$

$$\Rightarrow \quad \sin\theta\,\mathbf{N} = (lm' - l'm)\,H\mathbf{N} \qquad [\because \mathbf{r}_1 \times \mathbf{r}_2 = H\mathbf{N}]$$

$$\Rightarrow \quad \sin\theta = H(lm' - l'm). \qquad (2)$$

From (1) and (2), we have

$$\tan\theta = \frac{H(lm' - l'm)}{Ell' + F(lm' + l'm) + Gmm'}.$$

Note. If θ is such that $0 \le \theta \le \pi$, then

$$\cos\theta = Ell' + F(lm' + l'm) + Gmm',$$

$$\sin\theta = H|\, lm' - l'm\,|.$$

Corollary 1. If the directions having direction coefficients (l, m) and (l', m') are orthogonal, then $\cos\theta = 0$. Therefore, from (1), we have

$$Ell' + F(lm' + l'm) + Gmm' = 0$$

i.e., $$E\frac{l}{m}\frac{l'}{m'} + F\left(\frac{l}{m} + \frac{l'}{m'}\right) + G = 0.$$

Corollary 2. If the directions having direction coefficients (l, m) and (l', m') are coincident, then $\sin\theta = 0$. Therefore, from (2), we have

$$H(lm' - l'm) = 0$$

$$\Rightarrow \quad lm' - l'm = 0$$

$$\Rightarrow \quad \frac{l}{l'} = \frac{m}{m'}.$$

Direction Ratios. If (l, m) are the direction coefficients of a direction at a point P on the surface, then the numbers $\lambda, \mu,$ which are proportional to l, m respectively are called the direction ratios of that direction at P.

To find Direction Coefficients from Direction Ratios.

Let $P((u, v))$ be any point on the surface $\mathbf{r} = \mathbf{r}(u, v)$.

Let (l, m) be the direction coefficients of a direction at P and let (λ, μ) be the direction ratios of the direction at P determined by the direction coefficients (l, m). Then, by definition, we have

$$\frac{l}{\lambda} = \frac{m}{\mu} = k \text{ (say)}.$$

$$\therefore \quad l = \lambda k, m = \mu k. \tag{1}$$

Since (l, m) are direction coefficients,

$$El^2 + 2Flm + Gm^2 = 1$$

or $$(E\lambda^2 + 2F\lambda\mu + G\mu^2)k^2 = 1 \qquad \text{[using (1)]}$$

or $$k^2 = \frac{1}{E\lambda^2 + 2F\lambda\mu + G\mu^2}.$$

$$\therefore \quad k = \pm\frac{1}{\sqrt{(E\lambda^2 + 2F\lambda\mu + G\mu^2)}}.$$

Substituting this value of k in (1), we get

$$(l, m) = \frac{(\lambda, \mu)}{\sqrt{(E\lambda^2 + 2F\lambda\mu + G\mu^2)}}.$$

Direction Coefficients along the Parametric Curves.

Consider the parametric curve

$$v = \text{constant}.$$

We know that $\mathbf{r}_1$ is a vector along the tangent to this curve and we have

$$\mathbf{r}_1 = 1\,(\mathbf{r}_1) + 0\,(\mathbf{r}_2)$$

$\Rightarrow$ Direction ratios along the parametric curve $v = \text{constant}$ are $(1, 0)$.

$\Rightarrow$ Direction coefficients along the parametric curve $v = \text{constant}$ are $\left(\pm\frac{1}{\sqrt{E}}, 0\right)$.

Now, consider the parametric curve

$$u = \text{constant}.$$

We know that $\mathbf{r}_2$ is a vector along the tangent to the parametric curve $u = \text{constant}$ and we have

$$\mathbf{r}_2 = 0\,(\mathbf{r}_1) + 1\,(\mathbf{r}_2)$$

$\Rightarrow$ Direction ratios along the parametric curve $u = \text{constant}$ are $(0, 1)$.

$\Rightarrow$ Direction coefficients along the parametric curve $u = \text{constant}$ are $\left(0, \pm\frac{1}{\sqrt{G}}\right)$.

Angle between two Directions on a Surface at a Point whose Direction Ratios are known.

Let $P((u, v))$ be any point on the surface $\mathbf{r} = \mathbf{r}\,(u, v)$.

Let (λ, μ) and (λ', μ') be direction ratios of two directions on the surface at P. Then $\mathbf{a} = \lambda\mathbf{r}_1 + \mu\mathbf{r}_2$ and $\mathbf{a}' = \lambda'\mathbf{r}_1 + \mu'\mathbf{r}_2$ are vectors in the directions whose direction ratios are (λ, μ) and (λ', μ') respectively. Therefore if θ is the angle between these two directions, then

$$\mathbf{a}\,.\,\mathbf{a}' = (\lambda\mathbf{r}_1 + \mu\mathbf{r}_2)\,.\,(\lambda'\mathbf{r}_1 + \mu'\mathbf{r}_2)$$

$$\Rightarrow \quad |\mathbf{a}|\,|\mathbf{a}'|\cos\theta = E\lambda\lambda' + F\,(\lambda\mu' + \lambda'\mu) + G\mu\mu'$$

$$\Rightarrow \quad \cos\theta = \frac{E\lambda\lambda' + F\,(\lambda\mu' + \lambda'\mu) + G\mu\mu'}{|\mathbf{a}|\,|\mathbf{a}'|}$$

i.e.,

$$\cos\theta = \frac{E\lambda\lambda' + F\,(\lambda\mu' + \lambda'\mu) + G\mu\mu'}{\sqrt{(E\lambda^2 + 2F\lambda\mu + G\mu^2)}\sqrt{(E\lambda'^2 + 2F\lambda'\mu' + G\mu'^2)}}. \tag{1}$$

Similarly,

$$\mathbf{a}\times\mathbf{a}' = (\lambda\mathbf{r}_1 + \mu\mathbf{r}_2)\times(\lambda'\mathbf{r}_1 + \mu'\mathbf{r}_2)$$

$$\Rightarrow \quad |\mathbf{a}|\,|\mathbf{a}'|\sin\theta\,\mathbf{N} = (\lambda\mu' - \lambda'\mu)\ (\mathbf{r}_1\times\mathbf{r}_2)$$

$$\Rightarrow \quad |\mathbf{a}|\,|\mathbf{a}'|\sin\theta\,\mathbf{N} = (\lambda\mu' - \lambda'\mu)\,H\mathbf{N} \qquad [\because \mathbf{r}_1\times\mathbf{r}_2 = H\mathbf{N}]$$

$$\Rightarrow \quad |\mathbf{a}|\,|\mathbf{a}'|\sin\theta = H(\lambda\mu' - \lambda'\mu)$$

$$\Rightarrow \quad \sin\theta = \frac{H(\lambda\mu' - \lambda'\mu)}{|\mathbf{a}|\,|\mathbf{a}'|}$$

i.e.,
$$\sin\theta = \frac{H\,(\lambda\mu' - \lambda'\mu)}{\sqrt{(E\lambda^2 + 2F\lambda\mu + G\mu^2)}\sqrt{(E\lambda'^2 + 2F\lambda'\mu' + G\mu'^2)}}. \tag{2}$$

From (1) and (2), we have

$$\tan\theta = \frac{H\,(\lambda\mu' - \lambda'\mu)}{E\lambda\lambda' + F\,(\lambda\mu' + \lambda'\mu) + G\mu\mu'}.$$

Corollary 1. If the directions having direction coefficients (λ, μ) and (λ', μ') are orthogonal, then $\cos\theta = 0.$ Therefore, from (1), we have

$$E\lambda\lambda' + F\,(\lambda\mu' + \lambda'\mu) + G\mu\mu' = 0$$

i.e.,
$$E\frac{\lambda}{\mu}\frac{\lambda'}{\mu'} + F\left(\frac{\lambda}{\mu} + \frac{\lambda'}{\mu'}\right) + G = 0.$$

Corollary 2. If the directions having direction ratios (λ, μ) and (λ', μ') are coincident, then $\sin\theta = 0$. Therefore, from (2), we have

$$H(\lambda\mu' - \lambda'\mu) = 0$$

$$\Rightarrow \quad \lambda\mu' - \lambda'\mu = 0$$

$$\Rightarrow \quad \frac{\lambda}{\lambda'} = \frac{\mu}{\mu'}.$$

3.7 FAMILY OF CURVES

Definition. Let the equation of a surface be given by

$$\mathbf{r} = \mathbf{r}\,(u, v). \tag{1}$$

Let $\phi\,(u, v)$ be a single valued function of u and v which has continuous derivatives ϕ_1, ϕ_2 which do not vanish together. Then the equation

$$\phi\,(u, v) = c, \tag{2}$$

where c is a real parameter, gives a family of curves lying on the surface (1). For different values of c, we get different curves lying on the surface (1).

Theorem. Through every point of a surface, there passes one and only one member of the family of curves.

Proof. Let the equation of a surface be given by

$$\mathbf{r} = \mathbf{r}(u, v) \tag{1}$$

and let

$$\phi(u, v) = c \tag{2}$$

be a family of curves lying on the surface (1).

Let (u_0, v_0) be any point on the surface (1). Then (u_0, v_0) will also lie on (2) for some value of c, say c_1. Thus we have

$$\phi(u_0, v_0) = c_1.$$

Therefore $\phi(u, v) = c_1$ is a member of family of curves (2) passing through the point (u_0, v_0). Hence it follows that through every point (u_0, v_0) of a surface, there passes one and only one member of the family of curves.

Differential Equation of the Family of Curves.

Let the equation of a surface be given by

$$\mathbf{r} = \mathbf{r}(u, v) \tag{1}$$

and let

$$\phi(u, v) = c \tag{2}$$

be a family of curves lying on the surface (1).

Differentiating (2), we have

$$\frac{\partial \phi}{\partial u} du + \frac{\partial \phi}{\partial v} dv = 0$$

i.e., $$\phi_1 du + \phi_2 dv = 0. \tag{3}$$

Thus the integral of (3) is (2).

Let

$$\phi_1 = P(u, v), \phi_2 = Q(u, v),$$

where P and Q are functions of u, v and do not vanish together. Then (3) can be written as

$$P\,du + Q\,dv = 0. \tag{4}$$

Equation (4) is called the differential equation of the family of curves (2).

3.8 ORTHOGONAL TRAJECTORIES.

Definition. Let the equation of a family of curves lying on the surface $\mathbf{r} = \mathbf{r}(u, v)$ be

$$\phi(u, v) = c. \tag{1}$$

If there exists another family of curves

$$\psi (u, v) = k \tag{2}$$

lying on the same surface such that at every point of the surface, the two curves, one from each family are orthogonal, then the family of curves (2) is called the orthogonal trajectories of the family of curves (1).

Differential Equation of the Orthogonal Trajectories.

Case 1. When the family of curves is given by $\phi (u, v) = c$.

Let the equation of a surface be given by

$$\mathbf{r} = \mathbf{r} (u, v) \tag{1}$$

and let

$$\phi (u, v) = c \tag{2}$$

be the equation of a family of curves lying on the surface (1).

From (2), we have

$$\delta\phi = 0$$

$$\phi_1 \, \delta u + \phi_2 \, \delta v = 0$$

i.e.,
$$\frac{\delta u}{\phi_2} = \frac{\delta v}{-\phi_1} \tag{3}$$

i.e., $(\phi_2, -\phi_1)$ are the direction ratios of the tangent at the point (u, v) to the member of family (2).

Let (du, dv) be the direction ratios of the tangent at the point (u, v) of a member of the orthogonal trajectories of (2).

Then the two directions having direction rations (du, dv) and $(\delta u, \delta v)$ are orthogonal. Therefore

$$E \, du \, \delta u + F \, (du \, \delta v + dv \, \delta u) + G \, dv \, \delta v = 0$$

or
$$E \, du \, (\phi_2) + F \, [du \, (-\phi_1) + dv \, (\phi_2)] + G \, dv \, (-\phi_1) = 0 \qquad \text{[using (3)]}$$

or
$$(E\phi_2 - F\phi_1) \, du + (F\phi_2 - G\phi_1) \, dv = 0. \tag{4}$$

Equation (4) is differential equation of the orthogonal trajectories of (2).

Case 2. When the family of curves is given by $P \, \delta u + Q \, \delta v = 0$.

In this case,

$$\frac{\delta u}{Q} = \frac{\delta v}{-P}. \tag{5}$$

Therefore the differential equation of the orthogonal trajectories is obtained from (4) by putting $\phi_1 = P, \phi_2 = Q,$ which gives

$$(EQ - FP) \, du + (FQ - GP) \, dv = 0. \tag{6}$$

3.9 DOUBLE FAMILY OF CURVES

If P, Q, R are continuous functions of u and v which do not vanish together, then the quadratic differential equation

$$P\,du^2 + 2\,Q\,du\,dv + Q\,dv^2 = 0$$

represents two families of curves on the surface, provided $Q^2 - PR > 0$.

The differential equations for the separate families are obtained by solving the above equation as a quadratic in $\dfrac{du}{dv}$.

Condition for Orthogonality.

Determine the condition that the quadratic differential equation

$$P\,du^2 + 2\,Q\,du\,dv + R\,dv^2 = 0$$

represents two orthogonal families of curves.

Proof. The given quadratic differential equation is

$$P\,du^2 + 2\,Q\,du\,dv + R\,dv^2 = 0 \tag{1}$$

or
$$P\left(\frac{du}{dv}\right)^2 + 2\,Q\frac{du}{dv} + R = 0. \tag{1'}$$

Clearly, (1') is a quadratic equation in $\dfrac{du}{dv}$, giving two values of $\dfrac{du}{dv}$. Each value of $\dfrac{du}{dv}$ gives a family of curves (or directions on the surface). Hence it follows that the quadratic differential equation (1) or (1') represents two families of curves (or directions on the surface).

If the two directions given by (1') be $\dfrac{du}{dv}$ and $\dfrac{\delta u}{\delta v}$, then

$$\frac{du}{dv} + \frac{\delta u}{\delta v} = -\frac{2Q}{P}, \quad \frac{du}{dv}\cdot\frac{\delta u}{\delta v} = \frac{R}{P}. \tag{2}$$

We know that the directions du, dv and δu, δv are orthogonal if

$$E du\,\delta u + F\left(du\,\delta v + dv\,\delta u\right) + G dv\,\delta v = 0$$

i.e.,
$$E\frac{du}{dv}\frac{\delta u}{\delta v} + F\left(\frac{du}{dv} + \frac{\delta u}{\delta v}\right) + G = 0$$

i.e.,
$$E\left(\frac{R}{P}\right) + F\left(-\frac{2Q}{P}\right) + G = 0 \qquad \text{[using (2)]}$$

i.e.,
$$ER - 2FQ + GP = 0. \tag{3}$$

This is the condition for the two families of curves given by (1) to be orthogonal.

Solved Examples

Example 1. If θ is the angle at the point (u, v) between the two directions given by

$$P\,du^2 + 2Q\,du\,dv + R\,dv^2 = 0,$$

show that

$$\tan\theta = \frac{2H(Q^2 - PR)^{1/2}}{ER - 2FQ + GP}.$$

Solution. The given quadratic equation is

$$P\,du^2 + 2Q\,du\,dv + R\,dv^2 = 0 \tag{1}$$

or $$P\left(\frac{du}{dv}\right)^2 + 2Q\frac{du}{dv} + R = 0. \tag{1'}$$

Clearly, (1') is a quadratic equation in $\frac{du}{dv}$, giving two values of $\frac{du}{dv}$. Each value of $\frac{du}{dv}$ gives a family of curves (or directions on the surface). Hence it follows that the quadratic differential equation (1) or (1') represents two families of curves (or directions on the surface).

If the two directions given by (1') be $\frac{du}{dv}$ and $\frac{\delta u}{\delta v}$, then

$$\frac{du}{dv} + \frac{\delta u}{\delta v} = -\frac{2Q}{P}, \quad \frac{du}{dv}\cdot\frac{\delta u}{\delta v} = \frac{R}{P}. \tag{2}$$

We know that the angle θ between the directions du, dv and δu, δv is given by

$$\tan\theta = \frac{H(du\,\delta v - dv\,\delta u)}{E\,du\,\delta u + F(du\,\delta v + dv\,\delta u) + G\,dv\,\delta v}$$

or $$\tan\theta = \frac{H\left(\frac{du}{dv} - \frac{\delta u}{\delta v}\right)}{E\frac{du}{dv}\frac{\delta u}{\delta v} + F\left(\frac{du}{dv} + \frac{\delta u}{\delta v}\right) + G}$$

or $$\tan\theta = \frac{H\left[\left(\frac{du}{dv} + \frac{\delta u}{\delta v}\right)^2 - 4\frac{du}{dv}\frac{\delta u}{\delta v}\right]^{1/2}}{E\frac{du}{dv}\frac{\delta u}{\delta v} + F\left(\frac{du}{dv} + \frac{\delta u}{\delta v}\right) + G}$$

$$= \frac{H\left[\left(-\frac{2Q}{P}\right)^2 - 4\frac{R}{P}\right]^{1/2}}{E\left(\frac{R}{P}\right) + F\left(-\frac{2Q}{P}\right) + G} \qquad \text{[using (2)]}$$

$$= \frac{2H(Q^2 - PR)^{1/2}}{ER - 2FQ + GP}.$$

Example 2. If the parametric curves are orthogonal, show that the differential equation of lines on the surface cutting the curves $u = \text{constant}$ at a constant angle β is

$$\frac{du}{dv} = \tan\beta\sqrt{\frac{G}{E}}.$$

Solution. Since the parametric curves are orthogonal,

$$F = 0 \tag{1}$$

$$\Rightarrow \quad H^2 = EG - F^2 = EG$$

$$\Rightarrow \quad H = \sqrt{EG}. \tag{2}$$

We know that the direction ratios of the parametric curve $u = \text{constant}$ are (0, 1), i.e.,

$$(\delta u, \delta v) = (0,1). \tag{3}$$

Let (du, dv) be the direction ratios of the tangent to the curve which cuts the curve $u = \text{constant}$ at a constant angle β.

We know that the angle θ between the directions having direction ratios (du, dv) and (δu, δv) is given by

$$\tan\theta = \frac{H(du\,\delta v - dv\,\delta u)}{Edu\,\delta u + F(du\,\delta v + dv\,\delta u) + Gdv\,\delta v}.$$

Therefore, the angle β between the directions having direction ratios (du, dv) and (δu, δv) = (0, 1) is given by

$$\tan\beta = \frac{H(du\,\delta v - dv\,\delta u)}{Edu\,\delta u + F(du\,\delta v + dv\,\delta u) + Gdv\,\delta v}$$

i.e., $$\tan\beta = \frac{\sqrt{EG}(du - 0)}{E\,(0) + 0 + Gdv} \qquad \text{[using (1), (2) and (3)]}$$

$$= \sqrt{\frac{E}{G}}\frac{du}{dv}.$$

$$\therefore \quad \frac{du}{dv} = \tan\beta\sqrt{\frac{G}{E}}.$$

Example 3. Show that the curves bisecting the angle between the parametric curves are given by

$$E\,du^2 - G\,dv^2 = 0.$$

Solution. We know that the direction coefficients of the parametric curves $v =$ constant and $u =$ constant are $(1/\sqrt{E}, 0)$ and $(0, 1/\sqrt{G})$ respectively.

Let $\left(\frac{du}{ds}, \frac{dv}{ds}\right)$ be the direction coefficients of the direction which bisects the angle between the parametric curves. Let θ_1 and θ_2 be the angles which this direction makes with the parametric curves $v =$ constant and $u =$ constant respectively. Then

$$\cos\theta_1 = E\frac{1}{\sqrt{E}}\frac{du}{ds} + F\frac{1}{\sqrt{E}}\frac{dv}{ds} \tag{1}$$

and $$\cos\theta_2 = F\frac{1}{\sqrt{G}}\frac{du}{ds} + G\frac{1}{\sqrt{G}}\frac{dv}{ds}. \tag{2}$$

Since the direction $\left(\frac{du}{ds}, \frac{dv}{ds}\right)$ bisects the angle between the parametric curves,

$$\cos\theta_1 = \pm\cos\theta_2$$

or $$E\frac{1}{\sqrt{E}}\frac{du}{ds} + F\frac{1}{\sqrt{E}}\frac{dv}{ds} = \pm\left(F\frac{1}{\sqrt{G}}\frac{du}{ds} + G\frac{1}{\sqrt{G}}\frac{dv}{ds}\right)$$

[using (1) and (2)]

or $$E\sqrt{G}\,du + F\sqrt{G}\,dv = \pm(F\sqrt{E}\,du + G\sqrt{E}\,dv).$$

Squaring both sides, we get

$$(E\sqrt{G}\,du + F\sqrt{G}\,dv)^2 = (F\sqrt{E}\,du + G\sqrt{E}\,dv)^2$$

or $$E^2G\,du^2 + 2EFG\,du\,dv + F^2G\,dv^2 = F^2E\,du^2 + 2EFG\,du\,dv + G^2E\,dv^2$$

or $$E^2G\,du^2 + F^2G\,dv^2 = F^2E\,du^2 + G^2E\,dv^2$$

or $$E(EG - F^2)\,du^2 = G(EG - F^2)\,dv^2$$

or $$E\,du^2 = G\,dv^2 \qquad [\because EG - F^2 \neq 0]$$

or $$E\,du^2 - G\,dv^2 = 0.$$

Hence the result.

Example 4. Show that the curves $du^2 - (u^2 + c^2)\,dv^2 = 0$ form an orthogonal system on the right helicoid $\mathbf{r} = (u\cos v, u\sin v, cv)$.

Solution. We know that the two family of curves given by

$$P\,du^2 + 2Q\,du\,dv + R\,dv^2 = 0 \qquad (1)$$

are orthogonal if and only if

$$ER - 2FQ + GP = 0. \qquad (2)$$

Therefore we have to find the values of E, F, G, P, Q and R.
The equation of the given right helicoid is

$$\mathbf{r} = (u\cos v, u\sin v, cv). \qquad (3)$$

Taking u and v as the first and the second parameters, we have

$$\mathbf{r}_1 = (\cos v, \sin v, 0),$$

$$\mathbf{r}_2 = (-u\sin v, u\cos v, c).$$

$$\therefore \quad E = \mathbf{r}_1^{\,2} = \cos^2 v + \sin^2 v + 0 = 1,$$

$$F = \mathbf{r}_1 \,.\, \mathbf{r}_2 = -u\cos v\sin v + u\sin v\cos v + 0 = 0,$$

$$G = \mathbf{r}_2^{\,2} = u^2\sin^2 v + u^2\cos^2 v + c^2 = u^2 + c^2.$$

The given family of curves is

$$du^2 - (u^2 + c^2)\,dv^2 = 0. \qquad (4)$$

Comparing the given family (4) with (1), we get

$$P = 1,\ Q = 0,\ R = -(u^2 + c^2).$$

$$\therefore \quad ER - 2FQ + GP = 1.[-(u^2 + c^2)] - 0 + (u^2 + c^2).1 = 0.$$

Hence the condition (2) is satisfied for the given family (4). Therefore the two families of curves given by (4) form an orthogonal system on the given right helicoid (3).

Example 5. On the paraboloid $x^2 - y^2 = z$, find the orthogonal trajectories of the sections by the planes $z =$ constant.

Solution. The equation of the given paraboloid is

$$x^2 - y^2 = z.$$

Let $x = u, y = v, z = u^2 - v^2$.

Then the equation of the given paraboloid may be written in vector form as

$$\mathbf{r} = (u, v, u^2 - v^2).$$

$$\therefore \quad \mathbf{r}_1 = (1, 0, 2u), \quad \mathbf{r}_2 = (0, 1, -2v)$$

and $\quad E = \mathbf{r}_1^{\,2} = 1 + 4u^2, F = \mathbf{r}_1 \,.\, \mathbf{r}_2 = -4uv, G = \mathbf{r}_2^{\,2} = 1 + 4v^2. \qquad (1)$

The given family of curves is

$$z = \text{constant}$$

or $\quad u^2 - v^2 = \text{constant}. \qquad (2)$

Differentiating (2), we get

$$2u\,\delta u - 2v\,\delta v = 0$$

or $$\frac{\delta u}{v} = \frac{\delta v}{u} \qquad (3)$$

i.e., (v, u) are the direction ratios of the tangent at the point (u, v) to the member of family (2).

Let (du, dv) be the direction ratios of the tangent at the point (u, v) of a member of the orthogonal trajectories of (2).

Then the two directions having direction rations (du, dv) and $(\delta u, \delta v)$ are orthogonal. Therefore

$$E\,du\,\delta u + F\left(du\,\delta v + dv\,\delta u\right) + G\,dv\,\delta v = 0$$

or $$(1+4u^2)\,v\,du + (-4uv)\,(u\,du + v\,dv) + (1+4v^2)\,u\,dv = 0$$

[using (1) and (3)]

or $$v\,du + u\,dv = 0.$$

Integrating this, we get

$$uv = \text{constant.} \qquad (4)$$

Equation (4) is the equation of the orthogonal trajectories of (2).

But $u = x, v = y$. Hence the orthogonal trajectories of the given family of curves are the sections of the paraboloid by the hyperbolic cylinder xy = constant.

Example 6. A helicoid is generated by the screw motion of a straight line which meets the axis at an angle α. Find the orthogonal trajectories of the generators. Find also the metric of the surface referred to the generators and their orthogonal trajectories as parametric curves.

Solution. First part. Let the axis of the helicoid be taken as the z-axis. Let the generating line OA be initially in the zx-plane. The line OA makes an angle α with OZ.

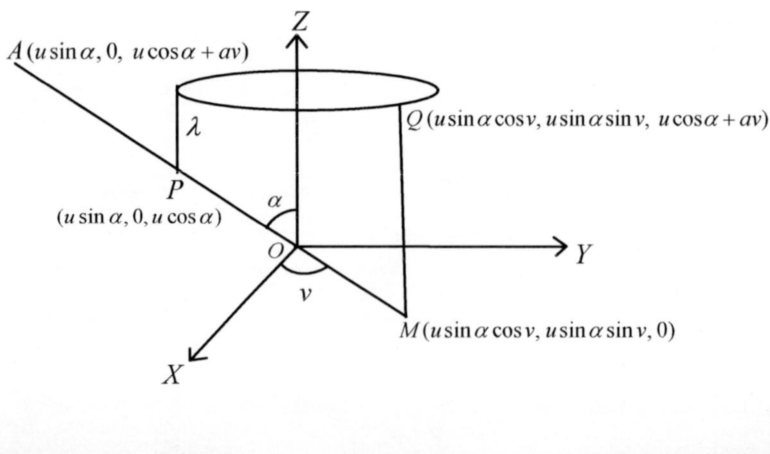

Let $OP = u$, where P is any point on the line OA. Then the coordinates of P are $(u\sin\alpha, 0, u\cos\alpha)$.

Now suppose that the line OA translates through a distance λ parallel to OZ and then revolves through an angle v about OZ, where

$$\frac{\lambda}{v} = a \text{ (constant)}.$$

Let Q be the final position of the point P. Since $\lambda = av$, coordinates of Q are $(u\sin\alpha\cos v, u\sin\alpha\sin v, u\cos\alpha + av)$. Thus the position vector of a current point Q on the helicoid is given by

$$\mathbf{r} = (u\sin\alpha\cos v, u\sin\alpha\sin v, u\cos\alpha + av).$$

$$\therefore \quad \mathbf{r}_1 = (\sin\alpha \cos v, \sin\alpha \sin v, \cos\alpha),$$

$$\mathbf{r}_2 = (-u \sin\alpha \sin v, u \sin\alpha \cos v, a)$$

and

$$\left.\begin{aligned} E &= \mathbf{r}_1^{\,2} = \sin^2\alpha\cos^2 v + \sin^2\alpha\sin^2 v + \cos^2\alpha = 1, \\ F &= \mathbf{r}_1 \cdot \mathbf{r}_2 = a\cos\alpha, \\ G &= \mathbf{r}_2^{\,2} = u^2\sin^2\alpha + a^2. \end{aligned}\right\} \tag{1}$$

The generators of the helicoid are the curves

$$v = \text{constant}. \tag{2}$$

We know that the direction ratios of the tangent to the curve given by (2) at the point (u, v) are $(1, 0)$, i.e.,

$$(\delta u, \delta v) = (1, 0). \tag{3}$$

Let (du, dv) be the direction ratios of the tangent at the point (u, v) of a member of the orthogonal trajectories of (2).

Then the two directions having direction ratios (du, dv) and $(\delta u, \delta v)$ are orthogonal. Therefore

$$E\, du\, \delta u + F\,(du\,\delta v + dv\,\delta u) + G\, dv\,\delta v = 0$$

or $\quad du + a\cos\alpha\,(0 + dv) + 0 = 0 \quad$ [using (1), (2) and (3)]

or $$du + a\cos\alpha\, dv = 0. \tag{4}$$

Equation (4) is the differential equation of the orthogonal trajectories of (2).

Integrating (4), we get

$$u + (a\cos\alpha)\, v = \text{constant}. \tag{5}$$

Equation (5) is the equation of the orthogonal trajectories of (2).

Second part. If the generators (2) and their orthogonal trajectories (5) are taken as the parametric curves, then the new parameters u', v' are given by

$$u' = u + (a\cos\alpha)\, v, \;\; v' = v.$$

$\therefore \qquad u = u' - (a\cos\alpha)\,v',\, v = v'$

$$\Rightarrow \qquad \frac{\partial u}{\partial u'} = 1, \frac{\partial u}{\partial v'} = a\cos\alpha, \frac{\partial v}{\partial u'} = 0, \frac{\partial v}{\partial v'} = 1. \tag{6}$$

Now, taking u' and v' as the first and the second parameters and using (6), we have

$$\mathbf{r}_1' = \frac{\partial \mathbf{r}}{\partial u'} = \frac{\partial \mathbf{r}}{\partial u}\frac{\partial u}{\partial u'} + \frac{\partial \mathbf{r}}{\partial v}\frac{\partial v}{\partial u'} = \frac{\partial \mathbf{r}}{\partial u} = \mathbf{r}_1,$$

$$\mathbf{r}_2' = \frac{\partial \mathbf{r}}{\partial v'} = \frac{\partial \mathbf{r}}{\partial u}\frac{\partial u}{\partial v'} + \frac{\partial \mathbf{r}}{\partial v}\frac{\partial v}{\partial v'} = (-a\cos\alpha)\,\mathbf{r}_1 + \mathbf{r}_2.$$

Therefore the new first fundamental coefficients E', F', G' are given by

$$E' = \mathbf{r}_1'^2 = \mathbf{r}_1^2 = 1,$$

$$F' = 0, \qquad \text{[since new parametric curves are orthogonal]}$$

$$\begin{aligned} G' &= \mathbf{r}_2'^2 = [-(a\cos\alpha)\,\mathbf{r}_1 + \mathbf{r}_2]^2 \\ &= (a^2\cos^2\alpha)\,\mathbf{r}_1^2 - (2a\cos\alpha)\,\mathbf{r}_1 . \mathbf{r}_2 + \mathbf{r}_2^2 \\ &= a^2\cos^2\alpha - (2a\cos\alpha)\,a\cos\alpha + u^2\sin^2\alpha + a^2 \qquad \text{[using (1)]} \\ &= \sin^2\alpha\,(u^2 + a^2) \\ &= \sin^2\alpha\,[\{u' - (a\cos\alpha)\,v'\}^2 + a^2]. \end{aligned}$$

Hence the metric of the surface referred to the new parameters is

$$\begin{aligned} ds^2 &= E'\,du'^2 + 2F'\,du'\,dv' + G'\,dv'^2 \\ &= du'^2 + \sin^2\alpha\,[\{u' - (a\cos\alpha)\,v'\}^2 + a^2]\,dv'^2. \end{aligned}$$

Example 7. The metric of a surface is $v^2du^2 + u^2dv^2$. Find the equation of the family of curves orthogonal to the curves $uv =$ constant and find the metric referred to the new parameters so that these two families are parametric curves.

Solution. **First part.** The metric of the given surface is

$$v^2du^2 + u^2dv^2 = ds^2. \tag{1}$$

Comparing (1) with the metric

$$E\,du^2 + 2F\,du\,dv + G\,dv^2 = ds^2,$$

we get

$$E = v^2, F = 0, G = u^2. \tag{2}$$

The equation of the given family of curves is

$$uv = \text{constant}. \tag{3}$$

Differentiating (3), we get

$$v\,\delta u + u\,\delta v = 0$$

or $$\frac{\delta u}{u}=\frac{\delta v}{-v} \tag{4}$$

i.e., $(u,-v)$ are the direction ratios of the tangent at the point (u, v) to the member of family (3).

Let (du, dv) be the direction ratios of the tangent at the point (u, v) of a member of the orthogonal trajectories of (3).

Then the two directions having direction rations (du, dv) and $(\delta u, \delta v)$ are orthogonal. Therefore

$$E\,du\,\delta u+F\left(du\,\delta v+dv\,\delta u\right)+G\,dv\,\delta v=0$$

or $$v^2u\,du+0-u^2v\,dv=0 \qquad \text{[using (2) and (4)]}$$

or $$\frac{du}{u}=\frac{dv}{v}.$$

Integrating this, we get

$$\log u=\log v+\log k$$

or $$\log\frac{u}{v}=\log k$$

or $$\frac{u}{v}=\text{constant}. \tag{5}$$

Equation (5) is the equation of the orthogonal trajectories of (3).

Second part. If the family of curves (3) and their orthogonal trajectories (5) are taken as the parametric curves, then the new parameters u', v' are given by

$$u'=\frac{u}{v},\ v'=uv.$$

$$\therefore \quad u^2=u'v',\ v^2=\frac{v'}{u'}$$

$$\Rightarrow \quad 2u\frac{\partial u}{\partial u'}=v',\ 2u\frac{\partial u}{\partial v'}=u',\ 2v\frac{\partial v}{\partial u'}=-\frac{v'}{u'^2},\ 2v\frac{\partial v}{\partial v'}=\frac{1}{u'}$$

$$\Rightarrow \quad \frac{\partial u}{\partial u'}=\frac{v'}{2u},\ \frac{\partial u}{\partial v'}=\frac{u'}{2u},\ \frac{\partial v}{\partial u'}=-\frac{v'}{2vu'^2},\ \frac{\partial v}{\partial v'}=\frac{1}{2vu'}. \tag{6}$$

Now, taking u' and v' as the first and the second parameters and using (6), we have

$$\mathbf{r}_1'=\frac{\partial\mathbf{r}}{\partial u'}=\frac{\partial\mathbf{r}}{\partial u}\frac{\partial u}{\partial u'}+\frac{\partial\mathbf{r}}{\partial v}\frac{\partial v}{\partial u'}=\frac{v'}{2u}\mathbf{r}_1-\frac{v'}{2vu'^2}\mathbf{r}_2,$$

$$\mathbf{r}_2'=\frac{\partial\mathbf{r}}{\partial v'}=\frac{\partial\mathbf{r}}{\partial u}\frac{\partial u}{\partial v'}+\frac{\partial\mathbf{r}}{\partial v}\frac{\partial v}{\partial v'}=\frac{u'}{2u}\mathbf{r}_1+\frac{1}{2vu'}\mathbf{r}_2.$$

Therefore the new first fundamental coefficients E', F', G' are given by

$$E' = \mathbf{r}_1'^2 = \left(\frac{v'}{2u}\mathbf{r}_1 - \frac{v'}{2vu'^2}\mathbf{r}_2\right)^2$$

$$= \frac{v'^2}{4u^2}\mathbf{r}_1^2 + \frac{v'^2}{4v^2u'^4}\mathbf{r}_2^2 \qquad [\because \mathbf{r}_1 . \mathbf{r}_2 = F = 0]$$

$$= E\frac{v'^2}{4u^2} + G\frac{v'^2}{4v^2u'^4}$$

$$= v^2\frac{v'^2}{4u^2} + u^2\frac{v'^2}{4v^2u'^4} \qquad \text{[using (2)]}$$

$$= \frac{v'^2}{4u'^2} + \frac{v'^2}{4u'^2} \qquad \left[\because u' = \frac{u}{v}\right]$$

$$= \frac{v'^2}{2u'^2},$$

$$F' = 0, \qquad \text{[since new parametric curves are orthogonal]}$$

$$G' = \mathbf{r}_2'^2 = \left(\frac{u'}{2u}\mathbf{r}_1 + \frac{1}{2vu'}\mathbf{r}_2\right)^2$$

$$= \frac{u'^2}{4u^2}\mathbf{r}_1^2 + \frac{1}{4v^2u'^2}\mathbf{r}_2^2$$

$$= E\frac{u'^2}{4u^2} + G\frac{1}{4v^2u'^2}$$

$$= v^2\frac{u'^2}{4u^2} + u^2\frac{1}{4v^2u'^2} \qquad \text{[using (2)]}$$

$$= \frac{1}{4} + \frac{1}{4} \qquad \left[\because u' = \frac{u}{v}\right]$$

$$= \frac{1}{2}.$$

Hence the metric of the surface referred to the new parameters is

$$ds^2 = E'\,du'^2 + 2F'\,du'\,dv' + G'\,dv'^2$$

$$= \frac{v'^2}{2u'^2}du'^2 + \frac{1}{2}dv'^2.$$

Example 8. Show that the parametric curves on the sphere

$$x = a\sin u\cos v,\ y = a\sin u\sin v,\ z = a\cos u$$

form an orthogonal system. Determine the two families of curves which meet the curves v = constant at angles $\pi/4$ and $3\pi/4$.

Solution. First part. The position vector **r** of any point on the given sphere is

$$\mathbf{r} = a(\sin u \cos v, \sin u \sin v, \cos u). \tag{1}$$

$$\therefore \quad \mathbf{r}_1 = a(\cos u \cos v, \cos u \sin v, -\sin u),$$

$$\mathbf{r}_2 = a(-\sin u \sin v, \sin u \cos v, 0)$$

and

$$E = \mathbf{r}_1^2 = a^2,\ F = \mathbf{r}_1 \cdot \mathbf{r}_2 = 0,\ G = \mathbf{r}_2^2 = a^2 \sin^2 u. \tag{2}$$

Since $F = 0$, it follows that the parametric curves on the sphere given by (1) are orthogonal.

Second part. We have

$$H^2 = EG - F^2 = a^4 \sin^2 u.$$

$$\therefore \quad H = a^2 \sin u. \tag{3}$$

Now, consider the given curve

$$v = \text{constant}. \tag{4}$$

We know that the direction ratios of the tangent to the curve given by (4) at the point (u, v) are $(1, 0)$, i.e.,

$$(\delta u, \delta v) = (1, 0). \tag{5}$$

Let (du, dv) be the direction ratios of the tangent at the point (u, v) of the curve which makes an angle $\pi/4$ with the curve (4). Then

$$\tan\frac{\pi}{4} = \frac{H\,|du\,\delta v - dv\,\delta u|}{E du\,\delta u + F(du\,\delta v + dv\,\delta u) + G dv\,\delta v}.$$

or $$1 = \frac{a^2 \sin u\,|0 - dv|}{a^2 du + 0 + 0} \qquad \text{[using (2), (3) and (5)]}$$

or $$\operatorname{cosec} u\, du = dv.$$

Integrating this, we get

$$\log\tan\frac{u}{2} = v + \log c_1$$

or $$\tan\frac{u}{2} = c_1 e^v$$

or $$e^v \cot\frac{u}{2} = c_1. \tag{6}$$

Equation (6) is the equation of the families of curves which make angle $\pi/4$ with the curves v = constant.

The differential equation of the families of curves which make angle $3\pi/4$ with the curves v = constant is given by

$$\tan\frac{3\pi}{4} = \frac{H\,|\,du\,\delta v - dv\,\delta u\,|}{E\,du\,\delta u + F(du\,\delta v + dv\,\delta u) + G\,dv\,\delta v}.$$

or $$-1 = \frac{a^2 \sin u\,|\,0 - dv\,|}{a^2 du + 0 + 0}$$ [using (2), (3) and (5)]

or $$-\operatorname{cosec} u\, du = dv.$$

Integrating this, we get

$$-\log\tan\frac{u}{2} = v + \log c_2$$

or $$\log\cot\frac{u}{2} = v + \log c_2$$

or $$\cot\frac{u}{2} = c_2 e^v$$

or $$e^v \tan\frac{u}{2} = c_2. \tag{7}$$

Equation (7) is the equation of the families of curves which make angle $3\pi/4$ with the curves $v = \text{constant}$.

Exercises

1. If ψ is the angle between the two directions

$$P\,du^2 + Q\,du\,dv + R\,dv^2 = 0,$$

show that

$$\tan\psi = \frac{H(Q^2 - 4PR)^{1/2}}{ER - FQ + GP}.$$

Hence find the condition that the two directions be orthogonal.

[**Ans.** $ER - FQ + GP = 0$.]

2. Find the direction which makes an angle $\pi/2$ with the direction whose coefficients are (l, m).

3. Show that on the right helicoid, the family of curves orthogonal to the curves $u\cos v = \text{constant}$ is the family

$$(u^2 + a^2)\sin^2 v = \text{constant}.$$

4. Show that the parametric curves are orthogonal on the surface

$$x = u\cos v,\ y = u\sin v,\ z = a\log\{u + (u^2 - a^2)^{1/2}\}.$$

4

Curves on a Surface

4.1 CURVATURE OF NORMAL SECTION OR NORMAL CURVATURE

Normal Section. Let $P(u, v)$ be a point on a surface. A section of the surface containing the normal to the surface at P is called a normal section of the surface at the point P. Such a section is a plane curve whose principal normal $\mathbf{n}$ is parallel to the surface normal $\mathbf{N}$. We adopt the convention that the vector $\mathbf{n}$ has the same direction as the vector $\mathbf{N}$.

Normal Curvature. The curvature of the normal section at $P(u, v)$ parallel to the direction (du, dv) is called the normal curvature at $P(u, v)$ in the direction (du, dv).

The reciprocal of the normal curvature is called the radius of normal curvature.

Formula for Normal Curvature in terms of Fundamental magnitudes.

Let the parametric equation of a surface in vector form be given by

$$\mathbf{r} = \mathbf{r}(u, v). \tag{1}$$

Take any point $P(u, v)$ on this surface and let $\mathbf{N}$ be the unit normal vector to this surface at the point P. Let Γ denote the normal section of the surface at P in the direction (du, dv) and let κ_n be the curvature of Γ at P in this direction. Then κ_n is positive when the curve is concave on the side toward which $\mathbf{N}$ is directed.

If $\mathbf{n}$ is the unit principal normal of Γ at P and if dashes denote differentiation with respect to the arc length s of the curve, then

$$\mathbf{r}'' = \kappa_n \mathbf{n} = \kappa_n \mathbf{N}. \qquad [\because \mathbf{n} = \mathbf{N}]$$

$$\therefore \quad \kappa_n = \mathbf{N} \cdot \mathbf{r}''. \tag{2}$$

But

$$\mathbf{r}' = \frac{d\mathbf{r}}{ds} = \frac{\partial \mathbf{r}}{\partial u}\frac{du}{ds} + \frac{\partial \mathbf{r}}{\partial v}\frac{dv}{ds} = \mathbf{r}_1 u' + \mathbf{r}_2 v'.$$

$$\therefore \quad \mathbf{r}'' = \frac{d\mathbf{r}_1}{ds}u' + \mathbf{r}_1 u'' + \frac{d\mathbf{r}_2}{ds}v' + \mathbf{r}_2 v''$$

$$= \left(\frac{\partial \mathbf{r}_1}{\partial u}\frac{du}{ds} + \frac{\partial \mathbf{r}_1}{\partial v}\frac{dv}{ds}\right)u' + \mathbf{r}_1 u'' + \left(\frac{\partial \mathbf{r}_2}{\partial u}\frac{du}{ds} + \frac{\partial \mathbf{r}_2}{\partial v}\frac{dv}{ds}\right)v' + \mathbf{r}_2 v''$$

$$= (\mathbf{r}_{11}u' + \mathbf{r}_{12}v')\,u' + \mathbf{r}_1 u'' + (\mathbf{r}_{21}u' + \mathbf{r}_{22}v')\,v' + \mathbf{r}_2 v''$$

i.e., $\mathbf{r}'' = \mathbf{r}_1 u'' + \mathbf{r}_2 v'' + \mathbf{r}_{11} u'^2 + 2\mathbf{r}_{12} u'v' + \mathbf{r}_{22} v'^2.$ (3)

Substituting this value of $\mathbf{r}''$ in (2) and using

$$\mathbf{N}\,.\,\mathbf{r}_1 = 0,\ \ \mathbf{N}\,.\,\mathbf{r}_2 = 0,\ \ \mathbf{N}\,.\,\mathbf{r}_{11} = L,\ \ \mathbf{N}\,.\,\mathbf{r}_{12} = M,\ \ \mathbf{N}\,.\,\mathbf{r}_{22} = N,$$

we get

$$\kappa_n = Lu'^2 + 2Mu'v' + Nv'^2$$

or $$\kappa_n = L\left(\frac{du}{ds}\right)^2 + 2M\left(\frac{du}{ds}\right)\left(\frac{dv}{ds}\right) + N\left(\frac{dv}{ds}\right)^2$$

or $$\kappa_n = \frac{L\,du^2 + 2M\,du\,dv + N\,dv^2}{ds^2}$$

or $$\kappa_n = \frac{L\,du^2 + 2M\,du\,dv + N\,dv^2}{E\,du^2 + 2F\,du\,dv + G\,dv^2}. \tag{4}$$

This gives the normal curvature at $P(u, v)$ in the direction (du, dv).

Meusnier's Theorem. If κ_n and κ are the curvature of the normal and oblique sections through the same tangent line and θ is the angle between the planes of these sections, then

$$\kappa_n = \kappa \cos\theta.$$

Proof. Let the parametric equation of a surface in vector form be given by

$$\mathbf{r} = \mathbf{r}(u, v). \tag{1}$$

Take any point $P(u, v)$ on this surface and let $\mathbf{N}$ be the unit normal vector to this surface at the point P. Then the unit principal normal to the normal section at P is $\mathbf{N}$.

Let $\mathbf{n}$ be the unit principal normal to an oblique section of surface (1) at P and let θ be the inclination of the plane of this section to the plane of the normal section at P. Then θ is the angle between the vectors $\mathbf{N}$ and $\mathbf{n}$. Therefore

$$\cos\theta = \mathbf{N}\,.\,\mathbf{n}$$

or $$\cos\theta = \mathbf{N}\,.\,\frac{\mathbf{r}''}{\kappa}, \tag{2}$$

$[\because \mathbf{r}'' = \kappa\mathbf{n}]$

where κ is the curvature of the oblique section at P.

But

$$\mathbf{r}' = \frac{d\mathbf{r}}{ds} = \frac{\partial\mathbf{r}}{\partial u}\frac{du}{ds} + \frac{\partial\mathbf{r}}{\partial v}\frac{dv}{ds} = \mathbf{r}_1 u' + \mathbf{r}_2 v'.$$

$\therefore$ $$\mathbf{r}'' = \frac{d\mathbf{r}_1}{ds}u' + \mathbf{r}_1 u'' + \frac{d\mathbf{r}_2}{ds}v' + \mathbf{r}_2 v''$$

or $$\mathbf{r}'' = \left(\frac{\partial \mathbf{r}_1}{\partial u}\frac{du}{ds} + \frac{\partial \mathbf{r}_1}{\partial v}\frac{dv}{ds}\right)u' + \mathbf{r}_1 u'' + \left(\frac{\partial \mathbf{r}_2}{\partial u}\frac{du}{ds} + \frac{\partial \mathbf{r}_2}{\partial v}\frac{dv}{ds}\right)v' + \mathbf{r}_2 v''$$

$$= (\mathbf{r}_{11}u' + \mathbf{r}_{12}v')\,u' + \mathbf{r}_1 u'' + (\mathbf{r}_{21}u' + \mathbf{r}_{22}v')\,v' + \mathbf{r}_2 v''$$

i.e., $$\mathbf{r}'' = \mathbf{r}_1 u'' + \mathbf{r}_2 v'' + \mathbf{r}_{11}u'^2 + 2\mathbf{r}_{12}u'v' + \mathbf{r}_{22}v'^2. \tag{3}$$

Substituting this value of $\mathbf{r}''$ in (2) and using

$$\mathbf{N}\cdot\mathbf{r}_1 = 0,\ \ \mathbf{N}\cdot\mathbf{r}_2 = 0,\ \ \mathbf{N}\cdot\mathbf{r}_{11} = L,\ \ \mathbf{N}\cdot\mathbf{r}_{12} = M,\ \ \mathbf{N}\cdot\mathbf{r}_{22} = N,$$

we get

$$\cos\theta = \frac{1}{\kappa}(Lu'^2 + 2Mu'v' + Nv'^2)$$

$$= \frac{1}{\kappa}\left[L\left(\frac{du}{ds}\right)^2 + 2M\left(\frac{du}{ds}\right)\left(\frac{dv}{ds}\right) + N\left(\frac{dv}{ds}\right)^2\right]$$

$$= \frac{1}{\kappa}\left(\frac{L\,du^2 + 2M\,du\,dv + N\,dv^2}{ds^2}\right)$$

$$= \frac{1}{\kappa}\left(\frac{L\,du^2 + 2M\,du\,dv + N\,dv^2}{E\,du^2 + 2F\,du\,dv + G\,dv^2}\right)$$

$$= \frac{\kappa_n}{\kappa}, \qquad \left[\because \frac{L\,du^2 + 2M\,du\,dv + N\,dv^2}{E\,du^2 + 2F\,du\,dv + G\,dv^2} = \kappa_n\right]$$

where κ_n is the curvature of the normal section at P parallel to the direction (du, dv).

$\therefore \qquad \kappa_n = \kappa\cos\theta.$

4.2 PRINCIPAL DIRECTIONS AND PRINCIPAL CURVATURES

Definition. Let S be a surface and let P be a point on it.

The normal curvatures of S at P in different directions are different. The directions on S in which the normal curvatures of S at P are maximum or minimum are known as the principal directions of S at P.

The normal curvatures of S at P in principal directions are known as the principal curvatures of S at P.

The reciprocals of the principal curvatures of S at P are known as the principal radii of curvature of S at P.

Equation giving the Principal Curvatures.

Let u and v denote the parametric coordinates of a point on a surface.

Take any point $P(u, v)$ on this surface and let κ be the normal curvature of the surface at P in the direction (du, dv).

We know that

$$\kappa = \frac{L\,du^2 + 2M\,du\,dv + N\,dv^2}{E\,du^2 + 2F\,du\,dv + G\,dv^2}$$

$$= \frac{L\,du^2 + 2M\,du\,dv + N\,dv^2}{ds^2}$$

$$= L\left(\frac{du}{ds}\right)^2 + 2M\left(\frac{du}{ds}\right)\left(\frac{dv}{ds}\right) + N\left(\frac{dv}{ds}\right)^2$$

i.e., $\kappa = Ll^2 + 2Mlm + Nm^2$, (1)

where

$$l = \frac{du}{ds},\ m = \frac{dv}{ds}.$$

Also,

$$E\,du^2 + 2F\,du\,dv + G\,dv^2 = ds^2$$

$\Rightarrow$ $E\left(\frac{du}{ds}\right)^2 + 2F\left(\frac{du}{ds}\right)\left(\frac{dv}{ds}\right) + G\left(\frac{dv}{ds}\right)^2 = 1$

$\Rightarrow$ $El^2 + 2Flm + Gm^2 = 1.$ (2)

Since E, F, G, L, M, N are fixed at a point of the surface and l, m vary in different directions at a point of the surface, it follows that the normal curvature κ given by (1) is function of two variables l and m. Further, these variables l and m are connected by the relation (2). We shall find the maximum and minimum values of κ by Lagrange's method of undetermined multipliers.

For a maximum or minimum value of κ, we have

$d\kappa = 0$

$\Rightarrow$ $d\,(Ll^2 + 2Mlm + Nm^2) = 0$ [using (1)]

$\Rightarrow$ $L\,(2l\,dl) + 2M(l\,dm + m\,dl) + N(2m\,dm) = 0$

$\Rightarrow$ $2\,[(Ll + Mm)\,dl + (Ml + Nm)\,dm] = 0$

$\Rightarrow$ $(Ll + Mm)\,dl + (Ml + Nm)\,dm = 0.$ (3)

Again, differentiating (2), we get

$d\,(El^2 + 2Flm + Gm^2) = 0$

$\Rightarrow$ $E\,(2l\,dl) + 2F(l\,dm + m\,dl) + G\,(2m\,dm) = 0$

$\Rightarrow$ $2\,[(El + Fm)\,dl + (Fl + Gm)\,dm] = 0$

$\Rightarrow$ $(El + Fm)\,dl + (Fl + Gm)\,dm = 0.$ (4)

Multiplying (4) by λ and adding it to (3), we get

$$[(Ll+Mm)+\lambda\,(El+Fm)]\,dl\ +[(Ml+Nm)+\lambda\,(Fl+Gm)]\,dm=0$$

$$\Rightarrow \quad (Ll+Mm)+\lambda\,(El+Fm)=0, \tag{5}$$

$$(Ml+Nm)+\lambda\,(Fl+Gm)=0. \tag{6}$$

Now, multiplying (5) and (6) by l and m respectively and adding, we get

$$(Ll^2+2Mlm+Nm^2)+\lambda\,(El^2+2Flm+Gm^2)=0$$

$$\Rightarrow \quad \kappa+\lambda=0 \qquad \text{[using (1) and (2)]}$$

$$\Rightarrow \quad \lambda=-\kappa.$$

Substituting this value of λ in (5) and (6), we get

$$(Ll+Mm)-\kappa\,(El+Fm)=0, \tag{7}$$

$$(Ml+Nm)-\kappa\,(Fl+Gm)=0. \tag{8}$$

Now, we shall eliminate l, m from (7) and (8).
From (7), we have

$$(L-\kappa E)\,l+(M-\kappa F)\,m=0. \tag{9}$$

From (8), we have

$$(M-\kappa F)\,l+(N-\kappa G)\,m=0. \tag{10}$$

From (9) by (10), we get

$$\frac{L-\kappa E}{M-\kappa F}=\frac{M-\kappa F}{N-\kappa G}$$

or $\quad (L-\kappa E)\,(N-\kappa G)=(M-\kappa F)^2$

or $\quad LN-\kappa\,(EN+GL)+\kappa^2 EG=M^2-2\kappa FM+\kappa^2F^2$

or $\quad (EG-F^2)\,\kappa^2-(EN-2FM+GL)\,\kappa+LN-M^2=0$

or $\quad H^2\kappa^2-(EN-2FM+GL)\,\kappa+T^2=0. \qquad (11)$

This is the required equation giving the maximum or minimum values of the normal curvature at P. The roots of this equation are the principal curvatures of the surface at P and are usually denoted by κ_a and κ_b.

$$\therefore \quad \kappa_a+\kappa_b=\frac{EN-2FM+GL}{H^2}$$

and $\quad \kappa_a\kappa_b=\dfrac{T^2}{H^2}.$

Some Definitions.

1. First Curvature. The first curvature of a surface at a point is the sum of the principal curvatures κ_a and κ_b of the surface at that point. It is denoted by J.

$$\therefore \qquad J = \kappa_a + \kappa_b = \frac{EN - 2FM + GL}{H^2}.$$

2. Mean Curvature or Mean Normal Curvature. The mean curvature or mean normal curvature of a surface at a point is the arithmetic mean of the principal curvatures κ_a and κ_b of the surface at that point. It is denoted by μ or B.

$$\therefore \qquad \mu \text{ (or } B) = \frac{1}{2}(\kappa_a + \kappa_b) = \frac{EN - 2FM + GL}{2H^2}.$$

3. Amplitude of Normal Curvature. The amplitude of the normal curvature of a surface at a point is denoted by A and is defined as

$$A = \frac{1}{2}(\kappa_a - \kappa_b),$$

where κ_a and κ_b are the maximum and the minimum normal curvatures of the surface at the point.

Obviously,

$$\kappa_a = B + A, \ \kappa_b = B - A.$$

4. Minimal Surface. If the mean curvature of a surface is zero at all the points, then the surface is called a minimal surface.

Thus a surface is a minimal surface if and only if

$$EN - 2FM + GL = 0.$$

at every point of the surface.

5. Principal Radii of Curvature. The principal radii of curvature of a surface at a point are the reciprocals of the principal curvatures of the surface at the point. These are usually denoted by R_a and R_b, i.e.,

$$R_a = \frac{1}{\kappa_a}, \ R_b = \frac{1}{\kappa_b}.$$

6. Gaussian Curvature or Specific Curvature or Second Curvature or Total Curvature. The Gaussian curvature of a surface at a point is defined as the product of the principal curvatures κ_a and κ_b of the surface at that point. It is denoted by K.

$$\therefore \qquad K = \kappa_a \kappa_b = \frac{T^2}{H^2} = \frac{LN - M^2}{EG - F^2}.$$

If B is the mean normal curvature of a surface at a point and A is the amplitude, then

$$K = \kappa_a \kappa_b = (B + A)(B - A) = B^2 - A^2.$$

Characterization of a Developable Surface in terms of Gaussian Curvature.

Theorem. A necessary and sufficient condition for a surface to be a developable surface is that its Gaussian curvature is zero.

Proof. Let the equation of a surface be

$$z = (x, y). \tag{1}$$

By definition, the Gaussian curvature K of a surface is given by

$$K = \frac{T^2}{H^2} = \frac{LN - M^2}{EG - F^2}. \tag{2}$$

First, we shall find K for the given surface (1).

If $\mathbf{r}$ is the position vector of any current point on the surface (1), then

$$\mathbf{r} = \mathbf{r}\,(x, y, z), \tag{3}$$

where z is a function of x and y.

We write

$$\frac{\partial z}{\partial x} = p,\ \frac{\partial z}{\partial y} = q,\ \frac{\partial^2 z}{\partial x^2} = r,\ \frac{\partial^2 z}{\partial x\,\partial y} = s,\ \frac{\partial^2 z}{\partial y^2} = t.$$

Let x and y be taken as the first and the second parameters. Then, from (3), we have

$$\mathbf{r}_1 = (1, 0, p),\quad \mathbf{r}_2 = (0, 1, q),$$
$$\mathbf{r}_{11} = (0, 0, r),\ \mathbf{r}_{12} = (0, 0, s),\ \mathbf{r}_{22} = (0, 0, t).$$

Therefore

$$E = \mathbf{r}_1^{\ 2} = 1 + p^2,\ F = \mathbf{r}_1 . \mathbf{r}_2 = pq,\ G = \mathbf{r}_2^{\ 2} = 1 + q^2,$$
$$H^2 = EG - F^2 = 1 + p^2 + q^2.$$

Again, we have

$$\mathbf{r}_1 \times \mathbf{r}_2 = (-p, -q, 1)$$

$$\Rightarrow \qquad \mathbf{N} = \frac{\mathbf{r}_1 \times \mathbf{r}_2}{H} = \frac{1}{H}(-p, -q, 1).$$

Therefore

$$L = \mathbf{r}_{11} . \mathbf{N} = \frac{r}{H},\ M = \mathbf{r}_{12} . \mathbf{N} = \frac{s}{H},\ N = \mathbf{r}_{22} . \mathbf{N} = \frac{t}{H};$$

$$T^2 = LN - M^2 = \frac{r}{H}\frac{t}{H} - \left(\frac{s}{H}\right)^2 = \frac{rt - s^2}{1 + p^2 + q^2}.$$

Substituting the values of H^2 and T^2 in (2), we get

$$K = \frac{rt - s^2}{(1 + p^2 + q^2)^2}. \tag{4}$$

The condition is necessary. Let the surface be developable.

To prove : The Gaussian curvature of the surface is zero.

Since the given surface (1) is developable, we have

$$rt = s^2 \qquad \left[\begin{array}{l}\because \text{A necessary and sufficient condition for the} \\ \text{surface (1) to be developable is that } rt = s^2\end{array}\right]$$

$\Rightarrow \quad K = 0$ [using (4)]

$\Rightarrow$ the Gaussian curvature of the surface (1) is zero.

Hence the condition is necessary.

The condition is sufficient. Now, we suppose that the Gaussian curvature K of the surface (1) is zero at all points of the surface.

To prove : The surface is developable.

Since $K = 0$, it follows from (4) that $rt = s^2$ and consequently, the surface is developable.

Differential Equation of Principal Directions.

We know that the normal curvature κ of a surface at a point $P(u, v)$ in the direction (du, dv) is given by

$$\kappa = \frac{L\,du^2 + 2M\,du\,dv + N\,dv^2}{E\,du^2 + 2F\,du\,dv + G\,dv^2}$$

$$= \frac{L\,du^2 + 2M\,du\,dv + N\,dv^2}{ds^2}$$

$$= L\left(\frac{du}{ds}\right)^2 + 2M\left(\frac{du}{ds}\right)\left(\frac{dv}{ds}\right) + N\left(\frac{dv}{ds}\right)^2$$

i.e., $\quad \kappa = Ll^2 + 2Mlm + Nm^2, \qquad (1)$

where

$$l = \frac{du}{ds}, m = \frac{dv}{ds}.$$

Also,

$$E\,du^2 + 2F\,du\,dv + G\,dv^2 = ds^2$$

$$\Rightarrow \quad E\left(\frac{du}{ds}\right)^2 + 2F\left(\frac{du}{ds}\right)\left(\frac{dv}{ds}\right) + G\left(\frac{dv}{ds}\right)^2 = 1$$

$\Rightarrow \quad El^2 + 2Flm + Gm^2 = 1. \qquad (2)$

If (du, dv) be the principal direction at P, then κ is either maximum or minimum. Therefore, we have

$$d\kappa = 0$$

$\Rightarrow \quad d\,(Ll^2 + 2Mlm + Nm^2) = 0$ [using (1)]

$$\Rightarrow \quad L\,(2l\,dl) + 2M\,(l\,dm + m\,dl) + N(2m\,dm) = 0$$

$$\Rightarrow \quad 2[(Ll + Mm)\,dl + (Ml + Nm)\,dm] = 0$$

$$\Rightarrow \quad (Ll + Mm)\,dl + (Ml + Nm)\,dm = 0. \tag{3}$$

Again, differentiating (2), we get

$$d(El^2 + 2Flm + Gm^2) = 0$$

$$\Rightarrow \quad E(2l\,dl) + 2F(l\,dm + m\,dl) + G(2m\,dm) = 0$$

$$\Rightarrow \quad 2[(El + Fm)\,dl + (Fl + Gm)\,dm] = 0$$

$$\Rightarrow \quad (El + Fm)\,dl + (Fl + Gm)\,dm = 0. \tag{4}$$

Multiplying (4) by λ and adding it to (3), we get

$$[(Ll + Mm) + \lambda(El + Fm)]\,dl + [(Ml + Nm) + \lambda(Fl + Gm)]\,dm = 0$$

$$\Rightarrow \quad (Ll + Mm) + \lambda(El + Fm) = 0, \tag{5}$$

$$(Ml + Nm) + \lambda(Fl + Gm) = 0. \tag{6}$$

Now, multiplying (5) and (6) by l and m respectively and adding, we get

$$(Ll^2 + 2Mlm + Nm^2) + \lambda(El^2 + 2Flm + Gm^2) = 0$$

$$\Rightarrow \quad \kappa + \lambda = 0 \qquad \text{[using (1) and (2)]}$$

$$\Rightarrow \quad \lambda = -\kappa.$$

Substituting this value of λ in (5) and (6), we get

$$(Ll + Mm) - \kappa(El + Fm) = 0,$$

$$(Ml + Nm) - \kappa(Fl + Gm) = 0$$

or

$$(L - \kappa E)\,l + (M - \kappa F)\,m = 0,$$

$$(M - \kappa F)\,l + (N - \kappa G)\,m = 0$$

or

$$(L - \kappa E)\,du + (M - \kappa F)\,dv = 0, \tag{7}$$

$$(M - \kappa F)\,du + (N - \kappa G)\,dv = 0. \tag{8}$$

From (7), we have

$$\frac{du}{dv} = -\frac{M - \kappa F}{L - \kappa E}. \tag{9}$$

From (8), we have

$$\frac{du}{dv} = -\frac{N - \kappa G}{M - \kappa F}. \tag{10}$$

From (9) and (10), it follows that the differential equation of principal direction corresponding to the principal curvature κ_a is given by

$$\frac{du}{dv} = -\frac{M - \kappa_a F}{L - \kappa_a E} \text{ or } -\frac{N - \kappa_a G}{M - \kappa_a F}. \tag{11}$$

Similarly, the differential equation of principal direction corresponding to the principal curvature κ_b is given by

$$\frac{du}{dv} = -\frac{M - \kappa_b F}{L - \kappa_b E} \text{ or } -\frac{N - \kappa_b G}{M - \kappa_b F}. \tag{12}$$

The combined differential equation of the two principal directions at P can be found by eliminating λ from (5) and (6) as under.

From (5) and (6), we get

$$\frac{Ll + Mm}{Ml + Nm} = \frac{El + Fm}{Fl + Gm}$$

or $\quad (Ll + Mm)(Fl + Gm) = (El + Fm)(Ml + Nm)$

or $\quad (EM - FL)l^2 + (EN - GL)lm + (FN - GM)m^2 = 0$

or $\quad (EM - FL)\left(\frac{du}{ds}\right)^2 + (EN - GL)\left(\frac{du}{ds}\right)\left(\frac{dv}{ds}\right) + (FN - GM)\left(\frac{dv}{ds}\right)^2 = 0$

or $\quad (EM - FL)\, du^2 + (EN - GL)\, du\, dv + (FN - GM)\, dv^2 = 0. \quad (13)$

The quadratic differential equation (13) is the differential equation of the principal directions at P.

Clearly, equation (13) can be regarded as a quadratic equation in $\dfrac{du}{dv}$.

Therefore there are two principal directions at each point of the surface.

Principal Directions at a point on a surface are orthogonal.

We know that the two directions given by

$$P\, du^2 + 2Q\, du\, dv + R\, dv^2 = 0 \tag{14}$$

are orthogonal if

$$ER - 2FQ + GP = 0. \tag{15}$$

Comparing (13) and (14), we get

$$P = EM - FL,\ Q = (EN - GL)/2,\ R = FN - GM.$$

$$\therefore \quad ER - 2FQ + GP = E(FN - GM) - F(EN - GL) + G(EM - FL)$$

$$= 0.$$

Hence the directions given by (13), i.e., the two principal directions at point P of the surface are orthogonal.

Umbilic. A point on a surface is called an umbilic or a naval point if

$$\frac{E}{L} = \frac{F}{M} = \frac{G}{N} \text{ at that point.}$$

We know that the normal curvature κ of a surface at a point $P(u, v)$ in the direction (du, dv) is given by

$$\kappa = \frac{L\,du^2 + 2M\,du\,dv + N\,dv^2}{E\,du^2 + 2F\,du\,dv + G\,dv^2}.$$

Therefore the normal curvature at an umbilic is the same in all directions.

4.3 LINES OF CURVATURE

Definition. A curve on a surface is called a line of curvature if the tangent at each point of it is along the principal direction at that point. Since there are two principal directions at each point of a surface cutting orthogonally, there are two orthogonal families of lines of curvature on a surface.

Differential Equation of the Lines of Curvature.

Since the direction of line of curvature at any point of a surface is along the principal direction at that point, the differential equation of the two system of lines of curvature is the same as the differential equation of the principal directions. Hence their differential equation is

$$(EM - FL)\,du^2 + (EN - GL)\,du\,dv + (FN - GM)\,dv^2 = 0.$$

The above equation can also be put in the form

$$\begin{vmatrix} dv^2 & -du\,dv & du^2 \\ E & F & G \\ L & M & N \end{vmatrix} = 0.$$

Some Results on Lines of Curvature.

Theorem 1. Rodrigue's Formula. A necessary and sufficient condition that a curve on a surface be a line of curvature is that

$$\kappa\,d\mathbf{r} + d\mathbf{N} = \mathbf{0}$$

at each of its points.

Proof. The condition is necessary. Let the equation of a surface be given by

$$\mathbf{r} = \mathbf{r}\,(u, v)$$

and let $\mathbf{r} = \mathbf{r}\,(s)$ be a curve lying on it. Also, let this curve be a line of curvature on the given surface.

To prove : $\kappa\,d\mathbf{r} + d\mathbf{N} = \mathbf{0}$ at each point of the curve $\mathbf{r} = \mathbf{r}\,(s)$.

Since the curve $\mathbf{r} = \mathbf{r}\,(s)$ is a line of curvature on the given surface, the direction (du, dv) of this curve at each point (u, v) is along the principal direction at that point. Therefore, along the curve $\mathbf{r}=\mathbf{r}(s)$, the differential equations of the lines of curvature are

$$(L-\kappa E)\,du+(M-\kappa F)\,dv=0 \tag{1}$$

and $$(M-\kappa F)\,du+(N-\kappa G)\,du=0. \tag{2}$$

We know that

$$L=-\mathbf{r}_1.\mathbf{N}_1,\ M=-\mathbf{r}_1.\mathbf{N}_2,\ E=\mathbf{r}_1.\mathbf{r}_1,\ F=\mathbf{r}_1.\mathbf{r}_2. \tag{3}$$

Substituting the values of L, M, E and F from (3) in (1), we get

$$(-\mathbf{r}_1.\mathbf{N}_1-\kappa\mathbf{r}_1.\mathbf{r}_1)\,du+(-\mathbf{r}_1.\mathbf{N}_2-\kappa\mathbf{r}_1.\mathbf{r}_2)\,dv=0$$

$\Rightarrow$ $$(\mathbf{r}_1.\mathbf{N}_1+\kappa\mathbf{r}_1.\mathbf{r}_1)\,du+(\mathbf{r}_1.\mathbf{N}_2+\kappa\mathbf{r}_1.\mathbf{r}_2)\,dv=0$$

$\Rightarrow$ $$\mathbf{r}_1.(\mathbf{N}_1\,du+\mathbf{N}_2\,dv)+\kappa\mathbf{r}_1.(\mathbf{r}_1\,du+\mathbf{r}_2\,dv)=0$$

$\Rightarrow$ $$\mathbf{r}_1.\left(\frac{\partial\mathbf{N}}{\partial u}du+\frac{\partial\mathbf{N}}{\partial v}dv\right)+\kappa\mathbf{r}_1.\left(\frac{\partial\mathbf{r}}{\partial u}du+\frac{\partial\mathbf{r}}{\partial v}dv\right)=0$$

$\Rightarrow$ $$\mathbf{r}_1.d\mathbf{N}+\kappa\mathbf{r}_1.d\mathbf{r}=0$$

$\Rightarrow$ $$\mathbf{r}_1.(d\mathbf{N}+\kappa\,d\mathbf{r})=0$$

$\Rightarrow$ $$\mathbf{r}_1.(\kappa\,d\mathbf{r}+d\mathbf{N})=0$$

$\Rightarrow$ Either $\mathbf{r}_1=\mathbf{0}$ or $\kappa\,d\mathbf{r}+d\mathbf{N}=\mathbf{0}$ or $\kappa\,d\mathbf{r}+d\mathbf{N}$ is perpendicular to $\mathbf{r}_1$.

$\Rightarrow$ Either $\kappa\,d\mathbf{r}+d\mathbf{N}=\mathbf{0}$ or $\kappa\,d\mathbf{r}+d\mathbf{N}$ is perpendicular to $\mathbf{r}_1$. (4)

$[\because \mathbf{r}_1\neq\mathbf{0}]$

Similarly,

$$M=-\mathbf{r}_2.\mathbf{N}_1,\ N=-\mathbf{r}_2.\mathbf{N}_2,\ F=\mathbf{r}_2.\mathbf{r}_1,\ G=\mathbf{r}_2.\mathbf{r}_2. \tag{5}$$

Substituting the values of M, N, F and G from (5) in (2), we get

$$(-\mathbf{r}_2.\mathbf{N}_1-\kappa\mathbf{r}_2.\mathbf{r}_1)\,du+(-\mathbf{r}_2.\mathbf{N}_2-\kappa\mathbf{r}_2.\mathbf{r}_2)\,dv=0$$

$\Rightarrow$ $$(\mathbf{r}_2.\mathbf{N}_1+\kappa\mathbf{r}_2.\mathbf{r}_1)\,du+(\mathbf{r}_2.\mathbf{N}_2+\kappa\mathbf{r}_2.\mathbf{r}_2)\,dv=0$$

$\Rightarrow$ $$\mathbf{r}_2.(\mathbf{N}_1\,du+\mathbf{N}_2\,dv)+\kappa\mathbf{r}_2.(\mathbf{r}_1\,du+\mathbf{r}_2\,dv)=0$$

$\Rightarrow$ $$\mathbf{r}_2.\left(\frac{\partial\mathbf{N}}{\partial u}du+\frac{\partial\mathbf{N}}{\partial v}dv\right)+\kappa\mathbf{r}_2.\left(\frac{\partial\mathbf{r}}{\partial u}du+\frac{\partial\mathbf{r}}{\partial v}dv\right)=0$$

$\Rightarrow$ $$\mathbf{r}_2.d\mathbf{N}+\kappa\mathbf{r}_2.d\mathbf{r}=0$$

$\Rightarrow$ $$\mathbf{r}_2.(d\mathbf{N}+\kappa\,d\mathbf{r})=0$$

$\Rightarrow$ $$\mathbf{r}_2.(\kappa\,d\mathbf{r}+d\mathbf{N})=0$$

$\Rightarrow$ Either $\mathbf{r}_2=\mathbf{0}$ or $\kappa\,d\mathbf{r}+d\mathbf{N}=\mathbf{0}$ or $\kappa\,d\mathbf{r}+d\mathbf{N}$ is perpendicular to $\mathbf{r}_2$.

$\Rightarrow$ Either $\kappa\,d\mathbf{r}+d\mathbf{N}=\mathbf{0}$ or $\kappa\,d\mathbf{r}+d\mathbf{N}$ is perpendicular to $\mathbf{r}_2$. (6)

$[\because \mathbf{r}_2\neq\mathbf{0}]$

If possible, let $\kappa\, d\mathbf{r} + d\mathbf{N} \neq \mathbf{0}$. Then it follows from (4) and (6) that $\kappa\, d\mathbf{r} + d\mathbf{N}$ is perpendicular to each of $\mathbf{r}_1$ and $\mathbf{r}_2$. Therefore

$\kappa\, d\mathbf{r} + d\mathbf{N}$ is parallel to $\mathbf{r}_1 \times \mathbf{r}_2$

i.e., $\kappa\, d\mathbf{r} + d\mathbf{N}$ is parallel to $\mathbf{N}$. (7)

$[\because \mathbf{r}_1 \times \mathbf{r}_2 = H\mathbf{N}]$

But

$$\mathbf{N}\,.\,\mathbf{N} = 1$$

$\Rightarrow$ $d(\mathbf{N}\,.\,\mathbf{N}) = 0$

$\Rightarrow$ $d\mathbf{N}\,.\,\mathbf{N} + \mathbf{N}\,.\,d\mathbf{N} = 0$

$\Rightarrow$ $d\mathbf{N}\,.\,\mathbf{N} + d\mathbf{N}\,.\,\mathbf{N} = 0$

$\Rightarrow$ $2d\mathbf{N}\,.\,\mathbf{N} = 0$

$\Rightarrow$ $d\mathbf{N}\,.\,\mathbf{N} = 0$

$\Rightarrow$ Either $d\mathbf{N} = \mathbf{0}$ or $\mathbf{N} = \mathbf{0}$ or $d\mathbf{N}$ is perpendicular to $\mathbf{N}$.

$\Rightarrow$ $d\mathbf{N}$ is perpendicular to $\mathbf{N}$. (8)

$[\because |\mathbf{N}| = 1 \Rightarrow \mathbf{N} \neq \mathbf{0},\ d\mathbf{N} \neq \mathbf{0}]$

Also,

$d\mathbf{r}$ is tangential to the surface.

$\Rightarrow$ $d\mathbf{r}$ is perpendicular to $\mathbf{N}$. (9)

From (8) and (9), it follows that $\kappa\, d\mathbf{r} + d\mathbf{N}$ is perpendicular to $\mathbf{N}$. But, this contradicts (7).

Hence our supposition that $\kappa\, d\mathbf{r} + d\mathbf{N} \neq \mathbf{0}$ is false.

Therefore, we must have

$$\kappa\, d\mathbf{r} + d\mathbf{N} = \mathbf{0}.$$

Hence the condition is necessary.

The condition is sufficient. Now, we suppose that

$$\kappa\, d\mathbf{r} + d\mathbf{N} = \mathbf{0}$$

at all the points of a curve lying on a surface.

To prove : The curve is a line of curvature.

By supposition,

$$\kappa\, d\mathbf{r} + d\mathbf{N} = \mathbf{0}$$

$\Rightarrow$ $\kappa\, d\mathbf{r} = -\,d\mathbf{N}$

$\Rightarrow$ $\kappa\,(\mathbf{r}_1 du + \mathbf{r}_2 dv) = -\,(\mathbf{N}_1 du + \mathbf{N}_2 dv)$

$\Rightarrow$ $\kappa\,(\mathbf{r}_1 du + \mathbf{r}_2 dv)\,.\,(\mathbf{r}_1 du + \mathbf{r}_2 dv) = -\,(\mathbf{N}_1 du + \mathbf{N}_2 dv)\,.\,(\mathbf{r}_1 du + \mathbf{r}_2 dv)$

$\Rightarrow$ $\kappa\,(E\,du^2 + 2F\,du\,dv + G\,dv^2) = L\,du^2 + 2M\,du\,dv + N\,dv^2$

[using (3) and (5)]

$\Rightarrow$ $$\kappa = \frac{L\,du^2 + 2M\,du\,dv + N\,dv^2}{E\,du^2 + 2F\,du\,dv + G\,dv^2}.$$

$\Rightarrow$ κ is the normal curvature of the surface at the point (u, v) in the direction (du, dv). (10)

Again,

$$\kappa\, d\mathbf{r} + d\mathbf{N} = \mathbf{0}$$

$$\Rightarrow \quad (\kappa\, d\mathbf{r} + d\mathbf{N})\cdot\mathbf{r}_1 = 0,$$

$$(\kappa\, d\mathbf{r} + d\mathbf{N})\cdot\mathbf{r}_2 = 0$$

$$\Rightarrow \quad [\kappa(\mathbf{r}_1\, du + \mathbf{r}_2\, dv) + (\mathbf{N}_1\, du + \mathbf{N}_2\, dv)]\cdot\mathbf{r}_1 = 0,$$

$$[\kappa(\mathbf{r}_1\, du + \mathbf{r}_2\, dv) + (\mathbf{N}_1\, du + \mathbf{N}_2\, dv)]\cdot\mathbf{r}_2 = 0$$

$$\Rightarrow \quad \kappa(E\, du + F\, dv) - (L\, du + M\, dv) = 0,$$

$$\kappa(F\, du + G\, dv) - (M\, du + N\, dv) = 0 \qquad \text{[using (3) and (5)]}$$

$$\Rightarrow \quad (L\, du + M\, dv) - \kappa(E\, du + F\, dv) = 0,$$

$$(M\, du + N\, dv) - \kappa(F\, du + G\, dv) = 0$$

$$\Rightarrow \quad (L - \kappa E)\, du + (M - \kappa F)\, dv = 0,$$

$$(M - \kappa F)\, du + (N - \kappa G)\, dv = 0$$

$\Rightarrow$ κ is the principal curvature of the surface at the point (u, v) in the direction (du, dv). [using (10)]

$\Rightarrow$ (du, dv) is the principal direction of the surface at the point (u, v).

Therefore the direction (du, dv) of the taken curve at each point (u, v) is along the principal direction at that point. Hence the taken curve is a line of curvature.

Theorem 2. Fundamental Normal Property of the Lines of Curvature. The normals to any surface at any two consecutive points of one of its lines of curvature intersect. Conversely, if the normals at any two consecutive points of a curve drawn on a surface intersect, then the curve is a line of curvature.

Proof. Let the equation of the surface be given by

$$\mathbf{r} = \mathbf{r}(u, v) \tag{1}$$

and let $P(\mathbf{r})$ and $Q(\mathbf{r} + d\mathbf{r})$ be any two consecutive points lying on one of the lines of curvature of this surface. Also, let $\mathbf{N}$ and $\mathbf{N} + d\mathbf{N}$ be unit vectors along the normal to this surface at P and Q respectively.

We have to prove that the normals $\mathbf{N}$ and $\mathbf{N} + d\mathbf{N}$ intersect.

Obviously, the normals will intersect if $\mathbf{N}$, $\mathbf{N} + d\mathbf{N}$ and $d\mathbf{r}$ are coplanar, i.e., if their scalar triple product vanishes.

We have

$$[\mathbf{N},\ \mathbf{N} + d\mathbf{N},\ d\mathbf{r}] = [\mathbf{N},\ \mathbf{N},\ d\mathbf{r}] + [\mathbf{N},\ d\mathbf{N},\ d\mathbf{r}]$$

$$= [\mathbf{N},\ d\mathbf{N},\ d\mathbf{r}]$$

$$= [\mathbf{N},\ -\kappa\, d\mathbf{r},\ d\mathbf{r}] \qquad \text{[by Rodrigue's formula]}$$

$$= -\kappa\ [\mathbf{N},\ d\mathbf{r},\ d\mathbf{r}]$$
$$= 0.$$

Therefore the normals $\mathbf{N}$ and $\mathbf{N}+d\mathbf{N}$ intersect.

Conversely, let the equation of a curve C lying on the surface (1) be

$$\mathbf{r} = \mathbf{r}\,(s)$$

and let $P(\mathbf{r})$ and $Q(\mathbf{r}+d\mathbf{r})$ be any two consecutive points lying on the curve C.

Also, let $\mathbf{N}$ and $\mathbf{N}+d\mathbf{N}$ be unit normals to the surface (1) at P and Q respectively which intersect at some point, say A.

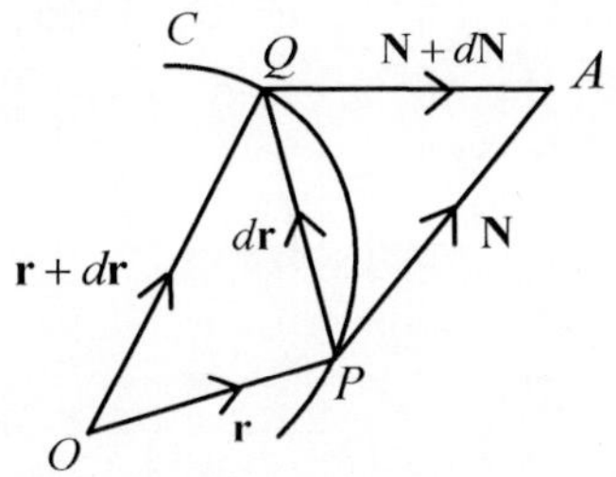

We have to prove that the curve C is a line of curvature.

Since $\mathbf{N}$ and $\mathbf{N}+d\mathbf{N}$ intersect, it follows that $\mathbf{N}, \mathbf{N}+d\mathbf{N}$ and $d\mathbf{r}$ are coplanar, i.e.,

$$[\mathbf{N},\ \mathbf{N}+d\mathbf{N},\ d\mathbf{r}] = 0$$

or $$[\mathbf{N},\ \mathbf{N},\ d\mathbf{r}] + [\mathbf{N},\ d\mathbf{N},\ d\mathbf{r}] = 0$$

or $$[\mathbf{N},\ d\mathbf{N},\ d\mathbf{r}] = 0. \tag{2}$$

Again, we have

$$d\mathbf{N} = \mathbf{N}_1\ du + \mathbf{N}_2\ dv,\ d\mathbf{r} = \mathbf{r}_1\ du + \mathbf{r}_2\ dv.$$

Substituting these values in (2), we get

$$[\mathbf{N},\ \mathbf{N}_1\ du + \mathbf{N}_2\ dv,\ \mathbf{r}_1\ du + \mathbf{r}_2\ dv] = 0$$

or $$[\mathbf{N},\ \mathbf{N}_1, \mathbf{r}_1]\,du^2 + \{[\mathbf{N},\ \mathbf{N}_1, \mathbf{r}_2] + [\mathbf{N},\ \mathbf{N}_2, \mathbf{r}_1]\}\,du\,dv + [\mathbf{N},\ \mathbf{N}_2,\ \mathbf{r}_2]\,dv^2 = 0. \tag{3}$$

Also, we know that

$$[\mathbf{N},\ \mathbf{N}_1,\ \mathbf{r}_1] = \frac{EM - FL}{H},$$

$$[\mathbf{N},\ \mathbf{N}_1,\ \mathbf{r}_2] = \frac{FM - GL}{H},$$

$$[\mathbf{N},\ \mathbf{N}_2,\ \mathbf{r}_1] = \frac{EN - FM}{H},$$

$$[\mathbf{N},\ \mathbf{N}_2,\ \mathbf{r}_2] = \frac{FN - GM}{H}.$$

Substituting these values in (3), we get

$$(EM - FL)\,du^2 + (EN - GL)\,du\,dv + (FN - GM)\,dv^2 = 0.$$

This is the differential equation of the lines of curvature. Therefore the curve C is a line of curvature.

Remark. The point of intersection of consecutive normals along a line of curvature at P is called a centre of curvature of the surface and its distance from P measured in the direction of the unit normal $\mathbf{N}$ is called a (principal) radius of curvature of the surface.

Theorem 3. Monge's Theorem. A necessary and sufficient condition that a curve on a surface be a line of curvature is that the surface normals along the curve form a developable.

Proof. Let the equation of a surface be given by

$$\mathbf{r} = \mathbf{r}(u, v) \tag{1}$$

and let the equation of a curve lying on this surface be given by

$$\mathbf{r} = \mathbf{r}(s) \tag{2}$$

Take any point P on the curve (2). Clearly, this point lies on the surface (1) also.

Let $\mathbf{N}$ be the unit normal to the surface (1) at P.

Now, take any point Q on this normal at P and let $\mathbf{R}$ be the position vector of Q. Then, we have

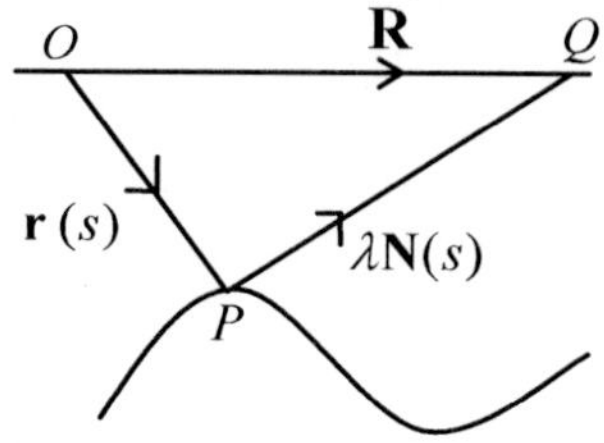

$$\mathbf{R} = \mathbf{r}(s) + \lambda \mathbf{N}(s). \tag{3}$$

Since the point P lies on the curve (2), the quantities $\mathbf{r}$ and $\mathbf{N}$ involved in (3) are functions of s. Hence the position vector of Q is a function of two independent variable parameters s and λ. Therefore the locus of Q is a surface, i.e., (3) represents a surface.

Thus the surface generated by the surface normals at the points on (2) is given by (3).

Let s and λ be taken as the first and the second parameters. Then, from (3), we have

$$\mathbf{R}_1 = \frac{\partial \mathbf{R}}{\partial s} = \frac{d\mathbf{r}}{ds} + \lambda \frac{d\mathbf{N}}{ds} = \mathbf{t} + \lambda \mathbf{N}', \quad \mathbf{R}_2 = \frac{\partial \mathbf{R}}{\partial \lambda} = \mathbf{N},$$

$$\mathbf{R}_{11} = \kappa \mathbf{n} + \lambda \mathbf{N}'', \quad \mathbf{R}_{12} = \mathbf{N}', \quad \mathbf{R}_{22} = \mathbf{0}.$$

Again, we have

$$\mathbf{R}_1 \times \mathbf{R}_2 = \mathbf{t} \times \mathbf{N} + \lambda \mathbf{N}' \times \mathbf{N}$$

$$\Rightarrow \quad \mathbf{N} = \frac{\mathbf{R}_1 \times \mathbf{R}_2}{H} = \frac{\mathbf{t} \times \mathbf{N} + \lambda \mathbf{N}' \times \mathbf{N}}{H}.$$

Therefore

$$L = \mathbf{R}_{11} \,.\, \mathbf{N} = (\kappa\mathbf{n} + \lambda\mathbf{N}'')\,.\,\frac{\mathbf{t}\times\mathbf{N} + \lambda\mathbf{N}'\times\mathbf{N}}{H},$$

$$M = \mathbf{R}_{12} \,.\, \mathbf{N} = \mathbf{N}' \,.\, \mathbf{N}$$

$$= \mathbf{N}'\,.\,\frac{\mathbf{t}\times\mathbf{N} + \lambda\mathbf{N}'\times\mathbf{N}}{H}$$

$$= \frac{1}{H}[\mathbf{N}',\mathbf{t},\mathbf{N}] + \frac{\lambda}{\mathrm{H}}[\mathbf{N}',\mathbf{N}',\mathbf{N}]$$

$$= \frac{[\mathbf{N}',\mathbf{t},\mathbf{N}]}{H}$$

$$= \frac{[\mathbf{t},\mathbf{N},\mathbf{N}']}{H},$$

$$N = \mathbf{R}_{22} \,.\, \mathbf{N} = 0.$$

$$\therefore \qquad T^2 = LN - M^2 = -\frac{[\mathbf{t},\mathbf{N},\mathbf{N}']^2}{H^2}\,. \tag{4}$$

We know that a necessary and sufficient condition for a surface to be developable is that its Gaussian curvature is zero, i.e.,

$$K = 0$$

$$\text{or} \qquad \frac{T^2}{H^2} = 0 \qquad \left[\because K = \frac{T^2}{H^2} = \frac{LN - M^2}{EG - F^2}\right]$$

$$\text{or} \qquad T = 0$$

$$\text{or} \qquad \frac{[\mathbf{t},\mathbf{N},\mathbf{N}']}{H} = 0 \qquad \text{[using (4)]}$$

$$\text{or} \qquad [\mathbf{t},\mathbf{N},\mathbf{N}'] = 0. \tag{5}$$

Thus, in order to prove the theorem, it is required to prove that the curve (2) on the surface (1) is a line of curvature if and only if relation (5) holds good.

The condition is necessary. Let the curve (2) lying on the surface (1) be a line of curvature.

To prove : Relation (5) is true.

Since the curve (2) is a line of curvature, we have

$$\kappa\, d\mathbf{r} + d\mathbf{N} = \mathbf{0} \qquad \text{[by Rodrigue's formula]}$$

$$\Rightarrow \qquad \kappa\frac{d\mathbf{r}}{ds} + \frac{d\mathbf{N}}{ds} = \mathbf{0}$$

$$\Rightarrow \qquad \kappa\mathbf{t} + \mathbf{N}' = \mathbf{0}$$

$$\Rightarrow \qquad \mathbf{N}' = -\kappa\mathbf{t}$$

$$\Rightarrow \qquad [\mathbf{t},\mathbf{N},\mathbf{N}'] = [\mathbf{t},\mathbf{N},-\kappa\mathbf{t}] = -\kappa[\mathbf{t},\mathbf{N},\mathbf{t}] = 0.$$

Therefore relation (5) is true.

Hence the condition is necessary.

The condition is sufficient. Now, we suppose that the relation (5) is true.

To prove : The curve (2) is a line of curvature.

From (5), we have

$$[\mathbf{t}, \mathbf{N}', \mathbf{N}] = 0$$

$\Rightarrow \quad \mathbf{t} \times \mathbf{N}' . \mathbf{N} = 0$

$\Rightarrow \quad$ Either $\mathbf{t} \times \mathbf{N}' = 0$ or $\mathbf{t} \times \mathbf{N}'$ is perpendicular to $\mathbf{N}$. (6)

We know that

$$\mathbf{N} . \mathbf{N} = 1$$

$\Rightarrow \quad \dfrac{d}{ds}(\mathbf{N} . \mathbf{N}) = 0$

$\Rightarrow \quad \mathbf{N}' . \mathbf{N} + \mathbf{N} . \mathbf{N}' = 0$

$\Rightarrow \quad \mathbf{N}' . \mathbf{N} + \mathbf{N}' . \mathbf{N} = 0$

$\Rightarrow \quad 2\mathbf{N}' . \mathbf{N} = 0$

$\Rightarrow \quad \mathbf{N}' . \mathbf{N} = 0$

$\Rightarrow \quad$ Either $\mathbf{N}' = \mathbf{0}$ or $\mathbf{N}'$ is perpendicular to $\mathbf{N}$.

$\Rightarrow \quad \mathbf{N}'$ is perpendicular to $\mathbf{N}$. $[\because \mathbf{N}' \neq \mathbf{0}]$

$\Rightarrow \quad \mathbf{N}'$ is tangential to the surface (1). (7)

Also,

$\mathbf{t}$ = unit tangent to curve (2) at P

$\Rightarrow \quad \mathbf{t}$ is tangential to the surface (1) at P. (8)

From (7) and (8), it follows that $\mathbf{t} \times \mathbf{N}'$ is a vector perpendicular to the tangent plane to the surface (1) at P. Therefore

$\mathbf{t} \times \mathbf{N}'$ is a vector parallel to $\mathbf{N}$.

$\Rightarrow \quad \mathbf{t} \times \mathbf{N}'$ is not perpendicular to $\mathbf{N}$.

$\Rightarrow \quad \mathbf{t} \times \mathbf{N}' = \mathbf{0}$ [using (6)]

$\Rightarrow \quad \mathbf{N}'$ is parallel to $\mathbf{t}$. $[\because \mathbf{t} \neq \mathbf{0}, \mathbf{N}' \neq \mathbf{0}]$

$\Rightarrow \quad \mathbf{N}' = -\kappa \mathbf{t}$ for some scalar function κ

$\Rightarrow \quad \mathbf{N}' + \kappa \mathbf{t} = \mathbf{0}$

$\Rightarrow \quad \dfrac{d\mathbf{N}}{ds} + \kappa \dfrac{d\mathbf{r}}{ds} = \mathbf{0}$

$\Rightarrow \quad d\mathbf{N} + \kappa\, d\mathbf{r} = \mathbf{0}$

$\Rightarrow \quad$ The curve (2) is a line of curvature. [by Rodrigue's formula]

Hence the condition is sufficient.

Theorem 4. The necessary and sufficient conditions that the parametric curves be lines of curvature are $F = 0, M = 0$.

Proof. Let the equation of a surface be given by

$$\mathbf{r} = \mathbf{r}\,(u, v),$$

where (u, v) are the parametric coordinates of a point on the surface.

Then the differential equation of the lines of curvature on the surface is given by

$$(EM - FL)\,du^2 + (EN - GL)\,du\,dv + (FN - GM)\,dv^2 = 0. \tag{1}$$

Also, the differential equation of the parametric curves on the surface is given by

$$du\,dv = 0. \tag{2}$$

The condition is necessary. Let the parametric curves be lines of curvature.

To prove : $F = 0, M = 0$.

Since the parametric curves are lines of curvature, equations (1) and (2) are identical. Hence, we have

$$EM - FL = 0, \tag{3}$$

$$FN - GM = 0, \tag{4}$$

$$EN - GL \neq 0. \tag{5}$$

Multiplying (3) and (4) by G and E respectively and adding, we get

$$(EN - GL)\,F = 0$$

$$\Rightarrow \quad F = 0. \tag{6}$$

[using (5)]

Now, multiplying (3) and (4) by N and L respectively and adding, we get

$$(EN - GL)\,M = 0$$

$$\Rightarrow \quad M = 0. \tag{7}$$

[using (5)]

From (6) and (7), we get

$$F = 0, M = 0.$$

Hence the condition is necessary.

The condition is sufficient. Now, let $F = 0, M = 0$.

To prove : The parametric curves are lines of curvature, i.e., equations (1) and (2) are identical.

Substituting $F = 0, M = 0$ in (1), we get

$$(EN - GL)\,du\,dv = 0$$

$$\Rightarrow \quad \text{Either } EN - GL = 0 \text{ or } du\,dv = 0. \tag{8}$$

If we take

$$EN - GL = 0,$$

then

$$\frac{E}{L} = \frac{G}{N}.$$

But this is true at an umbilic only. Therefore, we have

$$EN - GL \neq 0$$

$\Rightarrow$ $du\,dv = 0,$ [using (8)]

which is equation (2), i.e. the differential equation of the parametric curves.

Therefore equations (1) and (2) are identical, i.e., the parametric curves are lines of curvature.

Hence the condition is sufficient.

Theorem 5. Euler's Theorem. The normal curvature κ_n at a point P on a surface in any direction is given by

$$\kappa_n = \kappa_a \cos^2 \psi + \kappa_b \sin^2 \psi,$$

where κ_a and κ_b are the principal curvatures at P and ψ is the angle which the direction (du, dv) of the normal section at P makes with the principal direction through P in which the principal curvature is κ_a.

Proof. Let the equation of a surface be given by

$$\mathbf{r} = \mathbf{r}\,(u, v)$$

and let κ_n be the normal curvature at a point $P\,(u, v)$ on this surface in the direction (du, dv). Also, let κ_a and κ_b be the principal curvatures at $P\,(u, v)$.

We know that

$$\kappa_n = \frac{L\,du^2 + 2M\,du\,dv + N\,dv^2}{E\,du^2 + 2F\,du\,dv + G\,dv^2}. \tag{1}$$

Let the lines of curvature be taken as the parametric curves. Then

$$F = 0, M = 0. \tag{2}$$

Substituting the values of F and M from (2) in (1), we get

$$\kappa_n = \frac{L\,du^2 + N\,dv^2}{E\,du^2 + G\,dv^2}. \tag{3}$$

The principal curvature κ_a, being the normal curvature for the direction $dv = 0$, is expressed as

$$\kappa_a = \frac{L}{E}. \tag{4}$$

[using (3)]

Similarly, the principal curvature κ_b, being the normal curvature for the direction $du = 0$, is expressed as

$$\kappa_b = \frac{N}{G}. \tag{5}$$

[using (3)]

Now consider a normal section of the surface in the direction (du, dv). If this normal section makes an angle ψ with the direction $dv = 0$, then we know that

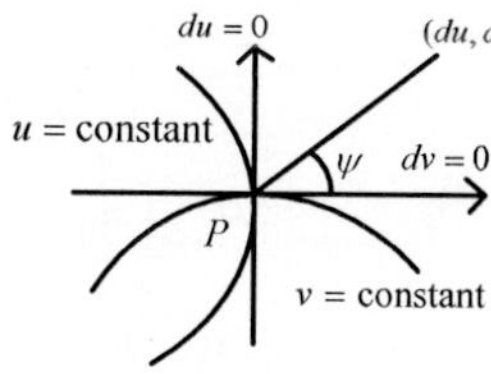

$$\cos\psi = \frac{1}{\sqrt{E}}\left(E\frac{du}{ds} + F\frac{dv}{ds}\right), \ \sin\psi = \frac{H}{\sqrt{E}}\frac{dv}{ds}$$

or $$\cos\psi = \sqrt{E}\frac{du}{ds}, \ \sin\psi = \sqrt{G}\frac{dv}{ds}. \tag{6}$$

[using (2) and $H^2 = EG - F^2$]

Since $ds^2 = E\,du^2 + 2F\,du\,dv + G\,dv^2 = E\,du^2 + G\,dv^2$ [using (2)], (3) can be written as

$$\kappa_n = \frac{L\,du^2 + N\,dv^2}{ds^2}.$$

or $$\kappa_n = L\left(\frac{du}{ds}\right)^2 + N\left(\frac{dv}{ds}\right)^2$$

or $$\kappa_n = L\left(\frac{\cos\psi}{\sqrt{E}}\right)^2 + N\left(\frac{\sin\psi}{\sqrt{G}}\right)^2$$ [using (6)]

or $$\kappa_n = \kappa_a \cos^2\psi + \kappa_b \sin^2\psi.$$ [using (4) and (5)]

Corollary. Dupin's Theorem. The sum of the normal curvatures at any point on a surface in two directions at right angles is constant and equal to the sum of the principal curvatures at the point.

Proof. Let κ_1, κ_2 be the normal curvatures in mutually perpendicular directions at a point $P(u, v)$ on a surface and let κ_a, κ_b be the principal curvatures of the surface at $P(u, v)$. Also, let ψ_1, ψ_2 be the inclinations of the directions of the normal curvatures κ_1, κ_2 respectively with the direction of the principal curvature κ_a.

By Euler's theorem,

$$\kappa_1 = \kappa_a \cos^2 \psi_1 + \kappa_b \sin^2 \psi_1, \tag{1}$$

$$\kappa_2 = \kappa_a \cos^2 \psi_2 + \kappa_b \sin^2 \psi_2. \tag{2}$$

But, $\psi_2 - \psi_1 = \dfrac{\pi}{2}$.

$$\therefore \quad \psi_2 = \frac{\pi}{2} + \psi_1. \tag{3}$$

Substituting the value of ψ_2 from (3) in (2), we get

$$\kappa_2 = \kappa_a \cos^2\left(\frac{\pi}{2} + \psi_1\right) + \kappa_b \sin^2\left(\frac{\pi}{2} + \psi_1\right)$$

or
$$\kappa_2 = \kappa_a \sin^2 \psi_1 + \kappa_b \cos^2 \psi_1. \tag{4}$$

Adding (1) and (4), we get

$$\kappa_1 + \kappa_2 = \kappa_a(\cos^2 \psi_1 + \sin^2 \psi_1) + \kappa_b(\sin^2 \psi_1 + \cos^2 \psi_1)$$

or
$$\kappa_1 + \kappa_2 = \kappa_a + \kappa_b.$$

Theorem 6. Joachimsthal's Theorem. If the curve of intersection of two surfaces is a line of curvature on both the surfaces, then the surfaces cut at a constant angle. Conversely, if two surfaces cut at a constant angle and the curve of intersection is a line of curvature on one of them, then it is also a line of curvature on the other.

Proof. Let C be the curve of intersection of two surfaces S_1 and S_2 and let P be any point on C. Let $\mathbf{N}_1$ and $\mathbf{N}_2$ be unit normal vectors to the surfaces S_1 and S_2 respectively at P. Also, let θ be the angle of intersection of the surfaces S_1 and S_2 at P.

First, let us suppose that the curve C is a line of curvature on S_1 and S_2 both.

We have to prove that θ is constant.

Clearly,

$$\cos\theta = \mathbf{N}_1 . \mathbf{N}_2 \cdot$$

$$\therefore \quad \frac{d}{ds}(\cos\theta) = \frac{d\mathbf{N}_1}{ds} . \mathbf{N}_2 + \mathbf{N}_1 . \frac{d\mathbf{N}_2}{ds}. \tag{1}$$

Since C is a line of curvature on S_1, we have

$$\kappa\, d\mathbf{r} + d\mathbf{N}_1 = \mathbf{0}, \tag{2}$$

[by Rodrigue's formula]

where $\mathbf{r}$ is the position vector of the point P and κ is the normal curvature of S_1 at P along C.

Differentiating (2) with respect to s, we get

$$\kappa \frac{d\mathbf{r}}{ds} + \frac{d\mathbf{N}_1}{ds} = \mathbf{0}$$

$\Rightarrow$ $\kappa \mathbf{t} + \dfrac{d\mathbf{N}_1}{ds} = \mathbf{0}$, where $\mathbf{t}$ is the unit tangent vector to C at P.

$\Rightarrow$ $\dfrac{d\mathbf{N}_1}{ds} = -\kappa \mathbf{t}$

$\Rightarrow$ $\dfrac{d\mathbf{N}_1}{ds} \cdot \mathbf{N}_2 = -\kappa \mathbf{t} \cdot \mathbf{N}_2 = 0.$ (3)

Similarly, since C is a line of curvature on S_2,

$$\mathbf{N}_1 \cdot \frac{d\mathbf{N}_2}{ds} = 0. \quad (4)$$

Using (3) and (4) in (1), we get

$$\frac{d}{ds}(\cos\theta) = 0$$

$\Rightarrow$ $\cos\theta$ = constant (along C)

$\Rightarrow$ θ = constant (along C)

$\Rightarrow$ The angle of intersection of S_1 and S_2 is constant.

Conversely, let that θ = constant (along C) and let C be a line of curvature on one of the surfaces, say S_1.

We have to prove that C is a line of curvature on S_2 as well.

By supposition,

θ = constant (along C)

$\Rightarrow$ $\cos\theta$ = constant (along C)

$\Rightarrow$ $\mathbf{N}_1 \cdot \mathbf{N}_2$ = constant (along C)

$\Rightarrow$ $\dfrac{d}{ds}(\mathbf{N}_1 \cdot \mathbf{N}_2) = 0$

$\Rightarrow$ $\dfrac{d\mathbf{N}_1}{ds} \cdot \mathbf{N}_2 + \mathbf{N}_1 \cdot \dfrac{d\mathbf{N}_2}{ds} = 0$

$\Rightarrow$ $\mathbf{N}_1 \cdot \dfrac{d\mathbf{N}_2}{ds} = 0$ $\left[\because \dfrac{d\mathbf{N}_1}{ds} \cdot \mathbf{N}_2 = 0, \text{as } C \text{ is a line of curvature on } S_1\right]$

$\Rightarrow$ $\dfrac{d\mathbf{N}_2}{ds}$ is perpendicular to $\mathbf{N}_1$. (5)

Also,

$\mathbf{N}_2 \cdot \mathbf{N}_2 = 1$

$\Rightarrow$ $\dfrac{d}{ds}(\mathbf{N}_2 \cdot \mathbf{N}_2) = 0$

$$\Rightarrow \quad \frac{d\mathbf{N}_2}{ds}\cdot\mathbf{N}_2 + \mathbf{N}_2\cdot\frac{d\mathbf{N}_2}{ds} = 0$$

$$\Rightarrow \quad \frac{d\mathbf{N}_2}{ds}\cdot\mathbf{N}_2 + \frac{d\mathbf{N}_2}{ds}\cdot\mathbf{N}_2 = 0$$

$$\Rightarrow \quad 2\frac{d\mathbf{N}_2}{ds}\cdot\mathbf{N}_2 = 0$$

$$\Rightarrow \quad \frac{d\mathbf{N}_2}{ds}\cdot\mathbf{N}_2 = 0$$

$$\Rightarrow \quad \frac{d\mathbf{N}_2}{ds} \text{ is perpendicular to } \mathbf{N}_2. \tag{6}$$

From (5) and (6), it follows that

$$\frac{d\mathbf{N}_2}{ds} \text{ is perpendicular to each of } \mathbf{N}_1 \text{ and } \mathbf{N}_2.$$

$$\Rightarrow \quad \frac{d\mathbf{N}_2}{ds} \text{ is parallel to } \mathbf{t}.$$

$$\Rightarrow \quad \frac{d\mathbf{N}_2}{ds} = -\kappa\mathbf{t}, \text{ where } \kappa \text{ is a scalar.}$$

$$\Rightarrow \quad \kappa\frac{d\mathbf{r}}{ds} + \frac{d\mathbf{N}_2}{ds} = \mathbf{0} \qquad \left[\because \mathbf{t} = \frac{d\mathbf{r}}{ds}\right]$$

$$\Rightarrow \quad \kappa\, d\mathbf{r} + d\mathbf{N}_2 = \mathbf{0}$$

$\Rightarrow$ C is a line of curvature on S_2. [using Rodrigue's formula]

4.4 ELLIPTIC, HYPERBOLIC AND PARABOLIC POINTS ON A SURFACE

Let $P(u, v)$ be any point on a surface S. Let E, F, G be the fundamental magnitudes of the first order and let L, M, N be the fundamental magnitudes of the second order of S at P.

We know that

$$K = \kappa_a\kappa_b = \frac{T^2}{H^2} = \frac{LN - M^2}{EG - F^2},$$

where K is the Gaussian curvature of S at P and κ_a, κ_b are the principal curvatures of S at P.

Elliptic Point. The point P is called an elliptic point if the Gaussian curvature K is positive at P, i.e., if $LN - M^2 > 0$.

Thus a point P is an elliptic point if and only if the two principal curvatures κ_a and κ_b are of the same sign.

Hyperbolic Point. The point P is called a hyperbolic point if the Gaussian curvature K is negative at P, i.e., if $LN - M^2 < 0$.

Thus a point P is a hyperbolic point if and only if the two principal curvatures κ_a and κ_b are of opposite signs.

Parabolic Point. The point P is called a parabolic point if the Gaussian curvature K is zero at P, i.e., if $LN - M^2 = 0$.

Thus a point P is an elliptic point if and only if at least one of the two principal curvatures κ_a and κ_b is zero.

Synclastic and Anticlastic Surfaces.

Those portions of the surface on, which the two principal curvatures κ_a and κ_b have the same sign, are said to be synclastic. Thus a portion of a surface is synclastic if and only if every point on this portion is an elliptic point. The surface of a sphere or of an ellipsoid is synclastic at all points.

If the two principal curvatures κ_a and κ_b have opposite signs on any part of the surface, then this part is said to be anticlastic. Thus a portion of a surface is anticlastic if and only if every point on this portion is a hyperbolic point. The surface of a hyperbolic paraboloid is anticlastic at all points.

4.5 DUPIN'S INDICATRIX

Definition. The Dupin's indicatrix at a point on a surface is the curve of intersection of the surface by a plane parallel and indefinitely close to the tangent plane at the point.

Equations of Dupin's Indicatrix. Let O be a point on the surface, the equation of Dupin's indicatrix at which is to be found.

Take any point $P(x, y, z)$ on the Dupin's indicatrix.

Then P is a point on the surface very near to O.

From the point P, draw PN perpendicular to the tangent plane at O. Let $PN = h$.

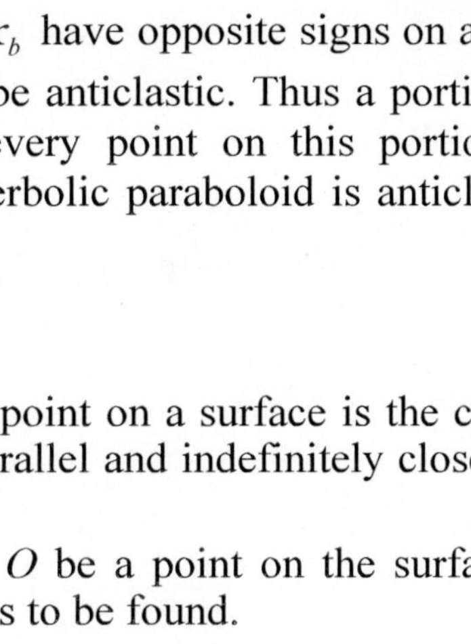

By the geometrical interpretation of the second fundamental form of a surface, we have

$$2h = L\,du^2 + 2M\,du\,dv + N\,dv^2. \quad (1)$$

where L, M, N are the fundamental magnitudes of the second order of the surface at O.

Let the lines of curvature be taken as the parametric curves. Then

$$F = 0, M = 0. \tag{2}$$

Substituting the value of M from (2) in (1), we get

$$2h = L\,du^2 + N\,dv^2. \tag{3}$$

Since the lines of curvature at each point of a surface are orthogonal, it follows that the parametric curves passing through O are also orthogonal.

Take O for origin, and let the axis of X and the axis of Y be taken along the parametric curves v = const. and u = const. respectively. The axis of Z is taken along the normal to the surface at O.

Let κ_n be the normal curvature of the surface at O in the direction (du, dv) and let κ_a, κ_b be the principal curvatures of the surface at O. Then

$$\kappa_n = \frac{L\,du^2 + 2M\,du\,dv + N\,dv^2}{E\,du^2 + 2F\,du\,dv + G\,dv^2}, \tag{4}$$

where E, F, G are the fundamental magnitudes of the first order of the surface at O.

Substituting the values of F and M from (2) in (4), we get

$$\kappa_n = \frac{L\,du^2 + N\,dv^2}{E\,du^2 + G\,dv^2}. \tag{5}$$

The principal curvature κ_a, being the normal curvature for the direction $dv = 0$, is expressed as

$$\kappa_a = \frac{L}{E}. \qquad \text{[using (5)]}$$

$$\therefore \quad L = \kappa_a E. \tag{6}$$

Similarly, the principal curvature κ_b, being the normal curvature for the direction $du = 0$, is expressed as

$$\kappa_b = \frac{N}{G}. \qquad \text{[using (5)]}$$

$$\therefore \quad N = \kappa_b G. \tag{7}$$

Substituting the values of L and N from (6) and (7) respectively in (3), we get

$$2h = \kappa_a E\,du^2 + \kappa_b G\,dv^2. \tag{8}$$

Let ds_1 and ds_2 denote elementary arc lengths along the parametric curves v = const. and u = const. respectively. Then, using the formula

$$ds^2 = E\,du^2 + 2F\,du\,dv + G\,dv^2,$$

we have

$$ds_1^{\,2} = E\,du^2,\; ds_2^{\,2} = G\,dv^2. \tag{9}$$

Using (9), (8) reduces to

$$2h = \kappa_a\, ds_1^{\,2} + \kappa_b\, ds_2^{\,2}. \tag{10}$$

Since the point P is very close to O, we can take $ds_1 = x, ds_2 = y$. Moreover, $z = h$. Therefore, from (10), we have

$$z = h,\; \kappa_a x^2 + \kappa_b y^2 = 2h$$

or
$$z = h,\; \frac{x^2}{\rho_1} + \frac{y^2}{\rho_2} = 2h, \tag{11}$$

where ρ_1 and ρ_2 are the principal radii of curvature at O, i.e., $\kappa_a = 1/\rho_1,\; \kappa_b = 1/\rho_2$.

Clearly, (11) represents a conic section and hence the Dupin's indicatrix is a conic section.

Now, we consider three cases according as the point O is elliptic, hyperbolic or parabolic.

Case I. If the point O be an elliptic point, then, at O, the principal radii of curvature ρ_1, ρ_2 are of the same sign. Hence the conic (11) is an ellipse with semi-axes of lengths $\sqrt{2h\rho_1}, \sqrt{2h\rho_2}$. The ellipse is real or imaginary according to the sign of h and thus the surface in the neighbourhood of O lies entirely on one side of the tangent plane at O.

Case II. If the point O be a hyperbolic point, then, at O, the principal radii of curvature ρ_1, ρ_2 are of opposite signs. Hence the indicatrix (11) is one of the two conjugate hyperbolas according to the sign of h and the surface in the neighbourhood of O lies on both sides of the tangent plane at O.

Case III. If the point O be a parabolic point, then, at O, one of the two principal curvatures is zero, i.e., either $\kappa_a = 0$ or $\kappa_b = 0$. If $\kappa_a = 0$, then the indicatrix reduces to $\kappa_b y^2 = 2h$. If $\kappa_b = 0$, then the indicatrix reduces to $\kappa_a x^2 = 2h$. In either case, the indicatrix is a pair of straight lines.

4.6 THIRD FUNDAMENTAL FORM

Definition. The third fundamental form of a surface, denoted by *III* is defined by

$$III = d\mathbf{N}\,.\,d\mathbf{N} = A\,du^2 + 2B\,du\,dv + C\,dv^2,$$

where $A = \mathbf{N}_1^{\,2}$, $B = \mathbf{N}_1 . \mathbf{N}_2$, $C = \mathbf{N}_2^{\,2}$.

Relation between the three fundamental forms. If the first, second and third fundamental forms are denoted by *I*, *II* and *III* respectively, then

$$KI - 2\mu II + III = 0,$$

where K is the Gaussian curvature and μ is the mean curvature.

Proof. By definition,

$$I = E\,du^2 + 2F\,du\,dv + G\,dv^2, \qquad (1)$$

$$II = L\,du^2 + 2M\,du\,dv + N\,dv^2. \qquad (2)$$

Let the lines of curvature be taken as the parametric curves. Then

$$F = 0, M = 0. \qquad (3)$$

Using (3), (1) and (2) reduce to

$$I = E\,du^2 + G\,dv^2 \qquad (4)$$

and $$II = L\,du^2 + N\,dv^2 \qquad (5)$$

respectively.

Let κ_n be the normal curvature of the surface at O in the direction (du, dv) and let κ_a, κ_b be the principal curvatures of the surface at O. Then

$$\kappa_n = \frac{L\,du^2 + 2M\,du\,dv + N\,dv^2}{E\,du^2 + 2F\,du\,dv + G\,dv^2}$$

or $$\kappa_n = \frac{L\,du^2 + N\,dv^2}{E\,du^2 + G\,dv^2}. \qquad (6)$$

[using (3)]

The principal curvature κ_a, being the normal curvature for the direction $dv = 0$, is expressed as

$$\kappa_a = \frac{L}{E}. \qquad \text{[using (6)]}$$

$$\therefore \quad L = \kappa_a E. \qquad (7)$$

Similarly, the principal curvature κ_b, being the normal curvature for the direction $du = 0$, is expressed as

$$\kappa_b = \frac{N}{G}. \qquad \text{[using (6)]}$$

$$\therefore \quad N = \kappa_b G. \qquad (8)$$

Using (7) and (8), (5) becomes

$$II = \kappa_a E\,du^2 + \kappa_b G\,dv^2. \qquad (9)$$

By Rodrigue's formula, we have

$$\kappa\, d\mathbf{r} + d\mathbf{N} = \mathbf{0}$$

or $$\kappa\,[\mathbf{r}_1 du + \mathbf{r}_2 dv] + [\mathbf{N}_1 du + \mathbf{N}_2 dv] = \mathbf{0}, \tag{10}$$

where κ is the normal curvature at the point $\mathbf{r}$ in the direction of line of curvature.

For the line of curvature $v =$ constant, we have

$$\kappa = \kappa_a,\ dv = 0.$$

Hence, by (10), we have

$$\kappa_a\, \mathbf{r}_1\, du + \mathbf{N}_1\, du = \mathbf{0}$$

$$\Rightarrow \quad (\kappa_a\, \mathbf{r}_1 + \mathbf{N}_1)\, du = \mathbf{0}$$

$$\Rightarrow \quad \kappa_a\, \mathbf{r}_1 + \mathbf{N}_1 = \mathbf{0}$$

$$\Rightarrow \quad \mathbf{N}_1 = -\kappa_a\, \mathbf{r}_1. \tag{11}$$

Similarly, for the line of curvature $u =$ constant, we have

$$\kappa = \kappa_b,\ du = 0.$$

Hence, by (10), we have

$$\kappa_b \mathbf{r}_2\, dv + \mathbf{N}_2\, dv = \mathbf{0}$$

$$\Rightarrow \quad (\kappa_b \mathbf{r}_2 + \mathbf{N}_2)\, dv = \mathbf{0}$$

$$\Rightarrow \quad \kappa_b \mathbf{r}_2 + \mathbf{N}_2 = \mathbf{0}$$

$$\Rightarrow \quad \mathbf{N}_2 = -\kappa_b \mathbf{r}_2. \tag{12}$$

Now, by definition,

$$\begin{aligned} III &= d\mathbf{N} \,.\, d\mathbf{N} = (\mathbf{N}_1\, du + \mathbf{N}_2\, dv)^2 \\ &= (\kappa_a \mathbf{r}_1\, du + \kappa_b \mathbf{r}_2\, dv)^2 \qquad \text{[using (11) and (12)]} \\ &= \kappa_a^{\ 2} E\, du^2 + 2\kappa_a \kappa_b\, F\, du\, dv + \kappa_b^{\ 2} G\, dv^2. \end{aligned}$$

But, by (2), $F = 0$. Therefore

$$III = \kappa_a^{\ 2} E\, du^2 + \kappa_b^{\ 2} G\, dv^2. \tag{13}$$

Eliminating $E\, du^2$ and $G\, dv^2$ from (4), (9) and (13), we get

$$\begin{vmatrix} I & 1 & 1 \\ II & \kappa_a & \kappa_b \\ III & \kappa_a^{\ 2} & \kappa_b^{\ 2} \end{vmatrix} = 0$$

or $$I(\kappa_a \kappa_b^{\ 2} - \kappa_a^{\ 2} \kappa_b) - II(\kappa_b^{\ 2} - \kappa_a^{\ 2}) + III(\kappa_b - \kappa_a) = 0$$

or $$(\kappa_b - \kappa_a)[I(\kappa_a \kappa_b) - II(\kappa_b + \kappa_a) + III] = 0.$$

Therefore

$$I(\kappa_a \kappa_b) - II(\kappa_b + \kappa_a) + III = 0$$

i.e., $$KI - 2\mu II + III = 0, \text{ where } K = \kappa_a \kappa_b \text{ and } \mu = \frac{\kappa_a + \kappa_b}{2}.$$

Solved Examples

Example 1. Find the curvature of normal section of a right helicoid.

Solution. We know that the curvature κ_n of the normal section is given by

$$\kappa_n = \frac{L\,du^2 + 2M\,du\,dv + N\,dv^2}{E\,du^2 + 2F\,du\,dv + G\,dv^2} \tag{1}$$

Now, we have to find the values of E, F, G, L, M, N for a right helicoid. The position vector $\mathbf{r}$ of any current point on the right helicoid is given by

$$\mathbf{r} = (u\cos\phi, u\sin\phi, c\phi).$$

Taking u and ϕ as the first and the second parameters, we have

$$\mathbf{r}_1 = (\cos\phi, \sin\phi, 0),\ \ \mathbf{r}_2 = (-u\sin\phi, u\cos\phi, c),$$
$$\mathbf{r}_{11} = (0, 0, 0),\ \ \mathbf{r}_{12} = (-\sin\phi, \cos\phi, 0),$$
$$\mathbf{r}_{22} = (-u\cos\phi, -u\sin\phi, 0).$$

Therefore

$$E = \mathbf{r}_1^{\ 2} = \cos^2\phi + \sin^2\phi + 0 = 1,$$
$$F = \mathbf{r}_1 \cdot \mathbf{r}_2 = -u\cos\phi\sin\phi + u\sin\phi\cos\phi + 0 = 0,$$
$$G = \mathbf{r}_2^{\ 2} = u^2\sin^2\phi + u^2\cos^2\phi + c^2 = u^2 + c^2,$$
$$H^2 = EG - F^2 = 1\,(u^2 + c^2) - 0 = u^2 + c^2.$$

Again, we have

$$\mathbf{r}_1 \times \mathbf{r}_2 = (c\sin\phi, -c\cos\phi, u)$$

$$\Rightarrow \qquad \mathbf{N} = \frac{\mathbf{r}_1 \times \mathbf{r}_2}{H} = \left(\frac{c\sin\phi}{\sqrt{u^2+c^2}}, -\frac{c\cos\phi}{\sqrt{u^2+c^2}}, \frac{u}{\sqrt{u^2+c^2}}\right).$$

Therefore

$$L = \mathbf{r}_{11} \cdot \mathbf{N} = (0, 0, 0) \cdot \left(\frac{c\sin\phi}{\sqrt{u^2+c^2}}, -\frac{c\cos\phi}{\sqrt{u^2+c^2}}, \frac{u}{\sqrt{u^2+c^2}}\right)$$
$$= 0,$$
$$M = \mathbf{r}_{12} \cdot \mathbf{N}$$
$$= (-\sin\phi, \cos\phi, 0) \cdot \left(\frac{c\sin\phi}{\sqrt{u^2+c^2}}, -\frac{c\cos\phi}{\sqrt{u^2+c^2}}, \frac{u}{\sqrt{u^2+c^2}}\right)$$
$$= -\frac{c}{\sqrt{u^2+c^2}},$$

$$N = \mathbf{r}_{22} \cdot \mathbf{N}$$

$$= (-u\cos\phi, -u\sin\phi, 0) \cdot \left(\frac{c\sin\phi}{\sqrt{u^2+c^2}}, -\frac{c\cos\phi}{\sqrt{u^2+c^2}}, \frac{u}{\sqrt{u^2+c^2}} \right)$$

$$= 0.$$

Substituting the values of E, F, G, L, M, N in (1), we get

$$\kappa_n = -\frac{\dfrac{2c}{\sqrt{(u^2+c^2)}}\, du\, dv}{du^2 + (u^2+c^2)\, dv^2}$$

$$= -\frac{2c\, du\, dv}{\sqrt{(u^2+c^2)}\left\{du^2 + (u^2+c^2)\, dv^2\right\}}.$$

Example 2. Find the principal curvatures and the lines of curvature on the right helicoid

$$x = u\cos\phi,\ y = u\sin\phi,\ z = c\phi.$$

Solution. We know that the principal curvatures κ of the surface are given by

$$H^2\kappa^2 - (EN - 2FM + GL)\,\kappa + LN - M^2 = 0. \tag{1}$$

Now, we have to find the values of E, F, G, L, M, N for the given surface.

The position vector $\mathbf{r}$ of any current point on the right helicoid is given by

$$\mathbf{r} = (u\cos\phi, u\sin\phi, c\phi).$$

Proceeding as in Example 1, we have

$$E = 1,\ F = 0,\ G = u^2 + c^2,\ H^2 = u^2 + c^2;$$

$$L = 0,\ M = -\frac{c}{\sqrt{u^2+c^2}},\ N = 0.$$

Substituting the above values in (1), we get

$$(u^2+c^2)\,\kappa^2 - \frac{c^2}{u^2+c^2} = 0.$$

$$\therefore \qquad \kappa = \pm\frac{c}{u^2+c^2}.$$

This gives the principal curvatures κ of the given surface.

Also, we know that the differential equation of the lines of curvature is given by

$$(EM - FL)\, du^2 + (EN - GL)\, du\, d\phi + (FN - GM)\, d\phi^2 = 0$$

or $$-\frac{c}{\sqrt{u^2+c^2}}\,du^2-(u^2+c^2)\left(-\frac{c}{\sqrt{u^2+c^2}}\right)d\phi^2=0$$

[substituting the values of E, F, G, L, M, N]

or $$-c\,du^2+c\,(u^2+c^2)\,d\phi^2=0$$

or $$d\phi=\pm\frac{du}{\sqrt{(u^2+c^2)}}.$$

Integrating this, we get

$$\phi=\pm\ \sinh^{-1}\left(\frac{u}{c}\right)+A,$$

where A is the constant of integration.

Example 3. Find the principal directions and the principal curvatures on the surface

$$x=a\,(u+v),\ y=b\,(u-v),\ z=uv.$$

Also, find the Gaussian curvature at a point of the above surface.

Solution. We know that the differential equation of the principal directions (or lines of curvature) is given by

$$(EM-FL)\,du^2+(EN-GL)\,du\,dv+(FN-GM)\,dv^2=0. \tag{1}$$

Now, we have to find the values of E, F, G, L, M, N for the given surface. The position vector $\mathbf{r}$ of any current point on the given surface is

$$\mathbf{r}=(a\,(u+v),b\,(u-v),u\,v).$$

Taking u and v as the first and the second parameters, we have

$$\mathbf{r}_1=(a,b,v),\ \ \mathbf{r}_2=(a,-b,u),$$

$$\mathbf{r}_{11}=(0,0,0),\ \ \mathbf{r}_{12}=(0,0,1),\ \ \mathbf{r}_{22}=(0,0,0).$$

Therefore

$$E=\mathbf{r}_1^{\ 2}=a^2+b^2+v^2,$$

$$F=\mathbf{r}_1\cdot\mathbf{r}_2=a^2-b^2+uv,$$

$$G=\mathbf{r}_2^{\ 2}=a^2+b^2+u^2,$$

$$H^2=EG-F^2$$

$$=(a^2+b^2+v^2)(a^2+b^2+u^2)-(a^2-b^2+uv)^2$$

$$=4a^2b^2+a^2(u-v)^2+b^2(u+v)^2.$$

Again, we have

$$\mathbf{r}_1\times\mathbf{r}_2=(b\,(u+v),a\,(v-u),-2ab)$$

$$\Rightarrow \qquad \mathbf{N} = \frac{\mathbf{r}_1 \times \mathbf{r}_2}{H} = \left(\frac{b\,(u+v)}{H}, \frac{a\,(v-u)}{H}, -\frac{2ab}{H} \right).$$

Therefore

$$L = \mathbf{r}_{11} \cdot \mathbf{N} = (0, 0, 0) \cdot \left(\frac{b\,(u+v)}{H}, \frac{a\,(v-u)}{H}, -\frac{2ab}{H} \right)$$

$$= 0,$$

$$M = \mathbf{r}_{12} \cdot \mathbf{N} = (0, 0, 1) \cdot \left(\frac{b\,(u+v)}{H}, \frac{a\,(v-u)}{H}, -\frac{2ab}{H} \right)$$

$$= -\frac{2ab}{H},$$

$$N = \mathbf{r}_{22} \cdot \mathbf{N} = (0, 0, 0) \cdot \left(\frac{b\,(u+v)}{H}, \frac{a\,(v-u)}{H}, -\frac{2ab}{H} \right)$$

$$= 0.$$

Substituting the above values in (1), we get

$$-(a^2 + b^2 + v^2) \frac{2ab}{H} du^2 + (a^2 + b^2 + u^2) \frac{2ab}{H} dv^2 = 0$$

or $$\frac{du}{\sqrt{(a^2 + b^2 + u^2)}} = \pm \frac{dv}{\sqrt{(a^2 + b^2 + v^2)}}.$$

Integrating this, we get

$$\sinh^{-1}\left(\frac{u}{\sqrt{a^2 + b^2}} \right) = \pm \sinh^{-1}\left(\frac{v}{\sqrt{a^2 + b^2}} \right) + C, \qquad (2)$$

where C is the constant of integration.

Equation (2) gives the principal directions of the given surface.

Also, we know that the equation giving the principal curvatures κ is

$$H^2 \kappa^2 - (EN - 2FM + GL)\,\kappa + LN - M^2 = 0$$

or $$H^2 \kappa^2 - \left\{ -2\,(a^2 - b^2 + uv) \left(-\frac{2ab}{H} \right) \right\} \kappa - \frac{4a^2b^2}{H^2} = 0$$

[substituting the values of E, F, G, L, M, N]

or $$H^4 \kappa^2 - 4abH\,(a^2 - b^2 + uv)\,\kappa - 4a^2b^2 = 0. \qquad (3)$$

It follows from (3) that the Gaussian curvature is

$$K = \kappa_a \kappa_b = -\frac{4a^2b^2}{H^4}.$$

Example 4. Show that the principal radii of curvature of the surface $y \cos(z/a) = x \sin(z/a)$

are equal to $\pm(x^2+y^2+a^2)/a$. Find the lines of curvature.

Solution. We know that the principal curvatures κ of the surface are given by

$$H^2\kappa^2-(EN-2FM+GL)\,\kappa+LN-M^2=0. \tag{1}$$

Now, we have to find the values of E, F, G, L, M, N for the given surface. The equation of the given surface is

$$y\cos(z/a)=x\sin(z/a)$$

or $\tan(z/a)=y/x$

or $z=a\tan^{-1}(y/x)$.

We can take parametric equations of the surface as

$$x=u\cos v,\ y=u\sin v,\ z=av.$$

Therefore the surface is a helicoid and the position vector $\mathbf{r}$ of any current point on the surface is

$$\mathbf{r}=(u\cos v, u\sin v, av).$$

Now, proceeding as in Example 1, we have

$$E=1,\ F=0,\ G=u^2+a^2,\ H^2=u^2+a^2\,;$$

$$L=0,\ M=-\frac{a}{\sqrt{u^2+a^2}},\ N=0.$$

Substituting the above values in (1), we get

$$(u^2+a^2)\,\kappa^2-\frac{a^2}{u^2+a^2}=0.$$

$$\therefore\qquad \kappa=\pm\frac{a}{u^2+a^2}=\pm\frac{a}{x^2+y^2+a^2}$$

and the principal radii of curvature are

$$\rho=\pm\frac{u^2+a^2}{a}=\pm\frac{x^2+y^2+a^2}{a}.$$

Also, we know that the differential equation of the lines of curvature is

$$(EM-FL)\,du^2+(EN-GL)\,du\,dv+(FN-GM)\,dv^2=0$$

or
$$-\frac{a}{\sqrt{u^2+a^2}}\,du^2-(u^2+a^2)\left(-\frac{a}{\sqrt{u^2+a^2}}\right)dv^2=0$$

[substituting the values of E, F, G, L, M, N]

or
$$-a\,du^2+a(u^2+a^2)\,dv^2=0$$

or
$$dv=\pm\frac{du}{\sqrt{(u^2+a^2)}}.$$

Integrating this, we get

$$v = \pm \sinh^{-1}\left(\frac{u}{a}\right) + C$$

or $$\tan^{-1}\left(\frac{y}{x}\right) = \pm \sinh^{-1}\left(\frac{\sqrt{(x^2+y^2)}}{a}\right) + C.$$

Example 5. Let $\kappa_1, \kappa_2, ..., \kappa_m$ denote the normal curvatures of m sections of a surface which make equal angles $2\pi/m$ with one another. If $m > 2$, prove that

$$\kappa_1 + \kappa_2 + ... + \kappa_m = m\mu.$$

Solution. Let κ_a, κ_b be principal curvatures at a point P of the given surface. Let κ_n be the normal curvature of the surface at P in a direction which makes an angle ψ with the principal direction through P in which normal curvature is κ_a. Then, by Euler's theorem,

$$\begin{aligned}\kappa_n &= \kappa_a \cos^2\psi + \kappa_b \sin^2\psi \\ &= \kappa_a\left(\frac{1+\cos 2\psi}{2}\right) + \kappa_b\left(\frac{1-\cos 2\psi}{2}\right) \\ &= \frac{1}{2}(\kappa_a + \kappa_b) + \frac{1}{2}(\kappa_a - \kappa_b)\cos 2\psi \\ &= \mu + A\cos 2\psi. \qquad \left[\because \mu = \frac{1}{2}(\kappa_a + \kappa_b),\ A = \frac{1}{2}(\kappa_a - \kappa_b)\right]\end{aligned}$$

Here, μ and A are constants for all sections through P.

Suppose θ is the angle which the section through P, in which normal curvature is κ_1 makes with the principal direction through P, in which normal curvature is κ_a. Then the other $m-1$ sections will make angles $\theta + \frac{2\pi}{m}, \theta + \frac{4\pi}{m}, ..., \theta + (m-1)\frac{2\pi}{m}$ with this direction. Therefore

$$\begin{aligned}\kappa_1 &= \mu + A\cos 2\theta, \\ \kappa_2 &= \mu + A\cos 2\left(\theta + \frac{2\pi}{m}\right), \\ &\dots\dots\dots\dots\dots\dots \\ &\dots\dots\dots\dots\dots\dots \\ \kappa_m &= \mu + A\cos 2\left\{\theta + (m-1)\frac{2\pi}{m}\right\}.\end{aligned}$$

$$\therefore \quad \kappa_1+\kappa_2+\ldots+\kappa_m = m\mu + \left[\cos 2\theta + \cos 2\left(\theta+\frac{2\pi}{m}\right)+\ldots + \cos 2\left\{\theta+(m-1)\frac{2\pi}{m}\right\}\right]. \qquad (1)$$

Let $C = \cos 2\theta + \cos 2\left(\theta+\frac{2\pi}{m}\right)+\ldots+\cos 2\left\{\theta+(m-1)\frac{2\pi}{m}\right\}$

and $S = \sin 2\theta + \sin 2\left(\theta+\frac{2\pi}{m}\right)+\ldots+\sin 2\left\{\theta+(m-1)\frac{2\pi}{m}\right\}$.

Then

$$\begin{aligned}
C+iS &= e^{i2\theta} + e^{i2\left(\theta+\frac{2\pi}{m}\right)}+\ldots+e^{i2\left\{\theta+(m-1)\frac{2\pi}{m}\right\}} \\
&= e^{i2\theta}\left[1+e^{i\left(\frac{4\pi}{m}\right)}+\ldots+e^{i(m-1)\frac{4\pi}{m}}\right] \\
&= e^{i2\theta}\left[\frac{1-\left(e^{i(4\pi/m)}\right)^m}{1-e^{i(4\pi/m)}}\right] \\
&= e^{i2\theta}\left[\frac{1-e^{4\pi i}}{1-e^{4\pi i/m}}\right] \\
&= e^{i2\theta}\left[\frac{1-(\cos 4\pi + i\sin 4\pi)}{1-e^{4\pi i/m}}\right] \\
&= e^{i2\theta}\left[\frac{1-(1+i0)}{1-e^{4\pi i/m}}\right] \\
&= 0.
\end{aligned}$$

$\therefore \quad C = 0$

i.e., $\quad \cos 2\theta + \cos 2\left(\theta+\frac{2\pi}{m}\right)+\ldots+\cos 2\left\{\theta+(m-1)\frac{2\pi}{m}\right\} = 0.$

Substituting the above value in (1), we get

$$\kappa_1+\kappa_2+\ldots+\kappa_m = m\mu.$$

Example 6. Show that the curvature κ at any point P of the curve of intersection of two surfaces is given by

$$\kappa^2 \sin^2\theta = \kappa_1^{\,2}+\kappa_2^{\,2} - 2\,\kappa_1\kappa_2\cos\theta,$$

where κ_1, κ_2 are the normal curvatures of the surfaces in the direction of the curve at P and θ is the angle between their normals at that point.

Solution. Let C be the curve of intersection of two surfaces S_1 and S_2 and let P be any point on C whose position vector is $\mathbf{r}$.

Let κ be the curvature of C at P and let κ_1, κ_2 be the normal curvatures of S_1, S_2 respectively at P in the direction of C.

Let $\mathbf{N}_1$ and $\mathbf{N}_2$ be unit normals to S_1 and S_2 respectively at P and let θ be the angle between $\mathbf{N}_1$ and $\mathbf{N}_2$. Then

$$\cos\theta = \mathbf{N}_1 . \mathbf{N}_2. \tag{1}$$

Also, $$\mathbf{r}'' = \mathbf{t}' = \kappa \mathbf{n}, \tag{2}$$

where $\mathbf{n}$ is the unit principal normal of C at P.

The vectors $\mathbf{N}_1, \mathbf{N}_2$ and $\mathbf{n}$ are all perpendicular to $\mathbf{t}$. Therefore they lie in the plane of $\mathbf{n}$ and $\mathbf{b}$, i.e., they are coplanar. Hence one of these can be expressed as a linear combination of the remaining two.

Writing $\mathbf{n}$ as a linear combination of $\mathbf{N}_1, \mathbf{N}_2$ and using (2), we have

$$\mathbf{r}'' = \lambda \mathbf{N}_1 + \mu \mathbf{N}_2. \tag{3}$$

Taking the dot product of both sides of (3) with $\mathbf{N}_1$ and using (1), we get

$$\mathbf{r}''. \mathbf{N}_1 = \lambda + \mu \cos\theta$$

or $$\kappa_1 = \lambda + \mu \cos\theta. \tag{4}$$

$$[\because \mathbf{r}''.\mathbf{N}_1 = \kappa_1]$$

Taking the dot product of both sides of (3) with $\mathbf{N}_2$ and using (1), we get

$$\mathbf{r}''. \mathbf{N}_2 = \lambda \cos\theta + \mu$$

or $$\kappa_2 = \lambda \cos\theta + \mu. \tag{5}$$

$$[\because \mathbf{r}''.\mathbf{N}_2 = \kappa_2]$$

Solving (4) and (5) for λ, μ we have

$$\lambda = \frac{\kappa_1 - \kappa_2 \cos\theta}{\sin^2\theta}, \quad \mu = \frac{\kappa_2 - \kappa_1 \cos\theta}{\sin^2\theta}.$$

Substituting the above values of λ and μ in (3), we have

$$\mathbf{r}'' = \frac{\kappa_1 - \kappa_2 \cos\theta}{\sin^2\theta} \mathbf{N}_1 + \frac{\kappa_2 - \kappa_1 \cos\theta}{\sin^2\theta} \mathbf{N}_2$$

or $$(\sin^2\theta)\, \kappa \mathbf{n} = (\kappa_1 - \kappa_2 \cos\theta)\, \mathbf{N}_1 + (\kappa_2 - \kappa_1 \cos\theta)\, \mathbf{N}_2. \qquad \text{[using (2)]}$$

Squaring both sides and using (1), we get

$$\kappa^2 \sin^4\theta = (\kappa_1 - \kappa_2 \cos\theta)^2 + (\kappa_2 - \kappa_1 \cos\theta)^2 + 2[\kappa_1 - \kappa_2 \cos\theta)(\kappa_2 - \kappa_1 \cos\theta]\cos\theta$$

$$= (\kappa_1^{\,2} + \kappa_2^{\,2})(1 + \cos^2\theta) - 4\kappa_1\kappa_2 \cos\theta + 2[\kappa_1\kappa_2 - (\kappa_1^{\,2} + \kappa_2^{\,2})\cos\theta + \kappa_1\kappa_2 \cos^2\theta]\cos\theta$$

$$= (\kappa_1^2 + \kappa_2^2)(1 - \cos^2\theta) - 2\kappa_1\kappa_2\cos\theta + 2\kappa_1\kappa_2\cos^3\theta$$
$$= (\kappa_1^2 + \kappa_2^2)\sin^2\theta - 2\kappa_1\kappa_2\cos\theta(1 - \cos^2\theta)$$
$$= (\kappa_1^2 + \kappa_2^2)\sin^2\theta - 2\kappa_1\kappa_2\cos\theta\sin^2\theta.$$

$$\therefore \quad \kappa^2\sin^2\theta = \kappa_1^2 + \kappa_2^2 - 2\,\kappa_1\kappa_2\cos\theta.$$

Example 7. Find the values of the two principal curvatures at any point of the surface of revolution

$$x = u\cos\phi,\ y = u\sin\phi,\ z = f(u).$$

Solution. We know that the principal curvatures κ of the surface are given by

$$H^2\kappa^2 - (EN - 2FM + GL)\kappa + LN - M^2 = 0. \tag{1}$$

Now, we have to find the values of E, F, G, L, M, N for the given surface.

The position vector $\mathbf{r}$ of any current point on the given surface of revolution is

$$\mathbf{r} = (u\cos\phi, u\sin\phi, f(u)).$$

Taking u and ϕ as the first and the second parameters, we have

$$\mathbf{r}_1 = (\cos\phi, \sin\phi, f_1),\ \ \mathbf{r}_2 = (-u\sin\phi, u\cos\phi, 0),$$
$$\mathbf{r}_{11} = (0, 0, f_{11}),\ \ \mathbf{r}_{12} = (-\sin\phi, \cos\phi, 0),$$
$$\mathbf{r}_{22} = (-u\cos\phi, -u\sin\phi, 0).$$

Therefore

$$E = \mathbf{r}_1^2 = \cos^2\phi + \sin^2\phi + f_1^2 = 1 + f_1^2,$$
$$F = \mathbf{r}_1 \cdot \mathbf{r}_2 = -u\cos\phi\sin\phi + u\sin\phi\cos\phi + 0 = 0,$$
$$G = \mathbf{r}_2^2 = u^2\sin^2\phi + u^2\cos^2\phi + 0 = u^2,$$
$$H^2 = EG - F^2 = u^2(1 + f_1^2).$$

Again, we have

$$\mathbf{r}_1 \times \mathbf{r}_2 = (-f_1 u\cos\phi, -f_1 u\sin\phi, u)$$

$$\Rightarrow \quad \mathbf{N} = \frac{\mathbf{r}_1 \times \mathbf{r}_2}{H} = \frac{1}{H}(-f_1 u\cos\phi, -f_1 u\sin\phi, u).$$

Therefore

$$L = \mathbf{r}_{11} \cdot \mathbf{N} = \frac{uf_{11}}{H},\ \ M = \mathbf{r}_{12} \cdot \mathbf{N} = 0,\ \ N = \mathbf{r}_{22} \cdot \mathbf{N} = \frac{u^2 f_1}{H}.$$

Substituting the above values in (1), we get

$$H^2\kappa^2 - \left\{(1 + f_1^2)\left(\frac{u^2 f_1}{H}\right) + u^2\left(\frac{uf_{11}}{H}\right)\right\}\kappa + \frac{u^3 f_1 f_{11}}{H^2} = 0$$

or $\quad H^4\kappa^2 - u^2 H\{(1+f_1^2) f_1 + uf_{11}\}\kappa + u^3 f_1 f_{11} = 0$

or $\quad u^4(1+f_1^2)^2\kappa^2 - u^3\sqrt{(1+f_1^2)}\{(1+f_1^2) f_1 + uf_{11}\}\kappa + u^3 f_1 f_{11} = 0$

or $\quad u(1+f_1^2)^2\kappa^2 - \sqrt{(1+f_1^2)}\{(1+f_1^2) f_1 + uf_{11}\}\kappa + f_1 f_{11} = 0$

or $\quad u\sqrt{(1+f_1^2)}\,\kappa\{(1+f_1^2)^{3/2}\kappa - f_{11}\} - f_1\{(1+f_1^2)^{3/2}\kappa - f_{11}\} = 0$

or $\quad \{u\sqrt{(1+f_1^2)}\,\kappa - f_1\}\{(1+f_1^2)^{3/2}\kappa - f_{11}\} = 0.$

Therefore the two principal curvatures κ_a and κ_b are given by

$$\kappa_a = \frac{f_1}{u\sqrt{(1+f_1^2)}} = \frac{\dfrac{df}{du}}{u\sqrt{\left\{1+\left(\dfrac{df}{du}\right)^2\right\}}},$$

$$\kappa_b = \frac{f_{11}}{(1+f_1^2)^{3/2}} = \frac{\dfrac{d^2f}{du^2}}{\left\{1+\left(\dfrac{df}{du}\right)^2\right\}^{3/2}}.$$

Example 8. The surface of revolution given by

$$x = u\cos v,\ y = u\sin v,\ z = a\log\left\{u+\sqrt{(u^2-a^2)}\right\}$$

is generated by revolving a catenary about its directrix. Prove that it is a minimal surface. Show also that it is the only minimal surface of revolution.

Solution. First Part. We know that a surface is minimal if

$$EN - 2FM + GL = 0.$$

For the given surface of revolution, we have

$$z = a\log\left\{u+\sqrt{(u^2-a^2)}\right\}$$

$$= a\log\left[\left\{\frac{u}{\mathrm{a}}+\sqrt{\left(\frac{u}{a}\right)^2-1}\right\}a\right]$$

$$= a\log\left\{\frac{u}{\mathrm{a}}+\sqrt{\left(\frac{u}{a}\right)^2-1}\right\}+a\log a$$

$$= a\cosh^{-1}\left(\frac{u}{\mathrm{a}}\right)+a\log a. \qquad \left[\because \cosh^{-1}\omega = \log\{\omega+\sqrt{(\omega^2-1)}\}\right]$$

$$\therefore \qquad z - a\log a = a\cosh^{-1}\left(\frac{u}{\mathrm{a}}\right)$$

or $$u = a\cosh\frac{z - a\log a}{a}.$$

Hence the given surface is generated by revolving the catenary

$$y = 0,\ x = a\cosh\frac{z - a\log a}{a}$$

about z-axis, i.e., its directrix.

Now, the position vector **r** of any current point on the given surface can be written as

$$\mathbf{r} = \left(u\cos v, u\sin v, a\log\left\{u + \sqrt{(u^2 - a^2)}\,\right\}\right).$$

Taking u and v as the first and the second parameters, we have

$$\mathbf{r}_1 = \left(\cos v, \sin v, \frac{a}{u + \sqrt{(u^2 - a^2)}}\left\{1 + \frac{1}{2\sqrt{(u^2 - a^2)}}2u\right\}\right)$$

$$= \left(\cos v, \sin v, \frac{a}{\sqrt{(u^2 - a^2)}}\right),$$

$$\mathbf{r}_2 = (-u\sin v, u\cos v, 0),$$

$$\mathbf{r}_{11} = \left(0, 0, -\frac{au}{(u^2 - a^2)^{3/2}}\right),$$

$$\mathbf{r}_{12} = (-\sin v, \cos v, 0),$$

$$\mathbf{r}_{22} = (-u\cos v, -u\sin v, 0).$$

Therefore

$$E = \mathbf{r}_1{}^2 = \cos^2 v + \sin^2 v + \frac{a^2}{u^2 - a^2} = \frac{u^2}{u^2 - a^2},$$

$$F = \mathbf{r}_1 \cdot \mathbf{r}_2 = -u\cos v\sin v + u\sin v\cos v + 0 = 0,$$

$$G = \mathbf{r}_2{}^2 = u^2\sin^2 v + u^2\cos^2 v + 0 = u^2,$$

$$H^2 = EG - F^2 = \frac{u^4}{u^2 - a^2}.$$

Again, we have

$$\mathbf{r}_1 \times \mathbf{r}_2 = \left(-\frac{au}{\sqrt{(u^2 - a^2)}}\cos v, -\frac{au}{\sqrt{(u^2 - a^2)}}\sin v, u\right)$$

$$\Rightarrow \qquad \mathbf{N} = \frac{\mathbf{r}_1 \times \mathbf{r}_2}{H} = \frac{1}{H}\left(-\frac{au}{\sqrt{(u^2 - a^2)}}\cos v, -\frac{au}{\sqrt{(u^2 - a^2)}}\sin v, u \right).$$

Therefore

$$L = \mathbf{r}_{11} \,.\, \mathbf{N} = -\frac{au^2}{(u^2 - a^2)^{3/2}}, \; M = \mathbf{r}_{12} \,.\, \mathbf{N} = 0,$$

$$N = \mathbf{r}_{22} \,.\, \mathbf{N} = \frac{au^2}{\sqrt{(u^2 - a^2)}}.$$

Now,

$$EN - 2FM + GL = \frac{u^2}{u^2 - a^2}\frac{au^2}{\sqrt{(u^2 - a^2)}} - 0 + u^2\left(-\frac{au^2}{(u^2 - a^2)^{3/2}} \right)$$

$$= 0.$$

Therefore the given surface is minimal.

Second Part. Let the equation of any surface of revolution be

$$x = u\cos v, y = u\sin v, z = f(u). \qquad (1)$$

Then the position vector **r** of any current point on this surface is

$$\mathbf{r} = (u\cos v, u\sin v, f(u)).$$

Therefore, proceeding as in Example 7, we have

$$E = 1 + f_1^2, \; F = 0, \; G = u^2, \; H^2 = u^2(1 + f_1^2),$$

$$L = \frac{uf_{11}}{H}, \; M = 0, \; N = \frac{u^2 f_1}{H}.$$

The surface (1) will be minimal if

$$EN - 2FM + GL = 0$$

or $$(1 + f_1^2)\frac{u^2 f_1}{H} - 0 + u^2\left(\frac{uf_{11}}{H}\right) = 0$$

or $$f_1(1 + f_1^2) + uf_{11} = 0$$

or $$u\frac{d^2 f}{du^2} + \frac{df}{du}\left\{1 + \left(\frac{df}{du}\right)^2\right\} = 0. \qquad (2)$$

Put $\frac{df}{du} = p$. Then (2) reduces to

$$u\frac{dp}{du} + p\,(1 + p^2) = 0$$

or $$\frac{dp}{p\,(1 + p^2)} + \frac{du}{u} = 0$$

or $$\left(\frac{1}{p}-\frac{p}{1+p^2}\right)dp+\frac{du}{u}=0.$$

Integrating, we get

$$\log p-\frac{1}{2}\log(1+p^2)+\log u=\text{constant}$$

or $$2\log p-\log(1+p^2)+2\log u=\text{constant}$$

or $$\log\left(\frac{u^2p^2}{1+p^2}\right)=\text{constant}$$

or $$\frac{u^2p^2}{1+p^2}=a^2,\ \text{where } a \text{ is a constant}$$

or $$p^2(u^2-a^2)=a^2$$

or $$p=\frac{a}{\sqrt{(u^2-a^2)}}$$

or $$\frac{df}{du}=\frac{a}{\sqrt{(u^2-a^2)}}.$$

Integrating, we get

$$z=f(u)=a\log\left\{u+\sqrt{(u^2-a^2)}\right\}+B. \tag{3}$$

Let $z=a\log a$ when $u=a$. Then, from (3), $B=0$.

Therefore (3) reduces to

$$z=f(u)=a\log\left\{u+\sqrt{(u^2-a^2)}\right\}.$$

Hence, except for position in space, the only minimal surface of revolution is

$$x=u\cos v,\ y=u\sin v,\ z=a\log\left\{u+\sqrt{(u^2-a^2)}\right\}.$$

Note. The surface of revolution formed by revolving a catenary about its directrix is called a catenoid. Thus we have proved that the catenoid is the only minimal surface of revolution.

Example 9. Prove that at a point of the conoid

$$x=u\cos\theta,\ y=u\sin\theta,\ z=f(\theta),$$

the principal curvatures are given by

$$u^2c^2(u^2+z'^2)\,\kappa^2-u^2cz''\kappa-z'^2=0,$$

where $z'=\dfrac{dz}{d\theta}$ etc. and $c^2=1+\dfrac{z'^2}{u^2}$.

Solution. We know that the equation giving the principal curvatures is

$$H^2\kappa^2 - (EN - 2FM + GL)\,\kappa + LN - M^2 = 0. \qquad (1)$$

We shall find the values of E, F, G, L, M, N for the given surface.
The position vector of any current point on the given surface is

$$\mathbf{r} = (u\cos\theta, u\sin\theta, f(\theta)).$$

Taking u and θ as the first and the second parameters, we have

$$\mathbf{r}_1 = (\cos\theta, \sin\theta, 0),$$

$$\mathbf{r}_2 = (-u\sin\theta, u\cos\theta, z'), \qquad \left[z' = f' = \frac{dz}{d\theta}\right]$$

$$\mathbf{r}_{11} = (0, 0, 0),$$

$$\mathbf{r}_{12} = (-\sin\theta, \cos\theta, 0),$$

$$\mathbf{r}_{22} = (-u\cos\theta, -u\sin\theta, z'').$$

Therefore

$$E = \mathbf{r}_1^{\,2} = \cos^2\theta + \sin^2\theta + 0 = 1,$$

$$F = \mathbf{r}_1 . \mathbf{r}_2 = -u\cos\theta\sin\theta + u\sin\theta\cos\theta + 0 = 0,$$

$$G = \mathbf{r}_2^{\,2} = u^2\sin^2\theta + u^2\cos^2\theta + z'^2 = u^2 + z'^2,$$

$$H^2 = EG - F^2 = u^2 + z'^2.$$

Again, we have

$$\mathbf{r}_1 \times \mathbf{r}_2 = (z'\sin\theta, -z'\cos\theta, u)$$

$$\Rightarrow \qquad \mathbf{N} = \frac{\mathbf{r}_1 \times \mathbf{r}_2}{H} = \frac{1}{H}(z'\sin\theta, -z'\cos\theta, u).$$

Therefore

$$L = \mathbf{r}_{11} . \mathbf{N} = 0,$$

$$M = \mathbf{r}_{12} . \mathbf{N} = \frac{1}{H}(-z'\sin^2\theta - z'\cos^2\theta + 0) = -\frac{z'}{H},$$

$$N = \mathbf{r}_{22} . \mathbf{N} = \frac{1}{H}(-uz'\cos\theta\sin\theta + uz'\sin\theta\cos\theta + uz'')$$

$$= \frac{uz''}{H}.$$

Substituting the above values in (1), we get

$$(u^2 + z'^2)\,\kappa^2 - \frac{uz''}{H}\kappa - \frac{z'^2}{H^2} = 0$$

or $\qquad (u^2 + z'^2)\,H^2\kappa^2 - uz''H\kappa - z'^2 = 0$

or $\qquad (u^2 + z'^2)(u^2 + z'^2)\,\kappa^2 - uz''\sqrt{(u^2 + z'^2)}\;\kappa - z'^2 = 0$

or $$u^2\left(1+\frac{z'^2}{u^2}\right)(u^2+z'^2)\,\kappa^2-u^2\sqrt{\left(1+\frac{z'^2}{u^2}\right)}\;z''\kappa-z'^2=0$$

or $$u^2c^2(u^2+z'^2)\,\kappa^2-u^2cz''\kappa-z'^2=0,$$

where $c^2=1+\dfrac{z'^2}{u^2}$.

Example 10. For the surface $x=u\cos\theta, y=u\sin\theta, z=f(\theta)$, prove that the angles that the lines of curvature make with the generators are given by

$$\tan^2\phi+\frac{f''}{f'}\frac{u}{\sqrt{(u^2+f'^2)}}\tan\phi-1=0.$$

Solution. We know that the differential equation of the lines of curvature is given by

$$(EM-FL)\,du^2+(EN-GL)\,du\,d\theta+(FN-GM)\,d\theta^2=0. \tag{1}$$

We shall find the values of *E*, *F*, *G*, *L*, *M*, *N* for the given surface.

The position vector of any current point on the given surface is

$$\mathbf{r}=(u\cos\theta, u\sin\theta, f(\theta)).$$

Now, proceeding as in Example 9, we have

$$E=1,\ F=0,\ G=u^2+f'^2,\ H^2=u^2+f'^2;$$

$$L=0,\ M=-\frac{f'}{H},\ N=\frac{uf''}{H}.$$

Substituting the above values in (1), we get

$$-\frac{f'}{H}du^2+\frac{uf''}{H}du\,d\theta+\left(u^2+f'^2\right)\frac{f'}{H}d\theta^2=0$$

or $$-f'+uf''\frac{d\theta}{du}+f'\left(u^2+f'^2\right)\left(\frac{d\theta}{du}\right)^2=0$$

or $$f'\left(u^2+f'^2\right)\left(\frac{d\theta}{du}\right)^2+uf''\frac{d\theta}{du}-f'=0. \tag{2}$$

The direction coefficients of the tangent are $\left(\dfrac{dx}{ds},\dfrac{dy}{ds},\dfrac{dz}{ds}\right)$,

where

$$\frac{dx}{ds}=\frac{\cos\theta\,du-u\sin\theta\,d\theta}{ds}=\frac{\cos\theta\,du-u\sin\theta\,d\theta}{\sqrt{E\,du^2+2F\,du\,d\theta+G\,d\theta^2}}$$

i.e.,
$$\frac{dx}{ds}=\frac{\cos\theta\,du-u\sin\theta\,d\theta}{\sqrt{du^2+(u^2+f'^2)\,d\theta^2}},$$

$$\frac{dy}{ds}=\frac{\sin\theta\,du+u\cos\theta\,d\theta}{ds}=\frac{\sin\theta\,du+u\cos\theta\,d\theta}{\sqrt{du^2+(u^2+f'^2)\,d\theta^2}},$$

$$\frac{dz}{ds}=\frac{f'd\theta}{ds}=\frac{f'd\theta}{\sqrt{du^2+(u^2+f'^2)\,d\theta^2}}.$$

The equations to the generators are

$$\frac{x}{\cos\theta}=\frac{y}{\sin\theta}=\frac{z-f(\theta)}{0}.$$

Therefore, if ϕ is the angle that the lines of curvature make with the generators, then

$$\cos\phi=\frac{dx}{ds}\cos\theta+\frac{dy}{ds}\sin\theta+\frac{dz}{ds}.0$$

$$=\frac{(\cos\theta\,du-u\sin\theta\,d\theta)\cos\theta}{\sqrt{du^2+(u^2+f'^2)\,d\theta^2}}+\frac{(\sin\theta\,du+u\cos\theta\,d\theta)\sin\theta}{\sqrt{du^2+(u^2+f'^2)\,d\theta^2}}$$

$$=\frac{du}{\sqrt{du^2+(u^2+f'^2)\,d\theta^2}}$$

$$\Rightarrow \qquad \tan\phi=\frac{\sqrt{(u^2+f'^2)\,d\theta^2}}{du}=\frac{\sqrt{(u^2+f'^2)}\;d\theta}{du}$$

$$\Rightarrow \qquad \frac{d\theta}{du}=\frac{1}{\sqrt{(u^2+f'^2)}}\tan\phi. \tag{3}$$

Substituting the value of $\frac{d\theta}{du}$ from (3) in (2), we get

$$f'\tan^2\phi+uf''\,\frac{1}{\sqrt{(u^2+f'^2)}}\tan\phi-f'=0$$

or
$$\tan^2\phi+\frac{f''}{f'}\frac{u}{\sqrt{(u^2+f'^2)}}\tan\phi-1=0.$$

Example 11. Show that the equation for the principal curvatures through a point of the surface $z=f(x,y)$ is

$$H^4\kappa^2-H\{(1+p^2)t-2pqs+(1+q^2)r\}\kappa+rt-s^2=0.$$

Also, obtain the differential equation of the lines of curvature on the surface $z = (x, y)$ and deduce that at an umbilic,

$$\frac{1+p^2}{r} = \frac{1+q^2}{t} = \frac{pq}{s}.$$

Solution. We know that the equation for the principal curvatures through a point of the surface $\mathbf{r} = \mathbf{r}(u, v)$ is

$$H^2\kappa^2 - (EN - 2FM + GL)\kappa + LN - M^2 = 0. \tag{1}$$

We shall find the values of E, F, G, L, M, N for the given surface.

The equation of the given surface is

$$z = f(x, y).$$

If $\mathbf{r}$ is the position vector of any current point on this surface, then

$$\mathbf{r} = (x, y, z), \tag{2}$$

where z is a function of x and y.

We write

$$\frac{\partial z}{\partial x} = p,\ \frac{\partial z}{\partial y} = q,\ \frac{\partial^2 z}{\partial x^2} = r,\ \frac{\partial^2 z}{\partial x\,\partial y} = s,\ \frac{\partial^2 z}{\partial y^2} = t.$$

Let x and y be taken as the first and the second parameters. Then, from (2), we have

$$\mathbf{r}_1 = (1, 0, p),\quad \mathbf{r}_2 = (0, 1, q),$$

$$\mathbf{r}_{11} = (0, 0, r),\ \mathbf{r}_{12} = (0, 0, s),\ \mathbf{r}_{22} = (0, 0, t).$$

Therefore

$$E = \mathbf{r}_1^{\,2} = 1 + p^2,\ F = \mathbf{r}_1 \cdot \mathbf{r}_2 = pq,\ G = \mathbf{r}_2^{\,2} = 1 + q^2,$$

$$H^2 = EG - F^2 = 1 + p^2 + q^2.$$

Again, we have

$$\mathbf{r}_1 \times \mathbf{r}_2 = (-p, -q, 1)$$

$$\Rightarrow \qquad \mathbf{N} = \frac{\mathbf{r}_1 \times \mathbf{r}_2}{H} = \frac{1}{H}(-p, -q, 1).$$

Therefore

$$L = \mathbf{r}_{11} \cdot \mathbf{N} = \frac{r}{H},\ M = \mathbf{r}_{12} \cdot \mathbf{N} = \frac{s}{H},\ N = \mathbf{r}_{22} \cdot \mathbf{N} = \frac{t}{H}.$$

Substituting the values of E, F, G, L, M, N in (1), we get

$$H^2\kappa^2 - \left(\frac{(1+p^2)t}{H} - \frac{2pqs}{H} + \frac{(1+q^2)r}{H}\right)\kappa + \frac{rt}{H^2} - \frac{s^2}{H^2} = 0$$

or $\qquad H^4\kappa^2 - H\left\{(1+p^2)t - 2pqs + (1+q^2)r\right\}\kappa + rt - s^2 = 0.$

This is the equation for the principal curvatures through a point of the given surface.
Also, we know that the differential equation of the lines of curvature is

$$(EM - FL)\,dx^2 + (EN - GL)\,dx\,dy + (FN - GM)\,dy^2 = 0$$

or
$$\left[\frac{(1+p^2)\,s}{H} - \frac{pqr}{H}\right] dx^2 + \left[\frac{(1+p^2)\,t}{H} - \frac{(1+q^2)\,r}{H}\right] dxdy + \left[\frac{pqt}{H} - \frac{(1+q^2)\,s}{H}\right] dy^2 = 0$$

or
$$[(1+p^2)\,s - pqr]\,dx^2 + [(1+p^2)\,t - (1+q^2)\,r]\,dxdy + [pqt - (1+q^2)\,s]\,dy^2 = 0,$$

which may be written in the determinant form as

$$\begin{vmatrix} dy^2 & -dx\,dy & dx^2 \\ 1+p^2 & pq & 1+q^2 \\ r & s & t \end{vmatrix} = 0.$$

At an umbilic, we have

$$\frac{E}{L} = \frac{F}{M} = \frac{G}{N}$$

i.e.,
$$\frac{1+p^2}{r} = \frac{pq}{s} = \frac{1+q^2}{t}.$$

Example 12. Obtain the differential equation of the lines of curvature through a point of the surface $F(x, y, z) = 0$.

Solution. We know that the differential equation of the lines of curvature is

$$(EM - FL)\,dx^2 + (EN - GL)\,dx\,dy + (FN - GM)\,dy^2 = 0. \tag{1}$$

We shall find the values of E, F, G, L, M, N for the given surface.
The equation of the given surface is

$$F(x,\ y,\ z) = 0. \tag{2}$$

If we regard z as a function of x and y, then the above equation can be transformed to the form $z = f(x, y)$. Therefore the position vector $\mathbf{r}$ of any current point on this surface can be written as

$$\mathbf{r} = (x, y, z), \tag{3}$$

where z is a function of x and y.
We write

$$\frac{\partial z}{\partial x} = p,\ \frac{\partial z}{\partial y} = q,\ \frac{\partial^2 z}{\partial x^2} = r,\ \frac{\partial^2 z}{\partial x\,\partial y} = s,\ \frac{\partial^2 z}{\partial y^2} = t.$$

Taking x and y be as the first and the second parameters, and proceeding as in Example 11, we get

$$E = 1 + p^2, F = pq,\ G = 1 + q^2;\ L = \frac{r}{H},\ M = \frac{s}{H},\ N = \frac{t}{H},$$

where

$$H = \sqrt{(1 + p^2 + q^2)}.$$

Substituting the values of E, F, G, L, M, N in (1), we get

$$[(1 + p^2)\, s - pqr]\, dx^2 + [(1 + p^2)\, t - (1 + q^2)\, r]\, dx\, dy + [pqt - (1 + q^2)\, s]\, dy^2 = 0. \qquad (4)$$

Now, we shall find the values of p, q, r, s, t from equation (2).

Differentiating (2) partially with respect to x, taking z as a function of x and y, we get

$$\frac{\partial F}{\partial x} + \frac{\partial F}{\partial z}\frac{\partial z}{\partial x} = 0$$

or $$F_x + F_z p = 0 \qquad \left[F_x = \frac{\partial F}{\partial x},\ F_z = \frac{\partial F}{\partial z}\right]$$

or $$p = -\frac{F_x}{F_z}. \qquad (5)$$

Similarly, differentiating (2) partially with respect to y, taking z as a function of x and y, we get

$$q = -\frac{F_y}{F_z}. \qquad (6)$$

Now, differentiating (5) partially with respect to x, taking z as a function of x and y, we get

$$r = \frac{\partial p}{\partial x}$$

$$= -\frac{\left\{\frac{\partial}{\partial x}(F_x) + \frac{\partial}{\partial z}(F_x)\frac{\partial z}{\partial x}\right\}F_z - \left\{\frac{\partial}{\partial x}(F_z) + \frac{\partial}{\partial z}(F_z)\frac{\partial z}{\partial x}\right\}F_x}{F_z^{\ 2}}$$

[using (5)]

or $$r = -\frac{(F_{xx} + F_{xz}\, p)\, F_z - (F_{zx} + F_{zz}\, p)\, F_x}{F_z^{\ 2}}$$

or $$r = -\frac{(F_{xx} F_z^{\ 2} - 2F_x F_z F_{xz} + F_{zz}\ F_x^{\ 2})}{F_z^{\ 3}}. \qquad (7)$$

[putting the value of p from (5)]

Similarly, differentiating (6) partially with respect to y, taking z as a function of x and y, we get

$$t = -\frac{(F_{yy} F_z^{\ 2} - 2F_y F_z F_{yz} + F_{zz}\ F_y^{\ 2})}{F_z^{\ 3}}. \qquad (8)$$

Again, differentiating (5) partially with respect to y, taking z as a function of x and y, we get

$$s=\frac{\partial p}{\partial y}$$

$$=-\frac{\left\{\frac{\partial}{\partial y}(F_x)+\frac{\partial}{\partial z}(F_x)\frac{\partial z}{\partial y}\right\}F_z-\left\{\frac{\partial}{\partial y}(F_z)+\frac{\partial}{\partial z}(F_z)\frac{\partial z}{\partial y}\right\}F_x}{F_z^{\,2}}$$

or $$s=-\frac{(F_{xy}+F_{xz}\,q)\,F_z-(F_{zy}+F_{zz}\,q)\,F_x}{F_z^{\,2}}$$

or $$s=-\frac{(F_{xy}F_z^{\,2}-F_{xz}F_yF_z-F_{zy}\,F_xF_z+F_{zz}\,F_yF_x)}{F_z^{\,3}}. \qquad (9)$$

[putting the value of q from (6)]

Now, putting the values of p, q, r, t, s from (5), (6), (7), (8), (9) respectively in (4), we get the required differential equation of lines of curvature.

Example 13. Show that the lines of curvature of the paraboloid $xy=az$ lies on the surfaces

$$\sinh^{-1}\frac{x}{a}\pm\sinh^{-1}\frac{y}{a}=\text{constant}.$$

Solution. We know that the differential equation of the lines of curvature is given by

$$(EM-FL)\,dx^2+(EN-GL)\,dx\,dy+(FN-GM)\,dy^2=0. \qquad (1)$$

We shall find the values of E, F, G, L, M, N for the given surface.

The equation of the given paraboloid is

$$xy=az$$

i.e., $$z=\frac{xy}{a}. \qquad (2)$$

Therefore the equation of the surface is in Monge's form.

If $\mathbf{r}$ is the position vector of any current point on this surface, then

$$\mathbf{r}=(x,y,z), \qquad (3)$$

where z is a function of x and y, given by (2).

Now, from (2), we have

$$p=\frac{\partial z}{\partial x}=\frac{y}{a},\ q=\frac{\partial z}{\partial y}=\frac{x}{a},$$

$$r = \frac{\partial^2 z}{\partial x^2} = 0,\ s = \frac{\partial^2 z}{\partial x\,\partial y} = \frac{1}{a},\ t = \frac{\partial^2 z}{\partial y^2} = 0.$$

Taking x and y be as the first and the second parameters, and proceeding as in Example 11, we get

$$E = 1 + p^2, F = pq,\ G = 1 + q^2;\ L = \frac{r}{H},\ M = \frac{s}{H},\ N = \frac{t}{H},$$

where $H = \sqrt{(1 + p^2 + q^2)}$.

Therefore, for the paraboloid (2), we have

$$E = 1 + p^2 = 1 + \frac{y^2}{a^2},\quad F = pq = \frac{xy}{a^2},\quad G = 1 + q^2 = 1 + \frac{x^2}{a^2},$$

$$L = \frac{r}{H} = 0,\ M = \frac{s}{H} = \frac{1}{aH},\quad N = \frac{t}{H} = 0.$$

Substituting the values of E, F, G, L, M, N in (1), we get

$$\left(1 + \frac{y^2}{a^2}\right)\frac{1}{aH}dx^2 - \left(1 + \frac{x^2}{a^2}\right)\frac{1}{aH}dy^2 = 0$$

or $$(a^2 + y^2)\,dx^2 - (a^2 + x^2)\,dy^2 = 0$$

or $$\frac{dx^2}{a^2 + x^2} = \frac{dy^2}{a^2 + y^2}$$

or $$\frac{dx}{\sqrt{(a^2 + x^2)}} \pm \frac{dy}{\sqrt{(a^2 + y^2)}} = 0.$$

Integrating this, we get

$$\sinh^{-1}\frac{x}{a} \pm \sinh^{-1}\frac{y}{a} = \text{constant}.$$

Example 14. Calculate the first and second curvatures of the helicoid
$x = u\cos v, y = u\sin v, z = f(u) + cv.$

Solution. The position vector $\mathbf{r}$ of any current point on the given helicoid is

$$\mathbf{r} = (u\cos v, u\sin v, f(u) + cv).$$

Taking u and v be as the first and the second parameters, we have

$$\mathbf{r}_1 = (\cos v, \sin v, f_1),\ \mathbf{r}_2 = (-u\sin v, u\cos v, c),$$

$$\mathbf{r}_{11} = (0, 0, f_{11}),\ \mathbf{r}_{12} = (-\sin v, \cos v, 0),$$

$$\mathbf{r}_{22} = (-u\cos v, -u\sin v, 0).$$

Therefore

$$E = \mathbf{r}_1^{\ 2} = \cos^2 v + \sin^2 v + f_1^{\ 2} = 1 + f_1^{\ 2},$$

$$F = \mathbf{r}_1 \,.\, \mathbf{r}_2 = -u\cos v \sin v + u \sin v \cos v + cf_1 = cf_1,$$
$$G = \mathbf{r}_2^{\,2} = u^2 \sin^2 v + u^2 \cos^2 v + c^2 = u^2 + c^2,$$
$$H^2 = EG - F^2 = u^2(1 + f_1^{\,2}) + c^2.$$

Again, we have

$$\mathbf{r}_1 \times \mathbf{r}_2 = (c \sin v - f_1 u \cos v, -c\cos v - f_1 u \sin v, u)$$

$$\Rightarrow \qquad \mathbf{N} = \frac{\mathbf{r}_1 \times \mathbf{r}_2}{H} = \frac{1}{H}(c \sin v - f_1 u \cos v, -c\cos v - f_1 u \sin v, u).$$

Therefore

$$L = \mathbf{r}_{11} \,.\, \mathbf{N} = \frac{uf_{11}}{H}, M = \mathbf{r}_{12} \,.\, \mathbf{N} = -\frac{c}{H}, N = \mathbf{r}_{22} \,.\, \mathbf{N} = \frac{u^2 f_1}{H}.$$

Hence the first curvature J is given by

$$J = \kappa_a + \kappa_b = \frac{EN - 2FM + GL}{H^2}$$
$$= \frac{(1 + f_1^{\,2})\, u^2 f_1 + 2c^2 f_1 + (u^2 + c^2)\, uf_{11}}{H^3}$$

and the second curvature K is given by

$$K = \kappa_a \kappa_b = \frac{LN - M^2}{H^2}$$
$$= \frac{u^3 f_1 f_{11} - c^2}{H^4}$$
$$= \frac{u^3 f_1 f_{11} - c^2}{\left[u^2 (1 + f_1^{\,2}) + c^2\right]^2}.$$

Example 15. Show that the lines of curvature of the helicoid in Example 14 are given by

$$c\,(1 + f_1^{\,2} + uf_1 f_{11})\, du^2 + [(u^2 + c^2)\, uf_{11} - (1 + f_1^{\,2})\, u^2 f_1]\, du\, dv$$
$$- c\,(u^2 + c^2 + u^2 f_1^{\,2})\, dv^2 = 0.$$

Hence prove that the meridians will be lines of curvature if

$$1 + f_1^{\,2} + uf_1 f_{11} = 0.$$

Solution. We know that the differential equation of the lines of curvature is given by

$$(EM - FL)\, du^2 + (EN - GL)\, du\, dv + (FN - GM)\, dv^2 = 0. \qquad (1)$$

Now, proceeding as in Example 14, we find the values of E, F, G, L, M, N for the given surface.

Substituting the above values in (1), we get

$$\frac{1}{H}[-c(1+f_1^2)-cf_1uf_{11}]\,du^2 + \frac{1}{H}[(1+f_1^2)u^2 f_1-(u^2+c^2)uf_{11}]\,du\,dv + \frac{1}{H}[cf_1u^2 f_1+(u^2+c^2)c]\,dv^2=0$$

or $$c(1+f_1^2+uf_1f_{11})\,du^2+[(u^2+c^2)uf_{11} - (1+f_1^2)u^2 f_1]\,du\,dv - c(u^2+c^2+u^2 f_1^2)\,dv^2=0. \tag{2}$$

This is the required equation of the lines of curvature.

If $1+f_1^2+uf_1f_{11}=0$, then the differential equation (2) of the lines of curvature reduces to

$$[(u^2+c^2)uf_{11}-(1+f_1^2)u^2 f_1]\,du\,dv-c(u^2+c^2+u^2 f_1^2)\,dv^2=0$$

$$\Rightarrow \quad dv[\{(u^2+c^2)uf_{11}-(1+f_1^2)u^2 f_1]\,du\}-\{c(u^2+c^2+u^2 f_1^2)\,dv\}]=0$$

$$\Rightarrow \quad dv=0$$

$$\Rightarrow \quad v=\text{ constant.}$$

Therefore the curves $v=$ constant are the lines of curvature, i.e. meridians are the lines of curvature.

Example 16. Find the principal radii at a point of the cylindroid $z(x^2+y^2)=2mxy$. Prove that at any point of the generator $x=y, z=m$, one principal radius is infinite and the other is $u^2/4m$, where u is the distance of the point from the axis.

Solution. We know that the principal curvatures κ of the surface are given by

$$H^2\kappa^2-(EN-2FM+GL)\kappa+LN-M^2=0. \tag{1}$$

Now, we have to find the values of E, F, G, L, M, N for the given surface. The equation of the given surface is

$$z(x^2+y^2)=2mxy.$$

We can take parametric equations of the surface as

$$x=u\cos\theta,\ y=u\sin\theta,\ z=m\sin 2\theta.$$

Therefore the position vector $\mathbf{r}$ of any current point on the surface is

$$\mathbf{r}=(u\cos\theta, u\sin\theta, m\sin 2\theta).$$

Taking u and θ as the first and the second parameters, we have

$$\mathbf{r}_1 = (\cos\theta, \sin\theta, 0),\ \ \mathbf{r}_2 = (-u\sin\theta, u\cos\theta, 2m\cos 2\theta),$$

$$\mathbf{r}_{11} = (0,0,0),\ \ \mathbf{r}_{12} = (-\sin\theta, \cos\theta, 0),$$

$$\mathbf{r}_{22} = (-u\cos\theta, -u\sin\theta, -4m\sin 2\theta).$$

Therefore

$$E = \mathbf{r}_1^{\ 2} = \cos^2\theta + \sin^2\theta + 0 = 1,$$

$$F = \mathbf{r}_1 \,.\, \mathbf{r}_2 = -u\cos\theta\sin\theta + u\sin\theta\cos\theta + 0 = 0,$$

$$G = \mathbf{r}_2^{\ 2} = u^2 + 4m^2\cos^2 2\theta,$$

$$H^2 = EG - F^2 = u^2 + 4m^2\cos^2 2\theta.$$

Again, we have

$$\mathbf{r}_1 \times \mathbf{r}_2 = (2m\sin\theta\cos 2\theta, -2m\cos\theta\cos 2\theta, u)$$

$$\Rightarrow \qquad \mathbf{N} = \frac{\mathbf{r}_1 \times \mathbf{r}_2}{H} = \frac{1}{H}(2m\sin\theta\cos 2\theta, -2m\cos\theta\cos 2\theta, u).$$

Therefore

$$L = \mathbf{r}_{11}\,.\,\mathbf{N} = 0,\quad M = \mathbf{r}_{12}\,.\,\mathbf{N} = -\frac{2m\cos 2\theta}{H},$$

$$N = \mathbf{r}_{22}\,.\,\mathbf{N} = -\frac{4mu\sin 2\theta}{H}.$$

Substituting the above values in (1), we get

$$(u^2 + 4m^2\cos^2 2\theta)\,\kappa^2 + \frac{4mu\sin 2\theta}{\sqrt{(u^2 + 4m^2\cos^2 2\theta)}}\kappa - \frac{4m^2\cos^2 2\theta}{u^2 + 4m^2\cos^2 2\theta} = 0. \qquad (2)$$

Putting $\kappa = 1/\rho$ in this equation, we get

$$(u^2 + 4m^2\cos^2 2\theta)\frac{1}{\rho^2} + \frac{4mu\sin 2\theta}{\sqrt{(u^2 + 4m^2\cos^2 2\theta)}}\frac{1}{\rho} - \frac{4m^2\cos^2 2\theta}{u^2 + 4m^2\cos^2 2\theta} = 0$$

or $$4m^2\rho^2\cos^2 2\theta - 4mu\rho\sqrt{(u^2 + 4m^2\cos^2 2\theta)}\sin 2\theta - (u^2 + 4m^2\cos^2 2\theta)^2 = 0. \qquad (3)$$

Equation (3) gives the principal radii at a point of the given cylindroid.

If $x = y, z = m,$ then $\theta = \pi/4$. Therefore, from (2), we have

$$u^2\kappa^2 + 4m\kappa - 0 = 0$$

$$\Rightarrow \quad \kappa\,(u^2\kappa + 4m) = 0$$

$$\Rightarrow \quad \text{Either } \kappa = 0 \text{ or } u^2\kappa + 4m = 0$$

$$\Rightarrow \quad \text{Either } \kappa = 0 \text{ or } \kappa = -\frac{4m}{u^2}$$

$$\Rightarrow \quad \text{Either } \rho = \infty \text{ or } \rho = -\frac{u^2}{4m}. \qquad \left[\because \kappa = \frac{1}{\rho}\right]$$

Example 17. For the hyperboloid $2z = 7x^2 + 6xy - y^2$, prove that the principal radii at the origin are $\frac{1}{8}, -\frac{1}{2}$ and the principal sections are

$$x = 3y, 3x = -y.$$

Solution. We know that the principal curvatures κ of the surface are given by

$$H^2\kappa^2 - (EN - 2FM + GL)\,\kappa + LN - M^2 = 0. \tag{1}$$

We shall find the values of E, F, G, L, M, N for the given surface.
The equation of the given hyperboloid is

$$2z = 7x^2 + 6xy - y^2$$

i.e.,
$$z = \frac{7x^2 + 6xy - y^2}{2}. \tag{2}$$

Therefore the equation of the surface is in Monge's form.
If $\mathbf{r}$ is the position vector of any current point on this surface, then

$$\mathbf{r} = (x, y, z), \tag{3}$$

where z is a function of x and y given by (2).

$$\therefore \quad p = \frac{\partial z}{\partial x} = 7x + 3y = 0 \text{ at the origin,}$$

$$q = \frac{\partial z}{\partial y} = 3x - y = 0 \text{ at the origin,}$$

$$r = \frac{\partial^2 z}{\partial x^2} = 7,\ s = \frac{\partial^2 z}{\partial x\,\partial y} = 3,\ t = \frac{\partial^2 z}{\partial y^2} = -1.$$

Taking x and y be as the first and the second parameters, and proceeding as in Example 11, we get

$$E = 1 + p^2, F = pq,\ G = 1 + q^2;\ L = \frac{r}{H},\ M = \frac{s}{H},\ N = \frac{t}{H},$$

where

$$H = \sqrt{(1 + p^2 + q^2)}.$$

Therefore, for the hyperboloid (2) at the origin, we have

$$E = 1 + p^2 = 1, F = pq = 0, G = 1 + q^2 = 1, H = \sqrt{1 + p^2 + q^2} = 1;$$

$$L = \frac{r}{H} = 7, \ M = \frac{s}{H} = 3, \ N = \frac{t}{H} = -1.$$

Substituting the values of E, F, G, L, M, N in (1), we get

$$\kappa^2 - 6\kappa - 16 = 0$$

or $(\kappa - 8)(\kappa + 2) = 0$

or $\kappa = 8, -2.$

Therefore the principal radii at the origin are given by

$$\rho = \frac{1}{8}, -\frac{1}{2}.$$

Also, we know that the differential equation of the lines of curvature is given by

$$(EM - FL)\,dx^2 + (EN - GL)\,dx\,dy + (FN - GM)\,dy^2 = 0$$

or $3\,dx^2 - 8\,dx\,dy - 3\,dy^2 = 0$

or $(3\,dx + dy)(dx - 3\,dy) = 0.$

Therefore the principal sections are given by

$$3\,dx + dy = 0, \ dx - 3\,dy = 0.$$

Integrating, we get

$$3x + y = c_1, \ x - 3y = c_2. \tag{4}$$

At origin, $x = 0, y = 0$. Therefore, from (4), we have

$$c_1 = 0, \ c_2 = 0.$$

Substituting the values of c_1 and c_2 in (4), we get

$$3x + y = 0, \ x - 3y = 0$$

i.e., $3x = -y, x = 3y.$

Example 18. Prove that at a point of curve of intersection of the paraboloid $xy = zc$ and the hyperboloid $x^2 + y^2 - z^2 + c^2 = 0$, the principal radii of the paraboloid are

$$(1 \pm \sqrt{2})\, z^2 / c.$$

Solution. We know that the principal curvatures κ of a surface are given by

$$H^2\kappa^2 - (EN - 2FM + GL)\,\kappa + LN - M^2 = 0. \tag{1}$$

We shall find the values of E, F, G, L, M, N for the given surface.
The equation of the given paraboloid is

$$xy = zc \tag{2}$$

and the equation of the given hyperboloid is

$$x^2 + y^2 - z^2 + c^2 = 0. \tag{3}$$

If $\mathbf{r}$ is the position vector of any current point on the surface (2), then

$$\mathbf{r} = (x, y, xy/c).$$

Taking x and y be as the first and the second parameters, we have

$$\mathbf{r}_1 = (1, 0, y/c), \quad \mathbf{r}_2 = (0, 1, x/c),$$

$$\mathbf{r}_{11} = (0, 0, 0), \quad \mathbf{r}_{12} = (0, 0, 1/c), \quad \mathbf{r}_{22} = (0, 0, 0).$$

Therefore

$$E = \mathbf{r}_1^{\ 2} = 1 + \frac{y^2}{c^2}, F = \mathbf{r}_1 \cdot \mathbf{r}_2 = \frac{xy}{c^2}, G = \mathbf{r}_2^{\ 2} = 1 + \frac{x^2}{c^2},$$

$$H^2 = EG - F^2 = 1 + \frac{x^2}{c^2} + \frac{y^2}{c^2} = \frac{c^2 + x^2 + y^2}{c^2}$$

$$= \frac{z^2}{c^2} \qquad \text{[using (3)]}$$

Again, we have

$$\mathbf{r}_1 \times \mathbf{r}_2 = \left(-\frac{y}{c}, -\frac{x}{c}, 1\right)$$

$$\Rightarrow \qquad \mathbf{N} = \frac{\mathbf{r}_1 \times \mathbf{r}_2}{H} = \frac{1}{H}\left(-\frac{y}{c}, -\frac{x}{c}, 1\right).$$

Therefore

$$L = \mathbf{r}_{11} \cdot \mathbf{N} = 0, \; M = \mathbf{r}_{12} \cdot \mathbf{N} = \frac{1}{cH}, \; N = \mathbf{r}_{22} \cdot \mathbf{N} = 0.$$

Substituting the values of E, F, G, L, M, N in (1), we get

$$H^2\kappa^2 + \frac{2xy}{c^3 H}\kappa - \frac{1}{c^2 H^2} = 0$$

or $$H^4 c^2 \kappa^2 + \frac{2xyH}{c}\kappa - 1 = 0$$

or $$H^4 c^2 \kappa^2 + 2zH\kappa - 1 = 0 \qquad \text{[using (2)]}$$

or $$\frac{z^4}{c^2}\kappa^2 + \frac{2z^2}{c}\kappa - 1 = 0 \qquad \left[\because H = \frac{z}{c}\right]$$

or $$z^4\kappa^2 + 2z^2 c\kappa - c^2 = 0$$

or $$\frac{z^4}{\rho^2} + \frac{2z^2 c}{\rho} - c^2 = 0 \qquad \left[\because \kappa = \frac{1}{\rho}\right]$$

or $$c^2\rho^2 - 2z^2 c\rho - z^4 = 0.$$

$$\therefore \qquad \rho = \frac{2z^2c \pm \sqrt{(4z^4c^2 + 4z^4c^2)}}{2c^2} = \frac{\left(1 \pm \sqrt{2}\right)z^2}{c}.$$

Example 19. If l, m, n are the direction cosines of the normal at a point to the surface $z = f(x, y)$, then prove that the equation for the principal radii can be written as

$$\frac{1}{\rho^2} + \frac{1}{\rho}\left(\frac{\partial l}{\partial x} + \frac{\partial m}{\partial y}\right) + \frac{\partial (l, m)}{\partial (x, y)} = 0.$$

Solution. We know that the equation for the principal curvatures through a point of the surface $\mathbf{r} = \mathbf{r}(u, v)$ is

$$H^2\kappa^2 - (EN - 2FM + GL)\,\kappa + LN - M^2 = 0.$$

Therefore the equation for the principal radii of curvature is

$$H^2\frac{1}{\rho^2} - (EN - 2FM + GL)\frac{1}{\rho} + LN - M^2 = 0. \tag{1}$$

We shall find the values of E, F, G, L, M, N for the given surface.
The equation of the given surface is

$$z = f(x, y).$$

If $\mathbf{r}$ is the position vector of any current point on this surface, then

$$\mathbf{r} = (x, y, z), \tag{2}$$

where z is a function of x and y.
We write

$$\frac{\partial z}{\partial x} = p, \quad \frac{\partial z}{\partial y} = q, \quad \frac{\partial^2 z}{\partial x^2} = r, \quad \frac{\partial^2 z}{\partial x\,\partial y} = s, \quad \frac{\partial^2 z}{\partial y^2} = t.$$

Taking x and y as the first and the second parameters, we have

$$\mathbf{r}_1 = (1, 0, p), \quad \mathbf{r}_2 = (0, 1, q),$$
$$\mathbf{r}_{11} = (0, 0, r), \ \mathbf{r}_{12} = (0, 0, s), \ \mathbf{r}_{22} = (0, 0, t).$$

Therefore

$$E = \mathbf{r}_1^{\,2} = 1 + p^2, \ F = \mathbf{r}_1 . \mathbf{r}_2 = pq, \ G = \mathbf{r}_2^{\,2} = 1 + q^2,$$
$$H^2 = EG - F^2 = 1 + p^2 + q^2.$$

Again, we have

$$\mathbf{r}_1 \times \mathbf{r}_2 = (-p, -q, 1)$$

$$\Rightarrow \qquad \mathbf{N} = \frac{\mathbf{r}_1 \times \mathbf{r}_2}{H} = \frac{1}{H}(-p, -q, 1).$$

Therefore

$$L = \mathbf{r}_{11} . \mathbf{N} = \frac{r}{H}, \ M = \mathbf{r}_{12} . \mathbf{N} = \frac{s}{H}, \ N = \mathbf{r}_{22} . \mathbf{N} = \frac{t}{H}.$$

Substituting the values of E, F, G, L, M, N in (1), we get

$$H^2 \frac{1}{\rho^2} - \left[(1+p^2)\frac{t}{H} - 2pq\frac{s}{H} + (1+q^2)\frac{r}{H}\right]\frac{1}{\rho} + \frac{rt-s^2}{H^2} = 0$$

or
$$\frac{1}{\rho^2} - \left[\frac{(1+p^2)t - 2pqs + (1+q^2)r}{H^3}\right]\frac{1}{\rho} + \frac{rt-s^2}{H^4} = 0. \tag{3}$$

Since l, m, n are the direction cosines of the normal at the point (x, y, z), we have

$$\mathbf{N} = (l, m, n)$$

$$\Rightarrow \quad \frac{1}{H}(-p, -q, 1) = (l, m, n)$$

$$\Rightarrow \quad l = -\frac{p}{H},\ m = -\frac{q}{H},\ n = \frac{1}{H}.$$

Now,

$$H^2 = 1 + p^2 + q^2$$

$$\Rightarrow \quad 2H\frac{\partial H}{\partial x} = 2p\frac{\partial p}{\partial x} + 2q\frac{\partial q}{\partial x} = 2pr + 2qs$$

$$\Rightarrow \quad \frac{\partial H}{\partial x} = \frac{pr+qs}{H}.$$

Similarly,

$$H^2 = 1 + p^2 + q^2$$

$$\Rightarrow \quad 2H\frac{\partial H}{\partial y} = 2p\frac{\partial p}{\partial y} + 2q\frac{\partial q}{\partial y} = 2ps + 2qt$$

$$\Rightarrow \quad \frac{\partial H}{\partial y} = \frac{ps+qt}{H}.$$

Also, $l = -\dfrac{p}{H}$

$$\Rightarrow \quad \frac{\partial l}{\partial x} = -\frac{\frac{\partial p}{\partial x}H - p\frac{\partial H}{\partial x}}{H^2} = -\frac{rH - p\left(\frac{pr+qs}{H}\right)}{H^2}$$

$$= -\frac{rH^2 - p(pr+qs)}{H^3}$$

$$= -\frac{r(1+p^2+q^2) - p(pr+qs)}{H^3}$$

$$= -\frac{r(1+q^2) - pqs}{H^3}.$$

Similarly,

$$l = -\frac{p}{H}$$

$$\Rightarrow \quad \frac{\partial l}{\partial y} = -\frac{\frac{\partial p}{\partial y}H - p\frac{\partial H}{\partial y}}{H^2} = -\frac{sH - p\left(\frac{ps+qt}{H}\right)}{H^2}$$

$$= -\frac{sH^2 - p\,(ps+qt)}{H^3}$$

$$= -\frac{s\,(1+p^2+q^2) - p\,(ps+qt)}{H^3}$$

$$= -\frac{s\,(1+q^2) - pqt}{H^3}.$$

Again, $m = -\frac{q}{H}$

$$\Rightarrow \quad \frac{\partial m}{\partial x} = -\frac{\frac{\partial q}{\partial x}H - q\frac{\partial H}{\partial x}}{H^2} = -\frac{sH - q\left(\frac{pr+qs}{H}\right)}{H^2}$$

$$= -\frac{sH^2 - q\,(pr+qs)}{H^3}$$

$$= -\frac{s\,(1+p^2+q^2) - q\,(pr+qs)}{H^3}$$

$$= -\frac{s\,(1+p^2) - pqr}{H^3}.$$

Similarly,

$$m = -\frac{q}{H}$$

$$\Rightarrow \quad \frac{\partial m}{\partial y} = -\frac{\frac{\partial q}{\partial y}H - q\frac{\partial H}{\partial y}}{H^2} = -\frac{tH - q\left(\frac{ps+qt}{H}\right)}{H^2}$$

$$= -\frac{tH^2 - q\,(ps+qt)}{H^3}$$

$$= -\frac{t\,(1+p^2+q^2) - q\,(ps+qt)}{H^3}$$

$$= -\frac{t\,(1+p^2) - pqs}{H^3}.$$

$$\therefore \quad \frac{\partial l}{\partial x}+\frac{\partial m}{\partial y}=-\left[\frac{(1+p^2)t-2pqs+(1+q^2)r}{H^3}\right] \tag{4}$$

and
$$\frac{\partial(l,m)}{\partial(x,y)}=\begin{vmatrix}\frac{\partial l}{\partial x} & \frac{\partial m}{\partial x}\\ \frac{\partial l}{\partial y} & \frac{\partial m}{\partial y}\end{vmatrix}$$

$$=\frac{\partial l}{\partial x}\frac{\partial m}{\partial y}-\frac{\partial l}{\partial y}\frac{\partial m}{\partial x}$$

$$=\frac{[r(1+q^2)-pqs][t(1+p^2)-pqs]}{H^6}$$

$$-\frac{[s(1+q^2)-pqt][s(1+p^2)-pqr]}{H^6}$$

i.e.,
$$\frac{\partial(l,m)}{\partial(x,y)}=\frac{rt-s^2}{H^4}. \tag{5}$$

Using (4) and (5), (3) gives

$$\frac{1}{\rho^2}+\frac{1}{\rho}\left(\frac{\partial l}{\partial x}+\frac{\partial m}{\partial y}\right)+\frac{\partial(l,m)}{\partial(x,y)}=0.$$

Example 20. Show that the indicatrix at every point of the helicoid

$$z=c\tan^{-1}(y/x)$$

is a rectangular hyperbola.

Solution. We know that the indicatrix at any point of a surface is a rectangular hyperbola if its first curvature J is zero at that point, i.e.,

$$\kappa_a+\kappa_b=0$$

i.e.,
$$\frac{EN-2FM+GL}{H^2}=0$$

i.e., $EN-2FM+GL=0$.

We shall find the values of E, F, G, L, M, N for the given surface.

The equation of the given helicoid is

$$z=c\tan^{-1}(y/x).$$

We can take parametric equations of the helicoid as

$$x=u\cos v,\ y=u\sin v,\ z=cv.$$

Therefore the position vector $\mathbf{r}$ of any current point on the helicoid is

$$\mathbf{r}=(u\cos v, u\sin v, cv).$$

Now, proceeding as in Example 1, we have

$$E = 1,\ F = 0,\ G = u^2 + c^2,\ H^2 = u^2 + c^2\ ;$$

$$L = 0,\ M = -\frac{c}{\sqrt{u^2 + c^2}},\ N = 0.$$

$\therefore$ $EN - 2FM + GL = 0$ at every point of the given helicoid.

Hence the indicatrix at every point of the given helicoid is a rectangular hyperbola.

Example 21. Show that the indicatrix at a point of the surface $z = f(x, y)$ is a rectangular hyperbola if

$$(1 + p^2)t - 2pqs + (1 + q^2)r = 0.$$

Solution. We know that the indicatrix at any point of a surface is a rectangular hyperbola if its first curvature J is zero at that point, i.e.,

$$\kappa_a + \kappa_b = 0$$

i.e., $$\frac{EN - 2FM + GL}{H^2} = 0$$

i.e., $$EN - 2FM + GL = 0. \quad (1)$$

We shall find the values of E, F, G, L, M, N for the given surface.

The equation of the given surface is

$$z = f(x, y).$$

If $\mathbf{r}$ is the position vector of any current point on this surface, then

$$\mathbf{r} = (x, y, z), \quad (2)$$

where z is a function of x and y.

We write

$$\frac{\partial z}{\partial x} = p,\ \frac{\partial z}{\partial y} = q,\ \frac{\partial^2 z}{\partial x^2} = r,\ \frac{\partial^2 z}{\partial x\, \partial y} = s,\ \frac{\partial^2 z}{\partial y^2} = t.$$

Taking x and y as the first and the second parameters, and proceeding as in Example 11, we have

$$E = 1 + p^2, F = pq,\ G = 1 + q^2;$$

$$L = \frac{r}{H},\ M = \frac{s}{H},\ N = \frac{t}{H},$$

where $H = \sqrt{(1 + p^2 + q^2)}$.

Hence, by (1), the indicatrix at any point of the given surface (2) is a rectangular hyperbola if

$$(1 + p^2)\frac{t}{\mathrm{H}} - 2pq\frac{s}{H} + (1 + q^2)\frac{r}{H} = 0$$

i.e., $$(1 + p^2)t - 2pqs + (1 + q^2)r = 0.$$

Example 22. Show that the differential equation of the lines of curvature can be put in the form

$$[\mathbf{N}, d\mathbf{N}, d\mathbf{r}] = 0.$$

Solution. We know that the differential equation of the lines of curvature is given by

$$(EM - FL)\, du^2 + (EN - GL)\, du\, dv + (FN - GM)\, dv^2 = 0. \quad (1)$$

We have to prove that this differential equation can be put in the form

$$[\mathbf{N}, d\mathbf{N}, d\mathbf{r}] = 0.$$

We have

$$d\mathbf{N} = \mathbf{N}_1\, du + \mathbf{N}_2\, dv,$$

$$d\mathbf{r} = \mathbf{r}_1\, du + \mathbf{r}_2\, dv.$$

$$\therefore \quad d\mathbf{N} \times d\mathbf{r} = (\mathbf{N}_1\, du + \mathbf{N}_2\, dv) \times (\mathbf{r}_1\, du + \mathbf{r}_2\, dv)$$

$$= (\mathbf{N}_1 \times \mathbf{r}_1)\, du^2 + \{(\mathbf{N}_1 \times \mathbf{r}_2) + (\mathbf{N}_2 \times \mathbf{r}_1)\}\, du\, dv + (\mathbf{N}_2 \times \mathbf{r}_2)\, dv^2.$$

$$\therefore \quad \mathbf{N}.(d\mathbf{N} \times d\mathbf{r}) = \mathbf{N}.(\mathbf{N}_1 \times \mathbf{r}_1)\, du^2 + \{\mathbf{N}.(\mathbf{N}_1 \times \mathbf{r}_2) + \mathbf{N}.(\mathbf{N}_2 \times \mathbf{r}_1)\}\, du\, dv + \mathbf{N}.(\mathbf{N}_2 \times \mathbf{r}_2)\, dv^2$$

i.e.,

$$[\mathbf{N}, d\mathbf{N}, d\mathbf{r}] = [\mathbf{N}, \mathbf{N}_1, \mathbf{r}_1]\, du^2 + \{[\mathbf{N}, \mathbf{N}_1, \mathbf{r}_2] + [\mathbf{N}, \mathbf{N}_2, \mathbf{r}_1]\}\, du\, dv + [\mathbf{N}, \mathbf{N}_2, \mathbf{r}_2]\, dv^2. \quad (2)$$

Now, by Weingarten equations, we have

$$\mathbf{N}_1 = \frac{FM - GL}{H^2}\mathbf{r}_1 + \frac{FL - EM}{H^2}\mathbf{r}_2,$$

$$\mathbf{N}_2 = \frac{FN - GM}{H^2}\mathbf{r}_1 + \frac{FM - EN}{H^2}\mathbf{r}_2.$$

Therefore

$$[\mathbf{N},\ \mathbf{N}_1,\ \mathbf{r}_1] = \mathbf{N}\,.\,\mathbf{N}_1 \times \mathbf{r}_1$$

$$= \mathbf{N}\,.\left\{\frac{FL - EM}{H^2}(\mathbf{r}_2 \times \mathbf{r}_1)\right\}$$

$$= \mathbf{N}\,.\left\{\frac{FL - EM}{H^2}(-H\mathbf{N})\right\}$$

$$= \frac{EM - FL}{H}.$$

Similarly,

$$[\mathbf{N},\ \mathbf{N}_1,\ \mathbf{r}_2] = \mathbf{N}\,.\,\mathbf{N}_1 \times \mathbf{r}_2$$

$$= \mathbf{N}\,.\left\{\frac{FM - GL}{H^2}(\mathbf{r}_1 \times \mathbf{r}_2)\right\}$$

$$= \mathbf{N} \cdot \left\{ \frac{FM - GL}{H^2} (H\mathbf{N}) \right\}$$

$$= \frac{FM - GL}{H},$$

$$[\mathbf{N},\ \mathbf{N}_2,\ \mathbf{r}_1] = \mathbf{N} \cdot \mathbf{N}_2 \times \mathbf{r}_1$$

$$= \mathbf{N} \cdot \left\{ \frac{FM - EN}{H^2} (\mathbf{r}_2 \times \mathbf{r}_1) \right\}$$

$$= \mathbf{N} \cdot \left\{ \frac{FM - EN}{H^2} (-H\mathbf{N}) \right\}$$

$$= \frac{EN - FM}{H},$$

$$[\mathbf{N},\ \mathbf{N}_2,\ \mathbf{r}_2] = \mathbf{N} \cdot \mathbf{N}_2 \times \mathbf{r}_2$$

$$= \mathbf{N} \cdot \left\{ \frac{FN - GM}{H^2} (\mathbf{r}_1 \times \mathbf{r}_2) \right\}$$

$$= \mathbf{N} \cdot \left\{ \frac{FN - GM}{H^2} (H\mathbf{N}) \right\}$$

$$= \frac{FN - GM}{H}.$$

Substituting the above values in (2), we get

$$[\mathbf{N}, d\mathbf{N}, d\mathbf{r}] = \left(\frac{EM - FL}{H} \right) du^2 + \left(\frac{EN - GL}{H} \right) du\, dv + \left(\frac{FN - GM}{H} \right) dv^2.$$

Hence $[\mathbf{N}, d\mathbf{N}, d\mathbf{r}] = 0$ if and only if

$$(EM - FL)\, du^2 + (EN - GL)\, du\, dv + (FN - GM)\, dv^2 = 0,$$

which is the differential equation of the lines of curvature given by (1). Therefore the differential equation of the lines of curvature given by (1) can be put in the form

$$[\mathbf{N}, d\mathbf{N}, d\mathbf{r}] = 0.$$

Example 23. The moment about origin of the unit normal $\mathbf{N}$ at a point $\mathbf{r}$ of the surface is $\mathbf{m} = \mathbf{r} \times \mathbf{N}$. Prove that the differential equation of the lines of curvature is

$$d\mathbf{m} \cdot d\mathbf{N} = 0.$$

Solution. We know that the differential equation of the lines of curvature is given by

$$[\mathbf{N}, d\mathbf{N}, d\mathbf{r}] = 0. \qquad (1)$$

Here, we have

$$\mathbf{m} = \mathbf{r} \times \mathbf{N}$$

$$\Rightarrow \quad d\mathbf{m} = d\mathbf{r} \times \mathbf{N} + \mathbf{r} \times d\mathbf{N}$$

$$\Rightarrow \quad d\mathbf{m} \,.\, d\mathbf{N} = (d\mathbf{r} \times \mathbf{N} + \mathbf{r} \times d\mathbf{N}) \,.\, d\mathbf{N}$$

$$= (d\mathbf{r} \times \mathbf{N} \,.\, d\mathbf{N}) + (\mathbf{r} \times d\mathbf{N} \,.\, d\mathbf{N})$$

$$= d\mathbf{r} \times \mathbf{N} \,.\, d\mathbf{N} \qquad [\because \mathbf{r} \times d\mathbf{N} \,.\, d\mathbf{N} = 0]$$

$$= \mathbf{N} \,.\, d\mathbf{N} \times d\mathbf{r}$$

$$= [\mathbf{N}, d\mathbf{N}, d\mathbf{r}].$$

Hence, by (1), the differential equation of the lines of curvature is

$$d\mathbf{m} \,.\, d\mathbf{N} = 0.$$

Example 24. Show that the surface $4a^2z^2 = (x^2 - 2a^2)(y^2 - 2a^2)$ has a line of umbilics lying on the sphere $x^2 + y^2 + z^2 = 4a^2$.

Solution. If we regard z as a function of x and y, then at an umbilic, we have

$$\frac{1 + p^2}{r} = \frac{1 + q^2}{t} = \frac{pq}{s}$$

or $\quad s(1 + p^2) = rpq. \qquad (1)$

The equation of the given surface is

$$4a^2z^2 = (x^2 - 2a^2)(y^2 - 2a^2) \qquad (2)$$

i.e., $\quad z = \dfrac{\sqrt{(x^2 - 2a^2)(y^2 - 2a^2)}}{2a}.$

Therefore

$$p = \frac{\partial z}{\partial x} = \frac{x}{2a}\sqrt{\frac{y^2 - 2a^2}{x^2 - 2a^2}}, \quad q = \frac{\partial z}{\partial y} = \frac{y}{2a}\sqrt{\frac{x^2 - 2a^2}{y^2 - 2a^2}},$$

$$r = \frac{\partial^2 z}{\partial x^2} = \frac{\partial}{\partial x}\left(\frac{x}{2a}\sqrt{\frac{y^2 - 2a^2}{x^2 - 2a^2}}\right)$$

$$= \frac{1}{2a}\sqrt{\frac{y^2 - 2a^2}{x^2 - 2a^2}} + \frac{x}{2a}\left(-\frac{x\sqrt{y^2 - 2a^2}}{(x^2 - 2a^2)^{3/2}}\right)$$

$$= \frac{1}{2a}\sqrt{\frac{y^2 - 2a^2}{x^2 - 2a^2}}\left(1 - \frac{x^2}{x^2 - 2a^2}\right)$$

$$= -a\frac{\sqrt{y^2-2a^2}}{(x^2-2a^2)^{3/2}},$$

$$s = \frac{\partial^2 z}{\partial x\,\partial y} = \frac{\partial}{\partial x}\left(\frac{y}{2a}\sqrt{\frac{x^2-2a^2}{y^2-2a^2}}\right)$$

$$= \frac{y}{2a\sqrt{y^2-2a^2}}\left(\frac{x}{\sqrt{x^2-2a^2}}\right)$$

$$= \frac{xy}{2a\sqrt{(x^2-2a^2)\,(y^2-2a^2)}},$$

$$t = \frac{\partial^2 z}{\partial y^2} = \frac{\partial}{\partial y}\left(\frac{y}{2a}\sqrt{\frac{x^2-2a^2}{y^2-2a^2}}\right)$$

$$= \frac{1}{2a}\sqrt{\frac{x^2-2a^2}{y^2-2a^2}} + \frac{y}{2a}\left(-\frac{y\sqrt{x^2-2a^2}}{(y^2-2a^2)^{3/2}}\right)$$

$$= \frac{1}{2a}\sqrt{\frac{x^2-2a^2}{y^2-2a^2}}\left(1-\frac{y^2}{y^2-2a^2}\right)$$

$$= -a\frac{\sqrt{x^2-2a^2}}{(y^2-2a^2)^{3/2}}.$$

Substituting the above values in (1), we get

$$\left[\frac{xy}{2a\sqrt{(x^2-2a^2)\,(y^2-2a^2)}}\right]\left[1+\frac{x^2}{4a^2}\left(\frac{y^2-2a^2}{x^2-2a^2}\right)\right]$$

$$= \left[-a\frac{\sqrt{y^2-2a^2}}{(x^2-2a^2)^{3/2}}\right]\frac{xy}{4a^2}$$

or $$1+\frac{x^2}{4a^2}\left(\frac{y^2-2a^2}{x^2-2a^2}\right) = -\frac{1}{2}\left(\frac{y^2-2a^2}{x^2-2a^2}\right)$$

or $$1+\left(\frac{x^2}{4a^2}+\frac{1}{2}\right)\left(\frac{y^2-2a^2}{x^2-2a^2}\right) = 0$$

or $$1+\left(\frac{x^2+2a^2}{4a^2}\right)\left(\frac{y^2-2a^2}{x^2-2a^2}\right) = 0$$

or $$1+\left(\frac{x^2-2a^2+4a^2}{4a^2}\right)\left(\frac{y^2-2a^2}{x^2-2a^2}\right)=0$$

or $$1+\left(\frac{x^2-2a^2}{4a^2}+1\right)\left(\frac{y^2-2a^2}{x^2-2a^2}\right)=0$$

or $$1+\left(\frac{y^2-2a^2}{4a^2}+\frac{y^2-2a^2}{x^2-2a^2}\right)=0$$

or $$(x^2-2a^2)+\left[\frac{(x^2-2a^2)(y^2-2a^2)}{4a^2}+(y^2-2a^2)\right]=0$$

or $$(x^2-2a^2)+[z^2+(y^2-2a^2)]=0 \qquad \text{[using (2)]}$$

or $$x^2+y^2+z^2=4a^2.$$

Example 25. Find the umbilics of the ellipsoid

$$\frac{x^2}{a^2}+\frac{y^2}{b^2}+\frac{z^2}{c^2}=1.$$

If P is an umbilic of this ellipsoid, prove that the curvature at P of any normal section through P is ac/b^3.

Solution. First Part. If we regard z as a function of x and y, then at an umbilic, we have

$$\frac{1+p^2}{r}=\frac{1+q^2}{t}=\frac{pq}{s}=\lambda \text{ (let)}. \tag{1}$$

The equation of the given ellipsoid is

$$\frac{x^2}{a^2}+\frac{y^2}{b^2}+\frac{z^2}{c^2}=1. \tag{2}$$

Differentiating (2) partially with respect to x, taking z as a function of x and y, we get

$$\frac{2x}{a^2}+\frac{2z}{c^2}\frac{\partial z}{\partial x}=0.$$

$$\therefore \quad p=\frac{\partial z}{\partial x}=-\frac{c^2x}{a^2z}. \tag{3}$$

Similarly, differentiating (2) partially with respect to y, taking z as a function of x and y, we get

$$\frac{2y}{b^2}+\frac{2z}{c^2}\frac{\partial z}{\partial y}=0.$$

$$\therefore \qquad q = \frac{\partial z}{\partial y} = -\frac{c^2 y}{b^2 z}. \tag{4}$$

Now

$$r = \frac{\partial p}{\partial x} = -\frac{c^2}{a^2}\left(\frac{1}{z} - \frac{x}{z^2} p\right) = -\frac{1}{z}\left(\frac{c^2}{a^2} + p^2\right), \tag{5}$$

$$s = \frac{\partial p}{\partial y} = -\frac{c^2}{a^2}\left(-\frac{x}{z^2} q\right) = -\frac{pq}{z}, \tag{6}$$

$$t = \frac{\partial q}{\partial y} = -\frac{c^2}{b^2}\left(\frac{1}{z} - \frac{y}{z^2} q\right) = -\frac{1}{z}\left(\frac{c^2}{b^2} + q^2\right). \tag{7}$$

At an umbilic, we have

$$\frac{1+p^2}{r} = \frac{pq}{s} \qquad \text{[From (1)]}$$

or $\qquad s(1+p^2) - rpq = 0$

or $\qquad -\dfrac{pq}{z}(1+p^2) - rpq = 0 \qquad$ [using (6)]

or $\qquad \dfrac{pq}{z}[(1+p^2) + rz] = 0$

or $\qquad pq[(1+p^2) + rz] = 0.$

$$\therefore \qquad \text{Either } p = 0 \text{ or } q = 0 \text{ or } (1+p^2) + rz = 0. \tag{8}$$

But, $\qquad (1+p^2) + rz = (1+p^2) - \left(\dfrac{c^2}{a^2} + p^2\right) = 1 - \dfrac{c^2}{a^2}$

$\Rightarrow \qquad (1+p^2) + rz \neq 0$ unless $c = a$

$$\Rightarrow \qquad \text{Either } p = 0 \text{ or } q = 0. \tag{9}$$

[using (8)]

Also, at an umbilic, we have

$$\frac{1+p^2}{r} = \frac{1+q^2}{t} \qquad \text{[From (1)]}$$

or $\qquad t(1+p^2) - r(1+q^2) = 0$

or $\qquad -\dfrac{1}{z}\left(\dfrac{c^2}{b^2} + q^2\right)(1+p^2) + \dfrac{1}{z}\left(\dfrac{c^2}{a^2} + p^2\right)(1+q^2) = 0$

[using (5) and (7)]

or $\qquad -\left(\dfrac{c^2 + b^2 q^2}{b^2}\right)(1+p^2) + \left(\dfrac{c^2 + a^2 p^2}{a^2}\right)(1+q^2) = 0$

or $\quad -a^2(c^2+b^2q^2)(1+p^2)+b^2(c^2+a^2p^2)(1+q^2)=0$

or $\quad a^2p^2(b^2-c^2)+b^2q^2(c^2-a^2)=c^2(a^2-b^2). \qquad (10)$

Now, let $a > b > c$.

Then, by (10), $p = 0$ gives imaginary values of q. Hence, by (9), we must have

$$q=0. \qquad (11)$$

Substituting $q=0$ in (10), we get

$$p=\pm\frac{c}{a}\sqrt{\left(\frac{a^2-b^2}{b^2-c^2}\right)}. \qquad (12)$$

From (3) and (12), we get

$$-\frac{c^2x}{a^2z}=\pm\frac{c}{a}\sqrt{\left(\frac{a^2-b^2}{b^2-c^2}\right)}$$

or $\quad \dfrac{c^2x}{a^2z}=\mp\dfrac{c}{a}\sqrt{\left(\dfrac{a^2-b^2}{b^2-c^2}\right)}$

or $\quad \dfrac{x/a}{\sqrt{(a^2-b^2)}}=\dfrac{z/c}{\mp\sqrt{(b^2-c^2)}}. \qquad (13)$

From (4) and (11), we get

$$-\frac{c^2y}{b^2z}=0$$

$\Rightarrow \quad y=0. \qquad (14)$

Hence, by (13) and (14), the umbilics are given by

$$\frac{x/a}{\sqrt{(a^2-b^2)}}=\frac{y/b}{0}=\frac{z/c}{\mp\sqrt{(b^2-c^2)}}=\pm\frac{\sqrt{(x/a)^2+(y/b)^2+(z/c)^2}}{\sqrt{(a^2-b^2)+0+(b^2-c^2)}}$$

or $\quad \dfrac{x/a}{\sqrt{(a^2-b^2)}}=\dfrac{y/b}{0}=\dfrac{z/c}{\mp\sqrt{(b^2-c^2)}}=\pm\dfrac{1}{\sqrt{(a^2-c^2)}}. \qquad (15)$

Second Part. If P is an umbilic of this ellipsoid, then, by definition, the curvature at P of any normal section through P is given by

$$\kappa=\frac{L\,du^2+2M\,du\,dv+N\,dv^2}{E\,du^2+2F\,du\,dv+G\,dv^2}. \qquad (16)$$

But,

$$E = 1 + p^2, F = pq,\ G = 1 + q^2\ ;\ L = \frac{r}{H},\ M = \frac{s}{H},\ N = \frac{t}{H}.$$

Therefore, by (16),

$$\kappa = \frac{\frac{r}{H} du^2 + 2\frac{s}{H} du\, dv + \frac{t}{H} dv^2}{(1+p^2)\, du^2 + 2pq\, du\, dv + (1+q^2)\, dv^2}$$

$$= \frac{\frac{r}{H} du^2 + 2\frac{s}{H} du\, dv + \frac{t}{H} dv^2}{(\lambda r)\, du^2 + 2(\lambda s)\, du\, dv + (\lambda t)\, dv^2} \qquad \text{[using (1)]}$$

$$= \frac{1}{H\lambda}$$

$$= \frac{t}{H(1+q^2)} \qquad \left[\because \lambda = \frac{1+q^2}{t}, \text{by (1)}\right]$$

$$= \frac{t}{H} \qquad [\because q = 0, \text{by (11)}]$$

$$= \frac{t}{\sqrt{1+p^2+q^2}} \qquad [\because H^2 = EG - F^2 = 1 + p^2 + q^2]$$

$$= \frac{t}{\sqrt{1+p^2}} \qquad [\because q = 0, \text{by (11)}]$$

$$= \frac{-\frac{1}{z}\left(\frac{c^2}{b^2} + q^2\right)}{\sqrt{1 + \frac{c^2}{a^2}\left(\frac{a^2-b^2}{b^2-c^2}\right)}} \qquad \text{[using (7) and (12)]}$$

$$= \frac{-c^2\sqrt{\frac{a^2(b^2-c^2)}{b^2(a^2-c^2)}}}{z b^2} \qquad [\because q = 0, \text{by (11)}]$$

$$= \frac{-c^2 a\sqrt{\frac{b^2-c^2}{a^2-c^2}}}{z b^3}$$

$$= \frac{-c^2 a}{z b^3}\left(-\frac{z}{c}\right) \qquad \text{[using (15)]}$$

$$= \frac{ac}{b^3}.$$

Example 26. Find the principal curvature and lines of curvature for the surface generated by the tangents to a twisted curve.

Solution. We know that the equation giving the principal curvatures κ_n is

$$H^2\kappa_n^2 - (EN - 2FM + GL)\,\kappa_n + LN - M^2 = 0. \qquad (1)$$

We shall find the values of E, F, G, L, M, N for the given surface.

Let $\mathbf{r} = \mathbf{r}(s)$ be a twisted curve and let $\mathbf{R}$ be the position vector of any current point on the surface generated by the tangents to the given curve. Then

$$\mathbf{R} = \mathbf{r}(s) + \lambda\mathbf{t}(s), \qquad (2)$$

where λ, s are the two independent parameters.

Let the suffixes 1 and 2 denote partial derivatives with respect to λ and s respectively. Then, from (2), we have

$$\mathbf{R}_1 = \frac{\partial\mathbf{R}}{\partial\lambda} = \mathbf{t},\ \mathbf{R}_2 = \frac{\partial\mathbf{R}}{\partial s} = \mathbf{t} + \lambda\kappa\mathbf{n},$$

$$\mathbf{R}_{11} = \mathbf{0},\ \mathbf{R}_{12} = \kappa\mathbf{n},$$

$$\mathbf{R}_{22} = \kappa\mathbf{n} + \lambda\,[\kappa'\mathbf{n} + \kappa\,(\tau\mathbf{b} - \kappa\mathbf{t})]$$

$$= -\lambda\kappa^2\,\mathbf{t} + (\kappa + \lambda\kappa')\,\mathbf{n} + \lambda\kappa\tau\,\mathbf{b}.$$

Therefore

$$E = \mathbf{R}_1^{\ 2} = 1,\ F = \mathbf{R}_1\mathbf{.}\mathbf{R}_2 = 1,\ G = \mathbf{R}_2^{\ 2} = 1 + \lambda^2\kappa^2,$$

$$H^2 = EG - F^2 = \lambda^2\kappa^2,\ \text{i.e.,}\ H = \lambda\kappa.$$

Again, we have

$$\mathbf{R}_1 \times \mathbf{R}_2 = \lambda\kappa\mathbf{b}$$

$$\Rightarrow \qquad \mathbf{N} = \frac{\mathbf{R}_1 \times \mathbf{R}_2}{H} = \frac{\lambda\kappa\mathbf{b}}{\lambda\kappa} = \mathbf{b}.$$

Therefore

$$L = \mathbf{R}_{11}\mathbf{.}\mathbf{N} = 0,\ M = \mathbf{R}_{12}\mathbf{.}\mathbf{N} = 0,\ N = \mathbf{R}_{22}\mathbf{.}\mathbf{N} = \lambda\kappa\tau.$$

Substituting the above values in (1), we get

$$\lambda^2\kappa^2\kappa_n^{\ 2} - \lambda\kappa\tau\kappa_n = 0$$

or $$\lambda\kappa\kappa_n(\lambda\kappa\kappa_n - \tau) = 0.$$

$$\therefore \qquad \kappa_n = 0 \text{ and } \kappa_n = \frac{\tau}{\lambda\kappa}.$$

Also, we know that the differential equation of the lines of curvature is

$$(EM - FL)\,d\lambda^2 + (EN - GL)\,d\lambda\,ds + (FN - GM)\,ds^2 = 0$$

or $$0\,d\lambda^2 + (\lambda\kappa\tau)\,d\lambda\,ds + (\lambda\kappa\tau)\,ds^2 = 0$$

or $\lambda\kappa\tau\, ds\,(d\lambda + ds) = 0.$

$\therefore$ $ds = 0$ and $d\lambda + ds = 0.$

Integrating these, we get

$$s = \text{constant and } \lambda + s = \text{constant},$$

which give the required lines of curvature.

Example 27. (a) Find the principal curvature and lines of curvature for the surface generated by the principal normals to a twisted curve.

Solution. We know that the equation giving the principal curvatures κ_n is

$$H^2\kappa_n{}^2 - (EN - 2FM + GL)\,\kappa_n + LN - M^2 = 0. \qquad (1)$$

We shall find the values of E, F, G, L, M, N for the given surface.

Let $\mathbf{r} = \mathbf{r}\,(s)$ be a twisted curve and let $\mathbf{R}$ be the position vector of any current point on the surface generated by the principal normals to the given curve. Then

$$\mathbf{R} = \mathbf{r}\,(s) + \lambda\mathbf{n}\,(s), \qquad (2)$$

where λ, s are the two independent parameters.

Let the suffixes 1 and 2 denote partial derivatives with respect to λ and s respectively. Then, from (2), we have

$$\mathbf{R}_1 = \frac{\partial\mathbf{R}}{\partial\lambda} = \mathbf{n},\;\; \mathbf{R}_2 = \frac{\partial\mathbf{R}}{\partial s} = \mathbf{t} + \lambda\,(\tau\mathbf{b} - \kappa\mathbf{t}) = (1 - \lambda\kappa)\,\mathbf{t} + \lambda\tau\mathbf{b},$$

$$\mathbf{R}_{11} = \mathbf{0},\;\; \mathbf{R}_{12} = \tau\mathbf{b} - \kappa\mathbf{t},$$

$$\mathbf{R}_{22} = (-\lambda\kappa')\,\mathbf{t} + (1 - \lambda\kappa)\,\kappa\mathbf{n} + \lambda\,[\tau'\mathbf{b} + \tau\,(-\tau\mathbf{n})]$$

$$= -\lambda\kappa'\,\mathbf{t} + (\kappa - \lambda\kappa^2 - \lambda\tau^2)\,\mathbf{n} + \lambda\tau'\mathbf{b}.$$

Therefore

$$E = \mathbf{R}_1{}^2 = 1,\quad F = \mathbf{R}_1 . \mathbf{R}_2 = 0,\quad G = \mathbf{R}_2{}^2 = (1 - \lambda\kappa)^2 + \lambda^2\tau^2,$$

$$H^2 = EG - F^2 = (1 - \lambda\kappa)^2 + \lambda^2\tau^2.$$

Again, we have

$$\mathbf{R}_1 \times \mathbf{R}_2 = \lambda\tau\mathbf{t} - (1 - \lambda\kappa)\,\mathbf{b}$$

$$\Rightarrow \quad \mathbf{N} = \frac{\mathbf{R}_1 \times \mathbf{R}_2}{H} = \frac{\lambda\tau\mathbf{t} - (1 - \lambda\kappa)\,\mathbf{b}}{H}.$$

Therefore

$$L = \mathbf{R}_{11} . \mathbf{N} = 0,\;\; M = \mathbf{R}_{12} . \mathbf{N} = -\frac{\tau}{H},$$

$$N = \mathbf{R}_{22} . \mathbf{N} = \frac{\lambda^2(\kappa\tau' - \kappa'\tau) - \lambda\tau'}{H}.$$

Substituting the above values in (1), we get

$$H^2\kappa_n^2 - \frac{\lambda^2(\kappa\tau' - \kappa'\tau) - \lambda\tau'}{H}\kappa_n - \frac{\tau^2}{H^2} = 0$$

or $$H^4\kappa_n^2 - [\lambda^2(\kappa\tau' - \kappa'\tau) - \lambda\tau']\,H\kappa_n - \tau^2 = 0$$

or $$[(1-\lambda\kappa)^2 + \lambda^2\tau^2]^2\,\kappa_n^2$$
$$-[\lambda^2(\kappa\tau' - \kappa'\tau) - \lambda\tau'][(1-\lambda\kappa)^2 + \lambda^2\tau^2]^{1/2}\,\kappa_n - \tau^2 = 0. \tag{3}$$

Equation (3) gives the principal curvatures of the given surface.

Also, the differential equation of the lines of curvature is

$$(EM - FL)\,d\lambda^2 + (EN - GL)\,d\lambda\,ds + (FN - GM)\,ds^2 = 0$$

or $$-\frac{\tau}{H}d\lambda^2 + \frac{\lambda^2(\kappa\tau' - \kappa'\tau) - \lambda\tau'}{H}\,d\lambda\,ds$$
$$+[(1-\lambda\kappa)^2 + \lambda^2\tau^2]\frac{\tau}{H}ds^2 = 0$$

or $$\tau\,d\lambda^2 - [\lambda^2(\kappa\tau' - \kappa'\tau) - \lambda\tau']\,d\lambda\,ds$$
$$-[(1-\lambda\kappa)^2 + \lambda^2\tau^2]\,\tau\,ds^2 = 0. \tag{4}$$

Equation (4) is the required differential equation of the lines of curvature.

Example 27. (b) A ruled surface is generated by the principal normals to a given curve. Prove that the principal radii at the point of a principal normal distant r from the curve are given by

$$\frac{R^2}{\sigma^2} + R\left\{\left(1-\frac{r}{\rho}\right)^2 + \frac{r^2}{\sigma^2}\right\}^{1/2}\left\{\frac{r\sigma'}{\sigma^2} - \frac{r^2}{\rho\sigma}\left(\frac{\sigma'}{\sigma} - \frac{\rho'}{\rho}\right)\right\}$$
$$-\left\{\left(1-\frac{r}{\rho}\right)^2 + \frac{r^2}{\sigma^2}\right\}^2 = 0.$$

Find also the principal radii at the points of the curve.

Solution. This is another form of Example 27 (a).

Here put

$$\kappa_n = \frac{1}{R},\ \lambda = r,\ \tau = \frac{1}{\sigma},\ \tau' = -\frac{\sigma'}{\sigma^2},\ \kappa = \frac{1}{\rho},\ \kappa' = -\frac{\rho'}{\rho^2},$$

$$H^2 = \left(1-\frac{r}{\rho}\right)^2 + \frac{r^2}{\sigma^2}.$$

Then, using (3) of Example 27 (a), we get the required result.

In order to find the principal radii at the points of the curve, we put $r = 0$. Then, we get

$$\frac{R^2}{\sigma^2} - 1 = 0.$$

$$\therefore \qquad R = \pm\sigma.$$

Example 28. (a) Find the principal curvature and lines of curvature for the surface generated by the binormals to a twisted curve.

Solution. We know that the equation giving the principal curvatures κ_n is

$$H^2\kappa_n^{\ 2} - (EN - 2FM + GL)\,\kappa_n + LN - M^2 = 0. \tag{1}$$

We shall find the values of E, F, G, L, M, N for the given surface.

Let $\mathbf{r} = \mathbf{r}(s)$ be a twisted curve and let $\mathbf{R}$ be the position vector of any current point on the surface generated by the binormals to the given curve. Then

$$\mathbf{R} = \mathbf{r}(s) + \lambda\mathbf{b}(s), \tag{2}$$

where λ, s are the two independent parameters.

Let the suffixes 1 and 2 denote partial derivatives with respect to λ and s respectively. Then, from (2), we have

$$\mathbf{R}_1 = \frac{\partial\mathbf{R}}{\partial\lambda} = \mathbf{b},\ \mathbf{R}_2 = \frac{\partial\mathbf{R}}{\partial s} = \mathbf{t} + \lambda(-\tau\mathbf{n}) = \mathbf{t} - \lambda\tau\mathbf{n},$$

$$\mathbf{R}_{11} = \mathbf{0},\ \mathbf{R}_{12} = -\tau\mathbf{n},$$

$$\begin{aligned}\mathbf{R}_{22} &= \kappa\mathbf{n} - \lambda[\tau'\mathbf{n} + \tau(\tau\mathbf{b} - \kappa\mathbf{t})]\\ &= \lambda\kappa\tau\mathbf{t} + (\kappa - \lambda\tau')\,\mathbf{n} - \lambda\tau^2\mathbf{b}.\end{aligned}$$

Therefore

$$E = \mathbf{R}_1^{\ 2} = 1,\ F = \mathbf{R}_1 . \mathbf{R}_2 = 0,\ G = \mathbf{R}_2^{\ 2} = 1 + \lambda^2\tau^2,$$

$$H^2 = EG - F^2 = 1 + \lambda^2\tau^2.$$

Again, we have

$$\mathbf{R}_1 \times \mathbf{R}_2 = \lambda\tau\mathbf{t} + \mathbf{n}$$

$$\Rightarrow \qquad \mathbf{N} = \frac{\mathbf{R}_1 \times \mathbf{R}_2}{H} = \frac{\lambda\tau\mathbf{t} + \mathbf{n}}{H}.$$

Therefore

$$L = \mathbf{R}_{11} . \mathbf{N} = 0,\ M = \mathbf{R}_{12} . \mathbf{N} = -\frac{\tau}{H},$$

$$N = \mathbf{R}_{22} . \mathbf{N} = \frac{\kappa - \lambda\tau' + \kappa\lambda^2\tau^2}{H}.$$

Substituting the above values in (1), we get

$$H^2\kappa_n^{\ 2} - \frac{\kappa - \lambda\tau' + \kappa\lambda^2\tau^2}{H}\kappa_n - \frac{\tau^2}{H^2} = 0$$

or $\quad H^4\kappa_n^{\ 2}-(\kappa-\lambda\tau'+\kappa\lambda^2\tau^2)\,H\kappa_n-\tau^2=0$

or $\quad (1+\lambda^2\tau^2)^2\kappa_n^{\ 2}-(\kappa-\lambda\tau'+\kappa\lambda^2\tau^2)\,(1+\lambda^2\tau^2)^{1/2}\kappa_n-\tau^2=0. \qquad (3)$

Equation (3) gives the principal curvatures of the given surface.

Also, we know that the differential equation of the lines of curvature is

$$(EM-FL)\,d\lambda^2+(EN-GL)\,d\lambda\,ds+(FN-GM)\,ds^2=0$$

or $$-\frac{\tau}{H}d\lambda^2+\frac{\kappa-\lambda\tau'+\kappa\lambda^2\tau^2}{H}d\lambda\,ds+(1+\lambda^2\tau^2)\frac{\tau}{H}ds^2=0$$

or $\quad \tau\,d\lambda^2-(\kappa-\lambda\tau'+\kappa\lambda^2\tau^2)\,d\lambda\,ds-(1+\lambda^2\tau^2)\,\tau\,ds^2=0. \qquad (4)$

Equation (4) is the differential equation of the lines of curvature.

Example 28. (b) A ruled surface is generated by binormals to a given curve. Show that the principal radii at a point distant r from the curve are given by

$$\frac{R^2}{\sigma^2}+R\left(1+\frac{r^2}{\sigma^2}\right)^{1/2}\left[\frac{1}{\rho}\left(1+\frac{r^2}{\sigma^2}\right)+\frac{r\sigma'}{\sigma^2}\right]-\left(1+\frac{r^2}{\sigma^2}\right)=0.$$

Solution. This is another form of Example 28 (a).

Here put

$$\kappa_n=\frac{1}{R},\ \lambda=r,\ \tau=\frac{1}{\sigma},\ \tau'=-\frac{\sigma'}{\sigma^2},\ \kappa=\frac{1}{\rho},\ \kappa'=-\frac{\rho'}{\rho^2},$$

$$H^2=1+\frac{r^2}{\sigma^2}.$$

Then, using (3) of Example 28 (a), we get the required result.

Exercises

1. Find the value of (i) Gaussian curvature, (ii) Mean curvature at any point of the right helicoid $\mathbf{r}=(u\cos v,u\sin v,av)$.

   Hence show that a right helicoid is a minimal surface.

   $$\left[\textbf{Ans.}\ \text{(i)}-\frac{a^2}{(u^2+a^2)^2}\quad \text{(ii)}\ 0.\right]$$

2. Show that the meridians and parallels of a surface of revolution are its lines of curvature.

3. Find the equation for the principal curvatures and the differential equation of the lines of curvature for the surface

$z = c \tan^{-1}(y/x).$

$$\left[\begin{array}{l}\textbf{Ans. } \kappa = \pm \dfrac{c}{\sqrt{(x^2+y^2+c^2)}},\ d\phi = \pm \dfrac{du}{\sqrt{(u^2+c^2)}}, \\ \text{where } x = u\cos\phi, y = u\sin\phi.\end{array}\right]$$

4. For the surface

$$x = u\cos\theta,\ y = u\sin\theta,\ z = c\log\left\{u + \sqrt{(u^2 - c^2)}\right\},$$

prove that $\rho_1 = -\rho_2$.

5. Show that the surface $e^x \cos x = \cos y$ is minimal.

6. Find the principal radii of the surface $a^2x^2 = z^2(x^2+y^2)$ at the points where $x = y = z$.

$$\left[\textbf{Ans. } \frac{a\sqrt{3}}{2}(1 \pm \sqrt{7}).\right]$$

7. Find the Gaussian curvature at the point (u, v) of the anchor ring

$$x = (b + a\cos u)\cos v,\ y = (b + a\cos u)\sin v, z = a\sin u,$$

where the domain of u, v is $0 < u < 2\pi$, $0 < v < 2\pi$. Verify that the total curvature of the whole surface is zero.

$$\left[\textbf{Ans. } K = \frac{\cos u}{a(b + a\cos u)}.\right]$$

8. If l_1, m_1, n_1 are the direction cosines of the tangent to a line of curvature and l, m, n are the direction cosines of the normal to the surface at the point, then prove that

$$\frac{dl}{l_1} = \frac{dm}{m_1} = \frac{dn}{n_1}.$$

9. Show that the value of the Gaussian curvature is independent of the particular parametric representation chosen.

10. Show that the umbilics of the surface

$$\left(\frac{x}{a}\right)^{2/3} + \left(\frac{y}{b}\right)^{2/3} + \left(\frac{z}{c}\right)^{2/3} = 1$$

lie on a sphere whose centre is the origin and whose radius is equal to $abc/(ab + bc + ca)$.

11. Show that the points of intersection of the surface

$$x^m + y^m + z^m = a^m$$

and the line $x = y = m$ are umbilics and the curvature at an umbilic is given by

$$\rho = \frac{a}{m-1}(3)^{(m-2)/2m}.$$

12. Show that at any point P, there are two directions on the surface, at right angles to each other, such that the normal at a consecutive point in either of these directions meets the normal at P.

4.7 CONJUGATE DIRECTIONS

Definition. Let P be any point on a surface. Let Q be a point on the surface adjacent to P and let PR be a direction at P parallel to the line of intersection of the tangent planes at P and Q. Then as Q tends to coincide with P, the limiting directions of the lines PQ and PR are said to be conjugate directions at P.

Analytical Expression for Two Directions to be Conjugate. Let the equation of a surface be

$$\mathbf{r} = \mathbf{r}\,(u, v). \tag{1}$$

Let $P\,(u, v)$ be any point on this surface. Let $\mathbf{r}$ be the position vector of P and let $\mathbf{N}$ be the unit normal to the surface at P.

Let $Q\,(u + du, v + dv)$ be a point in the neighbourhood of P in the direction (du, dv) and let $\mathbf{N} + d\mathbf{N}$ be the unit normal to the surface at Q.

Again, let $R\,(u + \delta u, v + \delta v)$ be a point adjacent to P in the direction $(\delta u, \delta v)$ and let $\mathbf{r} + \delta\mathbf{r}$ be the position vector of R. Then

$$\overrightarrow{PR} = (\mathbf{r} + \delta\mathbf{r}) - \mathbf{r} = \delta\mathbf{r}.$$

Let the direction PR, i.e., $(\delta u, \delta v)$ be parallel to the line of intersection of the tangent planes at P and Q. Then $\overrightarrow{PR} = \delta\mathbf{r}$ is parallel to each of the tangent planes at P and Q, i.e., $\delta\mathbf{r}$ is perpendicular to each of $\mathbf{N}$ and $\mathbf{N} + d\mathbf{N}$. Hence

$$\mathbf{N}\,.\,\delta\mathbf{r} = 0, \quad (\mathbf{N} + d\mathbf{N})\,.\,\delta\mathbf{r} = 0$$

$$\Rightarrow \quad d\mathbf{N}\,.\,\delta\mathbf{r} = 0$$

$$\Rightarrow \quad \left(\frac{\partial \mathbf{N}}{\partial u} du + \frac{\partial \mathbf{N}}{\partial v} dv\right) . \left(\frac{\partial \mathbf{r}}{\partial u} \delta u + \frac{\partial \mathbf{r}}{\partial v} \delta v\right) = 0$$

$$\Rightarrow \quad (\mathbf{N}_1 du + \mathbf{N}_2 dv)\,.\,(\mathbf{r}_1\,\delta u + \mathbf{r}_2\,\delta v) = 0$$

$$\Rightarrow \quad \mathbf{N}_1 . \mathbf{r}_1\, du\,\delta u + \mathbf{N}_1 . \mathbf{r}_2\, du\,\delta v + \mathbf{N}_2 . \mathbf{r}_1\, dv\,\delta u + \mathbf{N}_2 . \mathbf{r}_2\, dv\,\delta v = 0$$

$$\Rightarrow \qquad L\,du\,\delta u + M(du\,\delta v + dv\,\delta u) + N\,dv\,\delta v = 0$$

$$[\because \mathbf{N}_1 . \mathbf{r}_1 = -L,\ \mathbf{N}_1 . \mathbf{r}_2 = \mathbf{N}_2 . \mathbf{r}_1 = -M,\ \mathbf{N}_2 . \mathbf{r}_2 = -N\,]$$

$$\Rightarrow \qquad L\frac{du}{dv}\frac{\delta u}{\delta v} + M\left(\frac{du}{dv} + \frac{\delta u}{\delta v}\right) + N = 0. \tag{2}$$

This is the required condition that the directions (du, dv) and $(\delta u, \delta v)$ at point P of the given surface (1) are conjugate.

The symmetry of the relation (2) shows that if the direction $\delta u/\delta v$ is conjugate to the direction du/dv, then the direction du/dv is conjugate to the direction $\delta u/\delta v$, i.e., this property is a reciprocal one.

Moreover, equation (2) is linear in each of the ratios $du : dv$ and $\delta u : \delta v$ which implies that there is one and only one conjugate direction to a given direction.

Theorem 1. Condition for two directions to be Conjugate.

The two directions given by

$$P\,du^2 + Q\,du\,dv + R\,dv^2 = 0$$

are conjugate if

$$LR - MQ + NP = 0.$$

Proof. We know that the two directions (du, dv) and $(\delta u, \delta v)$ are conjugate if

$$L\frac{du}{dv}\frac{\delta u}{\delta v} + M\left(\frac{du}{dv} + \frac{\delta u}{\delta v}\right) + N = 0. \tag{1}$$

The given directions are

$$P\,du^2 + Q\,du\,dv + R\,dv^2 = 0$$

i.e.,
$$P\left(\frac{du}{dv}\right)^2 + Q\frac{du}{dv} + R = 0. \tag{2}$$

Equation (2) is a quadratic equation in du/dv. Therefore it has two roots corresponding to the two directions. If the two roots of (2) are du/dv and $\delta u/\delta v$, then

$$\frac{du}{dv} + \frac{\delta u}{\delta v} = -\frac{Q}{P},\quad \frac{du}{dv}\frac{\delta u}{\delta v} = \frac{R}{P}.$$

Substituting the above values in (1), we get

$$L\left(\frac{R}{P}\right) + M\left(-\frac{Q}{P}\right) + N = 0$$

or $\quad LR - MQ + NP = 0.$

Hence the result.

Theorem 2. Condition for Parametric Curves to have Conjugate Directions.

The necessary and sufficient condition that the parametric curves have conjugate directions is that $M = 0$.

Proof. We know that the condition for the two directions given by

$$P\,du^2 + Q\,du\,dv + R\,dv^2 = 0 \tag{1}$$

to be conjugate is

$$LR - MQ + NP = 0. \tag{2}$$

Also, the differential equation of the parametric curves on the surface is given by

$$du\,dv = 0. \tag{3}$$

The condition is necessary. Suppose that the parametric curves have conjugate directions, i.e., condition (2) is satisfied.

To prove : $M = 0$.

Comparing (3) with (1), we get

$$P = 0,\ Q = 1,\ R = 0.$$

Substituting the values of P, Q, R in (2), we get

$$M = 0.$$

Hence the condition is necessary.

The condition is sufficient. Now, we suppose that

$$M = 0. \tag{4}$$

To prove : The parametric curves have conjugate directions.

Since $P = 0,\ Q = 1,\ R = 0$ for the parametric curves and $M = 0$ [by (4)], therefore condition (2) is satisfied, i.e., we have

$$LR - MQ + NP = 0.$$

Hence the parametric curves have conjugate directions.

Theorem 3. The principal directions at a point are orthogonal conjugate.

Proof. Let the equation of a surface be given by

$$\mathbf{r} = \mathbf{r}\,(u, v),$$

where (u, v) are the parametric coordinates of a point on the surface. Then the parametric curves on the surface are given by

$$v = \text{constant},\ \ u = \text{constant}.$$

Let us take the parametric curves on the surface as the lines of curvature. Then, from the necessary and sufficient condition for the parametric curves to be lines of curvature, we have

$$F = 0, M = 0.$$

But $F = 0$ is the necessary and sufficient condition for the parametric curves to be orthogonal and $M = 0$ is the necessary and sufficient condition for the parametric curves to have conjugate directions.

Hence it follows that the principal directions at a point are orthogonal conjugate.

Theorem 4. A Characteristic Property of Conjugate Directions.

Two directions at a point on a surface are conjugate if they are parallel to the conjugate diameters of the indicatrix at that point.

Proof. Let the equation of a surface be given by

$$\mathbf{r} = \mathbf{r}(u, v). \tag{1}$$

The parametric curves on the surface are given by

$$v = \text{constant}, \quad u = \text{constant}.$$

Let us take the parametric curves on the surface as the lines of curvature. Then, from the necessary and sufficient condition for the parametric curves to be lines of curvature, we have

$$F = 0, M = 0. \tag{2}$$

Let (du, dv) and (δu, δv) be any two directions on the surface (1) which are conjugate. Then, we know that

$$L\frac{du}{dv}\frac{\delta u}{\delta v} + M\left(\frac{du}{dv} + \frac{\delta u}{\delta v}\right) + N = 0$$

or $$L\frac{du}{dv}\frac{\delta u}{\delta v} + N = 0 \qquad \text{[using (2)]}$$

or $$\frac{dv}{du}\frac{\delta v}{\delta u} = -\frac{L}{N}. \tag{3}$$

Let κ_n be the normal curvature at a point P (u, v) on this surface in the direction (du, dv). Also, let κ_a and κ_b be the principal curvatures at P (u, v).

We know that

$$\kappa_n = \frac{L\,du^2 + 2M\,du\,dv + N\,dv^2}{E\,du^2 + 2F\,du\,dv + G\,dv^2}$$

or $$\kappa_n = \frac{L\,du^2 + N\,dv^2}{E\,du^2 + G\,dv^2}. \tag{4}$$

[using (2)]

The principal curvature κ_a, being the normal curvature for the direction $dv = 0$, is expressed as

$$\kappa_a = \frac{L}{E}. \tag{5}$$

[using (4)]

Similarly, the principal curvature κ_b, being the normal curvature for the direction $du = 0$, is expressed as

$$\kappa_b = \frac{N}{G}. \tag{6}$$

[using (4)]

Now, if we take the point P as origin, x-axis along the principal direction $v =$ const. at P, y-axis along the principal direction $u =$ const. at P and z-axis along the surface normal at P, then we know that the Dupin's indicatrix at P is given by the equations

$$z = h,\ \kappa_a x^2 + \kappa_b y^2 = 2h$$

or

$$z = h,\ \frac{x^2}{R_a} + \frac{y^2}{R_b} = 2h, \tag{7}$$

where

$$R_a = \frac{1}{\kappa_a},\ \ R_b = \frac{1}{\kappa_b}.$$

Let the directions (du, dv) and $(\delta u, \delta v)$ at P be inclined to the parametric curve $v =$ const. at angles θ and θ' respectively. Then

$$\begin{aligned}\tan\theta &= \frac{H\,|\,du\,\delta v - dv\,\delta u\,|}{E\,du\,\delta u + F(du\,\delta v + dv\,\delta u) + G\,dv\,\delta v} \\ &= \frac{H}{E}\frac{dv}{du} \\ &= \frac{\sqrt{EG - F^2}}{E}\frac{dv}{du} \\ &= \sqrt{\frac{G}{E}}\frac{dv}{du}. \qquad [\because F = 0, \text{ by } (3)]\end{aligned}$$

$$\therefore \quad \frac{dv}{du} = \sqrt{\frac{E}{G}}\tan\theta. \tag{8}$$

Similarly,

$$\frac{\delta v}{\delta u} = \sqrt{\frac{E}{G}}\tan\theta'. \tag{9}$$

From (8) and (9), we have

$$\frac{dv}{du}\frac{\delta v}{\delta u}=\frac{E}{G}\tan\theta\tan\theta'. \tag{10}$$

Now, from (3) and (10), we have

$$-\frac{L}{N}=\frac{E}{G}\tan\theta\tan\theta'.$$

$$\therefore \qquad \tan\theta\tan\theta'=-\frac{L/E}{N/G}=-\frac{\kappa_a}{\kappa_b}=-\frac{R_b}{R_a}. \tag{11}$$

But, we know that if m_1, m_2 be the gradients of a pair of conjugate diameters of the ellipse $\frac{x^2}{a^2}+\frac{y^2}{b^2}=1$, then $m_1 m_2=-\frac{b^2}{a^2}$. Hence it follows from (11) that θ, θ' are inclinations of the conjugate diameters of the indicatrix (7) at P to the positive direction of x-axis.

Therefore the directions (du, dv) and $(\delta u, \delta v)$ at P are parallel to the conjugate diameters of the indicatrix at P.

4.8 CONJUGATE SYSTEMS

Definition. Let the equation of a surface be

$$\mathbf{r}=\mathbf{r}(u,v) \tag{1}$$

and let a family of curves on this surface be

$$\phi(u,v)=\text{constant}. \tag{2}$$

From (2), we have

$$\delta\phi=0$$

$$\Rightarrow \qquad \phi_1\,\delta u+\phi_2\,\delta v=0$$

$$\Rightarrow \qquad \frac{\delta u}{\delta v}=-\frac{\phi_2}{\phi_1}.$$

Therefore the direction $\delta u/\delta v$ of a member of the family (2) at the point (u, v) is given by

$$\frac{\delta u}{\delta v}=-\frac{\phi_2}{\phi_1}. \tag{3}$$

Let du/dv denote the direction conjugate to the direction $\delta u/\delta v$ at the point (u, v) on the surface (1). Then

$$L\frac{du}{dv}\frac{\delta u}{\delta v}+M\left(\frac{du}{dv}+\frac{\delta u}{\delta v}\right)+N=0$$

or $$L\frac{du}{dv}\left(-\frac{\phi_2}{\phi_1}\right)+M\left(\frac{du}{dv}-\frac{\phi_2}{\phi_1}\right)+N=0 \qquad \text{[using (3)]}$$

or $$\left(M - L\frac{\phi_2}{\phi_1}\right)\frac{du}{dv} + \left(N - M\frac{\phi_2}{\phi_1}\right) = 0$$

or $$(M\phi_1 - L\phi_2)\,du + (N\phi_1 - M\phi_2)\,dv = 0. \tag{4}$$

Equation (4) is a differential equation of first order and first degree. Therefore its solution

$$\psi\,(u, v) = \text{constant} \tag{5}$$

represents a one parameter family of curves.

The family of curves given by (2) and (5) are known as a pair of conjugate systems. At a point of intersection of two curves, one from each family, their directions are conjugate.

Theorem 1. Condition for two Families of Curves to form a Conjugate System.

The two families of curves given by

$$\phi\,(u, v) = \text{constant},$$
$$\psi\,(u, v) = \text{constant}$$

form a conjugate system if

$$L\phi_2\psi_2 - M(\phi_1\psi_2 + \phi_2\psi_1) + N\phi_1\psi_1 = 0.$$

Proof. We know that the two directions (δu, δv) and (du, dv) are conjugate if

$$L\,\frac{\delta u}{\delta v}\frac{du}{dv} + M\left(\frac{\delta u}{\delta v} + \frac{du}{dv}\right) + N = 0. \tag{1}$$

The given families of curves are

$$\phi\,(u, v) = \text{constant},$$
$$\psi\,(u, v) = \text{constant}.$$

The directions of the two curves, one from the first and the other from the second family through a point (u, v), are given by

$$\phi_1\,\delta u + \phi_2\,\delta v = 0,$$
$$\psi_1\,du + \psi_2\,dv = 0$$

i.e., $$\frac{\delta u}{\delta v} = -\frac{\phi_2}{\phi_1},\quad \frac{du}{dv} = -\frac{\psi_2}{\psi_1}.$$

$\therefore$ $$\frac{\delta u}{\delta v} + \frac{du}{dv} = -\frac{\phi_1\psi_2 + \phi_2\psi_1}{\phi_1\psi_1},\quad \frac{\delta u}{\delta v}\frac{du}{dv} = \frac{\phi_2\psi_2}{\phi_1\psi_1}.$$

Substituting the above values in (1), we get

$$L\left(\frac{\phi_2\psi_2}{\phi_1\psi_1}\right) + M\left(-\frac{\phi_1\psi_2 + \phi_2\psi_1}{\phi_1\psi_1}\right) + N = 0$$

or $\quad L\phi_2\psi_2 - M(\phi_1\psi_2 + \phi_2\psi_1) + N\phi_1\psi_1 = 0.$

Hence the result.

Theorem 2. Condition for Parametric Curves to form Conjugate Systems.

The necessary and sufficient condition that the parametric curves form conjugate systems is that $M = 0$.

Proof. We know that the two directions (δu, δv) and (du, dv) are conjugate if

$$L\,\delta u\,du + M(\delta u\,dv + \delta v\,du) + N\,\delta v\,dv = 0. \tag{1}$$

The directions of the parametric curves

$$v = \text{ constant}, \quad u = \text{ constant}$$

through a point (u, v) are given by

$$\delta v = 0, \; du = 0. \tag{2}$$

The condition is necessary. Suppose that the parametric curves form conjugate systems.

To prove : $M = 0$.

Since the parametric curves form conjugate systems, the directions of the parametric curves given by (2) satisfy condition (1). Substituting the values of δv and du from (2) in (1), we get

$$M = 0.$$

Hence the condition is necessary.

The condition is sufficient. Now, we suppose that

$$M = 0. \tag{3}$$

To prove : The parametric curves form conjugate systems.

Since $\delta v = 0, du = 0$ for the parametric curves and $M = 0$ [by (3)], therefore condition (1) is satisfied, i.e., we have

$$L\,\delta u\,du + M(\delta u\,dv + \delta v\,du) + N\,\delta v\,dv = 0.$$

Hence the parametric curves form conjugate systems.

Theorem 3. The lines of curvature form an orthogonal conjugate system.

Proof. Let the equation of a surface be given by

$$\mathbf{r} = \mathbf{r}\,(u, v),$$

where (u, v) are the parametric coordinates of a point on the surface.

Then the parametric curves on the surface are given by

$$v = \text{constant}, \quad u = \text{constant}.$$

Let us take the parametric curves on the surface as the lines of curvature. Then, from the necessary and sufficient condition for the parametric curves to be lines of curvature, we have

$$F = 0, M = 0.$$

But $F = 0$ is the necessary and sufficient condition for the parametric curves to be orthogonal and $M = 0$ is the necessary and sufficient condition for the parametric curves to form conjugate system.

Hence it follows that the lines of curvature form an orthogonal conjugate system.

4.9 ASYMPTOTIC LINES

Definition. A direction at a point on the surface is called asymptotic if it is self-conjugate. An asymptotic line is a curve whose direction at each point is self-conjugate.

Differential Equation of Asymptotic Lines

Let the equation of a surface be given by

$$\mathbf{r} = \mathbf{r}(u, v). \tag{1}$$

Let C denote an asymptotic line on the surface (1) and let P (u, v) be any point on C. Let (du, dv) denote the direction at P along C. Then (du, dv) is an asymptotic direction, i.e., the direction (du, dv) is self-conjugate.

We know that if the two directions (du, dv) and $(\delta u, \delta v)$ are conjugate, then

$$L\frac{du}{dv}\frac{\delta u}{\delta v} + M\left(\frac{du}{dv} + \frac{\delta u}{\delta v}\right) + N = 0. \tag{2}$$

Hence, for the self-conjugate direction (du, dv), we replace $\delta u/\delta v$ by du/dv in (2). Then, we get

$$L\left(\frac{du}{dv}\right)^2 + 2M\frac{du}{dv} + N = 0 \tag{3}$$

or $$L\,du^2 + 2M\,du\,dv + N\,dv^2 = 0 \tag{3'}$$

Equation (3) or (3') is the differential equation of the asymptotic lines. Since (3) is a quadratic equation in du/dv, it has two roots corresponding to the two directions. Hence it follows that there are two asymptotic directions through each point of the surface.

The two asymptotic directions given by (3) are real and different when $M^2 - LN$ is positive, i.e., when the Gaussian curvature K is negative and the point is hyperbolic. They are imaginary when K is positive and the point is elliptic. They are identical when K is zero and the point is parabolic. In the last case, the surface is developable, and the single asymptotic line through a point is the generator.

Remark. The differential equation of the asymptotic lines can also be written in an alternative form as follows.

Let the equation of an asymptotic line on the surface (1) be

$$\mathbf{r} = \mathbf{r}\,(s). \tag{4}$$

Let P (u, v) be any point on this asymptotic line and let $\mathbf{N}$ be the unit normal to the surface (1) at P. Then

$$\mathbf{N} = \mathbf{N}\,(s).$$

$$\therefore \qquad \frac{d\mathbf{N}}{ds} = \frac{\partial \mathbf{N}}{\partial u}\frac{du}{ds} + \frac{\partial \mathbf{N}}{\partial v}\frac{dv}{ds} = \mathbf{N}_1 u' + \mathbf{N}_2 v'. \tag{5}$$

Similarly,

$$\frac{d\mathbf{r}}{ds} = \frac{\partial \mathbf{r}}{\partial u}\frac{du}{ds} + \frac{\partial \mathbf{r}}{\partial v}\frac{dv}{ds} = \mathbf{r}_1 u' + \mathbf{r}_2 v'. \tag{6}$$

From (5) and (6), we have

$$\begin{aligned}\frac{d\mathbf{N}}{ds}\cdot\frac{d\mathbf{r}}{ds} &= (\mathbf{N}_1 u' + \mathbf{N}_2 v')\cdot(\mathbf{r}_1 u' + \mathbf{r}_2 v') \\ &= (\mathbf{N}_1\cdot\mathbf{r}_1)\,u'^2 + (\mathbf{N}_1\cdot\mathbf{r}_2 + \mathbf{N}_2\cdot\mathbf{r}_1)\,u'\,v' + (\mathbf{N}_2\cdot\mathbf{r}_2)\,v'^2 \\ &= -\,L\left(\frac{du}{ds}\right)^2 - 2M\left(\frac{du}{ds}\right)\left(\frac{dv}{ds}\right) - N\left(\frac{dv}{ds}\right)^2\end{aligned}$$

i.e.,

$$\frac{d\mathbf{N}}{ds}\cdot\frac{d\mathbf{r}}{ds} = -\frac{L\,du^2 + 2M\,du\,dv + N\,dv^2}{ds^2}$$

or

$$\frac{d\mathbf{N}}{ds}\cdot\frac{d\mathbf{r}}{ds} = 0. \tag{7}$$

[using (3')]

Equation (7) gives an alternative form of the differential equation of the asymptotic lines.

Theorem 1. Condition for Asymptotic Lines to be Orthogonal.

The asymptotic lines are orthogonal when the surface is a minimal surface.

Proof. We know that the differential equation of the asymptotic lines is given by

$$L\,du^2 + 2M\,du\,dv + N\,dv^2 = 0. \tag{1}$$

Also, we know that the two directions given by

$$P\,du^2 + 2Q\,du\,dv + R\,dv^2 = 0 \tag{2}$$

are orthogonal if

$$ER - 2FQ + GP = 0. \tag{3}$$

Comparing (1) and (2), we get

$$P = L,\ Q = M,\ R = N.$$

Hence, by (3), the asymptotic lines given by (1) are orthogonal if at each point of the surface,

$$EN - 2FM + GL = 0. \tag{4}$$

But, (4) is the condition for a surface to be minimal. Hence the surface is a minimal surface.

Theorem 2. The osculating plane at any point of an asymptotic line is the tangent plane to the surface.

Proof. Let the equation of a surface be given by

$$\mathbf{r} = \mathbf{r}\,(u, v) \tag{1}$$

and let the equation of an asymptotic line on this surface be given by

$$\mathbf{r} = \mathbf{r}\,(s). \tag{2}$$

Let $P\,(u, v)$ be any point on this asymptotic line and let $\mathbf{N}$ be the unit normal to the surface (1) at P. Let **t**, **n**, **b** be unit tangent, unit principal normal and unit binormal respectively to (2) at P, and let κ and τ denote the curvature and torsion of (2) at P.

Since the tangent **t** to the asymptotic line (2) is tangential to the surface (1), therefore

$$\mathbf{N} \text{ is perpendicular to } \mathbf{t}. \tag{3}$$

$$\Rightarrow \quad \mathbf{N}\,.\,\mathbf{t} = 0$$

$$\Rightarrow \quad \frac{d}{ds}(\mathbf{N}\,.\,\mathbf{t}) = 0$$

$$\Rightarrow \quad \frac{d\mathbf{N}}{ds}\,.\,\mathbf{t} + \mathbf{N}\,.\,\frac{d\mathbf{t}}{ds} = 0$$

$$\Rightarrow \quad \frac{d\mathbf{N}}{ds}\,.\,\frac{d\mathbf{r}}{ds} + \mathbf{N}\,.\,(\kappa\mathbf{n}) = 0$$

$$\Rightarrow \quad \mathbf{N}\,.\,\mathbf{n} = 0 \qquad \left[\because \frac{d\mathbf{N}}{ds}\,.\,\frac{d\mathbf{r}}{ds} = 0, \kappa \neq 0\right]$$

$$\Rightarrow \quad \mathbf{N} \text{ is perpendicular to } \mathbf{n}. \tag{4}$$

From (3) and (4), it follows that $\mathbf{N}$ is perpendicular to each of **t** and **n**. Therefore $\mathbf{N}$ is parallel to $\mathbf{t} \times \mathbf{n} = \mathbf{b}$. Hence, we can take

$$\mathbf{N} = \pm\,\mathbf{b}$$

i.e., at any point of a curved asymptotic line, the binomial is also normal to the surface.

For convenience, we take

$$\mathbf{N} = \mathbf{b}. \tag{5}$$

Now, the equation of the osculating plane to curve (2) at any point **r** on it is

$$(\mathbf{R}-\mathbf{r}) \cdot \mathbf{b}=0 \tag{6}$$

and the equation of the tangent plane to surface (1) at this point is

$$(\mathbf{R}-\mathbf{r}) \cdot \mathbf{N}=0. \tag{7}$$

It follows from (5), (6) and (7) that the osculating plane to curve (2) at any point **r** on it and the tangent plane to surface (1) at this point are coincident.

Hence the theorem is proved.

Theorem 3. Condition for Parametric Curves to be Asymptotic Lines.

The necessary and sufficient conditions that the parametric curves be asymptotic lines are $L=0, N=0, M \neq 0$.

Proof. We know that the differential equation of the asymptotic lines is given by

$$L\,du^2+2M\,du\,dv+N\,dv^2=0. \tag{1}$$

Also, the differential equation of the parametric curves on the surface is given by

$$du\,dv=0. \tag{2}$$

The condition is necessary. Suppose that the parametric curves are asymptotic lines.

To prove : $L=0, N=0, M \neq 0$.

Since the parametric curves are asymptotic lines, therefore (1) and (2) are identical. Comparing (2) with (1), we get

$$L=0, N=0, M \neq 0.$$

The condition is sufficient. Now, we suppose that

$$L=0, N=0, M \neq 0. \tag{3}$$

To prove : The parametric curves are asymptotic lines.

Substituting $L=0, N=0, M \neq 0$ in (1), we get

$$du\,dv=0,$$

which is the differential equation of the parametric curves given by (2).

Hence (1) and (2) are identical, i.e., the asymptotic lines are the parametric curves.

Corollary 1. The differential equation of the lines of curvature, if the asymptotic lines be taken as the parametric curves, is

$$E\,du^2-G\,dv^2=0.$$

Proof. Let the equation of a surface be given by

$$\mathbf{r} = \mathbf{r}(u, v). \tag{1}$$

We know that the differential equation of the lines of curvature on the surface (1) is given by

$$(EM - FL)\, du^2 + (EN - GL)\, du\, dv + (FN - GM)\, dv^2 = 0. \tag{2}$$

Let the equation of an asymptotic line on the surface (1) be

$$\mathbf{r} = \mathbf{r}(s) \tag{3}$$

and let asymptotic lines be taken as the parametric curves. Then

$$L = 0, N = 0, M \neq 0. \tag{4}$$

Substituting the values of L and N from (4) in (2), we get

$$EM\, du^2 - GM\, dv^2 = 0$$

or $$M(E\, du^2 - G\, dv^2) = 0$$

or $$E\, du^2 - G\, dv^2 = 0. \tag{5}$$

$[\because M \neq 0]$

Corollary 2. The equation for the principal curvatures, if the asymptotic lines be taken as the parametric curves, is

$$(EG - F^2)\, \kappa^2 + 2FM\kappa - M^2 = 0,$$

and, in this case,

$$J = -\frac{2FM}{H^2}, \quad K = -\frac{M^2}{H^2}.$$

Proof. Let the equation of a surface be given by

$$\mathbf{r} = \mathbf{r}(u, v). \tag{1}$$

We know that the differential equation of the lines of curvature on the surface (1) is given by

$$(EG - F^2)\, \kappa^2 - (EN - 2FM + GL)\, \kappa + LN - M^2 = 0. \tag{2}$$

Let the equation of an asymptotic line on the surface (1) be

$$\mathbf{r} = \mathbf{r}(s) \tag{3}$$

and let asymptotic lines be taken as the parametric curves. Then

$$L = 0, N = 0, M \neq 0. \tag{4}$$

Substituting the values of L and N from (4) in (2), we get

$$(EG - F^2)\, \kappa^2 + 2FM\kappa - M^2 = 0. \tag{5}$$

$$\therefore \quad J = -\frac{2FM}{EG - F^2} = -\frac{2FM}{H^2}, \quad K = -\frac{M^2}{EG - F^2} = -\frac{M^2}{H^2}.$$

Curvature and Torsion of an Asymptotic Line.

Let the equation of a surface be given by

$$\mathbf{r} = \mathbf{r}(u, v) \tag{1}$$

and let the equation of an asymptotic line on this surface be given by

$$\mathbf{r} = \mathbf{r}(s). \tag{2}$$

Let P (u, v) be any point on this asymptotic line. Let **t**, **n**, **b** be unit tangent, unit principal normal and unit binormal respectively to (2) at P and and let **N** be the unit normal to the surface (1) at P. Also, let κ and τ denote the curvature and torsion of (2) at P.

We know that the osculating plane to the curve (2) at P is the same as the tangent plane to surface (1) P. Hence **N** is parallel to **b**. Therefore we can take

$$\mathbf{N} = \pm\,\mathbf{b}.$$

For convenience, we take

$$\mathbf{N} = \mathbf{b}. \tag{3}$$

Differentiating (3) with respect to s, we get

$$\mathbf{N}' = -\tau\mathbf{n}$$

$$\Rightarrow \quad \mathbf{N}'.\mathbf{n} = -\tau\mathbf{n}\,.\,\mathbf{n} = -\tau$$

$$\Rightarrow \quad \tau = -\mathbf{N}'.\mathbf{n}$$

$$= -\mathbf{N}'.(\mathbf{b}\times\mathbf{t}) \qquad [\because \mathbf{n} = \mathbf{b}\times\mathbf{t}]$$

$$= \mathbf{N}'.(\mathbf{t}\times\mathbf{b})$$

$$= \mathbf{N}'.(\mathbf{t}\times\mathbf{N}) \qquad [\text{using (3)}]$$

$$= [\mathbf{N}', \mathbf{t}, \mathbf{N}]$$

$$= [\mathbf{N}, \mathbf{N}', \mathbf{t}]$$

i.e., $$\tau = [\mathbf{N}, \mathbf{N}', \mathbf{r}']. \tag{4}$$

Again, we have

$$\mathbf{r}' = \mathbf{t}$$

$$\Rightarrow \quad \mathbf{r}'' = \kappa\mathbf{n}$$

$$\Rightarrow \quad \mathbf{r}''.\mathbf{n} = \kappa\mathbf{n}\,.\,\mathbf{n} = \kappa$$

$$\Rightarrow \quad \kappa = \mathbf{r}''.(\mathbf{b}\times\mathbf{t}) \qquad [\because \mathbf{n} = \mathbf{b}\times\mathbf{t}]$$

$$= \mathbf{r}''.(\mathbf{N}\times\mathbf{t}) \qquad [\text{using (3)}]$$

$$= [\mathbf{r}'', \mathbf{N}, \mathbf{t}]$$

$$= [\mathbf{N}, \mathbf{t}, \mathbf{r}'']$$

i.e., $$\kappa = [\mathbf{N}, \mathbf{r}', \mathbf{r}'']. \tag{5}$$

Equations (4) and (5) give the torsion and curvature of the asymptotic line (2) lying on the surface (1).

Theorem 4. Theorem of Beltrami and Enneper.

The torsions of the two asymptotic lines through a point are equal in magnitude and opposite in sign and the square of either is the negative of the Gaussian curvature.

Proof. Let the equation of a surface be given by

$$\mathbf{r} = \mathbf{r}\,(u, v) \tag{1}$$

and let the equation of an asymptotic line on this surface be given by

$$\mathbf{r} = \mathbf{r}(s). \tag{2}$$

Let $P(u, v)$ be any point on this asymptotic line whose position vector is $\mathbf{r}$. Let $\mathbf{N}$ be the unit normal to the surface (1) at P. Also, let τ denote the torsion of (2) at P and let K be the Gaussian curvature of the surface (1) at P.

We know that

$$\tau = [\mathbf{N}, \mathbf{N}', \mathbf{r}']. \tag{3}$$

Now,

$$\mathbf{N} = \mathbf{N}(s)$$

$$\Rightarrow \quad \mathbf{N}' = \frac{d\mathbf{N}}{ds} = \frac{\partial \mathbf{N}}{\partial u}\frac{du}{ds} + \frac{\partial \mathbf{N}}{\partial v}\frac{dv}{ds} = \mathbf{N}_1 u' + \mathbf{N}_2 v'. \tag{4}$$

Similarly,

$$\mathbf{r}' = \frac{d\mathbf{r}}{ds} = \frac{\partial \mathbf{r}}{\partial u}\frac{du}{ds} + \frac{\partial \mathbf{r}}{\partial v}\frac{dv}{ds} = \mathbf{r}_1 u' + \mathbf{r}_2 v'. \tag{5}$$

From (4) and (5), we have

$$\begin{aligned} \mathbf{N}' \times \mathbf{r}' &= (\mathbf{N}_1 u' + \mathbf{N}_2 v') \times (\mathbf{r}_1 u' + \mathbf{r}_2 v') \\ &= (\mathbf{N}_1 \times \mathbf{r}_1) u'^2 + (\mathbf{N}_1 \times \mathbf{r}_2 + \mathbf{N}_2 \times \mathbf{r}_1) u' v' + (\mathbf{N}_2 \times \mathbf{r}_2) v'^2. \end{aligned}$$

$$\begin{aligned} \therefore \quad \mathbf{N} \cdot (\mathbf{N}' \times \mathbf{r}') = \mathbf{N} \cdot (\mathbf{N}_1 \times \mathbf{r}_1) u'^2 &+ \mathbf{N} \cdot (\mathbf{N}_1 \times \mathbf{r}_2) u' v' \\ &+ \mathbf{N} \cdot (\mathbf{N}_2 \times \mathbf{r}_1) u' v' + \mathbf{N} \cdot (\mathbf{N}_2 \times \mathbf{r}_2) v'^2 \end{aligned}$$

i.e.,

$$\begin{aligned} [\mathbf{N}, \mathbf{N}', \mathbf{r}'] = [\mathbf{N}, \mathbf{N}_1, \mathbf{r}_1] u'^2 &+ [\mathbf{N}, \mathbf{N}_1, \mathbf{r}_2] u' v' \\ &+ [\mathbf{N}, \mathbf{N}_2, \mathbf{r}_1] u' v' + [\mathbf{N}, \mathbf{N}_2, \mathbf{r}_2] v'^2. \end{aligned} \tag{6}$$

By Weingarten equations, we have

$$\left.\begin{aligned} H[\mathbf{N}, \mathbf{N}_1, \mathbf{r}_1] &= EM - FL, \\ H[\mathbf{N}, \mathbf{N}_1, \mathbf{r}_2] &= FM - GL, \\ H[\mathbf{N}, \mathbf{N}_2, \mathbf{r}_1] &= EN - FM, \\ H[\mathbf{N}, \mathbf{N}_2, \mathbf{r}_2] &= FN - GM. \end{aligned}\right\} \tag{7}$$

From (3), (6) and (7), we have

$$\tau = \frac{EM - FL}{H} u'^2 + \frac{EN - GL}{H} u' v' + \frac{FN - GM}{H} v'^2. \tag{8}$$

Let us take the asymptotic lines as the parametric curves. Then, we know that

$$L = 0, N = 0, M \neq 0. \tag{9}$$

Substituting the values of L and N from (9) in (8), we get

$$\tau = \frac{EM}{H}u'^2 - \frac{GM}{H}v'^2$$

or $$\tau = \frac{M}{H}(Eu'^2 - Gv'^2)$$

or $$\tau = \frac{M}{H}\left[E\left(\frac{du}{ds}\right)^2 - G\left(\frac{dv}{ds}\right)^2\right]. \tag{10}$$

For the asymptotic line $v =$ constant, we have

$$dv = 0, \quad ds^2 = E\,du^2.$$

Therefore, by (10), we have

$$\tau = \frac{M}{H}. \tag{11}$$

For the asymptotic line $u =$ constant, we have

$$du = 0, \quad ds^2 = G\,dv^2.$$

Therefore, by (10), we have

$$\tau = -\frac{M}{H}. \tag{12}$$

From (11) and (12), we have

$$\tau = \pm\frac{M}{H}. \tag{13}$$

Therefore the torsions of the two asymptotic lines through P are equal in magnitude and opposite in sign.

The Gaussian curvature K of the surface (1) at P is given by

$$K = \frac{LN - M^2}{H^2} = -\frac{M^2}{H^2} \qquad \text{[using (9)]}$$

$$\Rightarrow \qquad \frac{M^2}{H^2} = -K$$

$$\Rightarrow \qquad (\pm\tau)^2 = -K \qquad \text{[using (13)]}$$

$$\Rightarrow \qquad \tau^2 = -K.$$

Therefore the square of the torsion of each asymptotic line through P is equal to the negative of the Gaussian curvature of the surface through P.

Solved Examples

Example 1. Show that the parametric curves are conjugate on the following surfaces :

(i) a surface of revolution

$$x = u\cos v,\ y = u\sin v,\ z = f(u);$$

(ii) the surface generated by the tangents to a curve, on which $\mathbf{R} = \mathbf{r} + u\mathbf{t}$ (u, s are parameters) ;

(iii) the surface $z = f(x) + F(y)$, where x, y are parameters.

Solution. (i) We know that the necessary and sufficient condition for the parametric curves to be conjugate is $M = 0$. Therefore we shall find the value of M for the given surface.

The position vector $\mathbf{r}$ of any current point on the given surface is

$$\mathbf{r} = (u\cos v, u\sin v, f(u)).$$

Taking u and v as the first and second parameters, we have

$$\mathbf{r}_1 = (\cos v, \sin v, f_1),\ \mathbf{r}_2 = (-u\sin v, u\cos v, 0),$$

$$\mathbf{r}_{12} = (-\sin v, \cos v, 0).$$

Therefore

$$\mathbf{r}_1 \times \mathbf{r}_2 = (-f_1 u\cos v, -f_1 u\sin v, u)$$

$$\Rightarrow \quad \mathbf{N} = \frac{\mathbf{r}_1 \times \mathbf{r}_2}{H} = \frac{1}{H}(-f_1 u\cos v, -f_1 u\sin v, u)$$

$$\Rightarrow \quad M = \mathbf{r}_{12} \cdot \mathbf{N} = 0.$$

Hence the parametric curves are conjugate on the given surface.

(ii) Let the equation of the space curve be

$$\mathbf{r} = \mathbf{r}(s).$$

Then the position vector $\mathbf{R}$ of any current point on the surface generated by the tangents to this curve can be written as

$$\mathbf{R} = \mathbf{r} + u\mathbf{t},$$

where u and s are the parameters.

Taking u and s as the first and the second parameters, we have

$$\mathbf{R}_1 = \frac{\partial \mathbf{R}}{\partial u} = \mathbf{t},\ \mathbf{R}_2 = \frac{\partial \mathbf{R}}{\partial s} = \mathbf{t} + u(\kappa\mathbf{n}), \mathbf{R}_{12} = \kappa\mathbf{n}.$$

Therefore

$$\mathbf{R}_1 \times \mathbf{R}_2 = u\kappa\mathbf{b}$$

$$\Rightarrow \quad \mathbf{N} = \frac{\mathbf{R}_1 \times \mathbf{R}_2}{H} = \frac{u\kappa}{H}\mathbf{b}$$

$$\Rightarrow \quad M = \mathbf{R}_{12} \cdot \mathbf{N} = (\kappa\mathbf{n}) \cdot \left(\frac{u\kappa}{H}\mathbf{b}\right) = 0.$$

Hence the parametric curves are conjugate on the given surface.

(iii) The position vector $\mathbf{r}$ of any current point on the given surface is

$$\mathbf{r} = (x, y, f(x) + F(y)),$$

where x and y are the parameters.

Taking x and y as the first and the second parameters, we have

$$\mathbf{r}_1 = (1, 0, f'(x)),\ \mathbf{r}_2 = (0, 1, F'(y)),\ \mathbf{r}_{12} = (0, 0, 0).$$

Therefore

$$\mathbf{r}_1 \times \mathbf{r}_2 = (-f'(x), -F'(y), 1)$$

$$\Rightarrow \qquad \mathbf{N} = \frac{\mathbf{r}_1 \times \mathbf{r}_2}{H} = \frac{1}{H}(-f'(x), -F'(y), 1)$$

$$\Rightarrow \qquad M = \mathbf{r}_{12} . \mathbf{N} = 0 .$$

Hence the parametric curves are conjugate on the given surface.

Example 2. Prove that, at any point of the surface, the sum of the radii of normal curvatures in conjugate directions is constant.

Solution. Let the lines of curvature be taken as the parametric curves. Then

$$F = 0,\ M = 0. \tag{1}$$

Let (du, dv) and $(\delta u, \delta v)$ be any two conjugate directions through any point $P\ (u, v)$. Then

$$L\, du\, \delta u + M(du\, \delta v + dv\, \delta u) + N\, dv\, \delta v = 0$$

or $$L\, du\, \delta u + N\, dv\, \delta v = 0 \qquad \text{[using (1)]}$$

or $$\frac{\delta u}{N\, dv} = \frac{\delta v}{-L\, du} = \lambda \text{ (say)}. \tag{2}$$

Now, the normal curvature κ_n at P in direction (du, dv) is given by

$$\kappa_n = \frac{L\, du^2 + 2M\, dudv + N\, dv^2}{E\, du^2 + 2F\, dudv + G\, dv^2}$$

$$= \frac{L\, du^2 + N\, dv^2}{E\, du^2 + G\, dv^2}. \qquad \text{[using (1)]}$$

Let ρ_1, ρ_2 be the radii of normal curvature at P in the conjugate directions (du, dv) and $(\delta u, \delta v)$ respectively. Then

$$\rho_1 = \frac{E\, du^2 + G\, dv^2}{L\, du^2 + N\, dv^2}, \tag{3}$$

$$\rho_2 = \frac{E\, \delta u^2 + G\, \delta v^2}{L\, \delta u^2 + N\, \delta v^2}. \tag{4}$$

Substituting the values of δu, δv from (2) in (4), we get

$$\rho_2 = \frac{EN^2 dv^2 + GL^2 du^2}{LN^2 dv^2 + NL^2 du^2}$$

i.e., $$\rho_2 = \frac{EN^2 dv^2 + GL^2 du^2}{LN(N\, dv^2 + L\, du^2)}. \tag{5}$$

Adding (3) and (5), we get

$$\begin{aligned}\rho_1+\rho_2 &= \frac{E\,du^2+G\,dv^2}{L\,du^2+N\,dv^2}+\frac{EN^2dv^2+GL^2du^2}{LN(L\,du^2+N\,dv^2)}\\ &= \frac{LN(E\,du^2+G\,dv^2)+EN^2dv^2+GL^2du^2}{LN(L\,du^2+N\,dv^2)}\\ &= \frac{EN(L\,du^2+N\,dv^2)+GL\,(L\,du^2+N\,dv^2)}{LN(L\,du^2+N\,dv^2)}\\ &= \frac{(EN+GL)\,(L\,du^2+N\,dv^2)}{LN(L\,du^2+N\,dv^2)}\\ &= \frac{EN+GL}{LN},\end{aligned}$$

which is independent of the directions (du, dv) and $(\delta u, \delta v)$ and therefore constant.

Example 3. Show that the curvature of an asymptotic line may be expressed as

$$\frac{(\mathbf{r}_1\cdot\mathbf{r}')\,(\mathbf{r}_2\cdot\mathbf{r}'')-(\mathbf{r}_2\cdot\mathbf{r}')\,(\mathbf{r}_1\cdot\mathbf{r}'')}{H}.$$

Solution. The curvature of an asymptotic line is given by

$$\begin{aligned}\kappa &= [\mathbf{N},\mathbf{r}',\mathbf{r}'']\\ &= \mathbf{N}\cdot(\mathbf{r}'\times\mathbf{r}'')\\ &= \left(\frac{\mathbf{r}_1\times\mathbf{r}_2}{H}\right)\cdot(\mathbf{r}'\times\mathbf{r}'')\\ &= \frac{1}{H}\{(\mathbf{r}_1\times\mathbf{r}_2)\cdot(\mathbf{r}'\times\mathbf{r}'')\}\\ &= \frac{1}{H}\begin{vmatrix}\mathbf{r}_1\cdot\mathbf{r}' & \mathbf{r}_1\cdot\mathbf{r}''\\ \mathbf{r}_2\cdot\mathbf{r}' & \mathbf{r}_2\cdot\mathbf{r}''\end{vmatrix} \qquad \text{[by Lagrange's identity]}\\ &= \frac{1}{H}\{(\mathbf{r}_1\cdot\mathbf{r}')\,(\mathbf{r}_2\cdot\mathbf{r}'')-(\mathbf{r}_2\cdot\mathbf{r}')\,(\mathbf{r}_1\cdot\mathbf{r}'')\}.\end{aligned}$$

Example 4. Prove that the normal curvature in a direction perpendicular to an asymptotic line is twice the mean normal curvature.

Solution. Let κ_a, κ_b be the principal curvatures of a surface at the point P and let (du, dv) be the direction of an asymptotic line through P. Then

$$L\,du^2 + 2M\,du\,dv + N\,dv^2 = 0. \tag{1}$$

Let κ_n be the normal curvature of the surface at P in the direction (du, dv) and let κ_1 be the normal curvature at P in a direction perpendicular to (du, dv). Then

$$\kappa_n = \frac{L\,du^2 + 2M\,du\,dv + N\,dv^2}{E\,du^2 + 2F\,du\,dv + G\,dv^2}$$

or $$\kappa_n = 0. \tag{2}$$

[using (1)]

Let ψ be the angle which the direction (du, dv) of the normal section at P makes with the principal direction through P in which the principal curvature is κ_a. Then by Euler's theorem, we have

$$\kappa_n = \kappa_a \cos^2\psi + \kappa_b \sin^2\psi$$

and $$\kappa_1 = \kappa_a \cos^2\left(\frac{\pi}{2} + \psi\right) + \kappa_b \sin^2\left(\frac{\pi}{2} + \psi\right)$$

i.e., $$0 = \kappa_a \cos^2\psi + \kappa_b \sin^2\psi \tag{3}$$

[using (2)]

and $$\kappa_1 = \kappa_a \sin^2\psi + \kappa_b \cos^2\psi. \tag{4}$$

Adding (3) and (4), we get

$$\kappa_1 = \kappa_a + \kappa_b$$

$$= 2\left(\frac{\kappa_a + \kappa_b}{2}\right)$$

$= 2$ (mean normal curvature at P).

Example 5. Show that the parametric curves are asymptotic lines on the surface $x = u\cos\phi,\ y = u\sin\phi,\ z = c\phi$.

Solution. We know that the necessary and sufficient condition for the parametric curves to be asymptotic lines is

$$L = 0,\ N = 0,\ M \neq 0. \tag{1}$$

Therefore we shall find the values of L, M and N for the given surface.

The position vector $\mathbf{r}$ of any current point on the given surface is

$$\mathbf{r} = (u\cos\phi, u\sin\phi, c\phi).$$

Then, we have

$$L = 0,\ M = -\frac{c}{\sqrt{u^2 + c^2}} \neq 0,\ N = 0$$

i.e., the condition (1) is satisfied. Hence the parametric curves are asymptotic lines on the given surface.

Example 6. Show that the parametric curves are asymptotic lines on the surface $x = a(u+v), y = b(u-v), z = uv$.

Solution. We know that the necessary and sufficient condition for the parametric curves to be asymptotic lines is

$$L = 0, N = 0, M \neq 0. \tag{1}$$

Therefore we shall find the values of L, M and N for the given surface.

The position vector $\mathbf{r}$ of any current point on the given surface is

$$\mathbf{r} = (a(u+v), b(u-v), uv).$$

Then, we have

$$L = 0, \ M = -\frac{2ab}{H} \neq 0, \ N = 0$$

i.e., the condition (1) is satisfied. Hence the parametric curves are asymptotic lines on the given surface.

Example 7. Prove that on surface $z = f(x, y)$, the asymptotic lines are

$$r\,dx^2 + 2s\,dx\,dy + t\,dy^2 = 0$$

and their torsions are

$$\pm\sqrt{s^2 - rt}\big/(1 + p^2 + q^2).$$

Solution. We know that the differential equation of the asymptotic lines is given by

$$L\,dx^2 + 2M\,dx\,dy + N\,dy^2 = 0. \tag{1}$$

We shall find the values of L, M, N for the given surface.

The equation of the given surface is

$$z = f(x, y).$$

The position vector of any current point on this surface can be written as

$$\mathbf{r} = (x, y, z), \tag{2}$$

where z is a function of x and y.

Then, we have

$$E = 1 + p^2, F = pq, \ G = 1 + q^2, H^2 = 1 + p^2 + q^2;$$

$$L = \frac{r}{H}, \ M = \frac{s}{H}, \ N = \frac{t}{H}.$$

Substituting the values of L, M, N in (1), we get

$$\frac{r}{H}dx^2 + 2\frac{s}{H}dx\,dy + \frac{t}{H}dy^2 = 0$$

or $\quad r\,dx^2 + 2s\,dx\,dy + t\,dy^2 = 0.$

The Gaussian curvature K of the given surface is

$$K=\frac{LN-M^2}{H^2}=\frac{rt-s^2}{H^4}=\frac{rt-s^2}{(1+p^2+q^2)^2}.$$

Therefore the torsions of the asymptotic lines are given by

$$\tau=\pm\sqrt{-K}=\pm\frac{\sqrt{s^2-rt}}{(1+p^2+q^2}.$$

[by the theorem of Beltrami and Enneper]

Example 8. Prove that the asymptotic lines on the surface of revolution $x=u\cos v, y=u\sin v, z=f(u)$ are

$$f_{11}\,du^2+uf_1\,dv^2=0$$

and their torsions are

$$\pm\sqrt{-u^3f_1f_{11}}\Big/\{u^2(1+f_1^{\,2})\}.$$

Solution. We know that the differential equation of the asymptotic lines is given by

$$L\,du^2+2M\,du\,dv+N\,dv^2=0. \tag{1}$$

We shall find the values of L, M, N for the given surface.

The equation of the given surface is

$$x=u\cos v, y=u\sin v, z=f(u).$$

The position vector of any current point on this surface can be written as

$$\mathbf{r}=(u\cos v, u\sin v, f(u)).$$

Then, we have

$$E=1+f_1^{\,2},\ F=0,\ G=u^2,\ H^2=u^2(1+f_1^{\,2});$$

$$L=\frac{uf_{11}}{H},\ M=0,\ N=\frac{u^2f_1}{H}.$$

Substituting the values of L, M, N in (1), we get

$$\frac{uf_{11}}{H}du^2+\frac{u^2f_1}{H}dv^2=0$$

or $\quad f_{11}\,du^2+uf_1\,dv^2=0.$

The Gaussian curvature K of the given surface is

$$K=\frac{LN-M^2}{H^2}=\frac{u^3f_1f_{11}}{H^4}=\frac{u^3f_1f_{11}}{u^4(1+f_1^{\,2})^2}=\frac{u^3f_1f_{11}}{[u^2(1+f_1^{\,2})]^2}.$$

Therefore the torsions of the asymptotic lines are given by

$$\tau=\pm\sqrt{-K}=\pm\frac{\sqrt{-u^3f_1f_{11}}}{u^2(1+f_1^{\,2})}.$$

Example 9. Find the asymptotic lines and their torsions on the surface generated by the binormals to a twisted curve.

Solution. We know that the differential equation of the asymptotic lines is given by

$$L\,du^2 + 2M\,du\,ds + N\,ds^2 = 0. \tag{1}$$

We shall find the values of L, M, N for the given surface.

The position vector of a current point on the surface generated by the binormals to the twisted curve $\mathbf{r} = \mathbf{r}(s)$ is

$$\mathbf{R} = \mathbf{r}(s) + u\mathbf{b}(s), \tag{2}$$

where u and s are the parameters.

Taking u and s as the first and the second parameters, we have

$$\mathbf{R}_1 = \frac{\partial \mathbf{R}}{\partial u} = \mathbf{b},\ \mathbf{R}_2 = \frac{\partial \mathbf{R}}{\partial s} = \mathbf{t} + u(-\tau\mathbf{n}) = \mathbf{t} - u\tau\mathbf{n},$$

$$\mathbf{R}_{11} = \mathbf{0},\ \mathbf{R}_{12} = -\tau\mathbf{n},$$

$$\mathbf{R}_{22} = \kappa\mathbf{n} - u[\tau'\mathbf{n} + \tau(\tau\mathbf{b} - \kappa\mathbf{t})] = u\kappa\tau\mathbf{t} + (\kappa - u\tau')\mathbf{n} - u\tau^2\mathbf{b}.$$

Therefore

$$E = \mathbf{R}_1^{\,2} = 1,\ F = \mathbf{R}_1 . \mathbf{R}_2 = 0,\ G = \mathbf{R}_2^{\,2} = 1 + u^2\tau^2,$$

$$H^2 = EG - F^2 = 1 + u^2\tau^2.$$

Again, we have

$$\mathbf{R}_1 \times \mathbf{R}_2 = u\tau\mathbf{t} + \mathbf{n}$$

$$\Rightarrow \qquad \mathbf{N} = \frac{\mathbf{R}_1 \times \mathbf{R}_2}{H} = \frac{u\tau\mathbf{t} + \mathbf{n}}{H}.$$

Therefore

$$L = \mathbf{R}_{11} . \mathbf{N} = 0,\ M = \mathbf{R}_{12} . \mathbf{N} = -\frac{\tau}{H},$$

$$N = \mathbf{R}_{22} . \mathbf{N} = \frac{\kappa - u\tau' + \kappa u^2\tau^2}{H}.$$

Substituting the values of L, M, N in (1), we get

$$-2\frac{\tau}{H}\,du\,ds + \frac{u^2\tau^2\kappa + \kappa - u\tau'}{H}\,ds^2 = 0$$

or $$-2\tau\,du\,ds + (u^2\tau^2\kappa + \kappa - u\tau')\,ds^2 = 0,$$

which is the differential equation of the asymptotic lines.

The Gaussian curvature K of the given surface is

$$K = \frac{LN - M^2}{H^2} = -\frac{\tau^2}{H^4} = -\frac{\tau^2}{(1 + u^2\tau^2)^2}.$$

Therefore the torsions of the asymptotic lines are given by

$$\tau = \pm\sqrt{-K} = \pm\frac{\tau}{(1+u^2\tau^2)}.$$

Example 10. Find the asymptotic lines on the surface $z = y\sin x$.

Solution. We know that the differential equation of the asymptotic lines for the surface $z = f(x,, y)$ is given by

$$r\,dx^2 + 2s\,dx\,dy + t\,dy^2 = 0. \tag{1}$$

We shall find the values of r, s, t for the given surface.
The equation of the given surface is

$$z = y\sin x.$$

$$\therefore \qquad p = \frac{\partial z}{\partial x} = y\cos x,\ q = \frac{\partial z}{\partial y} = \sin x,$$

$$r = \frac{\partial^2 z}{\partial x^2} = -y\sin x,\ s = \frac{\partial^2 z}{\partial x\,\partial y} = \cos x,\ t = \frac{\partial^2 z}{\partial y^2} = 0.$$

Substituting the values of r, s, t in (1), we get

$$-y\sin x\,dx^2 + 2\cos x\,dx\,dy = 0$$

$$\Rightarrow \qquad dx\,(-y\sin x\,dx + 2\cos x\,dy) = 0$$

$$\Rightarrow \qquad dx = 0,\ \ -y\sin x\,dx + 2\cos x\,dy = 0.$$

Now,

$$dx = 0 \text{ gives } x = c_1$$

and

$$-y\sin x\,dx + 2\cos x\,dy = 0 \text{ gives}$$

$$\frac{\sin x}{\cos x}\,dx = 2\frac{dy}{y}$$

or $$2\frac{dy}{y} = \tan x\,dx.$$

Integrating this, we get

$$2\log y = -\log\cos x + \log c_2$$

or $$\log y^2 + \log\cos x = \log c_2$$

or $$\log(y^2\cos x) = \log c_2$$

or $$y^2\cos x = c_2.$$

Hence the asymptotic lines are given by

$$x = \text{constant},\ y^2\cos x = \text{constant}.$$

Example 11. Show that the asymptotic lines on the paraboloid $2z = \dfrac{x^2}{a^2} - \dfrac{y^2}{b^2}$ are

$$\frac{x}{a} \pm \frac{y}{b} = \text{constant.}$$

Solution. We know that the differential equation of the asymptotic lines for the surface $z = f(x,, y)$ is given by

$$r\,dx^2 + 2s\,dx\,dy + t\,dy^2 = 0. \qquad (1)$$

We shall find the values of r, s, t for the given surface.

The equation of the given surface is

$$z = \frac{1}{2}\left(\frac{x^2}{a^2} - \frac{y^2}{b^2}\right).$$

$$\therefore \qquad p = \frac{\partial z}{\partial x} = \frac{x}{a^2},\; q = \frac{\partial z}{\partial y} = -\frac{y}{b^2},$$

$$r = \frac{\partial^2 z}{\partial x^2} = \frac{1}{a^2},\; s = \frac{\partial^2 z}{\partial x\,\partial y} = 0,\; t = \frac{\partial^2 z}{\partial y^2} = -\frac{1}{b^2}.$$

Substituting the values of r, s, t in (1), we get

$$\frac{1}{a^2}dx^2 - \frac{1}{b^2}dy^2 = 0$$

$$\Rightarrow \qquad \left(\frac{dx}{a} + \frac{dy}{b}\right)\left(\frac{dx}{a} - \frac{dy}{b}\right) = 0$$

$$\Rightarrow \qquad \frac{dx}{a} + \frac{dy}{b} = 0,\; \frac{dx}{a} - \frac{dy}{b} = 0.$$

Now,

$$\frac{dx}{a} + \frac{dy}{b} = 0 \text{ gives } \frac{x}{a} + \frac{y}{b} = \text{constant}$$

and

$$\frac{dx}{a} - \frac{dy}{b} = 0 \text{ gives } \frac{x}{a} - \frac{y}{b} = \text{constant.}$$

Hence the asymptotic lines are given by

$$\frac{x}{a} \pm \frac{y}{b} = \text{constant.}$$

Example 12. Find the asymptotic lines of the conoid

$$x = u\cos v,\; y = u\sin v,\; z = f(v).$$

Solution. We know that the differential equation of the asymptotic lines is given by

$$L\,du^2 + 2M\,du\,dv + N\,dv^2 = 0. \qquad (1)$$

We shall find the values of L, M, N for the given surface.

The equation of the given surface is

$$x = u\cos v,\ y = u\sin v,\ z = f(v).$$

The position vector $\mathbf{r}$ of any current point on this surface can be written as

$$\mathbf{r} = (u\cos v, u\sin v, f(v)). \tag{2}$$

Taking u and v as the first and the second parameters, we have

$$\mathbf{r}_1 = (\cos v,\ \sin v, 0),\ \ \mathbf{r}_2 = (-u\sin v,\ u\cos v, f_1),\ \text{where } f_1 = \frac{df}{dv}$$

$$\mathbf{r}_{11} = (0, 0, 0),\ \ \mathbf{r}_{12} = (-\sin v, \cos v, 0),$$

$$\mathbf{r}_{22} = (-u\cos v, -u\sin v, f_{11}),\ \text{where } f_{11} = \frac{d^2 f}{dv^2}.$$

Again, we have

$$\mathbf{r}_1 \times \mathbf{r}_2 = (f_1 \sin v, -f_1 \cos v, u)$$

$$\Rightarrow \qquad \mathbf{N} = \frac{\mathbf{r}_1 \times \mathbf{r}_2}{H} = \frac{1}{H}(f_1 \sin v, -f_1 \cos v, u).$$

Therefore

$$L = \mathbf{r}_{11} \cdot \mathbf{N} = 0,\ M = \mathbf{r}_{12} \cdot \mathbf{N} = -\frac{f_1}{H},\ N = \mathbf{r}_{22} \cdot \mathbf{N} = \frac{uf_{11}}{H}.$$

Substituting the values of L, M, N in (1), we get

$$-2\frac{f_1}{H}\,du\,dv + \frac{uf_{11}}{H}\,dv^2 = 0$$

$$\Rightarrow \qquad dv\ (-2f_1\,du + uf_{11}\,dv) = 0$$

$$\Rightarrow \qquad dv = 0, -2f_1\,du + uf_{11}\,dv = 0.$$

Now,

$$dv = 0 \text{ gives } v = c_1.$$

and

$$-2f_1\,du + uf_{11}\,dv = 0 \text{ gives}$$

$$\frac{f_{11}}{f_1}\,dv = 2\frac{du}{u}$$

or

$$\frac{\dfrac{d^2 f}{dv^2}}{\dfrac{df}{dv}}\,dv = 2\frac{du}{u}.$$

Integrating this, we get

$$\log\left(\frac{df}{dv}\right) = 2\log u + \log c_2$$

or

$$\log f_1 - \log u^2 = \log c_2$$

or $\quad \log\left(\dfrac{f_1}{u^2}\right) = \log c_2$

or $\quad \dfrac{f_1}{u^2} = c_2.$

Hence the asymptotic lines are given by

$$v = \text{constant}, \ \frac{f_1}{u^2} = \text{constant}.$$

Example 13. Prove that the asymptotic lines on the surface

$$x = 3u\,(1+v^2) - u^3, \ y = 3v\,(1+u^2) - v^3, \ z = 3\,(u^2 - v^2),$$

are

$$u \pm v = \text{constant}.$$

Solution. We know that the differential equation of the asymptotic lines is given by

$$L\,du^2 + 2M\,du\,dv + N\,dv^2 = 0. \tag{1}$$

We shall find the values of L, M, N for the given surface.

The equation of the given surface is

$$x = 3u\,(1+v^2) - u^3, \ y = 3v\,(1+u^2) - v^3, \ z = 3\,(u^2 - v^2).$$

The position vector $\mathbf{r}$ of any current point on this surface can be written as

$$\mathbf{r} = (3u\,(1+v^2) - u^3, 3v\,(1+u^2) - v^3, 3\,(u^2 - v^2)). \tag{2}$$

Taking u and v as the first and the second parameters, we have

$$\mathbf{r}_1 = (3\,(1+v^2) - 3u^2, 6vu, 6u) = 3\,(1+v^2 - u^2, 2vu, 2u),$$

$$\mathbf{r}_2 = (6uv, 3\,(1+u^2) - 3v^2, -6v) = 3\,(2uv, 1+u^2 - v^2, -2v),$$

$$\mathbf{r}_{11} = 3\,(-2u, 2v, 2) = 6\,(-u, v, 1),$$

$$\mathbf{r}_{12} = 3\,(2v, 2u, 0) = 6\,(v, u, 0),$$

$$\mathbf{r}_{22} = 3\,(2u, -2v, -2) = 6\,(u, -v, -1).$$

Again, we have

$$\mathbf{r}_1 \times \mathbf{r}_2 = 9\,(1+u^2+v^2)\,(-2u, 2v, 1-u^2-v^2)$$

$$\Rightarrow \quad \mathbf{N} = \frac{\mathbf{r}_1 \times \mathbf{r}_2}{H} = \frac{9}{H}(1+u^2+v^2)\,(-2u, 2v, 1-u^2-v^2).$$

Therefore

$$L = \mathbf{r}_{11} \cdot \mathbf{N} = \frac{54}{H}(1+u^2+v^2)^2, \ M = \mathbf{r}_{12} \cdot \mathbf{N} = 0,$$

$$N = \mathbf{r}_{22} \cdot \mathbf{N} = -\frac{54}{H}(1+u^2+v^2)^2.$$

Substituting the values of L, M, N in (1), we get

$$\frac{54}{H}(1+u^2+v^2)^2\,du^2-\frac{54}{H}(1+u^2+v^2)^2\,dv^2=0$$

$\Rightarrow$ $du^2 - dv^2 = 0$

$\Rightarrow$ $(du+dv)(du-dv)=0$

$\Rightarrow$ $du+dv=0,\ du-dv=0.$

Now,

$$du+dv=0 \text{ gives } u+v=\text{constant}$$

and

$$du-dv=0 \text{ gives } u-v=\text{constant}.$$

Hence the asymptotic lines are given by

$$u \pm v = \text{constant}.$$

Example 14. Find the asymptotic lines on the catenoid of revolution

$$u = c\cos\text{h}\frac{z}{c}.$$

Solution. We know that the differential equation of the asymptotic lines is given by

$$L\,du^2 + 2M\,du\,dv + N\,dv^2 = 0. \tag{1}$$

We shall find the values of L, M, N for the given surface.

Let the parametric equations of the given catenoid of revolution be

$$x = u\cos v,\ y = u\sin v,\ z = c\cos\text{h}^{-1}\frac{u}{c}.$$

Then the position vector $\mathbf{r}$ of any current point on this surface is given by

$$\mathbf{r} = \left(u\cos v,\ u\sin v,\ c\cosh^{-1}\frac{u}{c}\right).$$

Taking u and v as the first and the second parameters, we have

$$\mathbf{r}_1 = \left(\cos v, \sin v, \frac{c}{\sqrt{u^2-c^2}}\right),\quad \mathbf{r}_2 = (-u\sin v, u\cos v, 0),$$

$$\mathbf{r}_{11} = \left(0, 0,\ -\frac{cu}{(u^2-c^2)^{3/2}}\right),\quad \mathbf{r}_{12} = (-\sin v, \cos v, 0),$$

$$\mathbf{r}_{22} = (-u\cos v, -u\sin v, 0).$$

Again, we have

$$\mathbf{r}_1\times\mathbf{r}_2 = \left(-\frac{uc\cos v}{\sqrt{u^2-c^2}}, -\frac{uc\ \sin v}{\sqrt{u^2-c^2}},\ u\right)$$

$$\Rightarrow \qquad \mathbf{N} = \frac{\mathbf{r}_1 \times \mathbf{r}_2}{H} = \frac{1}{H}\left(-\frac{uc\cos v}{\sqrt{u^2-c^2}}, -\frac{uc\sin v}{\sqrt{u^2-c^2}}, u\right).$$

Therefore

$$L = \mathbf{r}_{11}\cdot\mathbf{N} = -\frac{1}{H}\frac{cu^2}{(u^2-c^2)^{3/2}}, \quad M = \mathbf{r}_{12}\cdot\mathbf{N} = 0,$$

$$N = \mathbf{r}_{22}\cdot\mathbf{N} = \frac{1}{H}\frac{cu^2}{\sqrt{u^2-c^2}}.$$

Substituting the values of L, M, N in (1), we get

$$-\frac{1}{H}\frac{cu^2}{(u^2-c^2)^{3/2}}du^2 + \frac{1}{H}\frac{cu^2}{\sqrt{u^2-c^2}}dv^2 = 0$$

$$\Rightarrow \qquad \frac{du^2}{u^2-c^2} - dv^2 = 0$$

$$\Rightarrow \qquad \frac{dz^2}{c^2} - dv^2 = 0 \qquad \left[\because dz = \frac{c}{\sqrt{u^2-c^2}}du\right]$$

$$\Rightarrow \qquad dz^2 - c^2dv^2 = 0$$

$$\Rightarrow \qquad (dz + c\,dv)(dz - c\,dv) = 0$$

$$\Rightarrow \qquad dz + c\,dv = 0, \; dz - c\,dv = 0.$$

Now,

$$dz + c\,dv = 0 \text{ gives } z + cv = \text{constant}$$

and

$$dz - c\,dv = 0 \text{ gives } z - cv = \text{constant}.$$

Hence the asymptotic lines are given by

$$z \pm cv = \text{constant}.$$

Example 15. Find the asymptotic lines on the surface

$$x = a(1+\cos u)\cot v, \; y = a(1+\cos u), \; z = a\cos u \operatorname{cosec} v.$$

Solution. We know that the differential equation of the asymptotic lines is given by

$$L\,du^2 + 2M\,du\,dv + N\,dv^2 = 0. \tag{1}$$

We shall find the values of L, M, N for the given surface.

The equation of the given surface is

$$x = a(1+\cos u)\cot v, \; y = a(1+\cos u), \; z = a\cos u \operatorname{cosec} v.$$

The position vector $\mathbf{r}$ of any current point on this surface can be written as

$$\mathbf{r} = (a(1+\cos u)\cot v, a(1+\cos u), a\cos u \operatorname{cosec} v). \tag{2}$$

Taking u and v as the first and the second parameters, we have

$$\mathbf{r}_1 = (-a\sin u \cot v, -a\sin u, -a\sin u \operatorname{cosec} v)$$
$$= -a\sin u\,(\cot v,\ 1,\ \operatorname{cosec} v),$$
$$\mathbf{r}_2 = (-a\,(1+\cos u)\operatorname{cosec}^2 v, 0, -a\cos u \operatorname{cosec} v \cot v)$$
$$= -a \operatorname{cosec} v\,(\,(1+\cos u)\operatorname{cosec} v, 0, \cos u\ \cot v),$$
$$\mathbf{r}_{11} = (-a\cos u \cot v, -a\cos u, -a\cos u \operatorname{cosec} v)$$
$$= -a\cos u\,(\cot v,\ 1,\ \operatorname{cosec} v),$$
$$\mathbf{r}_{12} = (a\sin u \operatorname{cosec}^2 v, 0, a\sin u \operatorname{cosec} v \cot v)$$
$$= a\sin u \operatorname{cosec} v\,(\operatorname{cosec} v,\ 0,\ \cot v),$$
$$\mathbf{r}_{22} = (2a\,(1+\cos u)\operatorname{cosec}^2 v \cot v, 0,$$
$$a\cos u\,(\operatorname{cosec} v \cot^2 v + \operatorname{cosec}^3 v))$$
$$= a \operatorname{cosec} v\,(2\,(1+\cos u)\operatorname{cosec} v \cot v, 0,$$
$$\cos u\,(\,\cot^2 v + \operatorname{cosec}^2 v)).$$

Again, we have

$$\mathbf{r}_1 \times \mathbf{r}_2 = a^2 \sin u \operatorname{cosec} v\,(\,\cos u \cot v,\ \operatorname{cosec}^2 v + \cos u,$$
$$-\operatorname{cosec} v\,(1+\cos u))$$

$$\Rightarrow \quad \mathbf{N} = \frac{\mathbf{r}_1 \times \mathbf{r}_2}{H} = \frac{a^2 \sin u \operatorname{cosec} v}{H}(\,\cos u \cot v, \operatorname{cosec}^2 v + \cos u,$$
$$-\operatorname{cosec} v\,(1+\cos u)).$$

Therefore

$$L = \mathbf{r}_{11} \cdot \mathbf{N} = 0,$$

$$M = \mathbf{r}_{12} \cdot \mathbf{N} = -\frac{a^3}{H}\sin^2 u \operatorname{cosec}^3 v \cot v,$$

$$N = \mathbf{r}_{22} \cdot \mathbf{N} = -\frac{a^3}{H}\sin u\,(1+\cos u)\cos u \operatorname{cosec}^3 v.$$

Substituting the values of L, M, N in (1), we get

$$-2\frac{a^3}{H}\sin^2 u \operatorname{cosec}^3 v \cot v\, du\, dv$$
$$-\frac{a^3}{H}\sin u\,(1+\cos u)\cos u \operatorname{cosec}^3 v\, dv^2 = 0$$

$$\Rightarrow \quad -\frac{a^3}{H}\sin u \operatorname{cosec}^3 v\, dv\,[2\sin u \cot v\, du + (1+\cos u)\cos u\, dv] = 0$$

$$\Rightarrow \quad dv\ [2\sin u \cot v\, du + (1+\cos u)\cos u\, dv] = 0$$

$$\Rightarrow \quad dv = 0,\ 2\sin u \cot v\, du + (1+\cos u)\cos u\, dv = 0.$$

Now,

$$dv = 0 \text{ gives } v = c_1 \text{(constant)}$$

and

$$2\sin u \cot v\, du + (1+\cos u)\cos u\, dv = 0 \text{ gives}$$

$$\frac{2\sin u}{\cos u\,(1+\cos u)}du = -\frac{1}{\cot v}dv$$

or $$\frac{2\sin u\,[(1+\cos u)-\cos u]}{\cos u\,(1+\cos u)}du = -\tan v\, dv$$

or $$2\left[\frac{\sin u}{\cos u} - \frac{\sin u}{1+\cos u}\right]du = -\tan v\, dv.$$

Integrating this, we get

$$2[-\log\cos u + \log(1+\cos u)] = -\log\sec v + \log c_2$$

$$\Rightarrow \quad 2\log\left(\frac{1+\cos u}{\cos u}\right) = \log\left(\frac{c_2}{\sec v}\right)$$

$$\Rightarrow \quad \log\left(\frac{1+\cos u}{\cos u}\right)^2 = \log\left(\frac{c_2}{\sec v}\right)$$

$$\Rightarrow \quad \left(\frac{1+\cos u}{\cos u}\right)^2 = \frac{c_2}{\sec v}$$

$$\Rightarrow \quad \left(\frac{1+\cos u}{\cos u}\right)^2 \sec v = c_2.$$

Hence the asymptotic lines are given by

$$v = \text{constant}, \left(\frac{1+\cos u}{\cos u}\right)^2 \sec v = \text{constant}.$$

Exercises

1. Show that all straight lines on a surface are asymptotic lines.
2. Prove that the directions of lines of curvature bisect the asymptotic directions.
3. Prove that the asymptotic lines on the surface of revolution

   $$x = u\cos v,\ y = u\sin v,\ z = f(u),$$

   where

   $$u = a\sin\phi \text{ and } f(u) = a\left(\log\tan\frac{\phi}{2} + \cos\phi\right)$$

   are given by

$$d\theta = \pm \frac{d\phi}{\sin \phi}.$$

4. Show that the asymptotic lines of the helicoid

$$x = u \cos v, \; y = u \sin v, \; z = cv$$

consist of the generators and the curves of intersection with coaxial right circular cylinders.

5. Show that the projection on the xy-plane of an asymptotic line of the cylindroid $z\,(x^2 + y^2) = 2mxy$ is a lemniscate.

4.10 RULED SURFACES

Definition. A surface generated by the motion of a one parameter family of straight lines is called a called a ruled surface and the straight lines are called the generators of the surface.

If the consecutive generators intersect, the surface is a developable surface and if they do not intersect, it is a skew surface. Skew surfaces are also called scrolls.

Equation of the Ruled Surface. Let C be any curve drawn on the given ruled surface such that this curve cuts all the generators of the surface only once. Such a curve is called a base curve or directrix.

Obviously, many curves on the surface can be taken as the base curves and the ruled surface is determined by any base curve C and the directions of the generators at each point of C.

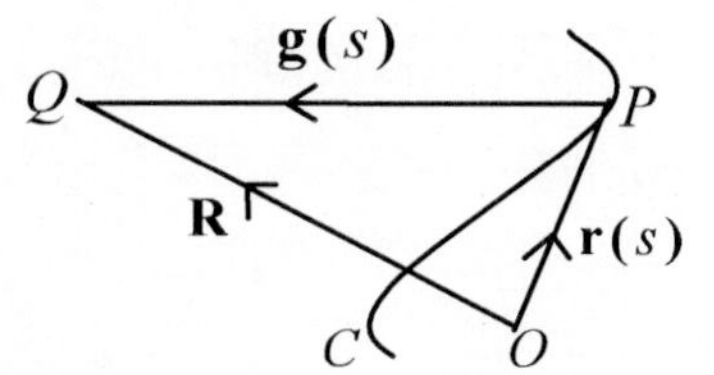

Let $\mathbf{r}\,(s)$ be the position vector of a current point P on the directrix C and let $\mathbf{g}(s)$ be a unit vector along the generator at P, where s is the arc length of C for P measured from some fixed on it. Then the position vector $\mathbf{R}$ of a current point Q on the ruled surface is given by

$$\mathbf{R} = \mathbf{r}\,(s) + v\mathbf{g}\,(s), \tag{1}$$

where v is a parameter which measures the distance of Q from the directrix in the direction of $\mathbf{g}$.

Equation (1) is the equation of the ruled surface with s, v as the parameters of the surface.

If u be taken as the parameter in place of s, i.e., if the equation of the directrix C be taken as $\mathbf{r} = \mathbf{r}\,(u)$ then the above equation of the ruled surface becomes

$$\mathbf{R} = \mathbf{r}(u) + v\mathbf{g}(u),$$

where u, v are the parameters of the surface.

The parametric curves $u =$ const. are the generators.

Fundamental Coefficients and Gaussian Curvature for a Ruled Surface. The position vector $\mathbf{R}$ of a current point Q on the ruled surface is given by

$$\mathbf{R} = \mathbf{r}(s) + v\mathbf{g}(s), \tag{1}$$

where s and v are the parameters.

Taking s and v as the first and the second parameters, we have

$$\mathbf{R}_1 = \frac{\partial \mathbf{R}}{\partial s} = \frac{d\mathbf{r}}{ds} + v\frac{d\mathbf{g}}{ds} = \mathbf{r}' + v\mathbf{g}', \quad \mathbf{R}_2 = \frac{\partial \mathbf{R}}{\partial v} = \mathbf{g},$$

$$\mathbf{R}_{11} = \frac{\partial}{\partial s}\left(\frac{\partial \mathbf{R}}{\partial s}\right) = \mathbf{r}'' + v\mathbf{g}'', \quad \mathbf{R}_{12} = \frac{\partial}{\partial v}\left(\frac{\partial \mathbf{R}}{\partial s}\right) = \mathbf{g}',$$

$$\mathbf{R}_{22} = \frac{\partial}{\partial v}\left(\frac{\partial \mathbf{R}}{\partial v}\right) = \mathbf{0}.$$

Therefore the fundamental magnitudes of the first order are

$$E = \mathbf{R}_1^{\,2} = (\mathbf{r}' + v\mathbf{g}')^2 = \mathbf{r}'^2 + 2v\mathbf{r}'.\mathbf{g}' + v^2\mathbf{g}'^2,$$

$$F = \mathbf{R}_1 \,.\, \mathbf{R}_2 = (\mathbf{r}' + v\mathbf{g}')\,.\,\mathbf{g} = \mathbf{r}'.\mathbf{g},$$

$$G = \mathbf{R}_2^{\,2} = \mathbf{g}^2 = 1.$$

Also, $H^2 = EG - F^2 = (\mathbf{r}'^2 + 2v\mathbf{r}'.\mathbf{g}' + v^2\mathbf{g}'^2) - (\mathbf{r}'.\mathbf{g})^2.$

Again, we have

$$\mathbf{R}_1 \times \mathbf{R}_2 = (\mathbf{r}' + v\mathbf{g}') \times \mathbf{g}$$

$$\Rightarrow \qquad \mathbf{N} = \frac{\mathbf{R}_1 \times \mathbf{R}_2}{H} = \frac{1}{H}\{(\mathbf{r}' + v\mathbf{g}') \times \mathbf{g}\}.$$

Therefore the fundamental magnitudes of the second order are

$$L = \mathbf{R}_{11}.\mathbf{N} = \frac{1}{H}[\mathbf{R}_{11},\ \mathbf{R}_1,\ \mathbf{R}_2]$$

$$= \frac{1}{H}[\mathbf{r}'' + v\mathbf{g}'',\ \mathbf{r}' + v\mathbf{g}',\ \mathbf{g}]$$

$$= \frac{1}{H}\{[\mathbf{r}'',\ \mathbf{r}',\ \mathbf{g}] + [\mathbf{g}'',\ \mathbf{r}',\ \mathbf{g}]\,v + [\mathbf{r}'',\ \mathbf{g}',\ \mathbf{g}]\,v + [\mathbf{g}'',\ \mathbf{g}',\ \mathbf{g}]\,v^2\},$$

$$M = \mathbf{R}_{12} . \mathbf{N} = \frac{1}{H}[\mathbf{R}_{12},\ \mathbf{R}_1,\ \mathbf{R}_2]$$
$$= \frac{1}{H}[\mathbf{g}',\ \mathbf{r}' + v\mathbf{g}',\ \mathbf{g}]$$
$$= \frac{1}{H}\{[\mathbf{g}',\ \mathbf{r}',\ \mathbf{g}] + [\mathbf{g}',\ \mathbf{g}',\ \mathbf{g}]\,v\}$$
$$= \frac{1}{H}[\mathbf{r}',\ \mathbf{g},\ \mathbf{g}']$$
$$= -\frac{1}{H}[\mathbf{r}',\ \mathbf{g}',\ \mathbf{g}],$$
$$N = \mathbf{R}_{22} . \mathbf{N} = \frac{1}{H}[\mathbf{R}_{22},\ \mathbf{R}_1,\ \mathbf{R}_2] = \frac{1}{H}[\mathbf{0},\ \mathbf{R}_1,\ \mathbf{R}_2] = 0.$$

The Gaussian curvature K has the value

$$K = \frac{LN - M^2}{H^2} = -\frac{M^2}{H^2}$$

i.e., $$K = -\frac{[\mathbf{r}',\ \mathbf{g}',\ \mathbf{g}]^2}{H^4} = -\frac{[\mathbf{t},\ \mathbf{g}',\ \mathbf{g}]^2}{H^4}.$$

Therefore, for a ruled surface, $K \leq 0$, i.e., K is never positive.

Asymptotic Lines on a Ruled Surface.

Let the equation of the ruled surface be given by

$$\mathbf{R} = \mathbf{r}(u) + v\mathbf{g}(u), \tag{1}$$

where u and v are the parameters.

First, we shall find the values of E, F, G, L, M, N.

Taking u and v as the first and the second parameters, we have

$$\mathbf{R}_1 = \frac{\partial \mathbf{R}}{\partial u} = \frac{d\mathbf{r}}{du} + v\frac{d\mathbf{g}}{du} = \dot{\mathbf{r}} + v\dot{\mathbf{g}},\quad \mathbf{R}_2 = \frac{\partial \mathbf{R}}{\partial v} = \mathbf{g},$$
$$\mathbf{R}_{11} = \frac{\partial}{\partial u}\left(\frac{\partial \mathbf{R}}{\partial u}\right) = \ddot{\mathbf{r}} + v\ddot{\mathbf{g}},\quad \mathbf{R}_{12} = \frac{\partial}{\partial v}\left(\frac{\partial \mathbf{R}}{\partial u}\right) = \dot{\mathbf{g}},$$
$$\mathbf{R}_{22} = \frac{\partial}{\partial v}\left(\frac{\partial \mathbf{R}}{\partial v}\right) = \mathbf{0}.$$

Therefore the fundamental magnitudes of the first order are

$$E = \mathbf{R}_1^{\ 2} = (\dot{\mathbf{r}} + v\dot{\mathbf{g}})^2 = \dot{\mathbf{r}}^2 + 2v\dot{\mathbf{r}}.\dot{\mathbf{g}} + v^2\dot{\mathbf{g}}^2,$$
$$F = \mathbf{R}_1 . \mathbf{R}_2 = (\dot{\mathbf{r}} + v\dot{\mathbf{g}}) . \mathbf{g} = \dot{\mathbf{r}} . \mathbf{g},$$
$$G = \mathbf{R}_2^{\ 2} = \mathbf{g}^2 = 1.$$

Also,

$$H^2 = EG - F^2 = (\dot{\mathbf{r}}^2 + 2v\dot{\mathbf{r}}.\dot{\mathbf{g}} + v^2\dot{\mathbf{g}}^2) - (\dot{\mathbf{r}}.\mathbf{g})^2.$$

Again, we have

$$\mathbf{R}_1 \times \mathbf{R}_2 = (\dot{\mathbf{r}} + v\dot{\mathbf{g}}) \times \mathbf{g}$$

$$\Rightarrow \qquad \mathbf{N} = \frac{\mathbf{R}_1 \times \mathbf{R}_2}{H} = \frac{1}{H}\{(\dot{\mathbf{r}} + v\dot{\mathbf{g}}) \times \mathbf{g}\}.$$

Therefore the fundamental magnitudes of the second order are

$$L = \mathbf{R}_{11}.\mathbf{N} = \frac{1}{H}[\mathbf{R}_{11},\ \mathbf{R}_1,\ \mathbf{R}_2]$$

$$= \frac{1}{H}[\ddot{\mathbf{r}} + v\ddot{\mathbf{g}},\ \dot{\mathbf{r}} + v\dot{\mathbf{g}},\ \mathbf{g}]$$

$$= \frac{1}{H}\{[\ddot{\mathbf{r}},\ \dot{\mathbf{r}},\ \mathbf{g}] + [\ddot{\mathbf{g}},\ \dot{\mathbf{r}},\ \mathbf{g}]\,v + [\ddot{\mathbf{r}},\ \dot{\mathbf{g}},\ \mathbf{g}]\,v + [\ddot{\mathbf{g}},\ \dot{\mathbf{g}},\ \mathbf{g}]\,v^2\},$$

$$M = \mathbf{R}_{12}.\mathbf{N} = \frac{1}{H}[\mathbf{R}_{12},\ \mathbf{R}_1,\ \mathbf{R}_2]$$

$$= \frac{1}{H}[\dot{\mathbf{g}},\ \dot{\mathbf{r}} + v\dot{\mathbf{g}},\ \mathbf{g}]$$

$$= \frac{1}{H}\{[\dot{\mathbf{g}},\ \dot{\mathbf{r}},\ \mathbf{g}] + [\dot{\mathbf{g}},\ \dot{\mathbf{g}},\ \mathbf{g}]\,v\}$$

$$= \frac{1}{H}[\dot{\mathbf{r}},\ \mathbf{g},\ \dot{\mathbf{g}}]$$

$$= -\frac{1}{H}[\dot{\mathbf{r}},\ \dot{\mathbf{g}},\ \mathbf{g}],$$

$$N = \mathbf{R}_{22}.\mathbf{N} = \frac{1}{H}[\mathbf{R}_{22},\ \mathbf{R}_1,\ \mathbf{R}_2] = \frac{1}{H}[\mathbf{0},\ \mathbf{R}_1,\ \mathbf{R}_2] = 0.$$

We know that the differential equation of the asymptotic lines on the surface $\mathbf{r} = \mathbf{r}(u,v)$ is given by

$$L\,du^2 + 2M\,du\,dv + N\,dv^2 = 0.$$

But, for a ruled surface, we have $N = 0$. Therefore the differential equation of the asymptotic lines for the ruled surface (1) becomes

$$L\,du^2 + 2M\,du\,dv = 0$$

or $$du\ (L\,du + 2M\,dv) = 0. \tag{2}$$

From (2), it follows that one family of asymptotic lines on a ruled surface is given by $du = 0$, i.e., $u =$ const. But the parametric curves $u =$ const. are the generators of the ruled surface. Hence the generators of a ruled surface constitute one family of asymptotic lines. Again, from (2), it follows that the other family of asymptotic lines (i.e., system of curved asymptotic lines) is given by

$$L\,du + 2M\,dv = 0$$

or $$\frac{dv}{du} = -\frac{L}{2M}$$

or $$\frac{dv}{du} = \frac{1}{2[\dot{\mathbf{r}},\ \dot{\mathbf{g}},\ \mathbf{g}]}\{[\ddot{\mathbf{r}},\ \dot{\mathbf{r}},\ \mathbf{g}] + [\ddot{\mathbf{g}},\ \dot{\mathbf{r}},\ \mathbf{g}]v + [\ddot{\mathbf{r}},\ \dot{\mathbf{g}},\ \mathbf{g}]v + [\ddot{\mathbf{g}},\ \dot{\mathbf{g}},\ \mathbf{g}]v^2\}. \tag{3}$$

This differential equation is of the Riccati type

$$\frac{dv}{du} = A + Bv + Cv^2,$$

where A, B, C are functions of u only. From the theory of differential equations, we know that the most general solution of this differential equation is of the form

$$v = \frac{cP + Q}{cR + S}, \tag{4}$$

where P, Q, R, S are functions of u, and c is an arbitrary constant.

Equation (4) is the required equation of the family of curved asymptotic lines and for different values of c, we have different members of the family.

Consider the intersections of four particular asymptotic lines c_1, c_2, c_3, c_4 with a given generator $u = u_0$. Let v_1, v_2, v_3, v_4 be the values of the parameter v at these four points. Then, from (4), we have

$$v_1 = \frac{c_1P + Q}{c_1R + S}, v_2 = \frac{c_2P + Q}{c_2R + S}, v_3 = \frac{c_3P + Q}{c_3R + S}, v_4 = \frac{c_4P + Q}{c_4R + S}.$$

$$\therefore \quad v_1 - v_2 = \frac{c_1P + Q}{c_1R + S} - \frac{c_2P + Q}{c_2R + S} = \frac{(c_1 - c_2)(PS - QR)}{(c_1R + S)(c_2R + S)}.$$

Similarly,

$$v_3 - v_4 = \frac{(c_3 - c_4)(PS - QR)}{(c_3R + S)(c_4R + S)},$$

$$v_1 - v_3 = \frac{(c_1 - c_3)(PS - QR)}{(c_1R + S)(c_3R + S)},$$

$$v_2 - v_4 = \frac{(c_2 - c_4)(PS - QR)}{(c_2R + S)(c_4R + S)}.$$

$$\therefore \quad \frac{(v_1 - v_2)(v_3 - v_4)}{(v_1 - v_3)(v_2 - v_4)} = \frac{(c_1 - c_2)(c_3 - c_4)}{(c_1 - c_3)(c_2 - c_4)}, \tag{5}$$

which is independent of u_0 and is therefore the same for all generators. But v is the distance measured along the generator from the directrix.

Therefore $v_1 - v_2$ is the distance between the points v_1 and v_2. Hence it follows from (5) that the cross-ratio of the four points, in which a generator is cut by four given curved asymptotic lines, is the same for all generators.

Condition for a Ruled Surface to be Developable.

Theorem 1. The necessary and sufficient condition for a ruled surface to be developable is that $[\mathbf{t}, \mathbf{g}, \mathbf{g}'] = 0$,

where $\mathbf{t}$ is the unit tangent vector at a point of the base curve and $\mathbf{g}$ is the unit vector along the generator through that point.

Proof. Let P and Q be two consecutive points on some base curve of a ruled surface. Let position vectors of P and Q with respect to some vector origin O be $\mathbf{r}(s)$ and $\mathbf{r}(s+ds) = \mathbf{r} + \mathbf{t}\, ds$ respectively.

Also, let $\mathbf{g}(s)$ and $\mathbf{g}(s+ds) = \mathbf{g} + \mathbf{g}'\, ds$ be the unit vectors along the generators through P and Q respectively.

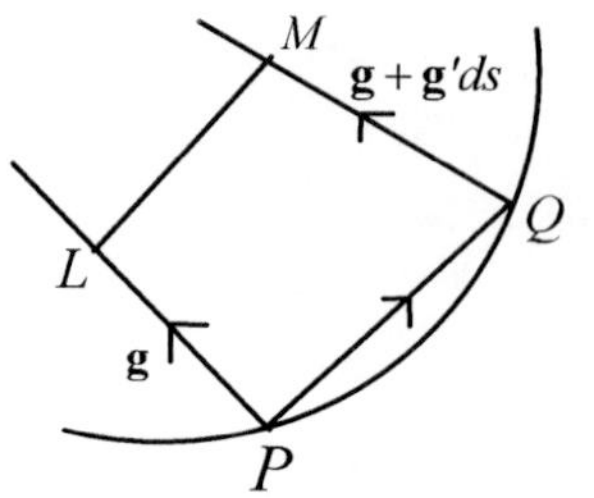

If LM be the shortest distance between the generators through P and Q, then LM is perpendicular to both these generators.

Also,

$$LM = \text{Projection of } PQ \text{ on } LM$$

i.e., $$LM = \overrightarrow{PQ} \cdot (\text{a unit vector along } LM). \tag{1}$$

Now,

$$\overrightarrow{PQ} = (\mathbf{r} + \mathbf{t}\, ds) - \mathbf{r} = \mathbf{t}\, ds. \tag{2}$$

Again,

LM is perpendicular to both the vectors $\mathbf{g}$ and $(\mathbf{g} + \mathbf{g}'\, ds)$.

$\Rightarrow$ LM is parallel to the vector $\mathbf{g} \times (\mathbf{g} + \mathbf{g}'\, ds)$.

$\Rightarrow$ LM is parallel to the vector $\mathbf{g} \times \mathbf{g}'$. $[\because \mathbf{g} \times (\mathbf{g} + \mathbf{g}'\, ds) = \mathbf{g} \times \mathbf{g}'\, ds]$

$\Rightarrow$ $\dfrac{\mathbf{g} \times \mathbf{g}'}{|\mathbf{g} \times \mathbf{g}'|}$ is a unit vector along LM.

$$\Rightarrow \qquad \frac{\mathbf{g}\times\mathbf{g}'}{|\mathbf{g}'|} = \text{a unit vector along } LM. \tag{3}$$

$$\left[\because |\mathbf{g}\times\mathbf{g}'| = |\mathbf{g}|\,|\mathbf{g}'|\sin\frac{\pi}{2} = |\mathbf{g}'|\right]$$

Using (2) and (3) in (1), we get

$$LM = \mathbf{t}\, ds \,.\, \frac{\mathbf{g}\times\mathbf{g}'}{|\mathbf{g}'|}$$

or $$LM = [\mathbf{t},\ \mathbf{g},\ \mathbf{g}']\ \frac{ds}{a}, \tag{4}$$

where $a = |\mathbf{g}'|$.

By definition, a ruled surface is developable if and only if the consecutive generators intersect. Hence it follows from (4) that a necessary and sufficient condition for a ruled surface to be developable is that $[\mathbf{t},\ \mathbf{g},\ \mathbf{g}'] = 0$.

Theorem 2. The necessary and sufficient condition for a ruled surface to be developable is that

$$a^2 \sin^2\theta = b^2,$$

where θ is the angle at which the generator $\mathbf{g}(s)$ of the ruled surface cuts the directrix, $a = |\mathbf{g}'|$ and $b = \mathbf{t}\,.\,\mathbf{g}'$.

Proof. Let the equation of the ruled surface be

$$\mathbf{R} = \mathbf{r}(s) + v\mathbf{g}(s), \tag{1}$$

where $\mathbf{r}(s)$ is the position vector of a current point P on the directrix, s is the arc length of the directrix for P measured from some fixed on it, $\mathbf{g}(s)$ is a unit vector along the generator at P and v is a parameter which measures the distance of a current point on the ruled surface from the directrix in the direction of $\mathbf{g}$.

Let θ be the angle at which the generator $\mathbf{g}(s)$ of the ruled surface cuts the directrix.

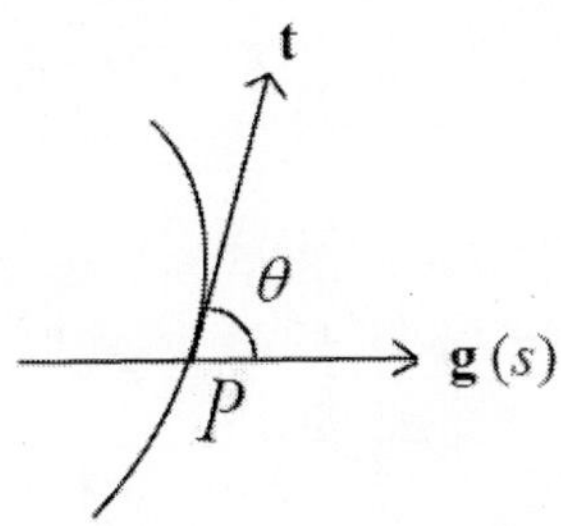

Taking s and v as the first and the second parameters, we have

$$\mathbf{R}_1 = \frac{\partial \mathbf{R}}{\partial s} = \frac{d\mathbf{r}}{ds} + v\frac{d\mathbf{g}}{ds} = \mathbf{t} + v\mathbf{g}', \quad \mathbf{R}_2 = \frac{\partial \mathbf{R}}{\partial v} = \mathbf{g}.$$

We know that a necessary and sufficient condition for a ruled surface to be developable is that

$$[\mathbf{t},\ \mathbf{g},\ \mathbf{g}'] = 0. \qquad (2)$$

Now,

$$[\mathbf{t},\ \mathbf{g},\ \mathbf{g}']^2 = \begin{vmatrix} \mathbf{t}^2 & \mathbf{t}.\mathbf{g} & \mathbf{t}.\mathbf{g}' \\ \mathbf{g}.\mathbf{t} & \mathbf{g}^2 & \mathbf{g}.\mathbf{g}' \\ \mathbf{g}'.\mathbf{t} & \mathbf{g}'.\mathbf{g} & \mathbf{g}'^2 \end{vmatrix} = \begin{vmatrix} 1 & \cos\theta & b \\ \cos\theta & 1 & 0 \\ b & 0 & a^2 \end{vmatrix}$$

$$= a^2 \sin^2\theta - b^2.$$

$$\therefore \qquad [\mathbf{t},\ \mathbf{g},\ \mathbf{g}'] = \pm\sqrt{(a^2 \sin^2\theta - b^2)}. \qquad (3)$$

It follows from (2) and (3) that a necessary and sufficient condition for a ruled surface to be developable is that $a^2 \sin^2\theta = b^2$.

Parameter of Distribution.

Definition. The parameter of distribution of a ruled surface is denoted by β and defined by

$$\beta = \frac{[\mathbf{r}',\ \mathbf{g},\ \mathbf{g}']}{a^2} = \frac{[\mathbf{t},\ \mathbf{g},\ \mathbf{g}']}{a^2} = \frac{D}{a^2}, \qquad (1)$$

where $a = |\mathbf{g}'|$, $D = [\mathbf{t},\ \mathbf{g},\ \mathbf{g}']$ and the position vector $\mathbf{R}$ of a general point on the ruled surface is given by

$$\mathbf{R} = \mathbf{r}(s) + v\mathbf{g}(s).$$

Properties of Parameter of Distribution.

(i) The parameter of distribution has a constant value at each point of a generator.

Proof. As the vectors $\mathbf{r}', \mathbf{g}, \mathbf{g}'$ are all functions of the parameter s only, it follows from (1) that the parameter of distribution β is a function of s only. But, along a generator, s is constant. Therefore β has a constant value at each point of a generator.

(ii) The parameter of distribution is independent of the particular base curve chosen.

Proof. By replacing $\mathbf{r}$ by $\mathbf{r} + w\mathbf{g}$ in (1), we get

$$\frac{[\mathbf{r}' + w\mathbf{g}',\ \mathbf{g},\ \mathbf{g}']}{a^2} = \frac{[\mathbf{r}',\ \mathbf{g},\ \mathbf{g}'] + w[\mathbf{g}',\ \mathbf{g},\ \mathbf{g}']}{a^2}$$

$$= \frac{[\mathbf{r}',\ \mathbf{g},\ \mathbf{g}']}{a^2}$$

$$= \beta.$$

Therefore the parameter of distribution is independent of the particular base curve chosen.

(iii) The parameter of distribution is independent of the choice of the parameter s.

Proof. If t be taken as the parameter instead of s, then

$$\frac{[\dot{\mathbf{r}},\ \mathbf{g},\ \dot{\mathbf{g}}]}{\dot{\mathbf{g}}^2} = \frac{\left[\dfrac{d\mathbf{r}}{dt},\ \mathbf{g},\ \dfrac{d\mathbf{g}}{dt}\right]}{\left(\dfrac{d\mathbf{g}}{dt}\right)^2}$$

$$= \frac{\left[\dfrac{d\mathbf{r}}{ds}\dfrac{ds}{dt},\ \mathbf{g},\ \dfrac{d\mathbf{g}}{ds}\dfrac{ds}{dt}\right]}{\left(\dfrac{d\mathbf{g}}{ds}\dfrac{ds}{dt}\right)^2}$$

$$= \frac{\left(\dfrac{ds}{dt}\right)^2\left[\dfrac{d\mathbf{r}}{ds},\ \mathbf{g},\ \dfrac{d\mathbf{g}}{ds}\right]}{\left(\dfrac{ds}{dt}\right)^2\left(\dfrac{d\mathbf{g}}{ds}\right)^2}$$

$$= \frac{\left[\dfrac{d\mathbf{r}}{ds},\ \mathbf{g},\ \dfrac{d\mathbf{g}}{ds}\right]}{\left(\dfrac{d\mathbf{g}}{ds}\right)^2}$$

$$= \frac{[\mathbf{r}',\ \mathbf{g},\ \mathbf{g}']}{a^2}$$

$$= \beta.$$

Therefore the parameter of distribution is independent of the choice of the parameter s.

Central point. Definition. Let P and Q be two consecutive points on some base curve of a ruled surface and let the common perpendicular to the generators through P and Q meet them at L and M respectively.

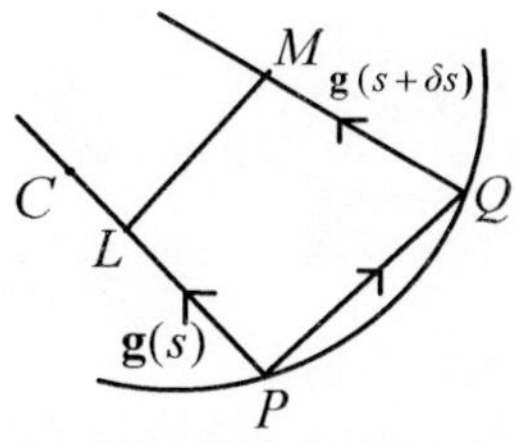

As $Q \to P$, the point L tends to a definite point C on the generator through P. This point C is called the central point of the generator through P.

Line of Striction. Definition. The locus of central points of all generators of a ruled surface is a definite curve lying on the ruled surface. This curve is called the line of striction of the ruled surface.

Equation of the Line of Striction. Let the equation of the ruled surface be

$$\mathbf{R} = \mathbf{r}(s) + v\mathbf{g}(s). \tag{1}$$

where s and v are the parameters.

Taking s and v as the first and the second parameters, we have

$$\mathbf{R}_1 = \frac{\partial \mathbf{R}}{\partial s} = \frac{d\mathbf{r}}{ds} + v\frac{d\mathbf{g}}{ds} = \mathbf{t} + v\mathbf{g}', \quad \mathbf{R}_2 = \frac{\partial \mathbf{R}}{\partial v} = \mathbf{g}.$$

$$\therefore \qquad H\mathbf{N} = \mathbf{R}_1 \times \mathbf{R}_2 = (\mathbf{t} + v\mathbf{g}') \times \mathbf{g}. \tag{2}$$

Let P and Q be two consecutive points on the directrix and let $\mathbf{g}(s)$ and $\mathbf{g}(s+\delta s)$ be the unit vectors along the generators through P and Q respectively. Let LM be the common perpendicular to the generators through P and Q. As $Q \to P$, the point L tends to a definite point C on the generator through P. Then C is the central point of the generator through P.

As $Q \to P$, the vector $\overrightarrow{LM}$ tends to be tangential to the surface at the point C. Therefore the limiting direction of the vector $\overrightarrow{LM}$ is perpendicular to $\mathbf{N}$, where $\mathbf{N}$ is normal to the surface at C. Also, the limiting direction of the vector $\overrightarrow{LM}$ is perpendicular to the generator through P, i.e., perpendicular to the vector $\mathbf{g}$. Therefore it is parallel to the vector $\mathbf{g} \times \mathbf{N}$. Again,

$\overrightarrow{LM}$ is perpendicular to both the generators through P and Q.

$$\Rightarrow \qquad \overrightarrow{LM} \cdot \mathbf{g}(s) = 0 \text{ and } \overrightarrow{LM} \cdot \mathbf{g}(s+\delta s) = 0$$

$$\Rightarrow \qquad \overrightarrow{LM} \cdot \left(\frac{\mathbf{g}(s+\delta s) - \mathbf{g}(s)}{\delta s} \right) = 0.$$

Taking the limit as $Q \to P$, we get

$$(\mathbf{g}\times\mathbf{N}).\mathbf{g}'=0$$

$$\Rightarrow \quad \mathbf{g}'.(\mathbf{g}\times\mathbf{N})=0$$

$$\Rightarrow \quad (\mathbf{g}'\times\mathbf{g}).\mathbf{N}=0$$

$$\Rightarrow \quad (\mathbf{g}'\times\mathbf{g}).\{(\mathbf{t}+v\mathbf{g}')\times\mathbf{g}\}=0 \qquad \text{[using (2)]}$$

$$\Rightarrow \quad (\mathbf{g}'\times\mathbf{g}).\{(\mathbf{t}\times\mathbf{g})+v(\mathbf{g}'\times\mathbf{g})\}=0$$

$$\Rightarrow \quad (\mathbf{g}'\times\mathbf{g}).(\mathbf{t}\times\mathbf{g})+v(\mathbf{g}'\times\mathbf{g})^2=0. \qquad (3)$$

Now,

$$(\mathbf{a}\times\mathbf{b}).(\mathbf{c}\times\mathbf{d})=\begin{vmatrix}\mathbf{a}.\mathbf{c} & \mathbf{a}.\mathbf{d}\\ \mathbf{b}.\mathbf{c} & \mathbf{b}.\mathbf{d}\end{vmatrix}.$$

$$\therefore \quad (\mathbf{g}'\times\mathbf{g}).(\mathbf{t}\times\mathbf{g})=\begin{vmatrix}\mathbf{g}'.\mathbf{t} & \mathbf{g}'.\mathbf{g}\\ \mathbf{g}.\mathbf{t} & \mathbf{g}.\mathbf{g}\end{vmatrix}=\begin{vmatrix} b & 0\\ \cos\theta & 1\end{vmatrix}=b, \qquad (4)$$

where $b=\mathbf{g}'.\mathbf{t}$.

Also,

$$(\mathbf{a}\times\mathbf{b})^2=\mathbf{a}^2\,\mathbf{b}^2-(\mathbf{a}.\mathbf{b})^2.$$

$$\therefore \quad (\mathbf{g}'\times\mathbf{g})^2=\mathbf{g}'^2\mathbf{g}^2-(\mathbf{g}'.\mathbf{g})^2=\mathbf{g}'^2=a^2, \qquad (5)$$

where $a=|\mathbf{g}'|$.

Using (4) and (5) in (3), we get

$$b+va^2=0$$

or

$$v=-\frac{b}{a^2}. \qquad (6)$$

Equation (6) gives the distance v of the central point C of the generator from the directrix.

If **R** is the position vector of C, then

$$\mathbf{R}=\mathbf{r}(s)+v\mathbf{g}(s), \text{ where } v=-\frac{b}{a^2}.$$

The parametric equation of the line of striction is

$$a^2v+b=0.$$

The vector equation of the line of striction is

$$\mathbf{R}=\mathbf{r}(s)+v\mathbf{g}(s), \text{ where } v=-\frac{b}{a^2}.$$

Corollary. The line of striction is directrix itself if and only if $b=0$.

Tangent Plane. The tangent plane to a developable is the same at all points of a given generator. But this is not so in case of a skew surface.

We shall now show that, as the point of contact moves along a generator from one end to the other, the tangent plane turns through an angle of 180°. For this, we shall find the inclination of the tangent plane at any point P of a generator to the tangent plane at the central point of the same generator. The tangent plane at the central point is called the central plane of the generator.

Let the equation of the ruled surface be

$$\mathbf{R} = \mathbf{r}(s) + v\mathbf{g}(s), \tag{1}$$

where s and v are the parameters.

Taking s and v as the first and the second parameters, we have

$$\mathbf{R}_1 = \frac{\partial \mathbf{R}}{\partial s} = \frac{d\mathbf{r}}{ds} + v\frac{d\mathbf{g}}{ds} = \mathbf{t} + v\mathbf{g}', \quad \mathbf{R}_2 = \frac{\partial \mathbf{R}}{\partial v} = \mathbf{g}.$$

$$\therefore \qquad H\mathbf{N} = \mathbf{R}_1 \times \mathbf{R}_2 = (\mathbf{t} + v\mathbf{g}') \times \mathbf{g}. \tag{2}$$

We know that the distance v of the central point of the generator from the directrix is given by

$$v = -\frac{b}{a^2} \tag{3}$$

and the equation of the line of striction is given by

$$a^2 v + b = 0, \tag{4}$$

where $a = |\mathbf{g}'|$, $b = \mathbf{g}'.\mathbf{t}$.

Let us take the line of striction as directrix. Then

$$b = 0 \tag{5}$$

and the central point of the generator is given by

$$v = 0. \tag{6}$$

Squaring (2), we get

$$H^2 = \{(\mathbf{t} + v\mathbf{g}') \times \mathbf{g}\}^2 = \{(\mathbf{t} \times \mathbf{g}) + (v\mathbf{g}' \times \mathbf{g})\}^2$$

or

$$H^2 = (\mathbf{t} \times \mathbf{g})^2 + 2v(\mathbf{t} \times \mathbf{g}).(\mathbf{g}' \times \mathbf{g}) + v^2(\mathbf{g}' \times \mathbf{g})^2. \tag{7}$$

Now,

$$(\mathbf{a} \times \mathbf{b})^2 = \mathbf{a}^2\mathbf{b}^2 - (\mathbf{a}.\mathbf{b})^2.$$

Therefore

$$\left.\begin{aligned} (\mathbf{t} \times \mathbf{g})^2 &= \mathbf{t}^2\mathbf{g}^2 - (\mathbf{t}.\mathbf{g})^2 = 1 - \cos^2\theta = \sin^2\theta, \\ (\mathbf{g}' \times \mathbf{g})^2 &= \mathbf{g}'^2\mathbf{g}^2 - (\mathbf{g}'.\mathbf{g})^2 = \mathbf{g}'^2 - 0 = a^2. \end{aligned}\right\} \tag{8}$$

Also, $(\mathbf{a} \times \mathbf{b}).(\mathbf{c} \times \mathbf{d}) = \begin{vmatrix} \mathbf{a}.\mathbf{c} & \mathbf{a}.\mathbf{d} \\ \mathbf{b}.\mathbf{c} & \mathbf{b}.\mathbf{d} \end{vmatrix}$.

$$\therefore \qquad (\mathbf{t} \times \mathbf{g}).(\mathbf{g}' \times \mathbf{g}) = \begin{vmatrix} \mathbf{t}.\mathbf{g}' & \mathbf{t}.\mathbf{g} \\ \mathbf{g}.\mathbf{g}' & \mathbf{g}.\mathbf{g} \end{vmatrix} = \begin{vmatrix} b & \cos\theta \\ 0 & 1 \end{vmatrix} = b$$

or $\quad (\mathbf{t}\times\mathbf{g})\,.\,(\mathbf{g}'\times\mathbf{g})=0.$ (9)

[using (5)]

Using (8) and (9), (7) becomes

$$H^2=a^2v^2+\sin^2\theta. \qquad 10)$$

Also, we have

$$D=[\,\mathbf{t},\ \mathbf{g},\ \mathbf{g}'\,]=\pm\sqrt{(a^2\sin^2\theta-b^2)}=\pm\,a\sin\theta. \qquad (11)$$

It follows from (6) and (10) that for the central point, we have

$$H=\sin\theta.$$

Hence the unit normal $\mathbf{N}_0$ at the central point is given by

$$\mathbf{N}_0=\frac{\mathbf{t}\times\mathbf{g}}{\sin\theta}. \qquad (12)$$

Let ϕ be the angle of rotation (in the sense which is positive for the direction **g**) from the central plane to the tangent plane at the point distant v from the central point. Then ϕ is equal to the angle of rotation from the normal $\mathbf{N}_0$ to the normal **N**. Therefore

$$\mathbf{N}_0\,.\,\mathbf{N}=\frac{1}{H\sin\theta}(\mathbf{t}\times\mathbf{g})\,.\{(\mathbf{t}\times\mathbf{g})+(v\mathbf{g}'\times\mathbf{g})\} \qquad \text{[using (2) and (12)]}$$

or $\quad |\,\mathbf{N}_0|\,|\,\mathbf{N}\,|\cos\phi=\dfrac{1}{H\sin\theta}\{(\mathbf{t}\times\mathbf{g})^2+v\,(\mathbf{t}\times\mathbf{g})\,.\,(\mathbf{g}'\times\mathbf{g})\}]$

or $\quad \cos\phi=\dfrac{1}{H\sin\theta}(\sin^2\theta+0)=\dfrac{\sin\theta}{H}$ [using (8) and (9)]

or $\quad \cos\phi=\dfrac{1}{H}\sqrt{H^2-a^2v^2}.$ [using (10)]

$\therefore \quad \tan\phi=\dfrac{av}{\sqrt{H^2-a^2v^2}}=\dfrac{av}{\sin\theta}$ [using (10)]

or $\quad \tan\phi=\dfrac{a^2v}{D}=\dfrac{v}{\beta},$ (13)

[using (11)]

where β is the parameter of distribution, which is constant along a generator.

From (13), it follows that the tangent of the angle through which the normal **N** rotates (i.e., $\tan\phi$) as the point P moves along a generator varies directly with the distance (v) moved from the central point.

As v increases from $-\infty$ to ∞, the angle ϕ increases from $-\dfrac{\pi}{2}$ to $\dfrac{\pi}{2}$ if $\beta>0$ and ϕ decreases from $\dfrac{\pi}{2}$ to $-\dfrac{\pi}{2}$ if $\beta<0$.

At the central point, $v = 0$. Therefore, from (13), we have $\phi = 0$.

Hence we can conclude that as the point of contact moves from one end of the generator to the other, the tangent plane turns through an angle of $180°$; and the tangent planes at the ends of the generator are perpendicular to the central plane.

Solved Examples

Example 1. Show that the ruled surface generated by the binormals of a skew curve has the curve itself as line of striction.

Also, find the parameter of distribution and the Gaussian curvature for this ruled surface.

Solution. First Part. Let the equation of the given skew curve be

$$\mathbf{r} = \mathbf{r}(s). \tag{1}$$

Let $\mathbf{t}$, $\mathbf{n}$, $\mathbf{b}$ be unit tangent, unit principal normal and unit binormal respectively at any point $P(\mathbf{r})$ of this curve.

For the ruled surface generated by the binormals of (1), take the curve (1) itself as the directrix.

We know that the line of striction is directrix itself if and only if $\mathbf{t} \cdot \mathbf{g}' = 0$. Therefore, in order to prove that the line of striction of the ruled surface generated by the binormals of curve (1) is the curve (1) itself, it is sufficient to prove that $\mathbf{t} \cdot \mathbf{g}' = 0$.

Since the binormals of curve (1) are the generators of the ruled surface, the unit vector $\mathbf{g}(s)$ along the generator of the ruled surface passing through the point $P(\mathbf{r})$ of the directrix is given by

$$\mathbf{g} = \mathbf{b}$$

$$\Rightarrow \quad \mathbf{g}' = \frac{d\mathbf{b}}{ds} = -\tau\mathbf{n}$$

$$\Rightarrow \quad \mathbf{t} \cdot \mathbf{g}' = \mathbf{t} \cdot (-\tau\mathbf{n}) = -\tau(\mathbf{t} \cdot \mathbf{n}) = 0.$$

Therefore the line of striction is directrix itself, i.e. the line of striction is the curve itself.

Second Part. The parameter of distribution β is given by

$$\beta = \frac{[\mathbf{r}', \mathbf{g}, \mathbf{g}']}{\mathbf{g}'^2} = \frac{[\mathbf{t}, \mathbf{b}, -\tau\mathbf{n}]}{(-\tau\mathbf{n})^2} = \frac{\mathbf{t} \times \mathbf{b} \cdot (-\tau\mathbf{n})}{\tau^2}$$

$$= \frac{(-\mathbf{n}) \cdot (-\tau\mathbf{n})}{\tau^2} = \frac{\tau(\mathbf{n} \cdot \mathbf{n})}{\tau^2} = \frac{1}{\tau} = \sigma,$$

where σ is the radius of torsion of the curve.

The Gaussian curvature K is given by

$$K = -\frac{[\mathbf{r}', \mathbf{g}, \mathbf{g}']^2}{H^4} = -\frac{\tau^2}{H^4}. \tag{2}$$

Now, we have to find the value of H.

The position vector $\mathbf{R}$ of a general point on the ruled surface is given by

$$\mathbf{R} = \mathbf{r}(s) + v\mathbf{g}(s) = \mathbf{r}(s) + v\mathbf{b}(s).$$

Taking s and v as the first and second parameters, we have

$$\mathbf{R}_1 = \frac{\partial \mathbf{R}}{\partial s} = \frac{d\mathbf{r}}{ds} + v\frac{d\mathbf{b}}{ds} = \mathbf{t} + v(-\tau\mathbf{n}) = \mathbf{t} - v\tau\mathbf{n},\ \mathbf{R}_2 = \frac{\partial \mathbf{R}}{\partial v} = \mathbf{b}.$$

Therefore

$$H\mathbf{N} = \mathbf{R}_1 \times \mathbf{R}_2 = (\mathbf{t} - v\tau\mathbf{n}) \times \mathbf{b} = -\mathbf{n} - v\tau\mathbf{t}$$

$$\Rightarrow \qquad H^2 = 1 + v^2\tau^2.$$

Substituting this value in (2), we get

$$K = -\frac{\tau^2}{(1 + v^2\tau^2)^2}.$$

Example 2. Show that the parameter of distribution of the ruled surface generated by the principal normals of a skew curve is equal to $\tau(\kappa^2 + \tau^2)^{-1}$, where κ and τ are the curvature and torsion of the curve. Also, find the distance of the central point from the curve and the Gaussian curvature for this ruled surface.

What is the value of the Gaussian curvature at a point on the curve itself ?

Solution. First Part. Let the equation of the given skew curve be

$$\mathbf{r} = \mathbf{r}(s). \tag{1}$$

Let $\mathbf{t}, \mathbf{n}, \mathbf{b}$ be unit tangent, unit principal normal and unit binormal respectively at any point $P(\mathbf{r})$ of this curve.

For the ruled surface generated by the principal normals of (1), take the curve (1) itself as the directrix.

Since the binormals of curve (1) are the generators of the ruled surface, the unit vector $\mathbf{g}(s)$ along the generator of the ruled surface passing through the point $P(\mathbf{r})$ of the directrix is given by

$$\mathbf{g} = \mathbf{n}$$

$$\Rightarrow \qquad \mathbf{g}' = \frac{d\mathbf{n}}{ds} = \tau\mathbf{b} - \kappa\mathbf{t}.$$

The parameter of distribution β is given by

$$\beta = \frac{[\mathbf{r}', \mathbf{g}, \mathbf{g}']}{\mathbf{g}'^2} = \frac{[\mathbf{t}, \mathbf{n}, \tau\mathbf{b} - \kappa\mathbf{t}]}{(\tau\mathbf{b} - \kappa\mathbf{t})^2} = \frac{\mathbf{t} \times \mathbf{n} . (\tau\mathbf{b} - \kappa\mathbf{t})}{\tau^2 + \kappa^2}$$

$$= \frac{\mathbf{b} . (\tau\mathbf{b} - \kappa\mathbf{t})}{\tau^2 + \kappa^2} = \frac{\tau (\mathbf{b} . \mathbf{b})}{\tau^2 + \kappa^2} = \frac{\tau}{\kappa^2 + \tau^2}.$$

Second Part. The distance v of the central point from the curve is given by

$$v = -\frac{\mathbf{g}' . \mathbf{r}'}{\mathbf{g}'^2} = -\frac{(\tau\mathbf{b} - \kappa\mathbf{t}) . \mathbf{t}}{(\tau\mathbf{b} - \kappa\mathbf{t})^2} = \frac{\kappa}{\kappa^2 + \tau^2}.$$

The Gaussian curvature K is given by

$$K = -\frac{[\mathbf{r}', \mathbf{g}, \mathbf{g}']^2}{H^4} = -\frac{\tau^2}{H^4}. \quad (2)$$

Now, we have to find the value of H.

The position vector $\mathbf{R}$ of a general point on the ruled surface is given by

$$\mathbf{R} = \mathbf{r}(s) + v\mathbf{g}(s) = \mathbf{r}(s) + v\mathbf{n}(s).$$

Taking s and v as the first and the second parameters, we have

$$\mathbf{R}_1 = \frac{\partial \mathbf{R}}{\partial s} = \frac{d\mathbf{r}}{ds} + v\frac{d\mathbf{n}}{ds} = \mathbf{t} + v(\tau\mathbf{b} - \kappa\mathbf{t}) = (1 - v\kappa)\mathbf{t} + v\tau\mathbf{b},$$

$$\mathbf{R}_2 = \frac{\partial \mathbf{R}}{\partial v} = \mathbf{n}.$$

$$\therefore \quad H\mathbf{N} = \mathbf{R}_1 \times \mathbf{R}_2 = \{(1 - v\kappa)\mathbf{t} + v\tau\mathbf{b}\} \times \mathbf{n} = (1 - v\kappa)\mathbf{b} - v\tau\mathbf{t}$$

$$\Rightarrow \quad H^2 = (1 - v\kappa)^2 + v^2\tau^2.$$

Substituting this value in (2), we get

$$K = -\frac{\tau^2}{\{(1 - v\kappa)^2 + v^2\tau^2\}^2}.$$

At a point on the curve itself, the Gaussian curvature is $-\tau^2$.

Example 3. A straight line cuts a twisted curve at a constant angle and lies in the rectifying plane. Show that, on the surface which it generates, the given curve is the line of striction. Find the parameter of distribution and the specific curvature.

Solution. First Part. Let the equation of the given skew curve be

$$\mathbf{r} = \mathbf{r}(s). \quad (1)$$

Let $\mathbf{t}$, $\mathbf{n}$, $\mathbf{b}$ be unit tangent, unit principal normal and unit binormal respectively at any point $P(\mathbf{r})$ of this curve and let the curve itself be taken as the directrix.

We know that the line of striction is directrix itself if and only if $\mathbf{t}\,.\,\mathbf{g}' = 0$. Therefore, in order to prove that the line of striction of the ruled surface generated by the given line is the curve (1) itself, it is sufficient to prove that $\mathbf{t}\,.\,\mathbf{g}' = 0$.

Since the generator of the ruled surface lies in the rectifying plane, the unit vector $\mathbf{g}(s)$ along the generator of the ruled surface passing through the point $P(\mathbf{r})$ of the directrix is given by

$$\mathbf{g} = \frac{\mathbf{t} + c\mathbf{b}}{\sqrt{1+c^2}}, \text{ where } c \text{ is a constant}$$

$$\Rightarrow \qquad \mathbf{g}' = \frac{\mathbf{t}' + c\mathbf{b}'}{\sqrt{1+c^2}} = \frac{\kappa\mathbf{n} + c(-\tau\mathbf{n})}{\sqrt{1+c^2}} = \frac{\kappa - c\tau}{\sqrt{1+c^2}}\mathbf{n}$$

$$\Rightarrow \qquad \mathbf{t}\,.\,\mathbf{g}' = \mathbf{t}\,.\left(\frac{\kappa - c\tau}{\sqrt{1+c^2}}\mathbf{n}\right) = 0.$$

Therefore the line of striction is directrix itself, i.e. the line of striction is the curve itself.

Second Part. The parameter of distribution β is given by

$$\beta = \frac{[\mathbf{r}', \mathbf{g}, \mathbf{g}']}{\mathbf{g}'^2} = \frac{\left[\mathbf{t}, \frac{\mathbf{t}+c\mathbf{b}}{\sqrt{1+c^2}}, \frac{\kappa - c\tau}{\sqrt{1+c^2}}\mathbf{n}\right]}{\left(\frac{\kappa - c\tau}{\sqrt{1+c^2}}\mathbf{n}\right)^2}$$

$$= \frac{\mathbf{t}\times\left(\frac{\mathbf{t}+c\mathbf{b}}{\sqrt{1+c^2}}\right).\frac{\kappa - c\tau}{\sqrt{1+c^2}}\mathbf{n}}{\left(\frac{\kappa - c\tau}{\sqrt{1+c^2}}\right)^2} = \frac{\left(-\frac{c\mathbf{n}}{\sqrt{1+c^2}}\right).\frac{\kappa - c\tau}{\sqrt{1+c^2}}\mathbf{n}}{\left(\frac{\kappa - c\tau}{\sqrt{1+c^2}}\right)^2}$$

$$= \frac{c}{c\tau - \kappa}.$$

The Gaussian curvature K is given by

$$K = -\frac{[\mathbf{r}', \mathbf{g}, \mathbf{g}']^2}{H^4} = -\frac{\left\{\frac{c(\kappa - c\tau)}{1+c^2}\right\}^2}{H^4} = -\frac{c^2(\kappa - c\tau)^2}{(1+c^2)^2 H^4}. \tag{2}$$

Now, we have to find the value of H.

The position vector $\mathbf{R}$ of a general point on the ruled surface is given by

$$\mathbf{R} = \mathbf{r}(s) + v\mathbf{g}(s) = \mathbf{r}(s) + v\frac{\mathbf{t}(s) + c\mathbf{b}(s)}{\sqrt{1+c^2}}.$$

Taking s and v as the first and the second parameters, we have

$$\mathbf{R}_1 = \frac{\partial \mathbf{R}}{\partial s} = \frac{d\mathbf{r}}{ds} + \frac{v}{\sqrt{1+c^2}}\left(\frac{d\mathbf{t}}{ds} + c\frac{d\mathbf{b}}{ds}\right)$$

$$= \mathbf{t} + \frac{v}{\sqrt{1+c^2}}\{\kappa\mathbf{n} + c(-\tau\mathbf{n})\}$$

$$= \mathbf{t} + \frac{v(\kappa - c\tau)}{\sqrt{1+c^2}}\mathbf{n},$$

$$\mathbf{R}_2 = \frac{\partial \mathbf{R}}{\partial v} = \frac{\mathbf{t} + c\mathbf{b}}{\sqrt{1+c^2}}.$$

Therefore

$$H\mathbf{N} = \mathbf{R}_1 \times \mathbf{R}_2 = \left\{\mathbf{t} + \frac{v(\kappa - c\tau)}{\sqrt{1+c^2}}\mathbf{n}\right\} \times \frac{\mathbf{t} + c\mathbf{b}}{\sqrt{1+c^2}}$$

$$= \frac{cv(\kappa - c\tau)}{1+c^2}\mathbf{t} - \frac{c}{\sqrt{1+c^2}}\mathbf{n} - \frac{v(\kappa - c\tau)}{1+c^2}\mathbf{b}$$

$$\Rightarrow \qquad H^2 = \frac{c^2v^2(\kappa - c\tau)^2}{(1+c^2)^2} + \frac{c^2}{1+c^2} + \frac{v^2(\kappa - c\tau)^2}{(1+c^2)^2}$$

$$= \frac{v^2(\kappa - c\tau)^2}{1+c^2} + \frac{c^2}{1+c^2}$$

$$= \frac{v^2(\kappa - c\tau)^2 + c^2}{1+c^2}.$$

Substituting this value in (2), we get

$$K = -\frac{c^2(\kappa - c\tau)^2}{\{v^2(\kappa - c\tau)^2 + c^2\}^2}.$$

Example 4. Obtain the formula

$$K = -\frac{p^2}{(p^2 + v^2)^2}$$

for the Gaussian curvature at the point distant v from the central point on a generator of parameter p. Hence show that on any one generator, the Gaussian curvature is greatest in absolute value at the central point; and at points equidistant from this, it has equal values.

Solution. Let the equation of the ruled surface be

$$\mathbf{R} = \mathbf{r}(s) + v\mathbf{g}(s), \tag{1}$$

where s and v are the parameters.

Taking s and v as the first and the second parameters, we have

$$\mathbf{R}_1 = \frac{\partial \mathbf{R}}{\partial s} = \frac{d\mathbf{r}}{ds} + v\frac{d\mathbf{g}}{ds} = \mathbf{t} + v\mathbf{g}', \quad \mathbf{R}_2 = \frac{\partial \mathbf{R}}{\partial v} = \mathbf{g}.$$

$$\therefore \quad H\mathbf{N} = \mathbf{R}_1 \times \mathbf{R}_2 = (\mathbf{t} + v\mathbf{g}') \times \mathbf{g}. \tag{2}$$

We know that the distance v of the central point of the generator from the directrix is given by

$$v = -\frac{b}{a^2} \tag{3}$$

and the equation of the line of striction is given by

$$a^2 v + b = 0, \tag{4}$$

where $a = |\mathbf{g}'|$, $b = \mathbf{g}'.\mathbf{t}$.

Let us take the line of striction as directrix. Then

$$b = 0 \tag{5}$$

and the central point of the generator is given by

$$v = 0. \tag{6}$$

Squaring (2), we get

$$H^2 = \{(\mathbf{t} + v\mathbf{g}') \times \mathbf{g}\}^2 = \{(\mathbf{t} \times \mathbf{g}) + (v\mathbf{g}' \times \mathbf{g})\}^2$$

or
$$H^2 = (\mathbf{t} \times \mathbf{g})^2 + 2v(\mathbf{t} \times \mathbf{g}) . (\mathbf{g}' \times \mathbf{g}) + v^2 (\mathbf{g}' \times \mathbf{g})^2. \tag{7}$$

Now,

$$(\mathbf{a} \times \mathbf{b})^2 = \mathbf{a}^2 \mathbf{b}^2 - (\mathbf{a} . \mathbf{b})^2.$$

Therefore

$$\left.\begin{aligned} (\mathbf{t} \times \mathbf{g})^2 &= \mathbf{t}^2\mathbf{g}^2 - (\mathbf{t} . \mathbf{g})^2 = 1 - \cos^2\theta = \sin^2\theta, \\ (\mathbf{g}' \times \mathbf{g})^2 &= \mathbf{g}'^2\mathbf{g}^2 - (\mathbf{g}' . \mathbf{g})^2 = \mathbf{g}'^2 - 0 = a^2. \end{aligned}\right\} \tag{8}$$

Also,

$$(\mathbf{a} \times \mathbf{b}).(\mathbf{c} \times \mathbf{d}) = \begin{vmatrix} \mathbf{a}.\mathbf{c} & \mathbf{a}.\mathbf{d} \\ \mathbf{b}.\mathbf{c} & \mathbf{b}.\mathbf{d} \end{vmatrix}.$$

$$\therefore \quad (\mathbf{t} \times \mathbf{g}).(\mathbf{g}' \times \mathbf{g}) = \begin{vmatrix} \mathbf{t}.\mathbf{g}' & \mathbf{t}.\mathbf{g} \\ \mathbf{g}.\mathbf{g}' & \mathbf{g}.\mathbf{g} \end{vmatrix} = \begin{vmatrix} b & \cos\theta \\ 0 & 1 \end{vmatrix} = b$$

or
$$(\mathbf{t} \times \mathbf{g}).(\mathbf{g}' \times \mathbf{g}) = 0. \tag{9}$$

[using (5)]

Using (8) and (9), (7) becomes

$$H^2 = a^2v^2 + \sin^2\theta. \tag{10}$$

Also, we have

$$D = [\mathbf{t}, \mathbf{g}, \mathbf{g}'] = \pm\sqrt{(a^2\sin^2\theta - b^2)} = \pm a\sin\theta. \tag{11}$$

The parameter of distribution p is defined by

$$p = \frac{[\mathbf{r}', \mathbf{g}, \mathbf{g}']}{a^2} = \frac{[\mathbf{t}, \mathbf{g}, \mathbf{g}']}{a^2} = \frac{D}{a^2}$$

or $$p = \pm\frac{a\sin\theta}{a^2} = \pm\frac{\sin\theta}{a}. \qquad \text{[using (11)]}$$

$\therefore$ $$\sin\theta = \pm\, ap. \qquad (12)$$

Now, the Gaussian curvature K is given by

$$K = -\frac{[\mathbf{r}', \mathbf{g}, \mathbf{g}']^2}{H^4} = -\frac{[\mathbf{t}, \mathbf{g}, \mathbf{g}']^2}{H^4}$$

$$= -\frac{a^2\sin^2\theta}{(a^2v^2 + \sin^2\theta)^2} \qquad \text{[using (10) and (11)]}$$

$$= -\frac{a^4p^2}{(a^2v^2 + a^2p^2)^2} \qquad \text{[using (12)]}$$

i.e., $$K = -\frac{p^2}{(p^2 + v^2)^2}. \qquad (13)$$

This proves the required formula.

Now, p has the same value at each point of a generator. Therefore it follows from (13) that K will be greatest at that point of the generator at which v^2 is least, i.e., at which $v = 0$, i.e., at the central point. Hence, on any one generator, the Gaussian curvature K is greatest in absolute value at the central point.

Also, at points equidistant from the central point, say, at points $v = \pm\, v_0$, we have

$$K = -\frac{p^2}{\{p^2 + (\pm\, v_0)^2\}^2} = -\frac{p^2}{(p^2 + v_0^{\,2})^2}. \qquad \text{[using (13)]}$$

Therefore the Gaussian curvature K has equal values.

Example 5. For the cylindroid $z(x^2 + y^2) = 2mxy$, prove that the parameter of distribution of the generator in the plane $x\sin\theta = y\cos\theta$ is $2m\cos 2\theta.$.

Solution. The equation of the cylindroid is

$$z(x^2 + y^2) = 2mxy. \qquad (1)$$

The equation of the given plane is

$$x\sin\theta = y\cos\theta. \qquad (2)$$

Therefore the equation of a generator of the cylindroid (1) in given plane (2) is

$$\frac{x-0}{\cos\theta}=\frac{y-0}{\sin\theta}=\frac{z-2m\sin\theta\cos\theta}{0}$$

or $$\frac{x-0}{\cos\theta}=\frac{y-0}{\sin\theta}=\frac{z-m\sin 2\theta}{0}=v \text{ (let)}. \tag{3}$$

Hence the position vector $\mathbf{R}$ of a general point on the given surface (1) can be written as

$$\mathbf{R}=\mathbf{r}(\theta)+v\mathbf{g}(\theta),$$

where

$$\mathbf{r}(\theta)=(0,0,m\sin 2\theta),\ \ \mathbf{g}(\theta)=(\cos\theta,\sin\theta,0).$$

The parameter of distribution β is given by

$$\beta=\frac{\left[\dfrac{d\mathbf{r}}{d\theta},\ \mathbf{g},\ \dfrac{d\mathbf{g}}{d\theta}\right]}{\left(\dfrac{d\mathbf{g}}{d\theta}\right)^2}=\frac{\left(\dfrac{d\mathbf{r}}{d\theta}\times\mathbf{g}\right)\cdot\dfrac{d\mathbf{g}}{d\theta}}{\left(\dfrac{d\mathbf{g}}{d\theta}\right)^2}. \tag{4}$$

Now,

$$\frac{d\mathbf{r}}{d\theta}=(0,0,2m\cos 2\theta),$$

$$\frac{d\mathbf{g}}{d\theta}=(-\sin\theta,\cos\theta,0).$$

Therefore

$$\frac{d\mathbf{r}}{d\theta}\times\ \mathbf{g}=(-2m\sin\theta\cos 2\theta,2m\cos\theta\cos 2\theta,0)$$

$$\Rightarrow\quad \left(\frac{d\mathbf{r}}{d\theta}\times\mathbf{g}\right)\cdot\frac{d\mathbf{g}}{d\theta}=2m\cos 2\theta\,(\sin^2\theta+\cos^2\theta)$$

i.e., $$\left(\frac{d\mathbf{r}}{d\theta}\times\mathbf{g}\right)\cdot\frac{d\mathbf{g}}{d\theta}=2m\cos 2\theta. \tag{5}$$

Again,

$$\left(\frac{d\mathbf{g}}{d\theta}\right)^2=(-\sin\theta)^2+\cos^2\theta+0=1. \tag{6}$$

Substituting the values of $\left(\dfrac{d\mathbf{r}}{d\theta}\times\mathbf{g}\right)\cdot\dfrac{d\mathbf{g}}{d\theta}$ and $\left(\dfrac{d\mathbf{g}}{d\theta}\right)^2$ from (5) and (6) respectively in (4), we get

$$\beta=2m\cos 2\theta.$$

Example 6. For the skew surface generated by the line

$$x+yt=3t\,(1+t^2),\ \ y+2zt=t^2(3+4t^2),$$

prove that the parameter of distribution of a generator is

$$\frac{3}{2}(1+2t^2)^2,$$

and that the line of striction is the curve

$$x = 3t,\ y = 3t^2,\ z = 2t^3.$$

Solution. The equation of a generator of the skew surface is

$$x + yt = 3t\,(1+t^2),\ y + 2zt = t^2(3+4t^2)$$

or $$x - 3t = -t\,(y - 3t^2),\ y - 3t^2 = -2t\,(z - 2t^3)$$

or $$\frac{x-3t}{t} = \frac{y-3t^2}{-1} = \frac{z-2t^3}{(1/2t)}$$

or $$\frac{x-3t}{\left(\dfrac{2t^2}{1+2t^2}\right)} = \frac{y-3t^2}{\left(-\dfrac{2t}{1+2t^2}\right)} = \frac{z-2t^3}{\left(\dfrac{1}{1+2t^2}\right)} = v\,(\text{let}). \tag{1}$$

Hence the position vector of a general point on the ruled surface generated by (1) can be written as

$$\mathbf{R} = \mathbf{r}\,(t) + v\mathbf{g}\,(t),$$

where

$$\mathbf{r}\,(t) = (3t, 3t^2, 2t^3),$$

$$\mathbf{g}\,(t) = \left(\frac{2t^2}{1+2t^2}, -\frac{2t}{1+2t^2}, \frac{1}{1+2t^2}\right).$$

The parameter of distribution β is given by

$$\beta = \frac{\left[\dfrac{d\mathbf{r}}{dt}, \mathbf{g}, \dfrac{d\mathbf{g}}{dt}\right]}{\left(\dfrac{d\mathbf{g}}{dt}\right)^2} = \frac{\left(\dfrac{d\mathbf{r}}{dt} \times \mathbf{g}\right) \cdot \dfrac{d\mathbf{g}}{dt}}{\left(\dfrac{d\mathbf{g}}{dt}\right)^2}. \tag{2}$$

Now,

$$\frac{d\mathbf{r}}{dt} = (3, 6t, 6t^2),$$

$$\frac{d\mathbf{g}}{dt} = \left(\frac{4t}{(1+2t^2)^2}, \frac{2\,(2t^2-1)}{(1+2t^2)^2}, -\frac{4t}{(1+2t^2)^2}\right).$$

Therefore

$$\frac{d\mathbf{r}}{dt} \times \mathbf{g} = (6t, 3\,(2t^2-1), -6t)$$

$$\Rightarrow \quad \left(\frac{d\mathbf{r}}{dt}\times\mathbf{g}\right)\cdot\frac{d\mathbf{g}}{dt}=\frac{24t^2}{(1+2t^2)^2}+\frac{6(2t^2-1)^2}{(1+2t^2)^2}+\frac{24t^2}{(1+2t^2)^2}$$

$$=\frac{6}{(1+2t^2)^2}[4t^2+(2t^2-1)^2+4t^2]$$

$$=\frac{6}{(1+2t^2)^2}[(2t^2-1)^2+8t^2]$$

i.e., $$\frac{d\mathbf{r}}{dt}\times\frac{d\mathbf{g}}{dt}\cdot\mathbf{g}=\frac{6}{(1+2t^2)^2}(1+2t^2)^2$$

or $$\frac{d\mathbf{r}}{dt}\times\frac{d\mathbf{g}}{dt}\cdot\mathbf{g}=6. \tag{3}$$

Again,

$$\left(\frac{d\mathbf{g}}{dt}\right)^2=\left\{\frac{4t}{(1+2t^2)^2}\right\}^2+\left\{\frac{2(2t^2-1)}{(1+2t^2)^2}\right\}^2+\left\{-\frac{4t}{(1+2t^2)^2}\right\}^2$$

$$=\frac{1}{(1+2t^2)^4}\{(4t)^2+4(2t^2-1)^2+(-4t)^2\}$$

$$=\frac{4}{(1+2t^2)^4}\{4t^2+(2t^2-1)^2+4t^2\}$$

$$=\frac{4}{(1+2t^2)^4}\{(2t^2-1)^2+8t^2\}$$

$$=\frac{4}{(1+2t^2)^4}(2t^2+1)^2$$

i.e., $$\left(\frac{d\mathbf{g}}{dt}\right)^2=\frac{4}{(1+2t^2)^2}. \tag{4}$$

Substituting the values of $\frac{d\mathbf{r}}{dt}\times\frac{d\mathbf{g}}{dt}\cdot\mathbf{g}$ and $\left(\frac{d\mathbf{g}}{dt}\right)^2$ from (3) and (4) respectively in (2), we get

$$\beta=\frac{3}{2}(1+2t^2)^2.$$

The vector equation of the line of striction is

$$\mathbf{R}=\mathbf{r}(t)+v\mathbf{g}(t), \tag{5}$$

where

$$v=-\frac{b}{a^2}=-\frac{\frac{d\mathbf{r}}{dt}\cdot\frac{d\mathbf{g}}{dt}}{\left(\frac{d\mathbf{g}}{dt}\right)^2}. \tag{6}$$

We have

$$\frac{d\mathbf{r}}{dt}\cdot\frac{d\mathbf{g}}{dt}=(3,6t,6t^2)\cdot\left(\frac{4t}{(1+2t^2)^2},\frac{2(2t^2-1)}{(1+2t^2)^2},-\frac{4t}{(1+2t^2)^2}\right)$$

i.e.,
$$\frac{d\mathbf{r}}{dt}\cdot\frac{d\mathbf{g}}{dt}=0. \tag{7}$$

Using (4) and (7), (6) gives

$$v=0. \tag{8}$$

Substituting this value of v in (5), we have

$$\mathbf{R}=\mathbf{r}\,(t)$$

or $(x,y,z)=(3t,\ 3t^2,\ 2t^3)$, where $\mathbf{R}=(x,y,z)$

or $$x=3t\,,\ y=3t^2,\ z=2t^3. \tag{9}$$

Equation (9) gives the equation of the line of striction.

Exercises

1. Prove that the line of striction on an hyperboloid of revolution of one sheet is the principal circular section.

2. If the line

$$\frac{x-\alpha}{l}=\frac{y-\beta}{m}=\frac{z-\gamma}{n}\ \text{(where } l^2+m^2+n^2=1\text{)}$$

generates a skew surface, prove that the parameter of distribution for the generator is

$$\begin{vmatrix}\delta\alpha & \delta\beta & \delta\gamma\\ \delta l & \delta m & \delta n\\ l & m & n\end{vmatrix}\div(\delta l^2+\delta m^2+\delta n^2).$$

3. If the line $x=az+\alpha, y=bz+\beta$, where a, b, α, β are functions of t, generates a skew surface, prove that the parameter of distribution of the generators is

$$\frac{(\alpha' b'-a'\beta')(1+a^2+b^2)}{a'^2+b'^2+(ab'-a'b)^2}.$$

4. Prove that the paraboloid which touches the helicoid $\dfrac{y}{x}=\dfrac{\tan z}{c}$ at all points of the generator $x\sin\theta=y\cos\theta, z=c\theta$ is

$$c\,(x\sin\theta-y\cos\theta)+(z-c\theta)\,(x\cos\theta+y\sin\theta)=0.$$

Prove also that the parameter of distribution of any generator is c and that the line of striction is the z-axis.

4.11 ISOMETRIC LINES

Let (u, v) denote the parametric coordinates of a point on a surface and let the square of the line element on the surface be given by

$$ds^2 = \lambda\,(du^2 + dv^2), \tag{1}$$

where λ is a function of u, v or is a constant.

We know that the square of the line element on the surface is given by

$$ds^2 = E\,du^2 + 2F\,du\,dv + G\,dv^2. \tag{2}$$

Comparing (1) with (2), we get

$$F = 0.$$

Hence the parametric curves on the surface are orthogonal.

Also, it follows from (1) that the length of an element of the parametric curve $v =$ const. is $\sqrt{\lambda}\,du$ and the length of an element of the parametric curve $u =$ const. is $\sqrt{\lambda}\,dv.$ Therefore the lengths of the elements of the two parametric curves are equal if $du = dv$.

Thus the parametric curves corresponding to the values u, $u + du$, v, $v + dv$ enclose a small square, provided $du = dv$. In this way, the surface may be mapped out into small squares by means of the parametric curves.

Now, we suppose that the square of the linear element has the form

$$ds^2 = \lambda\,(U\,du^2 + V\,dv^2), \tag{3}$$

where U is a function of u only, V is a function of v only and λ is a function of u, v or a constant.

Let us change the parameters u, v to ϕ, ψ by the transformation

$$d\phi = \sqrt{U}\,du,\ d\psi = \sqrt{V}\,dv. \tag{4}$$

Using (4), (3) reduces to

$$ds^2 = \lambda\,(d\phi^2 + d\psi^2), \tag{5}$$

which is of the same form as (1).

Thus the transformation (4) does not alter the parametric curves, since the parametric curves $u =$ const. and $v =$ const. are identical with the curves $\phi =$ const. and $\psi =$ const. respectively.

In such a case, the parametric curves are called the isometric lines and the parameters ϕ, ψ are called the isometric parameters.

Thus we have the following definition of the isometric lines.

Definition. Whenever the square of the line element of a surface has the form

$$ds^2 = \lambda\,(U\,du^2 + V\,dv^2),$$

so that without alteration of the parametric curves, it may be reduced to the form

$$ds^2 = \mu (d\phi^2 + d\psi^2),$$

then the parametric curves are called the isometric lines. In this case, the parameters ϕ, ψ are called the isometric parameters, and U is a function of u only, V is a function of v only, λ is a function of u, v or a constant and μ is a function of ϕ, ψ or a constant.

Sometimes the term isothermal or isothermic is used in place of isometric.

Theorem 1. The necessary and sufficient conditions for the parametric curves to be isometric lines are

$$F = 0, \frac{E}{G} = \frac{U}{V}.$$

Proof. The condition is necessary. Suppose that the parametric curves are isometric lines.

To prove : $F = 0, \dfrac{E}{G} = \dfrac{U}{V}.$

Since the parametric curves are isometric lines, the metric on the surface is given by

$$ds^2 = \lambda (U\, du^2 + V\, dv^2), \tag{1}$$

where U is a function of u only, V is a function of v only and λ is a function of u, v or a constant.

But, we know that the metric on the surface is given by

$$ds^2 = E\, du^2 + 2F\, du\, dv + G\, dv^2. \tag{2}$$

Comparing (1) with (2), we get

$$E = \lambda U,\ F = 0,\ G = \lambda V.$$

$$\therefore \qquad F = 0,\ \frac{E}{G} = \frac{U}{V} = \lambda.$$

The condition is sufficient. Now, we suppose that

$$F = 0,\ \frac{E}{G} = \frac{U}{V}. \tag{3}$$

To prove : The parametric curves are isometric lines.

From (3), we have

$$F = 0,\ E = \frac{U}{V} G.$$

$$\therefore \qquad ds^2 = \frac{U}{V} G\, du^2 + G\, dv^2 \qquad \text{[using (2)]}$$

or $$ds^2 = \frac{G}{V}(U\,du^2 + V\,dv^2) = \lambda\,(U\,du^2 + V\,dv^2),$$

where $\lambda = \frac{G}{V}$.

Hence the parametric curves are isometric lines.

Theorem 2. The necessary and sufficient conditions for the parametric curves to be isometric lines are

$$F = 0,\ \frac{\partial^2}{\partial u\,\partial v}\left(\log\frac{E}{G}\right) = 0.$$

Proof. The condition is necessary. Suppose that the parametric curves are isometric lines.

To prove : $F = 0,\ \frac{\partial^2}{\partial u\,\partial v}\left(\log\frac{E}{G}\right) = 0.$

Since the parametric curves are isometric lines, the metric on the surface is given by

$$ds^2 = \lambda\,(U\,du^2 + V\,dv^2), \tag{1}$$

where U is a function of u only, V is a function of v only and λ is a function of u, v or a constant.

But, we know that the metric on the surface is given by

$$ds^2 = E\,du^2 + 2F\,du\,dv + G\,dv^2. \tag{2}$$

Comparing (1) with (2), we get

$$E = \lambda U,\ F = 0,\ G = \lambda V.$$

$$\Rightarrow \quad F = 0,\ \frac{E}{G} = \frac{U}{V}$$

$$\Rightarrow \quad F = 0,\ \log\frac{E}{G} = \log\frac{U}{V}. \tag{3}$$

But, $$\log\frac{E}{G} = \log\frac{U}{V} = \log U - \log V$$

$$\Rightarrow \quad \frac{\partial}{\partial u}\left(\log\frac{E}{G}\right) = \frac{\partial}{\partial u}(\log U - \log V) = \frac{1}{U}\frac{\partial U}{\partial u}$$

[$\because V$ is a function of v only]

$$\Rightarrow \quad \frac{\partial^2}{\partial u\,\partial v}\left(\log\frac{E}{G}\right) = \frac{\partial}{\partial v}\left(\frac{1}{U}\frac{\partial U}{\partial u}\right) = 0. \tag{4}$$

[$\because U$ is a function of u only]

From (3) and (4), we get

$$F = 0, \quad \frac{\partial^2}{\partial u \, \partial v}\left(\log \frac{E}{G}\right) = 0.$$

The condition is sufficient. Now, we suppose that

$$F = 0, \quad \frac{\partial^2}{\partial u \, \partial v}\left(\log \frac{E}{G}\right) = 0. \tag{5}$$

To prove : The parametric curves are isometric lines.

From (5), we have

$$\frac{\partial^2}{\partial u \, \partial v}\left(\log \frac{E}{G}\right) = 0.$$

Integrating this with respect to v, we get

$$\frac{\partial}{\partial u}\left(\log \frac{E}{G}\right) = f(u), \text{ where } f(u) \text{ is a function of } u.$$

Again, integrating this with respect to u, we get

$$\log \frac{E}{G} = \int f(u)\, du + g(v)$$

$$\Rightarrow \quad \frac{E}{G} = e^{\int f(u)\, du + g(v)} = e^{\int f(u)\, du} e^{g(v)} = \frac{e^{\int f(u)\, du}}{e^{-g(v)}}$$

$$\Rightarrow \quad \frac{E}{G} = \frac{U}{V}, \tag{6}$$

where $U = e^{\int f(u)\, du}$ is a function of u only and $V = e^{-g(v)}$ is a function of v only.

From (5) and (6), we have

$$F = 0, \quad \frac{E}{G} = \frac{U}{V}.$$

Hence the parametric curves are isometric lines.

4.12 NULL LINES

Definition. The null lines (or minimal curves) on a surface are defined as the curves of zero length. They are therefore imaginary on a real surface and their importance is chiefly analytic.

Differential Equation of Null Lines. Let the parametric coordinates of a point on a surface be given by (u, v). Then the square of the length of an element on the surface is given by

$$ds^2 = E\, du^2 + 2F\, du\, dv + G\, dv^2. \tag{1}$$

Since the null lines on a surface are the curves of zero length, it follows from (1) that

$$E\,du^2 + 2F\,du\,dv + G\,dv^2 = 0. \tag{2}$$

Equation (2) is the required differential equation of null lines.

Theorem. The necessary and sufficient conditions that the parametric curves be null lines are

$$E = 0,\ G = 0,\ F \neq 0.$$

Proof. We know that the differential equation of the null lines is given by

$$E\,du^2 + 2F\,du\,dv + G\,dv^2 = 0. \tag{1}$$

Also, the differential equation of the parametric curves on the surface is given by

$$du\,dv = 0. \tag{2}$$

The condition is necessary. Suppose that the parametric curves are null lines.

To prove : $E = 0,\ G = 0,\ F \neq 0.$

Since the parametric curves are null lines, therefore (1) and (2) are identical. Comparing (2) with (1), we get

$$E = 0,\ G = 0,\ F \neq 0.$$

Hence the condition is necessary.

The condition is sufficient. Now, we suppose that

$$E = 0,\ G = 0,\ F \neq 0. \tag{3}$$

To prove : The parametric curves are asymptotic lines.

Substituting $E = 0,\ G = 0,\ F \neq 0$ in (1), we get

$$du\,dv = 0,$$

which is the differential equation of the parametric curves given by (2).

Hence (1) and (2) are identical, i.e., the null lines are the parametric curves.

Differential Equation of the Lines of Curvature when the Parametric Curves are taken as the Null Lines.

Let (u, v) be the parametric coordinates of a point on a surface. Then, we know that the differential equation of the lines of curvature is given by

$$(EM - FL)\,du^2 + (EN - GL)\,du\,dv + (FN - GM)\,dv^2 = 0. \tag{1}$$

Let the parametric curves be taken as the null lines. Then

$$E = 0,\ G = 0,\ F \neq 0. \tag{2}$$

Substituting the values of E and G from (2) in (1), we get

$$F\,(N\,dv^2 - L\,du^2) = 0$$

$$\Rightarrow \quad N\,dv^2 - L\,du^2 = 0$$

$$\Rightarrow \quad L\,du^2 - N\,dv^2 = 0. \tag{3}$$

$[\because F \neq 0]$

Equation (3) is the required differential equation of the lines of curvature when the parametric curves are taken as the null lines.

First Curvature and Gaussian Curvature when the Parametric Curves are taken as the Null Lines.

Let the parametric curves be taken as the null lines. Then

$$E = 0,\ G = 0,\ F \neq 0. \tag{1}$$

The first curvature J is given by

$$J = \frac{EN - 2FM + GL}{H^2} = \frac{EN - 2FM + GL}{EG - F^2} = \frac{2M}{F}, \text{ by (1)}$$

and the Gaussian curvature K is given by

$$K = \frac{LN - M^2}{H^2} = \frac{LN - M^2}{EG - F^2} = \frac{M^2 - LN}{F^2}, \text{ by (1).}$$

Solved Examples

Example 1. Show that the meridians and the parallels of a surface of revolution are isometric lines.

Solution. The position vector $\mathbf{r}$ of a current point on the surface of revolution is given by

$$\mathbf{r} = (u\cos\phi, u\sin\phi, f(u)). \tag{1}$$

$$\therefore \quad \mathbf{r}_1 = (\cos\phi, \sin\phi, f_1),\ \mathbf{r}_2 = (-u\sin\phi, u\cos\phi, 0)$$

$$\Rightarrow \quad E = \mathbf{r}_1^{\,2} = 1 + f_1^{\,2},\ F = \mathbf{r}_1 . \mathbf{r}_2 = 0,\ G = \mathbf{r}_2^{\,2} = u^2. \tag{2}$$

Now, the metric of the surface is given by

$$ds^2 = E\,du^2 + 2F\,du\,d\phi + G\,d\phi^2$$

or $$ds^2 = (1 + f_1^{\,2})\,du^2 + u^2\,d\phi^2 \qquad \text{[using (2)]}$$

or $$ds^2 = u^2\left(\frac{1 + f_1^{\,2}}{u^2}du^2 + d\phi^2\right). \tag{3}$$

The parametric curves are the meridians $\phi = \text{const.}$ and the parallels $u = \text{const.}$ If we make the transformation

$$d\psi = \frac{1}{u}\sqrt{1 + f_1^{\,2}}\,du, \tag{4}$$

then the curves $\psi =$ const. are the same as the parallels, and the square of the linear element given by (3) becomes

$$ds^2 = u^2\,(d\psi^2 + d\phi^2). \tag{5}$$

Therefore the parametric curves (which are the meridians and the parallels) are isometric lines.

Example 2. Show that the meridians and the parallels on a sphere form an isometric system and determine the isometric parameters.

Solution. The position vector **r** of a current point on a sphere is given by

$$\mathbf{r} = (a\sin u\cos v, a\sin u\sin v, a\cos u). \quad (1)$$

$$\therefore \quad \mathbf{r}_1 = (a\cos u\cos v, a\cos u\ \sin v, -a\sin u),$$

$$\mathbf{r}_2 = (-a\sin u\sin v, a\sin u\cos v, 0)$$

$$\Rightarrow \quad E = \mathbf{r}_1^{\ 2} = a^2,\ F = \mathbf{r}_1 . \mathbf{r}_2 = 0,\ G = \mathbf{r}_2^{\ 2} = a^2\sin^2 u. \quad (2)$$

Now, the metric of the surface is given by

$$ds^2 = E\,du^2 + 2F\,du\,dv + G\,dv^2$$

or $$ds^2 = a^2 du^2 + a^2\sin^2 u\,dv^2 \quad \text{[using (2)]}$$

or $$ds^2 = a^2\sin^2 u\,(\operatorname{cosec}^2 u\,du^2 + dv^2). \quad (3)$$

The parametric curves are the meridians $v=$ const. and the parallels $u=$ const. If we make the transformation

$$d\psi = \operatorname{cosec} u\,du, \quad (4)$$

then the curves $\psi =$ const. are the same as the parallels, and the square of the linear element given by (3) becomes

$$ds^2 = a^2\sin^2 u\,(d\psi^2 + dv^2). \quad (5)$$

Therefore the parametric curves (which are the meridians and the parallels) are isometric lines and the parameters ψ, v are the isometric parameters.

Exercises

1. Show that the lines of curvature on a surface of revolution form an isothermal system.
2. Show that a system of confocal ellipses and hyperbolas are isometric lines in the plane.

4.13 THE FUNDAMENTAL EQUATIONS OF SURFACE THEORY

In this section, we shall obtain relations between the fundamental coefficients E, F, G and L, M, N.

Gauss's Formulae for $\mathbf{r}_{11}, \mathbf{r}_{12}, \mathbf{r}_{22}$

If **r** is the position vector of a point P on the surface and if the suffixes 1 and 2 denote partial derivatives with respect to u and v respectively, then

$$\mathbf{r}_{11} = L\mathbf{N} + l\mathbf{r}_1 + \lambda\mathbf{r}_2,$$
$$\mathbf{r}_{12} = M\mathbf{N} + m\mathbf{r}_1 + \mu\mathbf{r}_2,$$
$$\mathbf{r}_{22} = N\mathbf{N} + n\mathbf{r}_1 + \nu\mathbf{r}_2.$$

where the coefficients $l,\ m,\ n,\ \lambda,\ \mu, \nu$ are known as Christoffel symbols or Gauss coefficients and are functions of E, F, G and their partial derivatives with respect to u and v.

Proof. We know that there are three non-coplanar vectors, $\mathbf{N}$, $\mathbf{r}_1$ and $\mathbf{r}_2$ at any point P of a surface. Hence every vector at P can be expressed as a linear combination of these vectors. Thus we have

$$\left.\begin{aligned} \mathbf{r}_{11} &= c_1\mathbf{N} + l\mathbf{r}_1 + \lambda\mathbf{r}_2, \\ \mathbf{r}_{12} &= c_2\mathbf{N} + m\mathbf{r}_1 + \mu\mathbf{r}_2, \\ \mathbf{r}_{22} &= c_3\mathbf{N} + n\mathbf{r}_1 + \nu\mathbf{r}_2. \end{aligned}\right\} \tag{1}$$

Taking the dot product of both sides of (1) with $\mathbf{N}$, we get

$$\left.\begin{aligned} \mathbf{r}_{11}\cdot\mathbf{N} &= L = c_1, \\ \mathbf{r}_{12}\cdot\mathbf{N} &= M = c_2, \\ \mathbf{r}_{22}\cdot\mathbf{N} &= N = c_3. \end{aligned}\right\}$$

Substituting the above values of c_1, c_2, c_3 in (1), we get

$$\left.\begin{aligned} \mathbf{r}_{11} &= L\mathbf{N} + l\mathbf{r}_1 + \lambda\mathbf{r}_2, \\ \mathbf{r}_{12} &= M\mathbf{N} + m\mathbf{r}_1 + \mu\mathbf{r}_2, \\ \mathbf{r}_{22} &= N\mathbf{N} + n\mathbf{r}_1 + \nu\mathbf{r}_2. \end{aligned}\right\} \tag{2}$$

Equations (2) are known as Gauss's formulae for $\mathbf{r}_{11}, \mathbf{r}_{12}, \mathbf{r}_{22}$.

Now, we shall find the values of the coefficients $l,\ m,\ n,\ \lambda,\ \mu, \nu$ occurring in (2).

Taking the dot product of both sides of the first equation in (2) with $\mathbf{r}_1$ and $\mathbf{r}_2$ successively, we get

$$\left.\begin{aligned} \mathbf{r}_1\cdot\mathbf{r}_{11} &= lE + \lambda F, \\ \mathbf{r}_2\cdot\mathbf{r}_{11} &= lF + \lambda G. \end{aligned}\right\} \tag{3}$$

Now,

$$\mathbf{r}_1\cdot\mathbf{r}_{11} = \frac{1}{2}\frac{\partial}{\partial u}(\mathbf{r}_1^{\ 2}) = \frac{1}{2}\frac{\partial}{\partial u}(E) = \frac{1}{2}E_1,$$

$$\mathbf{r}_2\cdot\mathbf{r}_{11} = \frac{\partial}{\partial u}(\mathbf{r}_2\cdot\mathbf{r}_1) - \mathbf{r}_{12}\cdot\mathbf{r}_1 = \frac{\partial}{\partial u}(\mathbf{r}_2\cdot\mathbf{r}_1) - \frac{1}{2}\frac{\partial}{\partial v}(\mathbf{r}_1^{\ 2})$$
$$= \frac{\partial}{\partial u}(F) - \frac{1}{2}\frac{\partial}{\partial v}(E) = F_1 - \frac{1}{2}E_2.$$

Putting these values of $\mathbf{r}_1 \cdot \mathbf{r}_{11}$ and $\mathbf{r}_2 \cdot \mathbf{r}_{11}$ in (3), we get

$$\left.\begin{aligned} \frac{1}{2}E_1 &= lE + \lambda F, \\ F_1 - \frac{1}{2}E_2 &= lF + \lambda G. \end{aligned}\right\}$$

Solving these for l and λ, we have

$$\left.\begin{aligned} l &= \frac{1}{2H^2}(GE_1 - 2FF_1 + FE_2), \\ \lambda &= \frac{1}{2H^2}(2EF_1 - EE_2 - FE_1). \end{aligned}\right\} \tag{4}$$

Similarly, to find the values of m and μ, we take the dot product of both sides of the second equation in (2) with $\mathbf{r}_1$ and $\mathbf{r}_2$ successively. Then we get

$$\left.\begin{aligned} \mathbf{r}_1 \cdot \mathbf{r}_{12} &= mE + \mu F, \\ \mathbf{r}_2 \cdot \mathbf{r}_{12} &= mF + \mu G. \end{aligned}\right\} \tag{5}$$

Now,

$$\mathbf{r}_1 \cdot \mathbf{r}_{12} = \frac{1}{2}\frac{\partial}{\partial v}(\mathbf{r}_1^{\ 2}) = \frac{1}{2}\frac{\partial}{\partial v}(E) = \frac{1}{2}E_2,$$

$$\mathbf{r}_2 \cdot \mathbf{r}_{12} = \frac{1}{2}\frac{\partial}{\partial u}(\mathbf{r}_2^{\ 2}) = \frac{1}{2}\frac{\partial}{\partial u}(G) = \frac{1}{2}G_1.$$

Putting these values of $\mathbf{r}_1 \cdot \mathbf{r}_{12}$ and $\mathbf{r}_2 \cdot \mathbf{r}_{12}$ in (5), we get

$$\left.\begin{aligned} \frac{1}{2}E_2 &= mE + \mu F, \\ \frac{1}{2}G_1 &= mF + \mu G. \end{aligned}\right\}$$

Solving these for m and μ, we have

$$\left.\begin{aligned} m &= \frac{1}{2H^2}(GE_2 - FG_1), \\ \mu &= \frac{1}{2H^2}(EG_1 - FE_2). \end{aligned}\right\} \tag{6}$$

Finally, to find the values of n and ν, we take the dot product of both sides of the third equation in (2) with $\mathbf{r}_1$ and $\mathbf{r}_2$ successively. Then

$$\left.\begin{aligned} \mathbf{r}_1 \cdot \mathbf{r}_{22} &= nE + \nu F, \\ \mathbf{r}_2 \cdot \mathbf{r}_{22} &= nF + \nu G. \end{aligned}\right\} \tag{7}$$

Now,

$$\mathbf{r}_1\cdot\mathbf{r}_{22}=\frac{\partial}{\partial v}(\mathbf{r}_1\cdot\mathbf{r}_2)-\mathbf{r}_{12}\cdot\mathbf{r}_2=\frac{\partial}{\partial v}(\mathbf{r}_1\cdot\mathbf{r}_2)-\frac{1}{2}\frac{\partial}{\partial u}(\mathbf{r}_2{}^2)$$
$$=\frac{\partial}{\partial v}(F)-\frac{1}{2}\frac{\partial}{\partial u}(G)=F_2-\frac{1}{2}G_1,$$
$$\mathbf{r}_2\cdot\mathbf{r}_{22}=\frac{1}{2}\frac{\partial}{\partial v}(\mathbf{r}_2{}^2)=\frac{1}{2}\frac{\partial}{\partial v}(G)=\frac{1}{2}G_2.$$

Putting these values of $\mathbf{r}_1\cdot\mathbf{r}_{22}$ and $\mathbf{r}_2\cdot\mathbf{r}_{22}$ in (7), we get

$$\left.\begin{aligned}F_2-\frac{1}{2}G_1&=nE+\nu F,\\ \frac{1}{2}G_2&=nF+\nu G.\end{aligned}\right\}$$

Solving these for n and ν, we have

$$\left.\begin{aligned}n&=\frac{1}{2H^2}(2GF_2-GG_1-FG_2),\\ \nu&=\frac{1}{2H^2}(EG_2-2FF_2+FG_1).\end{aligned}\right\}\qquad(8)$$

Thus the values of the coefficients l, λ ; m, μ ; n, ν occurring in (2) are given by (4), (6) and (8) respectively.

Corollary. If the parametric curves are orthogonal, then $F = 0$ and $H^2 = EG$. In this case, the values of l, λ ; m, μ ; n, ν given by (4), (6) and (8) respectively reduce to

$$\left.\begin{aligned}l&=\frac{E_1}{2E},\quad \lambda=-\frac{E_2}{2G};\\ m&=\frac{E_2}{2E},\quad \mu=\frac{G_1}{2G};\\ n&=-\frac{G_1}{2E},\quad \nu=\frac{G_2}{2G}.\end{aligned}\right\}$$

Hence, in this case, Gauss's formulae given by (2) reduce to

$$\left.\begin{aligned}\mathbf{r}_{11}&=L\mathbf{N}+\frac{E_1}{2E}\mathbf{r}_1-\frac{E_2}{2G}\mathbf{r}_2,\\ \mathbf{r}_{12}&=M\mathbf{N}+\frac{E_2}{2E}\mathbf{r}_1+\frac{G_1}{2G}\mathbf{r}_2,\\ \mathbf{r}_{22}&=N\mathbf{N}-\frac{G_1}{2E}\mathbf{r}_1+\frac{G_2}{2G}\mathbf{r}_2.\end{aligned}\right\}$$

Gauss Characteristic Equation.

If $l, m, n, \lambda, \mu, \nu$ denote the Gauss coefficients, then

$$LN - M^2 = \frac{1}{2}(2F_{12} - E_{22} - G_{11}) + (m^2 E + 2m\mu F + \mu^2 G)$$
$$- \{lnE + (lv + \lambda n)\, F + \lambda v G\}.$$

Proof. From Gauss's formulae, we have

$$\mathbf{r}_{11} = L\mathbf{N} + l\mathbf{r}_1 + \lambda \mathbf{r}_2,$$
$$\mathbf{r}_{12} = M\mathbf{N} + m\mathbf{r}_1 + \mu \mathbf{r}_2,$$
$$\mathbf{r}_{22} = N\mathbf{N} + n\mathbf{r}_1 + \nu \mathbf{r}_2.$$

Therefore

$$\mathbf{r}_{11} . \mathbf{r}_{22} = LN + lnE + (lv + \lambda n)\, F + \lambda v G,$$
$$\mathbf{r}_{12}^{\;2} = \mathbf{r}_{12} . \mathbf{r}_{12} = M^2 + m^2 E + 2m\mu F + \mu^2 G$$
$$\Rightarrow \quad \mathbf{r}_{11} . \mathbf{r}_{22} - \mathbf{r}_{12}^{\;2} = (LN - M^2) + \{lnE + (lv + \lambda n)\, F + \lambda v G\}$$
$$- (m^2 E + 2m\mu F + \mu^2 G)$$
$$\Rightarrow \quad LN - M^2 = (\mathbf{r}_{11} . \mathbf{r}_{22} - \mathbf{r}_{12}^2) + (m^2 E + 2m\mu F + \mu^2 G)$$
$$- \{lnE + (lv + \lambda n)\, F + \lambda v G\}. \qquad (1)$$

Now, we shall prove that

$$\mathbf{r}_{11} . \mathbf{r}_{22} - \mathbf{r}_{12}^{\;2} = \frac{1}{2}(2F_{12} - E_{22} - G_{11}).$$

We have

$$E = \mathbf{r}_1^{\;2}$$
$$\Rightarrow \quad E_2 = 2\,\mathbf{r}_1 . \mathbf{r}_{12}$$
$$\Rightarrow \quad E_{22} = 2\,(\mathbf{r}_{12} . \mathbf{r}_{12} + \mathbf{r}_1 . \mathbf{r}_{122}). \qquad (2)$$

Also,

$$G = \mathbf{r}_2^{\;2}$$
$$\Rightarrow \quad G_1 = 2\,\mathbf{r}_2 . \mathbf{r}_{12}$$
$$\Rightarrow \quad G_{11} = 2\,(\mathbf{r}_{12} . \mathbf{r}_{12} + \mathbf{r}_2 . \mathbf{r}_{112}). \qquad (3)$$

Again,

$$F = \mathbf{r}_1 . \mathbf{r}_2$$
$$\Rightarrow \quad F_1 = \mathbf{r}_{11} . \mathbf{r}_2 + \mathbf{r}_1 . \mathbf{r}_{12}$$
$$\Rightarrow \quad F_{12} = (\mathbf{r}_{112} . \mathbf{r}_2 + \mathbf{r}_{11} . \mathbf{r}_{22}) + (\mathbf{r}_{12} . \mathbf{r}_{12} + \mathbf{r}_1 . \mathbf{r}_{122})$$
$$\Rightarrow \quad 2F_{12} = 2\,(\mathbf{r}_{112} . \mathbf{r}_2 + \mathbf{r}_{11} . \mathbf{r}_{22}) + 2\,(\mathbf{r}_{12} . \mathbf{r}_{12} + \mathbf{r}_1 . \mathbf{r}_{122}). \qquad (4)$$

From (2), (3) and (4), we get

$$2F_{12} - E_{22} - G_{11} = 2\,(\mathbf{r}_{11} . \mathbf{r}_{22} - \mathbf{r}_{12}^2)$$
$$\Rightarrow \quad \mathbf{r}_{11} . \mathbf{r}_{22} - \mathbf{r}_{12}^{\;2} = \frac{1}{2}(2F_{12} - E_{22} - G_{11}). \qquad (5)$$

Substituting the value of $\mathbf{r}_{11}\cdot\mathbf{r}_{22}-\mathbf{r}_{12}^2$ from (5) in (1), we get

$$LN - M^2 = \frac{1}{2}(2F_{12} - E_{22} - G_{11}) + (m^2E + 2m\mu F + \mu^2 G) - \{lnE + (lv + \lambda n)\,F + \lambda v G\}. \qquad (6)$$

Equation (6) is known as the Gauss characteristic equation or simply the Gauss equation. It is sometimes expressed in the alternative form as follows.

$$LN - M^2 = \frac{1}{2}H\frac{\partial}{\partial u}\left(\frac{F}{EH}\frac{\partial E}{\partial v} - \frac{1}{H}\frac{\partial G}{\partial u}\right) + \frac{1}{2}H\frac{\partial}{\partial v}\left(\frac{2}{H}\frac{\partial F}{\partial u} - \frac{1}{H}\frac{\partial E}{\partial v} - \frac{F}{EH}\frac{\partial E}{\partial u}\right). \qquad (7)$$

Corollary 1. It follows from (7) that the Gaussian curvature, which is equal to $(LN - M^2)/H^2$, is expressible in terms of the fundamental magnitudes E, F, G and their derivatives of the first two orders. In this respect it differs from the first curvature.

Corollary 2. Surfaces which have the same first order magnitudes E, F, G (irrespective of the second order magnitudes L, M, N) have the same Gaussian curvature.

Mainardi-Codazzi Equations.

If l, m, n, λ, μ, v denote the Gauss coefficients, then

$$L_2 - M_1 = mL - (l - \mu)\,M - \lambda N,$$

$$M_2 - N_1 = nL - (m - v)\,M - \mu N.$$

Proof. From Gauss's formulae, we have

$$\left.\begin{aligned}\mathbf{r}_{11} &= L\mathbf{N} + l\mathbf{r}_1 + \lambda\mathbf{r}_2,\\ \mathbf{r}_{12} &= M\mathbf{N} + m\mathbf{r}_1 + \mu\mathbf{r}_2,\\ \mathbf{r}_{22} &= N\mathbf{N} + n\mathbf{r}_1 + v\mathbf{r}_2.\end{aligned}\right\} \qquad (1)$$

Now, consider the identity

$$\frac{\partial}{\partial v}(\mathbf{r}_{11}) = \frac{\partial}{\partial u}(\mathbf{r}_{12}). \qquad (2)$$

From (1), we have

$$\frac{\partial}{\partial v}(\mathbf{r}_{11}) = \frac{\partial}{\partial v}(L\mathbf{N} + l\mathbf{r}_1 + \lambda\mathbf{r}_2)$$

$$= L_2\mathbf{N} + L\mathbf{N}_2 + l_2\mathbf{r}_1 + l\mathbf{r}_{12} + \lambda_2\mathbf{r}_2 + \lambda\mathbf{r}_{22}$$

and $$\frac{\partial}{\partial u}(\mathbf{r}_{12}) = \frac{\partial}{\partial u}(M\mathbf{N} + m\mathbf{r}_1 + \mu\mathbf{r}_2)$$
$$= M_1\mathbf{N} + M\mathbf{N}_1 + m_1\mathbf{r}_1 + m\mathbf{r}_{11} + \mu_1\mathbf{r}_2 + \mu\mathbf{r}_{12}.$$

Substituting these values in (2), we get

$$L_2\mathbf{N} + L\mathbf{N}_2 + l_2\mathbf{r}_1 + l\mathbf{r}_{12} + \lambda_2\mathbf{r}_2 + \lambda\mathbf{r}_{22}$$
$$= M_1\mathbf{N} + M\mathbf{N}_1 + m_1\mathbf{r}_1 + m\mathbf{r}_{11} + \mu_1\mathbf{r}_2 + \mu\mathbf{r}_{12}. \tag{3}$$

From Weingarten's equations, we have

$$\left.\begin{aligned} \mathbf{N}_1 &= \frac{FM - GL}{H^2}\mathbf{r}_1 + \frac{FL - EM}{H^2}\mathbf{r}_2, \\ \mathbf{N}_2 &= \frac{FN - GM}{H^2}\mathbf{r}_1 + \frac{FM - EN}{H^2}\mathbf{r}_2. \end{aligned}\right\} \tag{4}$$

Substituting the values of $\mathbf{r}_{11}, \mathbf{r}_{12}$ and $\mathbf{r}_{22}$ from (1) and the values of $\mathbf{N}_1, \mathbf{N}_2$ from (4) in (3), we get

$$L_2\mathbf{N} + L\left(\frac{FN - GM}{H^2}\mathbf{r}_1 + \frac{FM - EN}{H^2}\mathbf{r}_2\right) + l_2\mathbf{r}_1$$
$$+ l\,(M\mathbf{N} + m\mathbf{r}_1 + \mu\mathbf{r}_2) + \lambda_2\mathbf{r}_2 + \lambda\,(N\mathbf{N} + n\mathbf{r}_1 + \nu\mathbf{r}_2)$$
$$= M_1\mathbf{N} + M\left(\frac{FM - GL}{H^2}\mathbf{r}_1 + \frac{FL - EM}{H^2}\mathbf{r}_2\right) + m_1\mathbf{r}_1$$
$$+ m\,(L\mathbf{N} + l\mathbf{r}_1 + \lambda\mathbf{r}_2) + \mu_1\mathbf{r}_2 + \mu\,(M\mathbf{N} + m\mathbf{r}_1 + \mu\mathbf{r}_2).$$

Equating the coefficients of $\mathbf{N}$ on both sides of this equation, we get

$$L_2 + lM + \lambda N = M_1 + mL + \mu M$$

or $$L_2 - M_1 = mL - (l - \mu)\,M - \lambda N, \tag{5}$$

which is the first Mainardi-Codazzi equation.

Now, consider the vector identity

$$\frac{\partial}{\partial v}(\mathbf{r}_{12}) = \frac{\partial}{\partial u}(\mathbf{r}_{22}). \tag{6}$$

From Gauss's formulae given by (1), we have

$$\frac{\partial}{\partial v}(\mathbf{r}_{12}) = \frac{\partial}{\partial v}(M\mathbf{N} + m\mathbf{r}_1 + \mu\mathbf{r}_2)$$
$$= M_2\mathbf{N} + M\mathbf{N}_2 + m_2\mathbf{r}_1 + m\mathbf{r}_{12} + \mu_2\mathbf{r}_2 + \mu\mathbf{r}_{22}$$

and $$\frac{\partial}{\partial u}(\mathbf{r}_{22}) = \frac{\partial}{\partial u}(N\mathbf{N} + n\mathbf{r}_1 + \nu\mathbf{r}_2)$$
$$= N_1\mathbf{N} + N\mathbf{N}_1 + n_1\mathbf{r}_1 + n\mathbf{r}_{11} + \nu_1\mathbf{r}_2 + \nu\mathbf{r}_{12}.$$

Substituting these values in (6), we get

$$M_2\mathbf{N} + M\mathbf{N}_2 + m_2\mathbf{r}_1 + m\mathbf{r}_{12} + \mu_2\mathbf{r}_2 + \mu\mathbf{r}_{22}$$
$$= N_1\mathbf{N} + N\mathbf{N}_1 + n_1\mathbf{r}_1 + n\mathbf{r}_{11} + \nu_1\mathbf{r}_2 + \nu\mathbf{r}_{12}. \tag{7}$$

Substituting the values of $\mathbf{r}_{11}$, $\mathbf{r}_{12}$ and $\mathbf{r}_{22}$ from (1) and the values of $\mathbf{N}_1, \mathbf{N}_2$ from (4) in (7), we get

$$M_2\mathbf{N} + M\left(\frac{FN - GM}{H^2}\mathbf{r}_1 + \frac{FM - EN}{H^2}\mathbf{r}_2\right) + m_2\mathbf{r}_1$$
$$+ m\,(M\mathbf{N} + m\mathbf{r}_1 + \mu\mathbf{r}_2) + \mu_2\mathbf{r}_2 + \mu\,(N\mathbf{N} + n\mathbf{r}_1 + \nu\mathbf{r}_2)$$
$$= N_1\mathbf{N} + N\left(\frac{FM - GL}{H^2}\mathbf{r}_1 + \frac{FL - EM}{H^2}\mathbf{r}_2\right) + n_1\mathbf{r}_1$$
$$+ n\,(L\mathbf{N} + l\mathbf{r}_1 + \lambda\mathbf{r}_2) + \nu_1\mathbf{r}_2 + \nu\,(M\mathbf{N} + m\mathbf{r}_1 + \mu\mathbf{r}_2).$$

Equating the coefficients of $\mathbf{N}$ on both sides of this equation, we get

$$M_2 + mM + \mu N = N_1 + nL + \nu M$$

or
$$M_2 - N_1 = nL - (m - \nu)\,M - \mu N, \tag{8}$$

which is the second Mainardi-Codazzi equation.

Alternative Form. Mainardi-Codazzi equations can also be expressed in a different form as follows.

$$\frac{\partial}{\partial v}\left(\frac{M}{H}\right) - \frac{\partial}{\partial u}\left(\frac{N}{H}\right) = \frac{nL - 2mM + lN}{H},$$

$$\frac{\partial}{\partial u}\left(\frac{M}{H}\right) - \frac{\partial}{\partial v}\left(\frac{L}{H}\right) = \frac{\nu L - 2\mu M + \lambda N}{H}.$$

Proof. We know that

$$\left.\begin{aligned} l &= \frac{1}{2H^2}(GE_1 - 2\,FF_1 + FE_2), \\ \lambda &= \frac{1}{2H^2}(2EF_1 - EE_2 - FE_1). \end{aligned}\right\} \tag{9}$$

$$\left.\begin{aligned} m &= \frac{1}{2H^2}(GE_2 - FG_1), \\ \mu &= \frac{1}{2H^2}(EG_1 - FE_2). \end{aligned}\right\} \tag{10}$$

$$\left.\begin{aligned} n &= \frac{1}{2H^2}(2GF_2 - GG_1 - FG_2), \\ \nu &= \frac{1}{2H^2}(EG_2 - 2FF_2 + FG_1). \end{aligned}\right\} \tag{11}$$

Now, $H^2 = EG - F^2$

$$\Rightarrow \quad \frac{\partial}{\partial u}(H^2) = \frac{\partial}{\partial u}(EG - F^2)$$

$$\Rightarrow \quad 2HH_1 = E_1G + EG_1 - 2FF_1$$

$$= 2H^2(l + \mu) \qquad \text{[using (9) and (10)]}$$

$$\Rightarrow \quad H_1 = H(l + \mu). \qquad (12)$$

Similarly,

$$H^2 = EG - F^2$$

$$\Rightarrow \quad \frac{\partial}{\partial v}(H^2) = \frac{\partial}{\partial v}(EG - F^2)$$

$$\Rightarrow \quad 2HH_2 = E_2G + EG_2 - 2FF_2$$

$$= 2H^2(m + \nu) \qquad \text{[using (10) and (11)]}$$

$$\Rightarrow \quad H_2 = H(m + \nu). \qquad (13)$$

Now, we have

$$\frac{\partial}{\partial v}\left(\frac{M}{H}\right) - \frac{\partial}{\partial u}\left(\frac{N}{H}\right) = \left(\frac{M_2}{H} - \frac{M}{H^2}H_2\right) - \left(\frac{N_1}{H} - \frac{N}{H^2}H_1\right)$$

$$= \frac{M_2 - N_1}{H} - \frac{M}{H^2}H_2 + \frac{N}{H^2}H_1$$

$$= \frac{M_2 - N_1}{H} - \frac{M}{H}(m + \nu) + \frac{N}{H}(l + \mu)$$

[using (12) and (13)]

$$= \frac{nL - 2mM + lN}{H} \qquad \text{[using (8)]}$$

i.e.,
$$\frac{\partial}{\partial v}\left(\frac{M}{H}\right) - \frac{\partial}{\partial u}\left(\frac{N}{H}\right) = \frac{nL - 2mM + lN}{H}, \qquad (14)$$

which is the first Mainardi-Codazzi equation.

Again,
$$\frac{\partial}{\partial u}\left(\frac{M}{H}\right) - \frac{\partial}{\partial v}\left(\frac{L}{H}\right) = \left(\frac{M_1}{H} - \frac{M}{H^2}H_1\right) - \left(\frac{L_2}{H} - \frac{L}{H^2}H_2\right)$$

$$= \frac{M_1 - L_2}{H} - \frac{M}{H^2}H_1 + \frac{L}{H^2}H_2$$

$$= \frac{M_1 - L_2}{H} - \frac{M}{H}(l + \mu) + \frac{L}{H}(m + \nu)$$

[using (12) and (13)]

$$= \frac{\nu L - 2\mu M + \lambda N}{H} \qquad \text{[using (5)]}$$

i.e., $$\frac{\partial}{\partial u}\left(\frac{M}{H}\right)-\frac{\partial}{\partial v}\left(\frac{L}{H}\right)=\frac{\nu L-2\mu M+\lambda N}{H}, \tag{15}$$

which is the second Mainardi-Codazzi equation.

Solved Examples

Example 1. Show that for the surface $z = f(x, y)$ with x, y as parameters,

$$l=\frac{pr}{H^2},\ m=\frac{ps}{H^2},\ n=\frac{pt}{H^2},\ \lambda=\frac{qr}{H^2},\ \mu=\frac{qs}{H^2}, \nu=\frac{qt}{H^2}.$$

Solution. The position vector of any current point on the given surface is

$$\mathbf{r}=(x, y, z),$$

where z is a function of x and y.

Then, we have

$$E=1+p^2,\ F=pq,\ G=1+q^2,\ H^2=1+p^2+q^2.$$

$$\therefore\quad E_1=2pr,\ E_2=2ps;\ F_1=rq+ps,\ F_2=sq+pt;$$

$$G_1=2qs, G_2=2qt.$$

Now,

$$l=\frac{1}{2H^2}(GE_1-2FF_1+FE_2)$$

$$=\frac{1}{2H^2}\{(1+q^2)\,2pr-2pq\,(rq+ps)+pq\,(2ps)\}$$

$$=\frac{pr}{H^2}.$$

Similarly,

$$m=\frac{1}{2H^2}(GE_2-FG_1)$$

$$=\frac{1}{2H^2}\{(1+q^2)\,2ps-pq\,(2qs)\}$$

$$=\frac{ps}{H^2}.$$

Again, $$n=\frac{1}{2H^2}(2GF_2-GG_1-FG_2)$$

$$=\frac{1}{2H^2}\{(2\,(1+q^2)\,(sq+pt)-(1+q^2)\,2qs-pq\,(2qt)\}$$

$$=\frac{pt}{H^2}.$$

Similarly, substituting the values in the expressions for λ, μ, ν, we get the required result.

Example 2. For the surface of revolution given by

$$x = u\cos v,\ y = u\sin v,\ z = f(u),$$

show that

$$l = u^2 f_1 f_{11}/H^2,\ m = 0,\ n = -u^3/H^2,\ \lambda = 0,\ \mu = 1/u,\ \nu = 0.$$

Solution. The position vector $\mathbf{r}$ of any current point on the given surface is

$$\mathbf{r} = (u\cos v, u\sin v, f(u)).$$

Then, we have

$$E = 1 + f_1^2, F = 0,\ G = u^2,\ H^2 = u^2(1 + f_1^2).$$

$$\therefore \quad E_1 = 2f_1 f_{11},\ E_2 = 0;\ F_1 = 0,\ F_2 = 0;\ G_1 = 2u, G_2 = 0.$$

Now,

$$l = \frac{1}{2H^2}(GE_1 - 2FF_1 + FE_2) = \frac{1}{2H^2}\{u^2(2f_1 f_{11})\} = \frac{u^2 f_1 f_{11}}{H^2},$$

$$m = \frac{1}{2H^2}(GE_2 - FG_1) = \frac{1}{2H^2}(0) = 0,$$

$$n = \frac{1}{2H^2}(2GF_2 - GG_1 - FG_2) = \frac{1}{2H^2}\{-u^2(2u)\} = -\frac{u^3}{H^2},$$

$$\lambda = \frac{1}{2H^2}(2EF_1 - EE_2 - FE_1) = \frac{1}{2H^2}(0) = 0,$$

$$\mu = \frac{1}{2H^2}(EG_1 - FE_2) = \frac{1}{2u^2(1 + f_1^2)}\{(1 + f_1^2)\,2u\} = \frac{1}{u},$$

$$\nu = \frac{1}{2H^2}(EG_2 - 2FF_2 + FG_1) = \frac{1}{2H^2}(0) = 0.$$

Example 3. For a surface whose linear element is given by

$$ds^2 = du^2 + D^2 dv^2,$$

show that

$$l = 0,\ m = 0,\ n = -DD_1,\ \lambda = 0,\ \mu = D_1/D,\ \nu = D_2/D.$$

Solution. The linear element for the surface is given by

$$ds^2 = du^2 + D^2 dv^2. \tag{1}$$

But, we know that

$$ds^2 = E\,du^2 + 2F\,du\,dv + G\,dv^2. \tag{2}$$

Comparing (1) with (2), we get

$$E = 1,\ F = 0,\ G = D^2.$$

Therefore

$$H^2 = EG - F^2 = D^2$$

and $\quad E_1 = 0,\ E_2 = 0;\ F_1 = 0,\ F_2 = 0;\ G_1 = 2DD_1, G_2 = 2DD_2.$

Now,

$$l = \frac{1}{2H^2}(GE_1 - 2FF_1 + FE_2) = \frac{1}{2H^2}(0) = 0,$$

$$m = \frac{1}{2H^2}(GE_2 - FG_1) = \frac{1}{2H^2}(0) = 0,$$

$$n = \frac{1}{2H^2}(2GF_2 - GG_1 - FG_2) = \frac{1}{2D^2}\{-D^2(2DD_1)\} = -DD_1,$$

$$\lambda = \frac{1}{2H^2}(2EF_1 - EE_2 - FE_1) = \frac{1}{2H^2}(0) = 0,$$

$$\mu = \frac{1}{2H^2}(EG_1 - FE_2) = \frac{1}{2D^2}\{2DD_1\} = \frac{D_1}{D},$$

$$\nu = \frac{1}{2H^2}(EG_2 - 2FF_2 + FG_1) = \frac{1}{2D^2}\{2DD_2\} = \frac{D_2}{D}.$$

Example 4. From the Gauss characteristic equation deduce that, when the parametric curves are orthogonal,

$$\kappa = -\frac{1}{\sqrt{EG}}\left[\frac{\partial}{\partial u}\left(\frac{1}{\sqrt{E}}\frac{\partial\sqrt{G}}{\partial u}\right) + \frac{\partial}{\partial v}\left(\frac{1}{\sqrt{G}}\frac{\partial\sqrt{E}}{\partial v}\right)\right].$$

Solution. We know that the Gauss characteristic equation is

$$LN - M^2 = \frac{1}{2}(2F_{12} - E_{22} - G_{11}) + (m^2E + 2m\mu F + \mu^2 G) - \{lnE + (l\nu + \lambda n)F + \lambda\nu G\}. \quad (1)$$

When the parametric curves are orthogonal, we have

$$F = 0. \quad (2)$$

Also, in this case,

$$\left.\begin{aligned} l &= \frac{E_1}{2E}, \quad m = \frac{E_2}{2E}, \quad n = -\frac{G_1}{2E}, \\ \lambda &= -\frac{E_2}{2G}, \quad \mu = \frac{G_1}{2G}, \quad \nu = \frac{G_2}{2G}. \end{aligned}\right\} \quad (3)$$

Moreover, since $F = 0$, we have

$$F_{12} = 0. \quad (4)$$

Using (2), (3) and (4), (1) reduces to

$$LN - M^2 = \frac{1}{2}(-E_{22} - G_{11}) + \left(\frac{E_2^{\ 2}}{4E} + \frac{G_1^{\ 2}}{4G}\right) + \left(\frac{E_1G_1}{4E} + \frac{E_2G_2}{4G}\right)$$

or $$LN - M^2 = -\left[\left(\frac{G_{11}}{2} - \frac{E_1G_1}{4E} - \frac{G_1^{\ 2}}{4G}\right) + \left(\frac{E_{22}}{2} - \frac{E_2G_2}{4G} - \frac{E_2^{\ 2}}{4E}\right)\right]. \tag{5}$$

Also,

$$H^2 = EG - F^2 = EG. \tag{6}$$

[using (2)]

Now, $\kappa = \dfrac{LN - M^2}{H^2}$

$$= -\frac{1}{EG}\left[\left(\frac{G_{11}}{2} - \frac{E_1G_1}{4E} - \frac{G_1^{\ 2}}{4G}\right) + \left(\frac{E_{22}}{2} - \frac{E_2G_2}{4G} - \frac{E_2^{\ 2}}{4E}\right)\right]$$

[using (5) and (6)]

$$= -\frac{1}{\sqrt{EG}}\left[\left(\frac{G_{11}}{2\sqrt{EG}} - \frac{E_1G_1}{4E\sqrt{EG}} - \frac{G_1^{\ 2}}{4G\sqrt{EG}}\right) + \left(\frac{E_{22}}{2\sqrt{EG}} - \frac{E_2G_2}{4G\sqrt{EG}} - \frac{E_2^{\ 2}}{4E\sqrt{EG}}\right)\right]$$

$$= -\frac{1}{\sqrt{EG}}\left[\left(\frac{1}{2\sqrt{EG}}G_{11} - \frac{G_1(GE_1 + EG_1)}{4\,(EG)^{3/2}}\right) + \left(\frac{1}{2\sqrt{EG}}E_{22} - \frac{E_2(EG_2 + GE_2)}{4\,(EG)^{3/2}}\right)\right]$$

$$= -\frac{1}{\sqrt{EG}}\left[\frac{\partial}{\partial u}\left(\frac{1}{2\sqrt{EG}}G_1\right) + \frac{\partial}{\partial v}\left(\frac{1}{2\sqrt{EG}}E_2\right)\right]$$

$$= -\frac{1}{\sqrt{EG}}\left[\frac{\partial}{\partial u}\left\{\frac{1}{\sqrt{E}}\left(\frac{1}{2\sqrt{G}}\frac{\partial G}{\partial u}\right)\right\} + \frac{\partial}{\partial v}\left\{\frac{1}{\sqrt{G}}\left(\frac{1}{2\sqrt{E}}\frac{\partial E}{\partial v}\right)\right\}\right]$$

$$= -\frac{1}{\sqrt{EG}}\left[\frac{\partial}{\partial u}\left(\frac{1}{\sqrt{E}}\frac{\partial\sqrt{G}}{\partial u}\right) + \frac{\partial}{\partial v}\left(\frac{1}{\sqrt{G}}\frac{\partial\sqrt{E}}{\partial v}\right)\right].$$

Alternative Method.

We know that the Gauss characteristic equation is

$$LN - M^2 = \frac{1}{2} H \frac{\partial}{\partial u}\left(\frac{F}{EH}\frac{\partial E}{\partial v} - \frac{1}{H}\frac{\partial G}{\partial u}\right)$$
$$+\frac{1}{2} H \frac{\partial}{\partial v}\left(\frac{2}{H}\frac{\partial F}{\partial u} - \frac{1}{H}\frac{\partial E}{\partial v} - \frac{F}{EH}\frac{\partial E}{\partial u}\right). \quad (1)$$

When the parametric curves are orthogonal, we have

$$F = 0. \quad (2)$$

Substituting the value of F from (2) in (1), we get

$$LN - M^2 = -\frac{1}{2} H\left[\frac{\partial}{\partial u}\left(\frac{1}{H}\frac{\partial G}{\partial u}\right) + \frac{\partial}{\partial v}\left(\frac{1}{H}\frac{\partial E}{\partial v}\right)\right]. \quad (3)$$

Also,

$$H^2 = EG - F^2 = EG. \quad (4)$$

[using (2)]

Now,

$$\kappa = \frac{LN - M^2}{H^2}$$
$$= -\frac{1}{2H}\left[\frac{\partial}{\partial u}\left(\frac{1}{H}\frac{\partial G}{\partial u}\right) + \frac{\partial}{\partial v}\left(\frac{1}{H}\frac{\partial E}{\partial v}\right)\right] \quad \text{[using (3)]}$$
$$= -\frac{1}{2\sqrt{EG}}\left[\frac{\partial}{\partial u}\left(\frac{1}{\sqrt{EG}}\frac{\partial G}{\partial u}\right) + \frac{\partial}{\partial v}\left(\frac{1}{\sqrt{EG}}\frac{\partial E}{\partial v}\right)\right] \quad \text{[using (4)]}$$
$$= -\frac{1}{\sqrt{EG}}\left[\frac{\partial}{\partial u}\left\{\frac{1}{\sqrt{E}}\left(\frac{1}{2\sqrt{G}}\frac{\partial G}{\partial u}\right)\right\} + \frac{\partial}{\partial v}\left\{\frac{1}{\sqrt{G}}\left(\frac{1}{2\sqrt{E}}\frac{\partial E}{\partial v}\right)\right\}\right]$$
$$= -\frac{1}{\sqrt{EG}}\left[\frac{\partial}{\partial u}\left(\frac{1}{\sqrt{E}}\frac{\partial\sqrt{G}}{\partial u}\right) + \frac{\partial}{\partial v}\left(\frac{1}{\sqrt{G}}\frac{\partial\sqrt{E}}{\partial v}\right)\right].$$

Example 5. Show that when the lines of curvature are chosen as parametric curves, the Codazzi relations expressed in terms of E, G, L, N and their derivatives are

$$L_2 = \frac{1}{2} E_2\left(\frac{L}{E} + \frac{N}{G}\right), \; N_1 = \frac{1}{2} G_1\left(\frac{L}{E} + \frac{N}{G}\right).$$

Show also that the equation of Gauss may be written as

$$\frac{LN}{\sqrt{EG}} + \frac{\partial}{\partial u}\left(\frac{1}{\sqrt{E}}\frac{\partial\sqrt{G}}{\partial u}\right) + \frac{\partial}{\partial v}\left(\frac{1}{\sqrt{G}}\frac{\partial\sqrt{E}}{\partial v}\right) = 0.$$

Solution. First Part. We know that the Codazzi relations are

$$L_2 - M_1 = mL - (l - \mu)\, M - \lambda N, \tag{1}$$
$$M_2 - N_1 = nL - (m - \nu)\, M - \mu N. \tag{2}$$

When the lines of curvature are taken as parametric curves, we know that

$$F = 0,\ M = 0. \tag{3}$$

Also, in this case,

$$\left.\begin{aligned} l &= \frac{E_1}{2E}, \quad m = \frac{E_2}{2E}, \quad n = -\frac{G_1}{2E}, \\ \lambda &= -\frac{E_2}{2G}, \quad \mu = \frac{G_1}{2G}, \quad \nu = \frac{G_2}{2G}. \end{aligned}\right\} \tag{4}$$

Moreover, since $M = 0$, we have

$$M_1 = 0,\ M_2 = 0. \tag{5}$$

Using (3), (4), (5), (1) gives

$$L_2 = \frac{E_2}{2E} L + \frac{E_2}{2G} N = \frac{1}{2} E_2 \left(\frac{L}{E} + \frac{N}{G} \right).$$

Using (3), (4), (5), (2) gives

$$-N_1 = -\frac{G_1}{2E} L - \frac{G_1}{2G} N.$$

$$\therefore \qquad N_1 = \frac{1}{2} G_1 \left(\frac{L}{E} + \frac{N}{G} \right).$$

Second Part. We know that the Gauss equation is

$$LN - M^2 = \frac{1}{2}(2F_{12} - E_{22} - G_{11}) + (m^2 E + 2m\mu F + \mu^2 G)$$
$$- \{lnE + (l\nu + \lambda n)\, F + \lambda \nu G\}. \tag{6}$$

Since $F = 0$, we have

$$F_{12} = 0. \tag{7}$$

Using (3), (4) and (7), (6) becomes

$$LN = \frac{1}{2}(-E_{22} - G_{11}) + \left(\frac{E_2{}^2}{4E} + \frac{G_1{}^2}{4G} \right) + \left(\frac{E_1 G_1}{4E} + \frac{E_2 G_2}{4G} \right)$$

or
$$LN + \left(\frac{1}{2} G_{11} - \frac{E_1 G_1}{4E} - \frac{G_1{}^2}{4G} \right) + \left(\frac{1}{2} E_{22} - \frac{E_2 G_2}{4G} - \frac{E_2{}^2}{4E} \right) = 0$$

or
$$\frac{LN}{\sqrt{EG}} + \left(\frac{G_{11}}{2\sqrt{EG}} - \frac{E_1 G_1}{4E\sqrt{EG}} - \frac{G_1{}^2}{4G\sqrt{EG}} \right)$$
$$+ \left(\frac{E_{22}}{2\sqrt{EG}} - \frac{E_2 G_2}{4G\sqrt{EG}} - \frac{E_2{}^2}{4E\sqrt{EG}} \right) = 0$$

or $$\frac{LN}{\sqrt{EG}}+\left[\frac{1}{2\sqrt{EG}}G_{11}-\frac{G_1(GE_1+EG_1)}{4\,(EG)^{3/2}}\right]$$
$$+\left[\frac{1}{2\sqrt{EG}}E_{22}-\frac{E_2(EG_2+GE_2)}{4\,(EG)^{3/2}}\right]=0$$

or $$\frac{LN}{\sqrt{EG}}+\frac{\partial}{\partial u}\left(\frac{1}{2\sqrt{EG}}G_1\right)+\frac{\partial}{\partial v}\left(\frac{1}{2\sqrt{EG}}E_2\right)=0$$

or $$\frac{LN}{\sqrt{EG}}+\frac{\partial}{\partial u}\left(\frac{1}{2\sqrt{EG}}\frac{\partial G}{\partial u}\right)+\frac{\partial}{\partial v}\left(\frac{1}{2\sqrt{EG}}\frac{\partial E}{\partial v}\right)=0$$

or $$\frac{LN}{\sqrt{EG}}+\frac{\partial}{\partial u}\left[\frac{1}{\sqrt{E}}\left(\frac{1}{2\sqrt{G}}\frac{\partial G}{\partial u}\right)\right]+\frac{\partial}{\partial v}\left[\frac{1}{\sqrt{G}}\left(\frac{1}{2\sqrt{E}}\frac{\partial E}{\partial v}\right)\right]=0$$

or $$\frac{LN}{\sqrt{EG}}+\frac{\partial}{\partial u}\left(\frac{1}{\sqrt{E}}\frac{\partial\sqrt{G}}{\partial u}\right)+\frac{\partial}{\partial v}\left(\frac{1}{\sqrt{G}}\frac{\partial\sqrt{E}}{\partial v}\right)=0.$$

Example 6. Show that when the linear element is of the form

$$ds^2=\phi\,(du^2+dv^2),$$

the Mainardi-Codazzi relations become

$$L_2-M_1=\frac{1}{2}(L+N)\frac{\phi_2}{\phi},$$
$$N_1-M_2=\frac{1}{2}(L+N)\frac{\phi_1}{\phi}$$

and the Gauss equation becomes

$$LN-M^2=\frac{1}{2\phi}({\phi_1}^2+{\phi_2}^2)-\frac{1}{2}(\phi_{11}+\phi_{22}).$$

Solution. First Part. We know that the Mainardi-Codazzi relations are

$$L_2-M_1=mL-(l-\mu)\,M-\lambda N, \quad (1)$$
$$M_2-N_1=nL-(m-\nu)\,M-\mu N. \quad (2)$$

The given line element is of the form

$$ds^2=\phi\,(du^2+dv^2). \quad (3)$$

But, we know that

$$ds^2=E\,du^2+2F\,du\,dv+G\,dv^2. \quad (4)$$

Comparing (3) with (4), we get

$$E=\phi,\ F=0,\ G=\phi. \quad (5)$$

Since $F = 0$, the Gauss coefficients reduce to

$$\left.\begin{aligned} l &= \frac{E_1}{2E} = \frac{\phi_1}{2\phi},\ m = \frac{E_2}{2E} = \frac{\phi_2}{2\phi},\ n = -\frac{G_1}{2E} = -\frac{\phi_1}{2\phi}, \\ \lambda &= -\frac{E_2}{2G} = -\frac{\phi_2}{2\phi}, \mu = \frac{G_1}{2G} = \frac{\phi_1}{2\phi}, \nu = \frac{G_2}{2G} = \frac{\phi_2}{2\phi}. \end{aligned}\right\} \tag{6}$$

Using (6), (1) and (2) give

$$L_2 - M_1 = \frac{\phi_2}{2\phi} L + \frac{\phi_2}{2\phi} N = \frac{1}{2}(L+N)\frac{\phi_2}{\phi}$$

and $$N_1 - M_2 = \frac{\phi_1}{2\phi} L + \frac{\phi_1}{2\phi} N = \frac{1}{2}(L+N)\frac{\phi_1}{\phi}$$

respectively.

Second Part. We know that the Gauss equation is

$$LN - M^2 = \frac{1}{2}(2F_{12} - E_{22} - G_{11}) + (m^2 E + 2m\mu F + \mu^2 G) - \{lnE + (l\nu + \lambda n) F + \lambda \nu G\}. \tag{7}$$

Since $F = 0$, we have

$$F_{12} = 0. \tag{8}$$

Using (5), (6) and (8), (7) gives

$$\begin{aligned} LN - M^2 &= \frac{1}{2}(-\phi_{22} - \phi_{11}) + \left(\frac{\phi_2^{\,2}}{4\phi^2}\phi + \frac{\phi_1^{\,2}}{4\phi^2}\phi\right) - \left(-\frac{\phi_1^{\,2}}{4\phi^2}\phi - \frac{\phi_2^{\,2}}{4\phi^2}\phi\right) \\ &= \frac{1}{2}(-\phi_{22} - \phi_{11}) + \frac{1}{4}\left(\frac{\phi_2^{\,2}}{\phi} + \frac{\phi_1^{\,2}}{\phi}\right) + \frac{1}{4}\left(\frac{\phi_1^{\,2}}{\phi} + \frac{\phi_2^{\,2}}{\phi}\right) \\ &= \frac{1}{2\phi}(\phi_1^{\,2} + \phi_2^{\,2}) - \frac{1}{2}(\phi_{11} + \phi_{22}). \end{aligned}$$

Example 7. If the asymptotic lines are taken as parametric curves, show that the Mainardi-Codazzi relations become

$$\frac{M_1}{M} = l - \mu,\quad \frac{M_2}{M} = \nu - m$$

and deduce that

$$2l = \frac{H_1}{H} + \frac{M_1}{M},\quad 2\mu = \frac{H_1}{H} - \frac{M_1}{M},$$

$$2m = \frac{H_2}{H} - \frac{M_2}{M},\quad 2\nu = \frac{H_2}{H} + \frac{M_2}{M}.$$

Solution. We know that the Mainardi-Codazzi relations are

$$L_2 - M_1 = mL - (l - \mu)\, M - \lambda N, \tag{1}$$

$$M_2 - N_1 = nL - (m - \nu)\, M - \mu N. \tag{2}$$

When the asymptotic lines are taken as parametric curves, we know that

$$L = 0,\, N = 0,\, M \neq 0. \tag{3}$$

$$\therefore \quad L_2 = 0,\, N_1 = 0. \tag{4}$$

Using (3) and (4), (1) becomes

$$-M_1 = -(l - \mu)\, M.$$

$$\therefore \quad \frac{M_1}{M} = l - \mu. \tag{5}$$

Similarly, using (3) and (4), (2) becomes

$$M_2 = -(m - \nu)\, M.$$

$$\therefore \quad \frac{M_2}{M} = \nu - m. \tag{6}$$

Deduction. We know that

$$l = \frac{1}{2H^2}(GE_1 - 2FF_1 + FE_2),\ \mu = \frac{1}{2H^2}(EG_1 - FE_2).$$

$$\therefore \quad l + \mu = \frac{1}{2H^2}(GE_1 + EG_1 - 2FF_1)$$

or $$l + \mu = \frac{1}{2H^2}(2HH_1)$$

$$[\because H^2 = EG - F^2 \Rightarrow 2HH_1 = E_1G + EG_1 - 2FF_1]$$

or $$l + \mu = \frac{H_1}{H}. \tag{7}$$

Adding (5) and (7), we get the value of l.

Subtracting (5) from (7), we get the value of μ.

Similarly, we have

$$m = \frac{1}{2H^2}(GE_2 - FG_1),\ \nu = \frac{1}{2H^2}(EG_2 - 2FF_2 + FG_1).$$

$$\therefore \quad \nu + m = \frac{1}{2H^2}(GE_2 + EG_2 - 2FF_2)$$

or $$\nu + m = \frac{1}{2H^2}(2HH_2)$$

$$[\because H^2 = EG - F^2 \Rightarrow 2HH_2 = E_2G + EG_2 - 2FF_2]$$

or $$\nu + m = \frac{H_2}{H}. \tag{8}$$

Adding (6) and (8), we get the value of ν.

Subtracting (6) from (8), we get the value of m.

Example 8. Show that the lines of curvature form an isothermal system on a minimal surface.

Solution. We know that a surface is minimal if

$$EN - 2FM + GL = 0. \tag{1}$$

Let us take the lines of curvature as the parametric curves. Then, we know that

$$F = 0, M = 0. \tag{2}$$

Using (2), (1) gives

$$EN + GL = 0$$

or $$\frac{L}{E} + \frac{N}{G} = 0. \tag{3}$$

When $F = 0$, $M = 0$, the Mainardi-Codazzi equations reduce to

$$L_2 = \frac{1}{2} E_2 \left(\frac{L}{E} + \frac{N}{G} \right), \tag{4}$$

$$N_1 = \frac{1}{2} G_1 \left(\frac{L}{E} + \frac{N}{G} \right). \tag{5}$$

Using (3), (4) gives

$$L_2 = 0$$

$\Rightarrow$ $$\frac{\partial L}{\partial v} = 0$$

$\Rightarrow$ L is a function of u only. (6)

Similarly, using (3), (5) gives

$$N_1 = 0$$

$\Rightarrow$ $$\frac{\partial N}{\partial u} = 0$$

$\Rightarrow$ N is a function of v only. (7)

Now, $$\frac{\partial^2}{\partial u\, \partial v} \log \frac{E}{G} = \frac{\partial^2}{\partial u\, \partial v} \log \left(-\frac{L}{N} \right) \qquad \text{[using (3)]}$$

$$= \frac{\partial^2}{\partial u\, \partial v} [\log (-L) - \log N]$$

$$= \frac{\partial}{\partial u} \left[\frac{\partial}{\partial v} \log (-L) \right] - \frac{\partial}{\partial v} \left[\frac{\partial}{\partial u} \log (N) \right]$$

$$= 0. \qquad \text{[using (6) and (7)]}$$

Thus, we have

$$F = 0, \quad \frac{\partial^2}{\partial u\, \partial v} \log \frac{E}{G} = 0.$$

Hence the parametric curves (which are the lines of curvature here) are isometric lines. Thus it follows that the lines of curvature form an isothermal system on a minimal surface.

Exercises

1. For the right helicoid

$$x = u\cos v,\ y = u\sin v,\ z = cv,$$

show that

$$l = 0,\ m = 0,\ n = -u,\ \lambda = 0,\ \mu = u/(u^2 + c^2),\ \nu = 0.$$

2. For the surface given by $ds^2 = \phi\,(du^2 + dv^2)$, show that

$$l = \frac{\phi_1}{2\phi}, \qquad m = \frac{\phi_2}{2\phi}, \qquad n = -\frac{\phi_1}{2\phi},$$

$$\lambda = -\frac{\phi_2}{2\phi}, \qquad \mu = \frac{\phi_1}{2\phi}, \qquad \nu = \frac{\phi_2}{2\phi}.$$

3. For any surface, prove that

$$\frac{\partial}{\partial u}(\log H) = l + \mu, \quad \frac{\partial}{\partial v}(\log H) = m + \nu.$$

4. If the null lines are taken as parametric curves, show that

$$l = F_1/F,\ m = 0,\ n = 0,\ \lambda = 0,\ \mu = 0,\ \nu = F_2/F.$$

5. When the parametric curves are null lines, show that the Mainardi-Codazzi relations may be expressed as

$$\frac{L_2}{M} = \frac{\partial}{\partial u}\log\frac{M}{F}, \quad \frac{N_1}{M} = \frac{\partial}{\partial v}\log\frac{M}{F}$$

and the Gauss characteristic equation as

$$LN - M^2 = F_{12} - \frac{F_1 F_2}{F}.$$

4.14 PARALLEL SURFACES

Definition. A surface $\bar{S}$, which is at a constant distance along the normal from another surface S, is said to be parallel to S. As the constant distance may be chosen arbitrarily, the number of such parallel surfaces is infinite. Two concentric spheres are parallel surfaces.

If $\mathbf{r}$ is the position vector of a current point P on the surface S and $\mathbf{N}$ is the unit normal to that surface at P, then the position vector $\bar{\mathbf{r}}$ of the corresponding point Q on the parallel surface $\bar{S}$ is given by

$$\bar{\mathbf{r}} = \mathbf{r} - c\mathbf{N}.$$

where c is the constant distance between the points P and Q along $\mathbf{N}$.

Fundamental Coefficients for the Parallel Surfaces when the Lines of Curvature on the Given Surface are taken as Parametric Curves.

Let $\mathbf{r}$ be the position vector of a current point P on the surface S and let $\mathbf{N}$ be the unit normal to that surface at P. Also, let $\bar{\mathbf{r}}$ be the position vector of the corresponding point Q on the parallel surface $\bar{S}$. Then

$$\bar{\mathbf{r}} = \mathbf{r} - c\mathbf{N}. \tag{1}$$

Differentiating (1) partially with respect to the parameters u and v, we get

$$\bar{\mathbf{r}}_1 = \mathbf{r}_1 - c\mathbf{N}_1, \tag{2}$$

$$\bar{\mathbf{r}}_2 = \mathbf{r}_2 - c\mathbf{N}_2. \tag{3}$$

When the lines of curvature are taken as parametric curves, we know that

$$F = 0,\ M = 0,\ H^2 = EG. \tag{4}$$

From Weingarten's equations, we have

$$\left.\begin{aligned} \mathbf{N}_1 &= \frac{FM - GL}{H^2}\mathbf{r}_1 + \frac{FL - EM}{H^2}\mathbf{r}_2, \\ \mathbf{N}_2 &= \frac{FN - GM}{H^2}\mathbf{r}_1 + \frac{FM - EN}{H^2}\mathbf{r}_2. \end{aligned}\right\} \tag{5}$$

Using (4), (5) reduces to

$$\mathbf{N}_1 = -\frac{L}{E}\mathbf{r}_1,$$

$$\mathbf{N}_2 = -\frac{N}{G}\mathbf{r}_2.$$

Substituting the above values of $\mathbf{N}_1$ and $\mathbf{N}_2$ in (2) and (3), we get

$$\bar{\mathbf{r}}_1 = \left(1 + \frac{cL}{E}\right)\mathbf{r}_1, \tag{6}$$

$$\bar{\mathbf{r}}_2 = \left(1 + \frac{cN}{G}\right)\mathbf{r}_2. \tag{7}$$

Let κ_a and κ_b be the principal curvatures of the surface S at P. Since the lines of curvature are taken as parametric curves, it follows that κ_a and κ_b are the normal curvatures of the surface S at P in the direction of parametric curves.

We know that the normal curvature κ_n at P in the direction (du, dv) is given by

$$\kappa_n = \frac{L\,du^2 + 2M\,du\,dv + N\,dv^2}{E\,du^2 + 2F\,du\,dv + G\,dv^2}$$

i.e., $$\kappa_n = \frac{L\,du^2 + N\,dv^2}{E\,du^2 + G\,dv^2}. \tag{8}$$

The principal curvature κ_a, being the normal curvature for the direction $dv = 0$, is expressed as

$$\kappa_a = \frac{L}{E}. \tag{9}$$

[using (8)]

Similarly, the principal curvature κ_b, being the normal curvature for the direction $du = 0$, is expressed as

$$\kappa_b = \frac{N}{G}. \tag{10}$$

[using (8)]

Using (9) and (10), (6) and (7) reduce to

$$\overline{\mathbf{r}}_1 = (1 + c\kappa_a)\,\mathbf{r}_1 \tag{11}$$

and $$\overline{\mathbf{r}}_2 = (1 + c\kappa_b)\,\mathbf{r}_2 \tag{12}$$

respectively.

Let $\overline{E}, \overline{F}, \overline{G}, \overline{L}, \overline{M}, \overline{N}$ denote the fundamental coefficients of $\overline{S}$. Then, we have

$$\overline{E} = \overline{\mathbf{r}}_1^{\,2} = (1 + c\kappa_a)^2\,\mathbf{r}_1^{\,2} = (1 + c\kappa_a)^2 E,$$

$$\overline{F} = \overline{\mathbf{r}}_1 \cdot \overline{\mathbf{r}}_2 = (1 + c\kappa_a)(1 + c\kappa_b)\,\mathbf{r}_1 \cdot \mathbf{r}_2 = (1 + c\kappa_a)(1 + c\kappa_b)\,F$$

$$= 0, \qquad \text{[using (4)]}$$

$$\overline{G} = \overline{\mathbf{r}}_2^{\,2} = (1 + c\kappa_b)^2\,\mathbf{r}_2^{\,2} = (1 + c\kappa_b)^2 G,$$

$$\overline{H}^2 = \overline{E}\,\overline{G} - \overline{F}^2 = \overline{E}\,\overline{G} = (1 + c\kappa_a)^2 (1 + c\kappa_b)^2 EG$$

$$= (1 + c\kappa_a)^2\,(1 + c\kappa_b)^2\,H^2$$

The unit normal $\overline{\mathbf{N}}$ to the parallel surface $\overline{S}$ at Q is given by

$$\overline{H}\,\overline{\mathbf{N}} = \overline{\mathbf{r}}_1 \times \overline{\mathbf{r}}_2 = (1 + c\kappa_a)(1 + c\kappa_b)\,\mathbf{r}_1 \times \mathbf{r}_2$$

or $$\overline{H}\,\overline{\mathbf{N}} = \{1 + c(\kappa_a + \kappa_b) + c^2\kappa_a\kappa_b\}\,H\mathbf{N}$$

or $$\overline{H}\,\overline{\mathbf{N}} = (1 + 2\mu c + Kc^2)\,H\mathbf{N}, \tag{13}$$

where μ is the mean curvature and K is the Gaussian curvature of S at P.
It follows from (13) that the normals to the two surfaces at corresponding points are parallel. Hence we may write

$$\overline{\mathbf{N}} = e\mathbf{N}, \tag{14}$$

where e is equal to 1 or -1 according as $1+2\mu c+Kc^2$ is positive or negative.

Again, we have

$$\overline{\mathbf{r}}_{11} = \left\{\frac{\partial}{\partial u}(c\kappa_a)\right\}\mathbf{r}_1 + (1+c\kappa_a)\,\mathbf{r}_{11},$$

$$\overline{\mathbf{r}}_{12} = \left\{\frac{\partial}{\partial v}(c\kappa_a)\right\}\mathbf{r}_1 + (1+c\kappa_a)\,\mathbf{r}_{12},$$

$$\overline{\mathbf{r}}_{22} = \left\{\frac{\partial}{\partial v}(c\kappa_b)\right\}\mathbf{r}_2 + (1+c\kappa_b)\,\mathbf{r}_{22}.$$

Therefore

$$\overline{L} = \overline{\mathbf{r}}_{11}\cdot\overline{\mathbf{N}} = \left[\left\{\frac{\partial}{\partial u}(c\kappa_a)\right\}\mathbf{r}_1 + (1+c\kappa_a)\,\mathbf{r}_{11}\right]\cdot e\mathbf{N}$$

$$= e\,(1+c\kappa_a)\,L,$$

$$\overline{M} = \overline{\mathbf{r}}_{12}\cdot\overline{\mathbf{N}} = \left[\left\{\frac{\partial}{\partial v}(c\kappa_a)\right\}\mathbf{r}_1 + (1+c\kappa_a)\,\mathbf{r}_{12}\right]\cdot e\mathbf{N}$$

$$= e\,(1+c\kappa_a)\,M$$

$$= 0, \qquad \text{[using (4)]}$$

$$\overline{N} = \overline{\mathbf{r}}_{22}\cdot\overline{\mathbf{N}} = \left[\left\{\frac{\partial}{\partial v}(c\kappa_b)\right\}\mathbf{r}_2 + (1+c\kappa_b)\,\mathbf{r}_{22}\right]\cdot e\mathbf{N}$$

$$= e\,(1+c\kappa_b)\,N.$$

Thus, we have $\overline{F} = \overline{M} = 0$. Hence it follows that the parametric curves are lines of curvature on the parallel surface $\overline{S}$ also.

Therefore we can conclude that the curves on the parallel surface $\overline{S}$ corresponding to the lines of curvature on S are lines of curvature on $\overline{S}$ also.

Gaussian Curvature and Mean Curvature for the Parallel Surface.

The Gaussian curvature $\overline{K}$ for the parallel surface $\overline{S}$ is given by

$$\overline{K} = \frac{\overline{L}\,\overline{N} - \overline{M}^2}{\overline{H}^2} = \frac{\{e\,(1+c\kappa_a)\,L\}\{e\,(1+c\kappa_b)\,N\} - 0}{(1+c\kappa_a)^2(1+c\kappa_b)^2\,H^2}$$

$$= \frac{LN}{(1+c\kappa_a)(1+c\kappa_b)H^2} \qquad [\because e = \pm 1 \Rightarrow e^2 = 1]$$

$$= \frac{K}{(1+c\kappa_a)(1+c\kappa_b)} \qquad \left[\because K = \frac{LN - M^2}{H^2} = \frac{LN}{H^2}\right]$$

$$= \frac{K}{1+2\mu c + Kc^2}.$$

If $\overline{\mu}$ is the mean curvature for the parallel surface $\overline{S}$, then

$$2\overline{\mu} = \overline{\kappa}_a + \overline{\kappa}_b = \frac{\overline{L}}{\overline{E}} + \frac{\overline{N}}{\overline{G}}$$

$$= \frac{e(1+c\kappa_a)L}{(1+c\kappa_a)^2 E} + \frac{e(1+c\kappa_b)N}{(1+c\kappa_b)^2 G}$$

$$= e\left\{\frac{\kappa_a}{1+c\kappa_a} + \frac{\kappa_b}{1+c\kappa_b}\right\} \qquad \left[\because \frac{L}{E} = \kappa_a, \frac{N}{G} = \kappa_b\right]$$

$$= e\left\{\frac{\kappa_a + \kappa_b + 2c\kappa_a\kappa_b}{(1+c\kappa_a)(1+c\kappa_b)}\right\}$$

$$= e\left\{\frac{2\mu + 2cK}{1+2\mu c + Kc^2}\right\}.$$

$$\therefore \qquad \overline{\mu} = e\left\{\frac{\mu + cK}{1+2\mu c + Kc^2}\right\}.$$

Bonnet's Theorem on Parallel Surfaces.

With every surface of constant second curvature $\frac{1}{a^2}$, there are associated two surfaces of constant first curvature $\pm\frac{1}{a}$, which are parallel to the former and distant a from it.

Similarly, with every surface of constant first curvature $\frac{1}{a}$, there is associated a parallel surface of constant second curvature $\frac{1}{a^2}$ at a distance a from it.

Proof. Case 1. Let a surface of constant second curvature $\frac{1}{a^2}$ be given.

To prove : There are associated two surfaces of constant first curvature $\pm\frac{1}{a}$, which are parallel to the former and distant a from it.

We know that the first curvature $\bar{J}$ of the parallel surface $\bar{S}$ is given by

$$\bar{J} = 2e\left\{\frac{\mu + cK}{1 + 2\mu c + Kc^2}\right\}, \tag{1}$$

where μ is the mean curvature and K is the Gaussian curvature (or the second curvature) of the given surface S.

Putting $K = \frac{1}{a^2}$ and $c = \pm a$ in (1), we have

$$\bar{J} = 2e\left\{\frac{\mu \pm (1/a)}{1 \pm 2\mu a + 1}\right\} = 2e\left\{\frac{\mu a \pm 1}{2a(1 \pm \mu a)}\right\}$$

$$= \pm e\frac{1}{a} = \pm\frac{1}{a}, \text{which is constant.}$$

Hence the result.

Case 2. Now, let a surface of constant first curvature $\frac{1}{a}$ be given.

To prove : There is associated a parallel surface of constant second curvature $\frac{1}{a^2}$ at a distance a from it.

We know that the second curvature $\bar{K}$ of the parallel surface $\bar{S}$ is given by

$$\bar{K} = \frac{K}{1 + 2\mu c + Kc^2}. \tag{2}$$

Putting $2\mu = \frac{1}{a}$ and $c = -a$ in (2), we get

$$\bar{K} = \frac{K}{1 + (1/a)(-a) + Ka^2} = \frac{1}{a^2} = \text{constant.}$$

Hence the proof is complete.

5
Geodesics

5.1 DEFINITION

A geodesic line or a geodesic on a surface may be defined as a curve whose osculating plane at each point contains the normal to the surface at that point.

Let S be a surface and let C denote a geodesic on it. Let P be any point on C and let $\mathbf{t}$, $\mathbf{n}$, $\mathbf{b}$ denote unit tangent, unit principal normal and unit binormal respectively to C at P. Also, let $\mathbf{N}$ be the unit normal to the surface S at P.

Then, $\mathbf{n}$ lies in the osculating plane of C at P and is perpendicular to $\mathbf{t}$.

Also, $\mathbf{N}$ lies in the osculating plane of C at P and is perpendicular to $\mathbf{t}$.

Therefore, it follows that

$$\mathbf{N} = \pm \mathbf{n}.$$

Without any loss of generality, we take

$$\mathbf{N} = \mathbf{n}$$

i.e., the normal to the surface coincides with the principal normal to the geodesic. This property is called the normal property of a geodesic.

Note 1. Clearly, the normal curvature of S at P is the curvature of C (geodesic) at P. Thus, if κ be the curvature of C at P, then κ is the normal curvature of S at P in the direction of $\mathbf{t}$. Therefore

$$\kappa = Lu'^2 + 2Mu'v' + Nv'^2,$$

where (u, v) are the parametric coordinates of the point P and dashes denote differentiation with respect to arc length s of C.

Note 2. If P and Q are two close points on S, then S will give infinitely many sections by planes passing through P and Q. Out of these infinitely many sections, the normal section gives the arcual distance PQ as the least one. Clearly, the osculating plane in this case containing the arcual element PQ contains the surface normal at P. This arc PQ is the arc of a geodesic. Thus it follows that the geodesic on a surface passing through two points on the surface is the path of least distance. More generally, the geodesic on a surface passing through two points on the surface is the curve of stationary length.

5.2 DIFFERENTIAL EQUATIONS OF GEODESICS

Let the parametric equation of a surface in vector form be given by

$$\mathbf{r} = \mathbf{r}(u, v), \tag{1}$$

where (u, v) denote the parametric coordinates of a point on the surface.

Let A and B be two points on this surface and let the equations of the arcs joining A and B be given by

$$u = u(t), v = v(t), \tag{2}$$

where $u(t)$ and $v(t)$ are functions of class two.

Without any loss of generality, we may suppose that for any such arc Γ, $t = 0$ at A and $t = 1$ at B. Thus Γ is given by $0 \le t \le 1$.

Now, we know that

$$\left(\frac{ds}{dt}\right)^2 = E\left(\frac{du}{dt}\right)^2 + 2F\frac{du}{dt}\frac{dv}{dt} + G\left(\frac{dv}{dt}\right)^2.$$

Therefore the whole length of the arc Γ is given by

$$s(\Gamma) = \int_0^1 (E\dot{u}^2 + 2F\dot{u}\dot{v} + G\dot{v}^2)^{1/2}\, dt. \tag{3}$$

Let the arc Γ be slightly deformed, keeping its end points A and B fixed. Let Γ' be the arc thus obtained. Then equations to represent Γ' may be written as

$$u'(t) = u(t) + \varepsilon g(t),\ v'(t) = v(t) + \varepsilon h(t), \tag{4}$$

where ε is small, and g and h are arbitrary functions of t of class two in $0 \le t \le 1$ such that $g = h = 0$ at $t = 0$ and $t = 1$.

The length of the arc Γ' is given by

$$s(\Gamma') = \int_0^1 (E\dot{u}'^2 + 2F\dot{u}'\dot{v}' + G\dot{v}'^2)^{1/2}\, dt. \tag{5}$$

The variation in the length of arc Γ is $s(\Gamma') - s(\Gamma)$. If this variation is a magnitude of order ε^2 for small variations in Γ, then we say that $s(\Gamma)$ is stationary and Γ is a geodesic.

Now, we shall use the method of calculus of variations to find the equations of geodesic.

Let $f = \sqrt{2T}$, where

$$2T(u, v, \dot{u}, \dot{v}) = E\dot{u}^2 + 2F\dot{u}\dot{v} + G\dot{v}^2.$$

Then

$$s(\Gamma') - s(\Gamma) = \int_0^1 [f(u + \varepsilon g, v + \varepsilon h, \dot{u} + \varepsilon\dot{g}, \dot{v} + \varepsilon\dot{h}) - f(u, v, \dot{u}, \dot{v})]\, dt. \tag{6}$$

By Taylor's theorem for several variables, we have

$$f(u+\varepsilon g, v+\varepsilon h, \dot{u}+\varepsilon \dot{g}, \dot{v}+\varepsilon \dot{h}) = f(u,v,\dot{u},\dot{v}) + \varepsilon g\frac{\partial f}{\partial u} + \varepsilon h\frac{\partial f}{\partial v} + \varepsilon \dot{g}\frac{\partial f}{\partial \dot{u}} + \varepsilon \dot{h}\frac{\partial f}{\partial \dot{v}} + O(\varepsilon^2).$$

Substituting the above value in (6), we get

$$s(\Gamma') - s(\Gamma) = \varepsilon \int_0^1 \left[g\frac{\partial f}{\partial u} + h\frac{\partial f}{\partial v} + \dot{g}\frac{\partial f}{\partial \dot{u}} + \dot{h}\frac{\partial f}{\partial \dot{v}} \right] dt + O(\varepsilon^2). \quad (7)$$

Now, integrating by parts,

$$\int_0^1 \dot{g}\frac{\partial f}{\partial \dot{u}}\, dt = \left[g\frac{\partial f}{\partial \dot{u}} \right]_0^1 - \int_0^1 g\frac{d}{dt}\left(\frac{\partial f}{\partial \dot{u}}\right) dt$$

$$= -\int_0^1 g\frac{d}{dt}\left(\frac{\partial f}{\partial \dot{u}}\right) dt. \qquad [\because g = 0 \text{ at } t = 0 \text{ and } t = 1.]$$

Similarly,

$$\int_0^1 \dot{h}\frac{\partial f}{\partial \dot{v}}\, dt = -\int_0^1 h\frac{d}{dt}\left(\frac{\partial f}{\partial \dot{v}}\right) dt.$$

Hence, from (7), we have

$$s(\Gamma') - s(\Gamma) = \varepsilon \int_0^1 (gL + hM)\, dt + O(\varepsilon^2), \quad (8)$$

where

$$L = \frac{\partial f}{\partial u} - \frac{d}{dt}\left(\frac{\partial f}{\partial \dot{u}}\right), \quad M = \frac{\partial f}{\partial v} - \frac{d}{dt}\left(\frac{\partial f}{\partial \dot{v}}\right).$$

By definition, Γ is a geodesic if $s(\Gamma)$ is stationary, i.e., the difference $s(\Gamma') - s(\Gamma)$ is of order ε^2. Hence it follows from (8) that Γ is a geodesic if and only if $u(t)$ and $v(t)$ are such that

$$\int_0^1 (gL + hM)\, dt = 0 \quad (9)$$

for all admissible values of g and h which satisfy $g = h = 0$ at $t = 0$ and $t = 1$.

Now, we shall prove that the condition (9) implies that $L = M = 0$.

For this, we shall state and prove the following lemma.

Lemma. If ϕ is a continuous function of t for $0 < t < 1$ and if

$$\int_0^1 \psi(t)\,\phi(t)\,dt = 0$$

for all admissible functions ψ as defined above, then

$$\phi(t) = 0.$$

Proof of the Lemma. Suppose, if possible, there exists a value t_0 of t between 0 and 1 such that $\phi(t_0) \neq 0$, say, $\phi(t_0) > 0$. Then, from continuity of ϕ, it follows that $\phi(t) > 0$ in some interval (a, b) such that $0 < a < t_0 < b < 1$.

Now, let us define the function ψ as

$$\psi(t) = \begin{cases} 0 \text{ for } 0 \leq t < a \text{ and } b < t \leq 1, \\ (t-a)^3 (b-t)^3 \text{ for } a \leq t \leq b. \end{cases}$$

Then $\psi(t)$ is admissible and

$$\int_0^1 \psi(t)\,\phi(t)\,dt = \int_0^a \psi(t)\,\phi(t)\,dt + \int_a^b \psi(t)\,\phi(t)\,dt + \int_b^1 \psi(t)\,\phi(t)\,dt$$

$$= \int_a^b (t-a)^3 (b-t)^3\, \phi(t)\,dt$$

$$> 0. \; [\because (t-a)^3 (b-t)^3 > 0 \text{ and } \phi(t) > 0 \text{ for } a < t < b]$$

But this contradicts the hypothesis that

$$\int_0^1 \psi(t)\,\phi(t)\,dt = 0$$

for all admissible functions ψ. Hence our supposition that $\phi(t_0) \neq 0$ is false. Therefore, we must have $\phi(t) = 0$.

Thus the lemma is proved.

Now, the functions L and M are continuous, since E, F, G are supposed to be of class one and $u(t)$, $v(t)$ of class two. Therefore we can apply the above lemma to (9).

Taking $h = 0$ in (9), we get

$$\int_0^1 gL\,dt = 0$$

$$\Rightarrow \quad L = 0. \tag{10}$$

[using the lemma]

Now, taking $g = 0$ in (9), we get

$$\int_0^1 hM\,dt = 0$$

$$\Rightarrow \qquad M = 0. \tag{11}$$

[using the lemma]

Equations (10) and (11) are the differential equations of the geodesics on the surface.

Since these equations do not involve the points A and B explicitly, these equations are same for all geodesics on the surface.

Now, we have

$$L = \frac{\partial f}{\partial u} - \frac{d}{dt}\left(\frac{\partial f}{\partial \dot{u}}\right)$$

$$= \frac{1}{\sqrt{2T}}\frac{\partial T}{\partial u} - \frac{d}{dt}\left(\frac{1}{\sqrt{2T}}\frac{\partial T}{\partial \dot{u}}\right) \qquad \left[\because f = \sqrt{2T}\right]$$

$$= \frac{1}{\sqrt{2T}}\frac{\partial T}{\partial u} - \frac{1}{\sqrt{2T}}\frac{d}{dt}\left(\frac{\partial T}{\partial \dot{u}}\right) + \frac{1}{(2T)^{3/2}}\frac{dT}{dt}\frac{\partial T}{\partial \dot{u}}.$$

Similarly,

$$M = \frac{\partial f}{\partial v} - \frac{d}{dt}\left(\frac{\partial f}{\partial \dot{v}}\right)$$

$$= \frac{1}{\sqrt{2T}}\frac{\partial T}{\partial v} - \frac{d}{dt}\left(\frac{1}{\sqrt{2T}}\frac{\partial T}{\partial \dot{v}}\right) \qquad \left[\because f = \sqrt{2T}\right]$$

$$= \frac{1}{\sqrt{2T}}\frac{\partial T}{\partial v} - \frac{1}{\sqrt{2T}}\frac{d}{dt}\left(\frac{\partial T}{\partial \dot{v}}\right) + \frac{1}{(2T)^{3/2}}\frac{dT}{dt}\frac{\partial T}{\partial \dot{v}}.$$

Hence, by (10) and (11), the equations of geodesics are given by

$$U \equiv \frac{d}{dt}\left(\frac{\partial T}{\partial \dot{u}}\right) - \frac{\partial T}{\partial u} = \frac{1}{2T}\frac{dT}{dt}\frac{\partial T}{\partial \dot{u}}, \tag{12}$$

$$V \equiv \frac{d}{dt}\left(\frac{\partial T}{\partial \dot{v}}\right) - \frac{\partial T}{\partial v} = \frac{1}{2T}\frac{dT}{dt}\frac{\partial T}{\partial \dot{v}}. \tag{13}$$

Eliminating $\dfrac{dT}{dt}$ between (12) and (13), we get

$$V\frac{\partial T}{\partial \dot{u}} - U\frac{\partial T}{\partial \dot{v}} = 0. \tag{14}$$

It can be seen that (14) is the necessary and sufficient condition for a curve on a surface to be geodesic.

Remark. The expressions U and V satisfy the identity

$$\dot{u}U + \dot{v}V = \frac{dT}{dt},$$

whether the curve $u = u\,(t), v = v\,(t)$ is a geodesic or not.

The above identity can be proved as follows.

We have

$$\dot{u}U+\dot{v}V=\dot{u}\left[\frac{d}{dt}\left(\frac{\partial T}{\partial \dot{u}}\right)-\frac{\partial T}{\partial u}\right]+\dot{v}\left[\frac{d}{dt}\left(\frac{\partial T}{\partial \dot{v}}\right)-\frac{\partial T}{\partial v}\right]$$

or $$\dot{u}U+\dot{v}V=\dot{u}\frac{d}{dt}\left(\frac{\partial T}{\partial \dot{u}}\right)+\dot{v}\frac{d}{dt}\left(\frac{\partial T}{\partial \dot{v}}\right)-\dot{u}\frac{\partial T}{\partial u}-\dot{v}\frac{\partial T}{\partial v}$$

or $$\dot{u}U+\dot{v}V=\frac{d}{dt}\left[\dot{u}\frac{\partial T}{\partial \dot{u}}+\dot{v}\frac{\partial T}{\partial \dot{v}}\right]-\ddot{u}\frac{\partial T}{\partial \dot{u}}-\ddot{v}\frac{\partial T}{\partial \dot{v}}-\dot{u}\frac{\partial T}{\partial u}-\dot{v}\frac{\partial T}{\partial v}. \tag{1}$$

Now T is a homogeneous function of degree two in $\dot{u},\dot{v}$. Therefore, by Euler's theorem on homogeneous functions,

$$\dot{u}\frac{\partial T}{\partial \dot{u}}+\dot{v}\frac{\partial T}{\partial \dot{v}}=2T. \tag{2}$$

Again, regarding T as a function of $u, v, \dot{u}, \dot{v}$, we have

$$\frac{dT}{dt}=\frac{\partial T}{\partial u}\dot{u}+\frac{\partial T}{\partial v}\dot{v}+\frac{\partial T}{\partial \dot{u}}\ddot{u}+\frac{\partial T}{\partial \dot{v}}\ddot{v}. \tag{3}$$

Using (2) and (3), (1) gives

$$\dot{u}U+\dot{v}V=\frac{d}{dt}(2T)-\frac{dT}{dt}=\frac{dT}{dt}.$$

Corollary 1. A necessary and sufficient condition that the curves $v=\text{const.}$ will be geodesics is $EE_2+FE_1-2EF_1=0$, and the curves $u=\text{const.}$ will be geodesics is $GG_1+FG_2-2GF_2=0$.

Proof. First, we consider the curve $v=c$ (constant).

For the curve $v=c$ (constant), u can be taken as the parameter, i.e., the equations of the curve $v=c$ (constant) can be written as

$$u=t, v=c$$

$$\therefore \quad \dot{u}=1, \dot{v}=0.$$

Now, we shall find the partial derivatives of T.

We have

$$T=\frac{1}{2}(E\dot{u}^2+2F\dot{u}\dot{v}+G\dot{v}^2).$$

$$\therefore \quad \frac{\partial T}{\partial u}=\frac{1}{2}(E_1\dot{u}^2+2F_1\dot{u}\dot{v}+G_1\dot{v}^2),$$

$$\frac{\partial T}{\partial v}=\frac{1}{2}(E_2\dot{u}^2+2F_2\dot{u}\dot{v}+G_2\dot{v}^2),$$

$$\frac{\partial T}{\partial \dot{u}}=\frac{1}{2}(2E\dot{u}+2F\dot{v})=E\dot{u}+F\dot{v},$$

$$\frac{\partial T}{\partial \dot{v}} = \frac{1}{2}(2F\dot{u} + 2G\dot{v}) = F\dot{u} + G\dot{v}.$$

At $\dot{u} = 1, \dot{v} = 0,$ we have

$$\frac{\partial T}{\partial u} = \frac{1}{2}E_1, \frac{\partial T}{\partial v} = \frac{1}{2}E_2, \frac{\partial T}{\partial \dot{u}} = E, \frac{\partial T}{\partial \dot{v}} = F.$$

Therefore

$$\begin{aligned} U &= \frac{d}{dt}\left(\frac{\partial T}{\partial \dot{u}}\right) - \frac{\partial T}{\partial u} = \frac{dE}{dt} - \frac{1}{2}E_1 \\ &= \left(\frac{\partial E}{\partial u}\frac{du}{dt} + \frac{\partial E}{\partial v}\frac{dv}{dt}\right) - \frac{1}{2}E_1 \\ &= (E_1\dot{u} + E_2\dot{v}) - \frac{1}{2}E_1 \\ &= (E_1.1 + E_2.0) - \frac{1}{2}E_1 \\ &= \frac{1}{2}E_1 \end{aligned}$$

and

$$\begin{aligned} V &= \frac{d}{dt}\left(\frac{\partial T}{\partial \dot{v}}\right) - \frac{\partial T}{\partial v} = \frac{dF}{dt} - \frac{1}{2}E_2 \\ &= \left(\frac{\partial F}{\partial u}\frac{du}{dt} + \frac{\partial F}{\partial v}\frac{dv}{dt}\right) - \frac{1}{2}E_2 \\ &= (F_1\dot{u} + F_2\dot{v}) - \frac{1}{2}E_2 \\ &= (F_1.1 + F_2.0) - \frac{1}{2}E_2 \\ &= F_1 - \frac{1}{2}E_2. \end{aligned}$$

Now, by (14), the necessary and sufficient condition for the curve $u = u(t), v = v(t)$ to be a geodesic is

$$V\frac{\partial T}{\partial \dot{u}} - U\frac{\partial T}{\partial \dot{v}} = 0.$$

Hence a necessary and sufficient condition for the curve $v = c$ to be a geodesic is

$$\left(F_1 - \frac{1}{2}E_2\right)E - \frac{1}{2}E_1F = 0$$

or $\quad 2EF_1 - EE_2 - FE_1 = 0$

or $\quad EE_2 + FE_1 - 2EF_1 = 0.$

Now, we consider the curve $u = c$ (constant).

For the curve $u = c$ (constant), v can be taken as the parameter, i.e., the equations of the curve $u = c$ (constant) can be written as

$$u = c, v = t.$$

$$\therefore \quad \dot{u} = 0, \dot{v} = 1.$$

Using the partial derivatives of T derived earlier, we have, at $\dot{u} = 0, \dot{v} = 1$,

$$\frac{\partial T}{\partial u} = \frac{1}{2}G_1, \frac{\partial T}{\partial v} = \frac{1}{2}G_2, \frac{\partial T}{\partial \dot{u}} = F, \frac{\partial T}{\partial \dot{v}} = G.$$

Therefore

$$U = \frac{d}{dt}\left(\frac{\partial T}{\partial \dot{u}}\right) - \frac{\partial T}{\partial u} = \frac{dF}{dt} - \frac{1}{2}G_1$$

$$= \left(\frac{\partial F}{\partial u}\frac{du}{dt} + \frac{\partial F}{\partial v}\frac{dv}{dt}\right) - \frac{1}{2}G_1$$

$$= (F_1\dot{u} + F_2\dot{v}) - \frac{1}{2}G_1$$

$$= (F_1.0 + F_2.1) - \frac{1}{2}G_1$$

$$= F_2 - \frac{1}{2}G_1$$

and

$$V = \frac{d}{dt}\left(\frac{\partial T}{\partial \dot{v}}\right) - \frac{\partial T}{\partial v} = \frac{dG}{dt} - \frac{1}{2}G_2$$

$$= \left(\frac{\partial G}{\partial u}\frac{du}{dt} + \frac{\partial G}{\partial v}\frac{dv}{dt}\right) - \frac{1}{2}G_2$$

$$= (G_1\dot{u} + G_2\dot{v}) - \frac{1}{2}G_2$$

$$= (G_1.0) + G_2.1) - \frac{1}{2}G_2$$

$$= \frac{1}{2}G_2.$$

Now, by (14), the necessary and sufficient condition for the curve $u = u(t), v = v(t)$ to be a geodesic is

$$V\frac{\partial T}{\partial \dot{u}} - U\frac{\partial T}{\partial \dot{v}} = 0.$$

Hence a necessary and sufficient condition for the curve $u = c$ to be a geodesic is

$$\frac{1}{2}G_2F - \left(F_2 - \frac{1}{2}G_1\right)G = 0$$

or $$FG_2 - 2GF_2 + GG_1 = 0$$

or $$GG_1 + FG_2 - 2GF_2 = 0.$$

Corollary 2. If the parametric curves are orthogonal, then the curves $v =$ const. will be geodesics provided E is a function of u only, and the curves $u =$ const. will be geodesics provided G is a function of v only.

Proof. If the parametric curves are orthogonal, then $F = 0$. Hence the result follows from Corollary 1.

5.3 CANONICAL GEODESIC EQUATIONS

Since the parameter t in equations (12) and (13) of Section 5.2 is arbitrary, we can take the arc length s of the curve measured from some fixed point on it as the parameter. Then, we get

$$\frac{d}{ds}\left(\frac{\partial T}{\partial u'}\right) - \frac{\partial T}{\partial u} = \frac{1}{2T}\frac{dT}{ds}\frac{\partial T}{\partial u'}, \tag{1}$$

$$\frac{d}{ds}\left(\frac{\partial T}{\partial v'}\right) - \frac{\partial T}{\partial v} = \frac{1}{2T}\frac{dT}{ds}\frac{\partial T}{\partial v'}, \tag{2}$$

where

$$T = \frac{1}{2}(Eu'^2 + 2Fu'v' + Gv'^2)$$

$$= \frac{1}{2}\left[E\left(\frac{du}{ds}\right)^2 + 2F\frac{du}{ds}\frac{dv}{ds} + G\left(\frac{dv}{ds}\right)^2\right]$$

$$= \frac{1}{2}\left[\frac{E\,du^2 + 2F\,du\,dv + G\,dv^2}{ds^2}\right]$$

$$= \frac{1}{2}. \qquad [\because E\,du^2 + 2F\,du\,dv + G\,dv^2 = ds^2]$$

Therefore $\frac{dT}{ds} = 0$ and equations (1) and (2) reduce to

$$\frac{d}{ds}\left(\frac{\partial T}{\partial u'}\right) - \frac{\partial T}{\partial u} = 0, \tag{3}$$

$$\frac{d}{ds}\left(\frac{\partial T}{\partial v'}\right) - \frac{\partial T}{\partial v} = 0. \tag{4}$$

Equations (3) and (4) are the canonical equations for geodesics.

5.4 NORMAL PROPERTY OF GEODESICS

Theorem. A curve on a surface is a geodesic if the principal normal at every point on the curve is normal to the surface.

Proof. Let the parametric equation of a surface in vector form be given by

$$\mathbf{r} = \mathbf{r}\,(u, v). \tag{1}$$

Let the equation of a geodesic on the surface (1) be given by

$$\mathbf{r} = \mathbf{r}\,(s). \tag{2}$$

Let $\mathbf{N}$ denote the unit normal vector to the surface (1) at P and let $\mathbf{n}$ denote the unit principal normal to geodesic (2) at P.

We have to prove that $\mathbf{n} = \mathbf{N}$.

Since (2) is a geodesic, its equations are given by

$$\frac{d}{ds}\left(\frac{\partial T}{\partial u'}\right) - \frac{\partial T}{\partial u} = 0, \tag{3}$$

$$\frac{d}{ds}\left(\frac{\partial T}{\partial v'}\right) - \frac{\partial T}{\partial v} = 0, \tag{4}$$

where

$$\begin{aligned} T &= \frac{1}{2}(Eu'^2 + 2Fu'v' + Gv'^2) \\ &= \frac{1}{2}(\mathbf{r}_1{}^2 u'^2 + 2\mathbf{r}_1 \cdot \mathbf{r}_2 u'v' + \mathbf{r}_2{}^2 v'^2) \\ &= \frac{1}{2}\left(\frac{\mathbf{r}_1{}^2 du^2 + 2\mathbf{r}_1 \cdot \mathbf{r}_2 du\, dv + \mathbf{r}_2{}^2 dv^2}{ds^2}\right) \\ &= \frac{1}{2}\frac{(\mathbf{r}_1 du + \mathbf{r}_2 dv)^2}{ds^2} \\ &= \frac{1}{2}\left(\frac{d\mathbf{r}}{ds}\right)^2 \end{aligned}$$

i.e.,

$$T = \frac{1}{2}\mathbf{r}'^2. \tag{5}$$

Also,

$$\mathbf{r}' = \frac{d\mathbf{r}}{ds} = \frac{\partial \mathbf{r}}{\partial u}\frac{du}{ds} + \frac{\partial \mathbf{r}}{\partial v}\frac{dv}{ds} = \mathbf{r}_1 u' + \mathbf{r}_2 v'. \tag{6}$$

From (6), we have

$$\left.\begin{aligned}&\frac{\partial \mathbf{r}'}{\partial u'}=\mathbf{r}_1,\ \frac{\partial \mathbf{r}'}{\partial v'}=\mathbf{r}_2,\\&\frac{\partial \mathbf{r}'}{\partial u}=\mathbf{r}_{11}u'+\mathbf{r}_{12}v',\\&\frac{\partial \mathbf{r}'}{\partial v}=\mathbf{r}_{12}u'+\mathbf{r}_{22}v'.\end{aligned}\right\}\tag{7}$$

Differentiating (5) partially with respect to u', v', u and v, and using (7), we get

$$\left.\begin{aligned}&\frac{\partial T}{\partial u'}=\mathbf{r}'.\frac{\partial \mathbf{r}'}{\partial u'}=\mathbf{r}'.\mathbf{r}_1,\\&\frac{\partial T}{\partial v'}=\mathbf{r}'.\frac{\partial \mathbf{r}'}{\partial v'}=\mathbf{r}'.\mathbf{r}_2,\\&\frac{\partial T}{\partial u}=\mathbf{r}'.\frac{\partial \mathbf{r}'}{\partial u}=\mathbf{r}'.(\mathbf{r}_{11}u'+\mathbf{r}_{12}v'),\\&\frac{\partial T}{\partial v}=\mathbf{r}'.\frac{\partial \mathbf{r}'}{\partial v}=\mathbf{r}'.(\mathbf{r}_{12}u'+\mathbf{r}_{22}v').\end{aligned}\right\}\tag{8}$$

Substituting the above values in (3), we get

$$\frac{d}{ds}(\mathbf{r}'.\mathbf{r}_1)-\mathbf{r}'.(\mathbf{r}_{11}u'+\mathbf{r}_{12}v')=0$$

or $$\mathbf{r}''.\mathbf{r}_1+\mathbf{r}'.\left(\frac{\partial \mathbf{r}_1}{\partial u}\frac{du}{ds}+\frac{\partial \mathbf{r}_1}{\partial v}\frac{dv}{ds}\right)-\mathbf{r}'.(\mathbf{r}_{11}u'+\mathbf{r}_{12}v')=0$$

or $$\mathbf{r}''.\mathbf{r}_1+\mathbf{r}'.(\mathbf{r}_{11}u'+\mathbf{r}_{12}v')-\mathbf{r}'.(\mathbf{r}_{11}u'+\mathbf{r}_{12}v')=0$$

or $$\mathbf{r}''.\mathbf{r}_1=0.\tag{9}$$

Again, substituting the above values in (4), we get

$$\frac{d}{ds}(\mathbf{r}'.\mathbf{r}_2)-\mathbf{r}'.(\mathbf{r}_{12}u'+\mathbf{r}_{22}v')=0$$

or $$\mathbf{r}''.\mathbf{r}_2+\mathbf{r}'.\left(\frac{\partial \mathbf{r}_2}{\partial u}\frac{du}{ds}+\frac{\partial \mathbf{r}_2}{\partial v}\frac{dv}{ds}\right)-\mathbf{r}'.(\mathbf{r}_{12}u'+\mathbf{r}_{22}v')=0$$

or $$\mathbf{r}''.\mathbf{r}_2+\mathbf{r}'.(\mathbf{r}_{12}u'+\mathbf{r}_{22}v')-\mathbf{r}'.(\mathbf{r}_{12}u'+\mathbf{r}_{22}v')=0$$

or $$\mathbf{r}''.\mathbf{r}_2=0.\tag{10}$$

From (9) and (10), we have

$$\mathbf{r}''.\mathbf{r}_1=0,\ \mathbf{r}''.\mathbf{r}_2=0.$$

$\Rightarrow$ The vector $\mathbf{r}''$ is perpendicular to each of $\mathbf{r}_1$ and $\mathbf{r}_2$.

$\Rightarrow$ The vector $\mathbf{r}''$ is parallel to $\mathbf{r}_1 \times \mathbf{r}_2$.

$\Rightarrow$ The vector $\mathbf{r}''=\kappa\mathbf{n}$ is parallel to $H\mathbf{N}$. $[\because \mathbf{r}_1 \times \mathbf{r}_2 = H\mathbf{N}]$

$\Rightarrow$ We can take $\mathbf{n}=\mathbf{N}$.

Hence the principal normal at every point on a geodesic is normal to the surface.

5.5 DIFFERENTIAL EQUATIONS OF GEODESICS USING NORMAL PROPERTY

Let the parametric equation of a surface in vector form be given by

$$\mathbf{r} = \mathbf{r}(u, v). \tag{1}$$

Let a geodesic on the surface (1) be given by

$$\mathbf{r} = \mathbf{r}(s). \tag{2}$$

Let $\mathbf{t}$, $\mathbf{n}$, $\mathbf{b}$ denote unit tangent, unit principal normal and unit binormal respectively to C at P and let $\mathbf{N}$ be the unit normal to the surface S at P. Then, we have

$$\mathbf{N} = \mathbf{n}. \tag{3}$$

We know that

$$\mathbf{r}'' = \kappa\mathbf{n}, \text{ where } \kappa \text{ is the curvature of } C \text{ at } P.$$

or $$\frac{d}{ds}(\mathbf{r}') = \kappa\mathbf{N} \qquad \text{[using (3)]}$$

or $$\frac{d}{ds}\left(\frac{d\mathbf{r}}{ds}\right) = \kappa\mathbf{N}$$

or $$\frac{d}{ds}\left(\frac{\partial\mathbf{r}}{\partial u}\frac{du}{ds} + \frac{\partial\mathbf{r}}{\partial v}\frac{dv}{ds}\right) = \kappa\mathbf{N}$$

or $$\frac{d}{ds}(\mathbf{r}_1 u' + \mathbf{r}_2 v') = \kappa\mathbf{N}$$

or $$\frac{d}{ds}(\mathbf{r}_1 u') + \frac{d}{ds}(\mathbf{r}_2 v') = \kappa\mathbf{N}$$

or $$\mathbf{r}_1 u'' + \frac{d\mathbf{r}_1}{ds}u' + \mathbf{r}_2 v'' + \frac{d\mathbf{r}_2}{ds}v' = \kappa\mathbf{N}$$

or $$\mathbf{r}_1 u'' + \left(\frac{\partial\mathbf{r}_1}{\partial u}\frac{du}{ds} + \frac{\partial\mathbf{r}_1}{\partial v}\frac{dv}{ds}\right)u' + \mathbf{r}_2 v'' + \left(\frac{\partial\mathbf{r}_2}{\partial u}\frac{du}{ds} + \frac{\partial\mathbf{r}_2}{\partial v}\frac{dv}{ds}\right)v' = \kappa\mathbf{N}$$

or $$\mathbf{r}_1 u'' + (\mathbf{r}_{11}u'^2 + \mathbf{r}_{12}u'v') + \mathbf{r}_2 v'' + (\mathbf{r}_{21}u'v' + \mathbf{r}_{22}v'^2) = \kappa\mathbf{N}$$

or $$\mathbf{r}_1 u'' + \mathbf{r}_2 v'' + \mathbf{r}_{11}u'^2 + 2\mathbf{r}_{12}u'v' + \mathbf{r}_{22}v'^2 = \kappa\mathbf{N}. \tag{4}$$

Taking the dot product of both sides of (4) with $\mathbf{r}_1$, we get

$$\mathbf{r}_1^2 u'' + \mathbf{r}_1 \cdot \mathbf{r}_2 v'' + \mathbf{r}_1 \cdot \mathbf{r}_{11}u'^2 + 2\mathbf{r}_1 \cdot \mathbf{r}_{12}u'v' + \mathbf{r}_1 \cdot \mathbf{r}_{22}v'^2 = \kappa\mathbf{r}_1 \cdot \mathbf{N}$$

or $$Eu'' + Fv'' + \mathbf{r}_1 \cdot \mathbf{r}_{11}u'^2 + 2\mathbf{r}_1 \cdot \mathbf{r}_{12}u'v' + \mathbf{r}_1 \cdot \mathbf{r}_{22}v'^2 = 0. \tag{5}$$

Now,

$$\mathbf{r}_1^2 = E$$

$$\Rightarrow \quad \frac{\partial}{\partial u}(\mathbf{r}_1^{\,2}) = E_1$$

$$\Rightarrow \quad 2\mathbf{r}_1 \cdot \mathbf{r}_{11} = E_1$$

$$\Rightarrow \quad \mathbf{r}_1 \cdot \mathbf{r}_{11} = \frac{E_1}{2}. \tag{6}$$

Also,

$$\mathbf{r}_1^{\,2} = E$$

$$\Rightarrow \quad \frac{\partial}{\partial v}(\mathbf{r}_1^{\,2}) = E_2$$

$$\Rightarrow \quad 2\mathbf{r}_1 \cdot \mathbf{r}_{12} = E_2$$

$$\Rightarrow \quad \mathbf{r}_1 \cdot \mathbf{r}_{12} = \frac{E_2}{2}. \tag{7}$$

Again,

$$\mathbf{r}_1 \cdot \mathbf{r}_2 = F$$

$$\Rightarrow \quad \frac{\partial}{\partial v}(\mathbf{r}_1 \cdot \mathbf{r}_2) = F_2$$

$$\Rightarrow \quad \mathbf{r}_{12} \cdot \mathbf{r}_2 + \mathbf{r}_1 \cdot \mathbf{r}_{22} = F_2$$

$$\Rightarrow \quad \frac{1}{2}\frac{\partial}{\partial u}(\mathbf{r}_2^{\,2}) + \mathbf{r}_1 \cdot \mathbf{r}_{22} = F_2$$

$$\Rightarrow \quad \frac{1}{2}G_1 + \mathbf{r}_1 \cdot \mathbf{r}_{22} = F_2$$

$$\Rightarrow \quad \mathbf{r}_1 \cdot \mathbf{r}_{22} = F_2 - \frac{1}{2}G_1. \tag{8}$$

Substituting the values of $\mathbf{r}_1 \cdot \mathbf{r}_{11}$, $\mathbf{r}_1 \cdot \mathbf{r}_{12}$ and $\mathbf{r}_1 \cdot \mathbf{r}_{22}$ from (6), (7) and (8) respectively in (5), we get

$$Eu'' + Fv'' + \frac{1}{2}E_1 u'^2 + E_2 u'v' + \left(F_2 - \frac{1}{2}G_1\right)v'^2 = 0. \tag{9}$$

Now, taking the dot product of both sides of (4) with $\mathbf{r}_2$, we get

$$\mathbf{r}_2 \cdot \mathbf{r}_1 u'' + \mathbf{r}_2^{\,2} v'' + \mathbf{r}_2 \cdot \mathbf{r}_{11} u'^2 + 2\mathbf{r}_2 \cdot \mathbf{r}_{12} u'v' + \mathbf{r}_2 \cdot \mathbf{r}_{22} v'^2 = \kappa \mathbf{r}_2 \cdot \mathbf{N}$$

or

$$Fu'' + Gv'' + \mathbf{r}_2 \cdot \mathbf{r}_{11} u'^2 + 2\mathbf{r}_2 \cdot \mathbf{r}_{12} u'v' + \mathbf{r}_2 \cdot \mathbf{r}_{22} v'^2 = 0. \tag{10}$$

Now,

$$\mathbf{r}_2 \cdot \mathbf{r}_1 = F$$

$$\Rightarrow \quad \frac{\partial}{\partial u}(\mathbf{r}_2 \cdot \mathbf{r}_1) = F_1$$

$$\Rightarrow \quad \mathbf{r}_{21} \cdot \mathbf{r}_1 + \mathbf{r}_2 \cdot \mathbf{r}_{11} = F_1$$

$$\Rightarrow \quad \frac{1}{2}\frac{\partial}{\partial v}(\mathbf{r}_1{}^2) + \mathbf{r}_2 \cdot \mathbf{r}_{11} = F_1$$

$$\Rightarrow \quad \frac{1}{2}E_2 + \mathbf{r}_2 \cdot \mathbf{r}_{11} = F_1$$

$$\Rightarrow \quad \mathbf{r}_2 \cdot \mathbf{r}_{11} = F_1 - \frac{1}{2}E_2 . \tag{11}$$

Again,

$$\mathbf{r}_2{}^2 = G$$

$$\Rightarrow \quad \frac{\partial}{\partial u}(\mathbf{r}_2{}^2) = G_1$$

$$\Rightarrow \quad 2\mathbf{r}_2 \cdot \mathbf{r}_{21} = G_1$$

$$\Rightarrow \quad 2\mathbf{r}_2 \cdot \mathbf{r}_{12} = G_1$$

$$\Rightarrow \quad \mathbf{r}_2 \cdot \mathbf{r}_{12} = \frac{G_1}{2}. \tag{12}$$

Also,

$$\mathbf{r}_2{}^2 = G$$

$$\Rightarrow \quad \frac{\partial}{\partial v}(\mathbf{r}_2{}^2) = G_2$$

$$\Rightarrow \quad 2\mathbf{r}_2 \cdot \mathbf{r}_{22} = G_2$$

$$\Rightarrow \quad \mathbf{r}_2 \cdot \mathbf{r}_{22} = \frac{G_2}{2}. \tag{13}$$

Substituting the values of $\mathbf{r}_2 \cdot \mathbf{r}_{11}$, $\mathbf{r}_2 \cdot \mathbf{r}_{12}$ and $\mathbf{r}_2 \cdot \mathbf{r}_{22}$ from (11), (12) and (13) respectively in (10), we get

$$Fu'' + Gv'' + \left(F_1 - \frac{1}{2}E_2\right)u'^2 + G_1u'v' + \frac{1}{2}G_2v'^2 = 0. \tag{14}$$

Equations (9) and (14) are the general differential equations of geodesics on a surface.

These equations can be put in an alternate form as follows.

We have

$$\frac{d}{ds}(Eu' + Fv')$$

$$= \frac{d}{ds}(Eu') + \frac{d}{ds}(Fv')$$

$$= Eu'' + \frac{dE}{ds}u' + Fv'' + \frac{dF}{ds}v'$$

$$= Eu'' + \left(\frac{\partial E}{\partial u}u' + \frac{\partial E}{\partial v}v'\right)u' + Fv'' + \left(\frac{\partial F}{\partial u}u' + \frac{\partial F}{\partial v}v'\right)v'$$

$$= Eu'' + (E_1u'^2 + E_2u'v') + Fv'' + (F_1u'v' + F_2v'^2)$$

$$= Eu'' + Fv'' + E_1u'^2 + (E_2 + F_1)u'v' + F_2v'^2$$

$$= \left[Eu'' + Fv'' + \frac{1}{2}E_1u'^2 + E_2u'v' + \left(F_2 - \frac{1}{2}G_1\right)v'^2\right] + \frac{1}{2}E_1u'^2 + F_1u'v' + \frac{1}{2}G_1v'^2$$

$$= \frac{1}{2}E_1u'^2 + F_1u'v' + \frac{1}{2}G_1v'^2 \qquad \text{[using (9)]}$$

i.e., $$\frac{d}{ds}(Eu' + Fv') = \frac{1}{2}(E_1u'^2 + 2F_1u'v' + G_1v'^2). \tag{15}$$

Similarly,

$$\frac{d}{ds}(Fu' + Gv')$$

$$= \frac{d}{ds}(Fu') + \frac{d}{ds}(Gv')$$

$$= Fu'' + \frac{dF}{ds}u' + Gv'' + \frac{dG}{ds}v'$$

$$= Fu'' + \left(\frac{\partial F}{\partial u}u' + \frac{\partial F}{\partial v}v'\right)u' + Gv'' + \left(\frac{\partial G}{\partial u}u' + \frac{\partial G}{\partial v}v'\right)v'$$

$$= Fu'' + (F_1u'^2 + F_2u'v') + Gv'' + (G_1u'v' + G_2v'^2)$$

$$= Fu'' + Gv'' + F_1u'^2 + (F_2 + G_1)u'v' + G_2v'^2$$

$$= \left[Fu'' + Gv'' + \left(F_1 - \frac{1}{2}E_2\right)u'^2 + G_1u'v' + \frac{1}{2}G_2v'^2\right] + \frac{1}{2}E_2u'^2 + F_2u'v' + \frac{1}{2}G_2v'^2$$

$$= \frac{1}{2}E_2u'^2 + F_2u'v' + \frac{1}{2}G_2v'^2 \qquad \text{[using (14)]}$$

i.e., $$\frac{d}{ds}(Fu' + Gv') = \frac{1}{2}(E_2u'^2 + 2F_2u'v' + G_2v'^2). \tag{16}$$

Equations (15) and (16) give another form of differential equations of geodesics on a surface.

A third form may be found as follows by solving (9) and (14) for u'' and v''.

From (14), we have

$$v'' = -\frac{1}{G}\left[Fu'' + \left(F_1 - \frac{1}{2}E_2 \right)u'^2 + G_1u'v' + \frac{1}{2}G_2v'^2 \right].$$

Putting this value of v'' in (9), we get

$$Eu'' - \frac{F}{G}\left[Fu'' + \left(F_1 - \frac{1}{2}E_2 \right)u'^2 + G_1u'v' + \frac{1}{2}G_2v'^2 \right]$$
$$+ \frac{1}{2}E_1u'^2 + E_2u'v' + \left(F_2 - \frac{1}{2}G_1 \right)v'^2 = 0$$

or
$$\frac{EG - F^2}{G}u'' + \left[\frac{1}{2}E_1 - \frac{F}{G}\left(F_1 - \frac{1}{2}E_2 \right) \right]u'^2$$
$$+ \left(E_2 - \frac{FG_1}{G} \right)u'v' + \left[\left(F_2 - \frac{1}{2}G_1 \right) - \frac{FG_2}{2G} \right]v'^2 = 0.$$

Multiplying both sides by $G/(EG - F^2)$ and using $H^2 = EG - F^2$, we have

$$u'' + \frac{1}{2H^2}(GE_1 - 2FF_1 + FE_2)u'^2$$
$$+ \frac{1}{H^2}(GE_2 - FG_1)u'v' + \frac{1}{2H^2}(2GF_2 - GG_1 - FG_2)v'^2 = 0$$

or
$$u'' + lu'^2 + 2mu'v' + nv'^2 = 0, \tag{17}$$

where

$$l = \frac{1}{2H^2}(GE_1 - 2FF_1 + FE_2),$$
$$m = \frac{1}{2H^2}(GE_2 - FG_1),$$
$$n = \frac{1}{2H^2}(2GF_2 - GG_1 - FG_2).$$

Again, from (14), we have

$$u'' = -\frac{1}{F}\left[Gv'' + \left(F_1 - \frac{1}{2}E_2 \right)u'^2 + G_1u'v' + \frac{1}{2}G_2v'^2 \right]$$

Putting this value of u'' in (9), we get

$$-\frac{E}{F}\left[Gv'' + \left(F_1 - \frac{1}{2}E_2 \right)u'^2 + G_1u'v' + \frac{1}{2}G_2v'^2 \right]$$
$$+ Fv'' + \frac{1}{2}E_1u'^2 + E_2u'v' + \left(F_2 - \frac{1}{2}G_1 \right)v'^2 = 0$$

or $$-\left(\frac{EG-F^2}{F}\right)v''-\left[\frac{E}{F}\left(F_1-\frac{1}{2}E_2\right)-\frac{1}{2}E_1\right]u'^2$$

$$-\left(\frac{EG_1}{F}-E_2\right)u'v'-\left[\frac{EG_2}{2F}-\left(F_2-\frac{1}{2}G_1\right)\right]v'^2=0.$$

Multiplying both sides by $\left[-\left(\frac{F}{EG-F^2}\right)\right]$ and using $H^2=EG-F^2$, we have

$$v''+\frac{1}{2H^2}(2EF_1-EE_2-FE_1)\,u'^2$$

$$+\frac{1}{H^2}(EG_1-FE_2)\,u'v'+\frac{1}{2H^2}(EG_2-2FF_2+FG_1)\,v'^2=0$$

or $$v''+\lambda u'^2+2\mu u'v'+\nu v'^2=0, \tag{18}$$

where

$$\lambda=\frac{1}{2H^2}(2EF_1-EE_2-FE_1),$$

$$\mu=\frac{1}{2H^2}(EG_1-FE_2),$$

$$\nu=\frac{1}{2H^2}(EG_2-2FF_2+FG_1).$$

Equations (17) and (18) give the third form of the differential equations of the geodesics.

The above pair of differential equations may be replaced by a single relation between the parameters u, v as follows.

Multiplying (18) by $1/u'^2$ and subtracting it from the product of (17) with v'/u'^3, we get

$$\left(\frac{u''v'}{u'^3}-\frac{v''}{u'^2}\right)+\left(\frac{lv'}{u'}-\lambda\right)+2\left(\frac{mv'^2}{u'^2}-\frac{\mu v'}{u'}\right)+\left(\frac{nv'^3}{u'^3}-\frac{\nu v'^2}{u'^2}\right)=0$$

or $$\frac{\dfrac{du'}{ds}\dfrac{dv}{du}-\dfrac{dv'}{ds}}{u'^2}+n\left(\frac{dv}{du}\right)^3+(2m-\nu)\left(\frac{dv}{du}\right)^2+(l-2\mu)\frac{dv}{du}-\lambda=0$$

$$\left[\because \frac{v'}{u'}=\frac{dv}{du}\right]$$

or $$\frac{\frac{dv'}{du}\frac{du}{ds}-\frac{du'}{du}\frac{du}{ds}\frac{dv}{du}}{u'^2}=n\left(\frac{dv}{du}\right)^3+(2m-\nu)\left(\frac{dv}{du}\right)^2+(l-2\mu)\frac{dv}{du}-\lambda$$

or $$\frac{\frac{dv'}{du}u'-\frac{du'}{du}v'}{u'^2}=n\left(\frac{dv}{du}\right)^3+(2m-\nu)\left(\frac{dv}{du}\right)^2+(l-2\mu)\frac{dv}{du}-\lambda$$

or $$\frac{d}{du}\left(\frac{v'}{u'}\right)=n\left(\frac{dv}{du}\right)^3+(2m-\nu)\left(\frac{dv}{du}\right)^2+(l-2\mu)\frac{dv}{du}-\lambda$$

or $$\frac{d}{du}\left(\frac{dv}{du}\right)=n\left(\frac{dv}{du}\right)^3+(2m-\nu)\left(\frac{dv}{du}\right)^2+(l-2\mu)\frac{dv}{du}-\lambda$$

or $$\frac{d^2v}{du^2}=n\left(\frac{dv}{du}\right)^3+(2m-\nu)\left(\frac{dv}{du}\right)^2+(l-2\mu)\frac{dv}{du}-\lambda, \tag{19}$$

which is the single differential equation of geodesics.

Remark. A single geodesic passes through each point of the surface in each direction.

Proof. It follows from the theory of differential equations that there exists a unique integral v of equation (19) which takes a given value v_0 when $u=u_0$, and whose derivative $\frac{dv}{du}$ also takes a given value when $u=u_0$. Hence a single geodesic passes through each point of the surface in each direction.

Corollary 1. A necessary and sufficient condition that the curves $v=\text{const.}$ will be geodesics is $EE_2+FE_1-2EF_1=0$, and the curves $u=\text{const.}$ will be geodesics is $GG_1+FG_2-2GF_2=0$.

Proof. It follows from (19) that the parametric curves $v=$ const. will be geodesics if and only if

$$\lambda=0$$

or $$\frac{1}{2H^2}(2EF_1-EE_2-FE_1)=0$$

or $$2EF_1-EE_2-FE_1=0$$

or $$EE_2+FE_1-2EF_1=0.$$

Again, (19) may be written as

$$\frac{d^2u}{dv^2} - \lambda\left(\frac{du}{dv}\right)^3 + (l - 2\mu)\left(\frac{du}{dv}\right)^2 + (2m - \nu)\frac{du}{dv} + n = 0. \tag{20}$$

It follows from (20) that the parametric curves $u =$ const. will be geodesics if and only if

$$n = 0$$

or $$\frac{1}{2H^2}(2GF_2 - GG_1 - FG_2) = 0$$

or $$2GF_2 - GG_1 - FG_2 = 0$$

or $$GG_1 + FG_2 - 2GF_2 = 0.$$

Corollary 2. If the parametric curves are orthogonal, then the curves $v =$ const. will be geodesics provided E is a function of u only, and the curves $u =$ const. will be geodesics provided G is a function of v only.

Proof. If the parametric curves are orthogonal, then $F = 0$. Hence the result follows from Corollary 1.

5.6 GEODESICS ON A SURFACE OF REVOLUTION

We know that the equations of the geodesics are given by

$$u'' + lu'^2 + 2mu'v' + nv'^2 = 0, \tag{1}$$

$$v'' + \lambda u'^2 + 2\mu u'v' + \nu v'^2 = 0. \tag{2}$$

Now, we shall find the values of λ, μ, ν for the given surface.
The position vector $\mathbf{r}$ of any current point on the surface of revolution is

$$\mathbf{r} = (u\cos v, u\sin v, f(u)).$$

Now, proceeding as in Section 3.4, Chapter 3, we have

$$E = 1 + f_1^2,\ F = 0,\ G = u^2, H^2 = u^2(1 + f_1^2).$$

$\therefore$ $$\lambda = \frac{1}{2H^2}(2EF_1 - EE_2 - FE_1) = 0,$$

$$\mu = \frac{1}{2H^2}(EG_1 - FE_2) = \frac{1}{u},$$

$$\nu = \frac{1}{2H^2}(EG_2 - 2FF_2 + FG_1) = 0.$$

Substituting the values of λ, μ, ν in (2), we have

$$v'' + \frac{2u'v'}{u} = 0$$

or $$u^2v'' + 2uu'v' = 0.$$

This is an exact differential equation and its solution is given by

$$u^2 v' = h, \tag{3}$$

where h is an arbitrary constant.

The geodesics on the surface of revolution are given by (3).

If $h = 0$, then (3) reduces to

$$u^2 v' = 0$$

$\Rightarrow \quad v' = 0$

$\Rightarrow \quad v = c$ (constant).

Hence every meridian is a geodesic on a surface of revolution.

To find the complete integral of (3), we proceed as follows.

We have

$$ds^2 = E\,du^2 + 2F\,du\,dv + G\,dv^2$$

or $\quad ds^2 = (1 + f_1^2)\,du^2 + u^2\,dv^2. \tag{4}$

Now, (3) may be written as

$$u^2 \frac{dv}{ds} = h$$

or $\quad u^2 dv = h\,ds.$

Squaring, we get

$$u^4 dv^2 = h^2\,ds^2$$

or $\quad u^4 dv^2 = h^2\,\{(1 + f_1^2)\,du^2 + u^2\,dv^2\}$ [using (4)]

or $\quad u^2(u^2 - h^2)\,dv^2 = h^2\,(1 + f_1^2)\,du^2 \tag{5}$

or $\quad dv = \pm \frac{h}{u}\sqrt{\frac{1 + f_1^2}{u^2 - h^2}}du$ if $u^2 \neq h^2$.

Integrating, we get

$$v = C \pm h \int \frac{1}{u}\sqrt{\frac{1 + f_1^2}{u^2 - h^2}}\,du, \tag{6}$$

where h and C are arbitrary constants.

Equation (6) gives the complete integral of the equation of geodesics on a surface of revolution.

If $u^2 = h^2$, i.e., $u = \pm h \neq 0$, then (5) gives

$$du = 0$$

$\Rightarrow \quad u = c$ (constant)

$\Rightarrow \quad G = u^2 = c^2$

$\Rightarrow \quad G_1 = 0$

$\Rightarrow \quad G$ is a function of v only. (7)

Also, we have

$$F = 0$$

$\therefore$ The parametric curves are orthogonal. (8)

We know that if the parametric curves are orthogonal, then the curves $u =$const. will be geodesics provided G is a function of v only. Hence, it follows from (7) and (8) that the curve $u = c$ (parallel) is a geodesic if the constant value of u is not zero. This constant value of u is the radius of the parallel.

Hence a parallel is a geodesic on a surface of revolution if and only if the radius of the parallel is constant.

5.7 CLAIRAUT'S THEOREM

If a geodesic on a surface of revolution cuts the meridian through any point P on it at an angle ψ, then $u \sin\psi$ is constant, where u is the distance of the point P from the axis.

Proof. Let the geodesic be a parallel. We know that on a surface of revolution, a parallel cuts each meridian at an angle $\pi/2$ and the distance of each point of a parallel from the axis is constant. Thus each of u and ψ is a constant and therefore $u \sin\psi$ is constant.

Now, let us prove the result for geodesics other than the parallels.

We know that the equations of the geodesics are given by

$$u'' + lu'^2 + 2mu'v' + nv'^2 = 0, \tag{1}$$

$$v'' + \lambda u'^2 + 2\mu u'v' + \nu v'^2 = 0. \tag{2}$$

Now, we shall find the values of λ, μ, ν for the given surface of revolution.

Let the equation of the surface of revolution be

$$\mathbf{r} = (u \cos v, u \sin v, f(u)).$$

Then, proceeding as in Section 5.6, we have

$$E = 1 + f_1^2,\ F = 0,\ G = u^2, H^2 = u^2(1 + f_1^2),$$

$$\lambda = 0,\ \mu == \frac{1}{u},\ \nu = 0.$$

Substituting the values of λ, μ, ν in (2), we have

$$v'' + \frac{2u'v'}{u} = 0$$

or $$u^2 v'' + 2uu'v' = 0.$$

This is an exact differential equation and its solution is given by

$$u^2 v' = h, \tag{3}$$

where h is an arbitrary constant.

The geodesics on the surface of revolution are given by (3).

Now, let P be any point on a geodesic and let this geodesic cut the meridian $v = \text{const.}$ through P at an angle ψ. The direction coefficients of the geodesic are $\left(\frac{du}{ds}, \frac{dv}{ds}\right)$ and the direction coefficients of the meridian are $(1/\sqrt{E}, 0)$. Therefore

$$\sin\psi = H\left|\frac{du}{ds}.0 - \frac{1}{\sqrt{E}}\frac{dv}{ds}\right|$$

[using the formula $\sin\theta = H\,|lm' - l'm|$]

$$= \sqrt{EG}\frac{1}{\sqrt{E}}\frac{dv}{ds} \qquad [\because F = 0 \Rightarrow H^2 = EG]$$

$$= \sqrt{G}\frac{dv}{ds}$$

$$= u\frac{dv}{ds}. \qquad [\because G = u^2]$$

$$\therefore \quad u\sin\psi = u^2\frac{dv}{ds} = u^2 v'$$

or $\quad u\sin\psi = h.$ [using (3)]

Deductions.

(i) The geodesics on a right circular cylinder are helices.

Proof. From Clairaut's theorem, we know that if a geodesic on a surface of revolution cuts the meridian through any point P on it at an angle ψ, then

$$u\sin\psi = h\,(\text{const.}), \qquad (1)$$

where u is the distance of the point P from the axis.

If the surface of revolution is a right circular cylinder, then the meridians are the generators of the cylinder and the distance of each point from the axis is constant, say, equal to a. Hence substituting $u = a$ in (1), we get

$$a\sin\psi = h$$

or $\quad \sin\psi = \frac{h}{a}$

or $\quad \psi = \sin^{-1}\left(\frac{h}{a}\right) = \text{const.}$

Thus a geodesic on a right circular cylinder cuts each generator at a constant angle and so it is a helix.

(ii) If a geodesic on a surface of revolution cuts the meridians at a constant angle, then the surface is a right cylinder.

Proof. Let the equation of the surface of revolution be

$$x = u\cos v,\ y = u\sin v,\ z = f(u).$$

From Clairaut's theorem, we know that if a geodesic on a surface of revolution cuts the meridian through any point P on it at an angle ψ, then

$$u\sin\psi = h\,(\text{const.}), \tag{1}$$

where $u = \sqrt{x^2 + y^2}$ is the distance of the point P from the axis.

Now, it is given that a geodesic cuts the meridians at a constant angle. Therefore ψ is constant and hence, from (1), we have

$$u = \frac{h}{\sin\psi} = \text{const.} = a, \text{ say.}$$

$\therefore \quad u^2 = a^2$

i.e., $\quad x^2 + y^2 = a^2,$

which is the equation of a right cylinder.

Solved Examples

Example 1. Prove that the curves of the family $\dfrac{v^3}{u^2} = \text{constant}$ are geodesics on a surface with metric

$$v^2 du^2 - 2uv\,du\,dv + 2u^2\,dv^2 \quad (u > 0, v > 0).$$

Solution. We know that a necessary and sufficient condition for the curve $u = u(t), v = v(t)$ to be a geodesic on the surface $\mathbf{r} = \mathbf{r}(u, v)$ is

$$V\frac{\partial T}{\partial \dot{u}} - U\frac{\partial T}{\partial \dot{v}} = 0. \tag{1}$$

The equation of the given family of curves is

$$\frac{v^3}{u^2} = \text{constant}.$$

This equation can be written in parametric form as

$$u = ct^3,\ v = ct^2.$$

$\therefore \quad \dot{u} = 3ct^2,\ \dot{v} = 2ct.$

The metric of the surface is given by

$$ds^2 = v^2 du^2 - 2uv\,du\,dv + 2u^2 dv^2.$$

Comparing this with

$$ds^2 = E\,du^2 + 2F\,du\,dv + G\,dv^2,$$

we get

$$E = v^2,\ F = -uv,\ G = 2u^2.$$

Now, we have

$$T = \frac{1}{2}(E\dot{u}^2 + 2F\dot{u}\dot{v} + G\dot{v}^2) = \frac{1}{2}(v^2\dot{u}^2 - 2uv\dot{u}\dot{v} + 2u^2\dot{v}^2).$$

$$\therefore \qquad \frac{\partial T}{\partial u} = -v\dot{u}\dot{v} + 2u\dot{v}^2 = 2c^3t^5,$$

$$\frac{\partial T}{\partial v} = v\dot{u}^2 - u\dot{u}\dot{v} = 3c^3t^6,$$

$$\frac{\partial T}{\partial \dot{u}} = v^2\dot{u} - uv\dot{v} = c^3t^6,$$

$$\frac{\partial T}{\partial \dot{v}} = -uv\dot{u} + 2u^2\dot{v} = c^3t^7.$$

Now,

$$U = \frac{d}{dt}\left(\frac{\partial T}{\partial \dot{u}}\right) - \frac{\partial T}{\partial u} = \frac{d}{dt}(c^3t^6) - 2c^3t^5 = 4c^3t^5,$$

$$V = \frac{d}{dt}\left(\frac{\partial T}{\partial \dot{v}}\right) - \frac{\partial T}{\partial v} = \frac{d}{dt}(c^3t^7) - 3c^3t^6 = 4c^3t^6.$$

Therefore

$$V\frac{\partial T}{\partial \dot{u}} - U\frac{\partial T}{\partial \dot{v}} = (4c^3t^6)(c^3t^6) - (4c^3t^5)(c^3t^7) = 0$$

i.e., the curves of the given family satisfy condition (1). Hence they are geodesics on the given surface.

Example 2. Prove that the curves of the family $u + v = \text{constant}$ are geodesics on a surface with metric

$$(1+u^2)\,du^2 - 2uv\,du\,dv + (1+v^2)\,dv^2.$$

Solution. We know that a necessary and sufficient condition for the curve $u = u(t), v = v(t)$ to be a geodesic on the surface $\mathbf{r} = \mathbf{r}(u, v)$ is

$$V\frac{\partial T}{\partial \dot{u}} - U\frac{\partial T}{\partial \dot{v}} = 0. \qquad (1)$$

The equation of the given family of curves is

$$u + v = c \text{ (constant)}.$$

This equation can be written in parametric form as

$$u = t, \ v = c - t.$$

$$\therefore \quad \dot{u} = 1, \ \dot{v} = -1.$$

The metric of the surface is given by

$$ds^2 = (1+u^2)\,du^2 - 2uv\,du\,dv + (1+v^2)\,dv^2.$$

Comparing this with

$$ds^2 = E\,du^2 + 2F\,du\,dv + G\,dv^2,$$

we get

$$E = 1+u^2, \ F = -uv, \ G = 1+v^2.$$

Now, we have

$$T = \frac{1}{2}(E\dot{u}^2 + 2F\dot{u}\dot{v} + G\dot{v}^2)$$

$$= \frac{1}{2}[(1+u^2)\,\dot{u}^2 - 2uv\dot{u}\dot{v} + (1+v^2)\,\dot{v}^2].$$

$$\therefore \quad \frac{\partial T}{\partial u} = u\dot{u}^2 - v\dot{u}\dot{v} = t - (c-t)(-1) = c,$$

$$\frac{\partial T}{\partial v} = -u\dot{u}\dot{v} + v\dot{v}^2 = -t\ (-1) + (c-t) = c,$$

$$\frac{\partial T}{\partial \dot{u}} = (1+u^2)\,\dot{u} - uv\dot{v} = (1+t^2) - t\,(c-t)(-1) = 1 + ct,$$

$$\frac{\partial T}{\partial \dot{v}} = -uv\dot{u} + (1+v^2)\,\dot{v} = -t\,(c-t) + [1 + (c-t)^2](-1)$$

$$= ct - 1 - c^2.$$

Now,

$$U = \frac{d}{dt}\left(\frac{\partial T}{\partial \dot{u}}\right) - \frac{\partial T}{\partial u} = \frac{d}{dt}(1+ct) - c = 0,$$

$$V = \frac{d}{dt}\left(\frac{\partial T}{\partial \dot{v}}\right) - \frac{\partial T}{\partial v} = \frac{d}{dt}(ct - 1 - c^2) - c = 0.$$

Therefore

$$V\frac{\partial T}{\partial \dot{u}} - U\frac{\partial T}{\partial \dot{v}} = (0)\,(1+ct) - (0)\,(ct - 1 - c^2) = 0$$

i.e., the curves of the given family satisfy condition (1). Hence they are geodesics on the given surface.

Example 3. Find the equations to the geodesics on the right helicoid

$$x = u\cos\phi, \ y = u\sin\phi, \ z = c\phi.$$

Solution. We know that the equations of the geodesics are given by

$$u'' + lu'^2 + 2mu'\phi' + n\phi'^2 = 0, \tag{1}$$

$$\phi'' + \lambda u'^2 + 2\mu u'\phi' + \nu\phi'^2 = 0. \tag{2}$$

Now, we shall find the values of λ, μ, ν for the right helicoid.

The position vector **r** of any current point on the right helicoid is

$$\mathbf{r} = (u\cos\phi, u\sin\phi, c\phi).$$

Now, proceeding as in Section 3.4, Chapter 3, we have

$$E = 1,\ F = 0,\ G = u^2 + c^2,\ H^2 = u^2 + c^2.$$

$$\therefore \quad \lambda = \frac{1}{2H^2}(2EF_1 - EE_2 - FE_1) = 0,$$

$$\mu = \frac{1}{2H^2}(EG_1 - FE_2) = \frac{u}{u^2 + c^2},$$

$$\nu = \frac{1}{2H^2}(EG_2 - 2FF_2 + FG_1) = 0.$$

Substituting the values of λ, μ, ν in (2), we have

$$\phi'' + \frac{2u}{u^2 + c^2}u'\phi' = 0$$

or $$(u^2 + c^2)\,\phi'' + 2uu'\phi' = 0.$$

This is an exact differential equation and its solution is given by

$$(u^2 + c^2)\,\phi' = h, \tag{3}$$

where h is an arbitrary constant.

Again, we have

$$ds^2 = E\,du^2 + 2F\,du\,d\phi + G\,d\phi^2$$

or $$ds^2 = du^2 + (u^2 + c^2)\,d\phi^2. \tag{4}$$

Now, (3) may be written as

$$(u^2 + c^2)\frac{d\phi}{ds} = h$$

or $$(u^2 + c^2)\,d\phi = h\,ds.$$

Squaring, we get

$$(u^2 + c^2)^2\,d\phi^2 = h^2\,ds^2$$

or $$(u^2 + c^2)^2\,d\phi^2 = h^2\,\{du^2 + (u^2 + c^2)\,d\phi^2\} \qquad \text{[using (4)]}$$

or $$(u^2 + c^2)^2\,d\phi^2 - h^2\,(u^2 + c^2)\,d\phi^2 = h^2du^2$$

or $$(u^2 + c^2)\,(u^2 + c^2 - h^2)\,d\phi^2 = h^2du^2.$$

$$\therefore \quad d\phi = \pm\frac{h}{\sqrt{(u^2 + c^2)\,(u^2 + c^2 - h^2)}}du. \tag{5}$$

Equation (5) gives the first integral of the differential equation of geodesics on a right helicoid.

Example 4. Find the equations of the geodesics for the surface in Monge's form $z = f(x, y)$.

Solution. We know that the equation of the geodesics with x, y as parameters is given by

$$\frac{d^2y}{dx^2} = n\left(\frac{dy}{dx}\right)^3 + (2m - \nu)\left(\frac{dy}{dx}\right)^2 + (l - 2\mu)\frac{dy}{dx} - \lambda. \tag{1}$$

Now, we shall find the values of $l, m, n, \lambda, \mu, \nu$ for the given surface.

If $\mathbf{r}$ is the position vector of any current point on the surface in Monge's form $z = f(x, y)$, then

$$\mathbf{r} = (x, y, z),$$

where x and y are taken as the independent variables and z is a function of x and y.

Therefore, we have

$$E = 1 + p^2, \quad F = pq, \; G = 1 + q^2, \; H^2 = 1 + p^2 + q^2$$

and $\quad l = \dfrac{pr}{H^2}, \; m = \dfrac{ps}{H^2}, \; n = \dfrac{pt}{H^2}, \; \lambda = \dfrac{qr}{H^2}, \; \mu = \dfrac{qs}{H^2}, \nu = \dfrac{qt}{H^2}.$

Substituting the values of $l, m, n, \lambda, \mu, \nu$ in (1), we have

$$\frac{d^2y}{dx^2} = \frac{pt}{H^2}\left(\frac{dy}{dx}\right)^3 + \frac{2ps - qt}{H^2}\left(\frac{dy}{dx}\right)^2 + \left(\frac{pr - 2qs}{H^2}\right)\frac{dy}{dx} - \frac{qr}{H^2}$$

or

$$H^2\frac{d^2y}{dx^2} = pt\left(\frac{dy}{dx}\right)^3 + (2ps - qt)\left(\frac{dy}{dx}\right)^2 + (pr - 2qs)\frac{dy}{dx} - qr$$

$$= p\frac{dy}{dx}\left\{t\left(\frac{dy}{dx}\right)^2 + 2s\frac{dy}{dx} + r\right\} - q\left\{t\left(\frac{dy}{dx}\right)^2 + 2s\frac{dy}{dx} + r\right\}$$

$$= \left(p\frac{dy}{dx} - q\right)\left\{t\left(\frac{dy}{dx}\right)^2 + 2s\frac{dy}{dx} + r\right\}$$

or

$$(1 + p^2 + q^2)\frac{d^2y}{dx^2} = \left(p\frac{dy}{dx} - q\right)\left\{t\left(\frac{dy}{dx}\right)^2 + 2s\frac{dy}{dx} + r\right\}. \tag{2}$$

Equation (2) gives the equation of geodesics for the surface in Monge's form $z = f(x, y)$.

Example 5. In the case of a right circular cone of semi-vertical angle α, show that the equation for geodesics is equivalent to

$$u = h \sec(\phi \sin\alpha + \beta),$$

where h and β are constants.

Solution. We know that the equations of the geodesics are given by

$$u'' + lu'^2 + 2mu'\phi' + n\phi'^2 = 0, \tag{1}$$

$$\phi'' + \lambda u'^2 + 2\mu u'\phi' + \nu\phi'^2 = 0. \tag{2}$$

Now, we shall find the values of λ, μ, ν for the right circular cone.

The position vector $\mathbf{r}$ of any current point on the right circular cone of semi-vertical angle α is given by

$$\mathbf{r} = (u\cos\phi, u\sin\phi, u\cot\alpha).$$

Taking u and ϕ as the first and the second parameters, we have

$$\mathbf{r}_1 = (\cos\phi,\ \sin\phi, \cot\alpha),\ \ \mathbf{r}_2 = (-u\sin\phi,\ u\cos\phi, 0).$$

Therefore

$$E = \mathbf{r}_1^{\,2} = \cos^2\phi + \sin^2\phi + \cot^2\alpha = 1 + \cot^2\alpha = \operatorname{cosec}^2\alpha,$$

$$F = \mathbf{r}_1 . \mathbf{r}_2 = -u\cos\phi\sin\phi + u\sin\phi\cos\phi + 0 = 0,$$

$$G = \mathbf{r}_2^{\,2} = u^2\sin^2\phi + u^2\cos^2\phi = u^2,$$

$$H^2 = EG - F^2 = u^2\operatorname{cosec}^2\alpha.$$

$$\therefore \qquad \lambda = \frac{1}{2H^2}(2EF_1 - EE_2 - FE_1) = 0,$$

$$\mu = \frac{1}{2H^2}(EG_1 - FE_2) = \frac{1}{u},$$

$$\nu = \frac{1}{2H^2}(EG_2 - 2FF_2 + FG_1) = 0.$$

Substituting the values of λ, μ, ν in (2), we have

$$\phi'' + \frac{2u'\phi'}{u} = 0$$

or

$$u^2\phi'' + 2uu'\phi' = 0.$$

This is an exact differential equation and its solution is given by

$$u^2\phi' = h, \tag{3}$$

where h is an arbitrary constant.

Again, we have

$$ds^2 = E\,du^2 + 2F\,du\,d\phi + G\,d\phi^2$$

or $\quad ds^2 = \text{cosec}^2\alpha\, du^2 + u^2 d\phi^2.$ (4)

Now, (3) may be written as

$$u^2 \frac{d\phi}{ds} = h$$

or $\quad u^2 d\phi = h\, ds.$

Squaring, we get

$$u^4 d\phi^2 = h^2 ds^2$$

or $\quad u^4 d\phi^2 = h^2(\text{cosec}^2\alpha\, du^2 + u^2 d\phi^2)$ [using (4)]

or $\quad u^2(u^2 - h^2)d\phi^2 = h^2\text{cosec}^2\alpha\, du^2$

or $\quad d\phi = \pm \frac{h}{u}\frac{\text{cosec}\alpha}{\sqrt{u^2 - h^2}} du.$

Integrating, we get

$$\phi = C \pm h\,\text{cosec}\,\alpha \int \frac{1}{u\sqrt{u^2 - h^2}} du$$

or $\quad \phi \sin\alpha = C\sin\alpha \pm h \int \frac{1}{u\sqrt{u^2 - h^2}} du$

or $\quad \phi \sin\alpha - C\sin\alpha = \pm h \int \frac{1}{u\sqrt{u^2 - h^2}} du$

or $\quad \phi \sin\alpha + \beta = \pm h \int \frac{1}{u\sqrt{u^2 - h^2}} du$, where $\beta = -C\sin\alpha$

or $\quad \phi \sin\alpha + \beta = \pm \sec^{-1}\left(\frac{u}{h}\right)$

or $\quad \frac{u}{h} = \sec[\pm(\phi\sin\alpha + \beta)] = \sec(\phi\sin\alpha + \beta)$

or $\quad u = h\sec(\phi\sin\alpha + \beta).$ (5)

Equation (5) gives the equation of geodesics for a right circular cone of semi-vertical angle α.

Example 6. Find the geodesics on the catenoid of revolution

$$u = c\cosh(z/c).$$

Solution. We know that the equations of the geodesics are given by

$$u'' + lu'^2 + 2mu'\phi' + n\phi'^2 = 0, \quad (1)$$

$$\phi'' + \lambda u'^2 + 2\mu u'\phi' + \nu\phi'^2 = 0. \quad (2)$$

Now, we shall find the values of λ, μ, ν for the catenoid of revolution.

The position vector **r** of any current point on the catenoid of revolution is given by

$$\mathbf{r} = (u\cos\phi, u\sin\phi, c\cosh^{-1}(u/c)).$$

Taking u and ϕ as the first and the second parameters, we have

$$\mathbf{r}_1 = \left(\cos\,\phi,\ \sin\phi,\ \frac{c}{\sqrt{u^2-c^2}}\right),$$

$$\mathbf{r}_2 = (-u\sin\phi,\ u\cos\phi, 0).$$

Therefore

$$E = \mathbf{r}_1^{\,2} = \cos^2\phi + \sin^2\phi + \frac{c^2}{u^2-c^2} = \frac{u^2}{u^2-c^2},$$

$$F = \mathbf{r}_1 \,.\, \mathbf{r}_2 = -u\cos\phi\sin\phi + u\sin\phi\cos\phi + 0 = 0,$$

$$G = \mathbf{r}_2^{\,2} = u^2\sin^2\phi + u^2\cos^2\phi = u^2,$$

$$H^2 = EG - F^2 = \frac{u^4}{u^2-c^2}.$$

$$\therefore \qquad \lambda = \frac{1}{2H^2}(2EF_1 - EE_2 - FE_1) = 0,$$

$$\mu = \frac{1}{2H^2}(EG_1 - FE_2) = \frac{1}{u},$$

$$\nu = \frac{1}{2H^2}(EG_2 - 2FF_2 + FG_1) = 0.$$

Substituting the values of λ, μ, ν in (2), we have

$$\phi'' + \frac{2u'\phi'}{u} = 0$$

or $$u^2\phi'' + 2uu'\phi' = 0.$$

This is an exact differential equation and its solution is given by

$$u^2\phi' = h, \tag{3}$$

where h is an arbitrary constant.

Again, we have

$$ds^2 = E\,du^2 + 2F\,du\,d\phi + G\,d\phi^2$$

or $$ds^2 = \frac{u^2}{u^2-c^2}du^2 + u^2 d\phi^2. \tag{4}$$

Now, (3) may be written as

$$u^2\frac{d\phi}{ds} = h$$

or $$u^2 d\phi = h\,ds.$$

Squaring, we get

$$u^4 d\phi^2 = h^2 ds^2$$

or $$u^4 d\phi^2 = h^2\left(\frac{u^2}{u^2 - c^2}du^2 + u^2 d\phi^2\right)$$ [using (4)]

or $$(u^2 - h^2)\, d\phi^2 = \frac{h^2}{u^2 - c^2}du^2$$

or $$d\phi = \pm\frac{h}{\sqrt{(u^2 - h^2)(u^2 - c^2)}}du.$$

Integrating, we get

$$\phi = C \pm h\int\frac{1}{\sqrt{(u^2 - h^2)(u^2 - c^2)}}du, \tag{5}$$

where h and C are arbitrary constants.

Equation (5) gives the required equation of geodesics for the catenoid of revolution.

Example 7. Prove that every helix on a cylinder is a geodesic. Conversely, geodesics on a cylinder are helices.

Solution. **Case 1.** Let **a** be a unit vector parallel to the generator of the cylinder and let C be a helix on the cylinder.

To prove : C is a geodesic.

Take any point P on C. Let **t**, **n**, **b** denote unit tangent, unit principal normal and unit binormal respectively to C at P and let **N** be the unit normal to the cylinder at P. Then, we have

$$\mathbf{t} \cdot \mathbf{a} = \text{constant}$$ [$\because C$ is a helix]

$$\Rightarrow \quad \frac{d}{ds}(\mathbf{t} \cdot \mathbf{a}) = 0$$

$$\Rightarrow \quad \frac{d\mathbf{t}}{ds} \cdot \mathbf{a} + \mathbf{t} \cdot \frac{d\mathbf{a}}{ds} = 0$$

$$\Rightarrow \quad \kappa\mathbf{n} \cdot \mathbf{a} = 0$$ $\left[\because \frac{d\mathbf{a}}{ds} = \mathbf{0}, \text{ as } \mathbf{a} \text{ is a constant vector.}\right]$

$$\Rightarrow \quad \mathbf{n} \cdot \mathbf{a} = 0. \tag{1}$$

[$\because \kappa \neq 0$]

But, we know that

$$\mathbf{n} \cdot \mathbf{t} = 0. \tag{2}$$

From (1) and (2), it follows that

n is perpendicular to each of **a** and **t**.

$\Rightarrow$ **n** is perpendicular to the tangent plane to the cylinder at P.
[$\because$ **a** and **t** are tangential to the cylinder]

$\Rightarrow$ **n** is parallel to **N**.

$\Rightarrow$ We can take $\mathbf{n} = \mathbf{N}$.

$\Rightarrow$ The helix C is a geodesic on the cylinder.

Converse. Now, we suppose that C is a geodesic on the cylinder.

To prove : C is a helix.

Since C is a geodesic on the cylinder, we have

$$\mathbf{n} = \mathbf{N}$$

$\Rightarrow$ $\mathbf{n} \cdot \mathbf{a} = \mathbf{N} \cdot \mathbf{a}$

$\Rightarrow$ $\mathbf{n} \cdot \mathbf{a} = 0$ [$\because$ **N** is perpendicular to **a**]

$\Rightarrow$ $\kappa \mathbf{n} \cdot \mathbf{a} = 0$

$\Rightarrow$ $\dfrac{d\mathbf{t}}{ds} \cdot \mathbf{a} = 0$

$\Rightarrow$ $\dfrac{d\mathbf{t}}{ds} \cdot \mathbf{a} + \mathbf{t} \cdot \dfrac{d\mathbf{a}}{ds} = 0$

$\Rightarrow$ $\dfrac{d}{ds}(\mathbf{t} \cdot \mathbf{a}) = 0$

$\Rightarrow$ $\mathbf{t} \cdot \mathbf{a} = \text{constant}$

$\Rightarrow$ The geodesic C cuts the generators of the cylinder at a constant angle.

$\Rightarrow$ C is a helix.

Example 8. Discuss the nature of geodesics on a sphere.

Or

Show that a curve on a sphere is a geodesic if and only if it is a great circle.

Solution. Case 1. Let the curve Γ be a geodesic on a sphere.

To prove : Γ is a great circle on the sphere.

Let C be the centre of the sphere and let P be any point on Γ.

Let **t**, **n**, **b** denote unit tangent, unit principal normal and unit binormal respectively to Γ at P and let **N** be the unit normal to the sphere at P.

Since Γ is a geodesic on the sphere,

$$\mathbf{N} = \mathbf{n}. \tag{1}$$

We know that the normal at each point of the sphere passes through its centre. Therefore it follows from (1) that the principal normal at each point of Γ passes through the centre of the sphere.

If the position vectors of the points P and C be **r** and **a** respectively, then

$$\mathbf{r} = \mathbf{a} + \overrightarrow{CP}$$

or $\quad \mathbf{r} = \mathbf{a} + \lambda\mathbf{N}$, where $\lambda =$ radius of the sphere

or $\quad \mathbf{r} = \mathbf{a} + \lambda\mathbf{n}$. [using (1)]

Differentiating with respect to s, we get

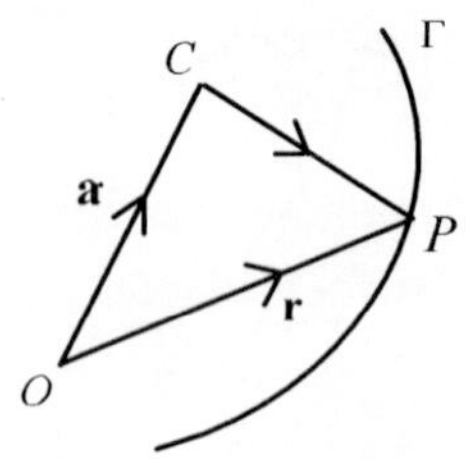

$$\frac{d\mathbf{r}}{ds} = \frac{d\mathbf{a}}{ds} + \frac{d\lambda}{ds}\mathbf{n} + \lambda\frac{d\mathbf{n}}{ds}$$

$\Rightarrow \quad \mathbf{t} = \lambda\,(\tau\mathbf{b} - \kappa\mathbf{t})$

$\Rightarrow \quad \mathbf{t}\,.\,\mathbf{b} = \lambda\,(\tau\mathbf{b} - \kappa\mathbf{t})\,.\,\mathbf{b}$

$\Rightarrow \quad 0 = \lambda\tau$

$\Rightarrow \quad \tau = 0 \qquad [\because \lambda \neq 0]$

$\Rightarrow \quad \Gamma$ is a plane curve.

Thus Γ is a plane curve on the sphere such that the normal at every point of Γ passes through the centre of the sphere. Therefore Γ is a section of the sphere by a plane passing through the centre of the sphere. Hence Γ is a great circle on the sphere.

Converse. Now, we suppose that Γ is a great circle on the sphere.

To prove : Γ is a geodesic on the sphere.

Since Γ is a great circle, it follows that the principal normal to Γ at each point of it coincides with the surface normal to the sphere. Hence, at each point of Γ, we have

$$\mathbf{n} = \mathbf{N}.$$

Therefore Γ is a geodesic on the sphere.

Example 9. Two surfaces touch each other along a curve. If the curve is a geodesic on one surface, prove that it is a geodesic on the other also.

Solution. Let S and $\overline{S}$ be two surfaces which touch each other along a curve C. Also, let the curve C be a geodesic on S.

We have to prove that C is a geodesic on $\overline{S}$.

Let P be any point on C. Let $\mathbf{n}$ denote unit principal normal to C at P, and let $\mathbf{N}$ and $\overline{\mathbf{N}}$ denote the unit normal vectors to S and $\overline{S}$ respectively at P.

Since the surfaces S and $\overline{S}$ touch each other along C,

$$\mathbf{N} = \overline{\mathbf{N}}. \tag{1}$$

Since C is a geodesic on S, we have

$$\mathbf{n} = \mathbf{N}. \tag{2}$$

From (1) and (2), we have

$$\mathbf{n} = \overline{\mathbf{N}}.$$

Therefore C is a geodesic on $\overline{S}$.

Example 10. Show that every space curve is geodesic on its rectifying developable.

Solution. Let the equation of a space curve be given by

$$\mathbf{r} = \mathbf{r}\,(s). \tag{1}$$

Let P be any point on this curve and let **t, n, b** denote unit tangent, unit principal normal and unit binormal respectively to this space curve at P.

The equation of the rectifying plane to this curve at P is given by

$$(\mathbf{R}-\mathbf{r})\,.\,\mathbf{n} = 0, \tag{2}$$

where **R** is the position vector of any current point on the rectifying plane.

Clearly, the left hand side of (2) is a function of a single variable parameter s. Therefore (2) represents a one parameter family of planes. The envelope of this family is known as the rectifying developable of (1).

Differentiating (2) with respect to s, we get

$$(\mathbf{R}-\mathbf{r})\,.\,(\tau\mathbf{b}-\kappa\mathbf{t})-\mathbf{t}\,.\,\mathbf{n} = 0$$

or $$(\mathbf{R}-\mathbf{r})\,.\,(\tau\mathbf{b}-\kappa\mathbf{t}) = 0. \tag{3}$$

The line of intersection of planes (2) and (3) is the generator of the rectifying developable of (1).

From (2) and (3), we get

$$(\mathbf{R}-\mathbf{r})\,.\,\mathbf{n} = 0 \text{ and } (\mathbf{R}-\mathbf{r})\,.\,(\tau\mathbf{b}-\kappa\mathbf{t}) = 0$$

$\Rightarrow$ $(\mathbf{R}-\mathbf{r})$ is perpendicular to each of $\mathbf{n}$ and $(\tau\mathbf{b}-\kappa\mathbf{t})$.

$\Rightarrow$ $(\mathbf{R}-\mathbf{r})$ is parallel to $\mathbf{n}\times(\tau\mathbf{b}-\kappa\mathbf{t}) = \tau\mathbf{t}+\kappa\mathbf{b}$.

$\Rightarrow$ We can take

$$\mathbf{R}-\mathbf{r} = u\,(\tau\mathbf{t}+\kappa\mathbf{b})$$

$$\Rightarrow \quad \mathbf{R} = \mathbf{r}+u\,(\tau\mathbf{t}+\kappa\mathbf{b}), \tag{4}$$

which gives the vector equation of the rectifying developable of (1). In this equation, u and s are parameters.

Taking u and s as the first and the second parameters, we have

$$\mathbf{R}_1 = \frac{\partial \mathbf{R}}{\partial u} = \tau\mathbf{t}+\kappa\mathbf{b},$$

$$\mathbf{R}_2 = \frac{\partial \mathbf{R}}{\partial s} = \mathbf{t}+u\,(\tau'\mathbf{t}+\tau\kappa\mathbf{n}+\kappa'\mathbf{b}-\kappa\tau\mathbf{n})$$

$$= (1+u\tau')\,\mathbf{t}+u\kappa'\mathbf{b}.$$

$$\therefore \qquad \mathbf{R}_1 \times \mathbf{R}_2 = (\tau\mathbf{t} + \kappa\mathbf{b}) \times [(1 + u\tau')\,\mathbf{t} + u\kappa'\mathbf{b}]$$
$$= \kappa[(1 + u\tau') - u\tau\kappa']\,\mathbf{n}.$$

On the curve (1), we have $u = 0$. Therefore, on this curve,

$$\mathbf{R}_1 \times \mathbf{R}_2 = \kappa\mathbf{n}$$

$\Rightarrow$ $\quad H\mathbf{N} = \kappa\mathbf{n}$

$\Rightarrow$ $\quad$ We can take $\mathbf{N} = \mathbf{n}$.

$\Rightarrow$ $\quad$ The space curve (1) is a geodesic on the surface (4).

Therefore every space curve is geodesic on its rectifying developable.

Example 11. Prove that if a geodesic is either a plane curve or a line of curvature, it is both.

Or

Prove that a necessary and sufficient condition that a geodesic be a plane curve is that it is a line of curvature.

Solution. Case 1. Let C be a plane curve and let it be a geodesic on a surface.

To prove : C is a line of curvature on the surface.

Let P be any point on C.

Let $\mathbf{t}$, $\mathbf{n}$, $\mathbf{b}$ denote unit tangent, unit principal normal and unit binormal respectively to C at P and let $\mathbf{N}$ be the unit normal to the surface at P.

Since C is a geodesic on the surface, we have

$$\mathbf{N} = \mathbf{n}.$$

$$\Rightarrow \qquad \frac{d\mathbf{N}}{ds} = \frac{d\mathbf{n}}{ds}$$

$$\Rightarrow \qquad \frac{d\mathbf{N}}{ds} = \tau\mathbf{b} - \kappa\mathbf{t}$$

$$\Rightarrow \qquad \frac{d\mathbf{N}}{ds} = -\kappa\mathbf{t} \qquad [\because \tau = 0, \text{ as } C \text{ is a plane curve.}]$$

$$\Rightarrow \qquad \frac{d\mathbf{N}}{ds} + \kappa\mathbf{t} = \mathbf{0}$$

$$\Rightarrow \qquad \frac{d\mathbf{N}}{ds} + \kappa\frac{d\mathbf{r}}{ds} = \mathbf{0}$$

$$\Rightarrow \qquad d\mathbf{N} + \kappa\, d\mathbf{r} = \mathbf{0}.$$

$\Rightarrow$ $\quad$ The curve C is a line of curvature. $\qquad$ [by Rodrigue's formula]

Converse. Now, we suppose that a curve C is a geodesic and a line of curvature on the surface.

To prove: C is a plane curve.

Since the curve C is a line of curvature, we have

$$d\mathbf{N} + \kappa\, d\mathbf{r} = \mathbf{0} \qquad \text{[by Rodrigue's formula]}$$

$$\Rightarrow \quad \frac{d\mathbf{N}}{ds} + \kappa \frac{d\mathbf{r}}{ds} = \mathbf{0}$$

$$\Rightarrow \quad \frac{d\mathbf{n}}{ds} + \kappa \frac{d\mathbf{r}}{ds} = \mathbf{0} \qquad [\because \mathbf{N} = \mathbf{n}, \text{ as } C \text{ is a geodesic.}]$$

$$\Rightarrow \quad \frac{d\mathbf{n}}{ds} + \kappa \mathbf{t} = \mathbf{0}$$

$$\Rightarrow \quad (\tau \mathbf{b} - \kappa \mathbf{t}) + \kappa \mathbf{t} = \mathbf{0}$$

$$\Rightarrow \quad \tau\, \mathbf{b} = \mathbf{0}$$

$$\Rightarrow \quad \tau = 0$$

$\Rightarrow$ C is a plane curve.

Example 12. Prove that any space curve is a geodesic on the surface generated by its binormals, and an asymptotic line on the surface generated by its principal normals.

Solution. First Part. Let the equation of a space curve be given by

$$\mathbf{r} = \mathbf{r}\,(s). \tag{1}$$

Let P be any point on this curve and let $\mathbf{t}, \mathbf{n}, \mathbf{b}$ denote unit tangent, unit principal normal and unit binormal respectively to this space curve at P.

If $\mathbf{R}$ be the position vector of a current point on the surface generated by the binormals of the space curve (1), then

$$\mathbf{R} = \mathbf{r} + u\mathbf{b}. \tag{2}$$

Clearly, the parameters of the surface (2) are u and s.

Taking u and s as the first and the second parameters, we have

$$\mathbf{R}_1 = \frac{\partial \mathbf{R}}{\partial u} = \mathbf{b},$$

$$\mathbf{R}_2 = \frac{\partial \mathbf{R}}{\partial s} = \mathbf{t} + u\,(-\tau \mathbf{n}) = \mathbf{t} - u\tau \mathbf{n}.$$

$$\therefore \quad \mathbf{R}_1 \times \mathbf{R}_2 = \mathbf{b} \times (\mathbf{t} - u\tau \mathbf{n}) = \mathbf{n} + u\tau \mathbf{t}.$$

On the curve (1), we have $u = 0$. Therefore, on this curve,

$$\mathbf{R}_1 \times \mathbf{R}_2 = \mathbf{n}$$

$\Rightarrow$ $H\mathbf{N} = \mathbf{n}$, where $\mathbf{N}$ is the unit normal vector to the surface (2) at P.

$\Rightarrow$ We can take $\mathbf{N} = \mathbf{n}$.

$\Rightarrow$ The space curve (1) is a geodesic on the surface (2).

Therefore every space curve is geodesic on the surface generated by its binormals.

Second Part. Now, we shall prove that every space curve is an asymptotic line on the surface generated by its principal normals.

If **R** be the position vector of a current point on the surface generated by the principal normals of the space curve (1), then

$$\mathbf{R} = \mathbf{r} + \lambda \mathbf{n}. \tag{3}$$

Clearly, the parameters of the surface (2) are λ and s.

Taking λ and s as the first and the second parameters, we have

$$\mathbf{R}_1 = \frac{\partial \mathbf{R}}{\partial \lambda} = \mathbf{n},$$

$$\mathbf{R}_2 = \frac{\partial \mathbf{R}}{\partial s} = \mathbf{t} + \lambda\,(\tau \mathbf{b} - \kappa \mathbf{t}) = (1 - \lambda\kappa)\,\mathbf{t} + \lambda\tau\mathbf{b}.$$

$\therefore$ $$\mathbf{R}_1 \times \mathbf{R}_2 = \mathbf{n} \times [(1-\lambda\kappa)\,\mathbf{t} + \lambda\tau\mathbf{b}] = (\lambda\kappa - 1)\,\mathbf{b} + \lambda\tau\mathbf{t}.$$

On the curve (1), we have $\lambda = 0$. Therefore, on this curve,

$$\mathbf{R}_1 \times \mathbf{R}_2 = -\mathbf{b}$$

$\Rightarrow$ $H\mathbf{N} = -\mathbf{b}$, where $\mathbf{N}$ is the unit normal vector to the surface (2) at P.

$\Rightarrow$ **N** is parallel to **b**.

$\Rightarrow$ The space curve (1) is an asymptotic line on (3).

Therefore every space curve is an asymptotic line on the surface generated by its principal normals.

Example 13. Prove that straight lines on a surface are the only asymptotic lines which are geodesics.

Solution. We know that every straight line on a surface is an asymptotic line as well as a geodesic. Therefore we have only to prove that if a curve C lying on a surface S is an asymptotic line as well as a geodesic, then it must be a straight line.

Let P be any point on C. Let $\mathbf{t}, \mathbf{n}, \mathbf{b}$ denote unit tangent, unit principal normal and unit binormal respectively of C at P and let **N** be the unit normal vector to the surface (3) at P.

Since C is a geodesic, we have

$$\mathbf{N} = \mathbf{n}. \tag{1}$$

Since C is an asymptotic line,

$$\mathbf{N} = \mathbf{b}. \tag{2}$$

From (1) and (2), we get

$$\mathbf{n} = \mathbf{b}$$

$$\Rightarrow \quad \frac{d\mathbf{n}}{ds} = \frac{d\mathbf{b}}{ds}$$

$$\Rightarrow \quad \tau\mathbf{b} - \kappa\mathbf{t} = -\tau\mathbf{n}$$

$$\Rightarrow \quad (\tau\mathbf{b} - \kappa\mathbf{t}) \cdot \mathbf{t} = -\tau\mathbf{n} \cdot \mathbf{t}$$

$$\Rightarrow \quad -\kappa = 0$$

$$\Rightarrow \quad \kappa = 0$$

$\Rightarrow$ The curve C is a straight line.

Example 14. Prove that a curve for which κ/τ is constant is a geodesic on a cylinder, and a curve for which $\frac{d}{ds}(\tau/\kappa)$ is constant is a geodesic on a cone.

Solution. First Part. Let C be a curve lying on a cylinder such that κ/τ is constant for C.

We have to prove that C is a geodesic.

Let P be any point on C. Let $\mathbf{t}, \mathbf{n}, \mathbf{b}$ denote unit tangent, unit principal normal and unit binormal respectively of C at P and let $\mathbf{N}$ be the unit normal vector to the cylinder at P.

By supposition, we have

$$\frac{\kappa}{\tau} = \text{const.} = c \text{ (say)}.$$

$$\therefore \quad \kappa = c\tau. \tag{1}$$

Now,

$$\frac{d\mathbf{t}}{ds} = \kappa\mathbf{n}$$

$$\Rightarrow \quad \frac{d\mathbf{t}}{ds} - \kappa\mathbf{n} = \mathbf{0}$$

$$\Rightarrow \quad \frac{d\mathbf{t}}{ds} - c\tau\mathbf{n} = \mathbf{0} \qquad \text{[using (1)]}$$

$$\Rightarrow \quad \frac{d\mathbf{t}}{ds} + c\frac{d\mathbf{b}}{ds} = \mathbf{0} \qquad \left[\because \frac{d\mathbf{b}}{ds} = -\tau\mathbf{n}\right]$$

$$\Rightarrow \quad \frac{d}{ds}(\mathbf{t} + c\mathbf{b}) = \mathbf{0}$$

$$\Rightarrow \quad \mathbf{t} + c\mathbf{b} = \mathbf{a} \text{ (a constant vector)}$$

$$\Rightarrow \quad (\mathbf{t} + c\mathbf{b}) \times \mathbf{t} = \mathbf{a} \times \mathbf{t}$$

$$\Rightarrow \quad c\mathbf{n} = \mathbf{a} \times \mathbf{t}. \tag{2}$$

Let us suppose that the generators of the cylinder are parallel to **a**. Then **N** is perpendicular to **a**. Therefore

$$\mathbf{N} \cdot \mathbf{a} = 0. \tag{3}$$

Also, we have

$$\mathbf{N} \cdot \mathbf{t} = 0. \tag{4}$$

From (3) and (4), it follows that

N is perpendicular to each of **a** and **t**.

$\Rightarrow$ **N** is parallel to $\mathbf{a} \times \mathbf{t}$.

$\Rightarrow$ **N** is parallel to **n**. [using (2)]

$\Rightarrow$ We can take $\mathbf{N} = \mathbf{n}$.

$\Rightarrow$ The curve C is a geodesic on the cylinder.

Second Part. Let C be a curve lying on a cone such that $\frac{d}{ds}(\tau/\kappa)$ is constant for C.

We have to prove that C is a geodesic.

Let P be any point on C and let $\mathbf{t}, \mathbf{n}, \mathbf{b}$ and **N** have their usual meaning.

By supposition, we have

$$\frac{d}{ds}(\tau/\kappa) = \text{const.} = c \text{ (say)}.$$

$$\therefore \quad \tau = \kappa\,(cs + A), \tag{5}$$

where A is a constant.

Now,

$$\frac{d\mathbf{b}}{ds} = -\tau\mathbf{n}$$

$$\Rightarrow \quad \frac{d\mathbf{b}}{ds} + \tau\mathbf{n} = \mathbf{0}$$

$$\Rightarrow \quad \frac{d\mathbf{b}}{ds} + (cs + A)\,\kappa\mathbf{n} = \mathbf{0} \qquad \text{[using (5)]}$$

$$\Rightarrow \quad \frac{d\mathbf{b}}{ds} + (cs + A)\frac{d\mathbf{t}}{ds} = \mathbf{0}. \qquad \left[\because \frac{d\mathbf{t}}{ds} = \kappa\mathbf{n}\right]$$

Integrating this equation, we get

$$\int \frac{d\mathbf{b}}{ds}\,ds + \int \left[(cs + A)\frac{d\mathbf{t}}{ds}\right] ds = \text{ constant vector}$$

or $\quad \mathbf{b} + (cs + A)\,\mathbf{t} - \int c\mathbf{t}\,ds = \text{ constant vector}$

or $\quad \mathbf{b} + (cs + A)\,\mathbf{t} - \int c\,\frac{d\mathbf{r}}{ds}\,ds = \text{constant vector} = -c\mathbf{a} \text{ (say)}$

or $\quad \mathbf{b} + (cs + A)\,\mathbf{t} - c\mathbf{r} = -c\mathbf{a}$

or $\quad \mathbf{b}+(cs+A)\,\mathbf{t}=c\,(\mathbf{r}-\mathbf{a}).$

Taking cross product of both sides with **t**, we get

or $\quad [\mathbf{b}+(cs+A)\,\mathbf{t}]\times\mathbf{t}=c\,(\mathbf{r}-\mathbf{a})\times\mathbf{t}$

or $\quad \mathbf{n}=c\,(\mathbf{r}-\mathbf{a})\times\mathbf{t}.$ (6)

Let us suppose that the position vector of the vertex of the cone is **a**. Then generators of the cone will be parallel to $(\mathbf{r}-\mathbf{a})$ and **N** will be perpendicular to $(\mathbf{r}-\mathbf{a})$. Therefore

$$\mathbf{N}\,.\,(\mathbf{r}-\mathbf{a})=0. \tag{7}$$

Also, we have

$$\mathbf{N}\,.\,\mathbf{t}=0.$$

From (7) and (8), it follows that

N is perpendicular to each of $(\mathbf{r}-\mathbf{a})$ and $\mathbf{t}$.

$\Rightarrow$ N is parallel to $(\mathbf{r}-\mathbf{a})\times\mathbf{t}$.

$\Rightarrow$ N is parallel to **n**. [using (6)]

$\Rightarrow$ We can take $\mathbf{N}=\mathbf{n}$.

$\Rightarrow$ The curve C is a geodesic on the cone.

Exercises

1. Prove that the meridians of a right helicoid are geodesics.
2. A particle is constrained to move on a smooth surface under no forces except the normal reaction. Show that its path is s geodesic.
3. Prove that the projections of the xy-plane of the geodesics on the catenoid $u=c\cosh(z/c)$ are given by

$$d\phi=\frac{a}{\sqrt{(u^2-c^2)(u^2-h^2)}}\,du.$$

4. Prove that if a curve on a surface has any two of the following three properties, then it also has the third:
   (i) It is a geodesic,
   (ii) It is an asymptotic line,
   (iii) It is a straight line.

5.8 TORSION AND CURVATURE OF A GEODESIC

I. Formula for Torsion of a Geodesic.

Let the parametric equation of a surface in vector form be given by

$$\mathbf{r}=\mathbf{r}\,(u,v), \tag{1}$$

where u,v are the parametric coordinates of a point on the surface.

Let the vector equation of a geodesic on the surface (1) be given by

$$\mathbf{r} = \mathbf{r}(s), \tag{2}$$

where s is the arc length of a point on the geodesic measured from a fixed point on it.

Let P be any point on the geodesic (2) whose position vector is $\mathbf{r}$.

Let $\mathbf{t}$, $\mathbf{n}$, $\mathbf{b}$ denote the fundamental unit vectors of (2) at P and let $\mathbf{N}$ be the unit normal vector to the surface (1) at P. Also, let κ and τ be the curvature and torsion respectively of (2) at P.

Since (2) is a geodesic, we have

$$\mathbf{N} = \mathbf{n}. \tag{3}$$

Now,

$$\mathbf{b} = \mathbf{t} \times \mathbf{n}$$

$$= \mathbf{t} \times \mathbf{N} \qquad \text{[using (3)]}$$

$$\Rightarrow \quad \frac{d\mathbf{b}}{ds} = \frac{d\mathbf{t}}{ds} \times \mathbf{N} + \mathbf{t} \times \frac{d\mathbf{N}}{ds}$$

$$\Rightarrow \quad -\tau\mathbf{n} = \kappa\mathbf{n} \times \mathbf{N} + \mathbf{t} \times \mathbf{N}'$$

$$\Rightarrow \quad -\tau\mathbf{N} = \kappa\mathbf{N} \times \mathbf{N} + \mathbf{t} \times \mathbf{N}' \qquad \text{[using (3)]}$$

$$\Rightarrow \quad -\tau\mathbf{N} = \mathbf{r}' \times \mathbf{N}' \qquad [\because \mathbf{N} \times \mathbf{N} = \mathbf{0} \text{ and } \mathbf{t} = \mathbf{r}']$$

$$\Rightarrow \quad \tau\mathbf{N} = \mathbf{N}' \times \mathbf{r}'$$

$$\Rightarrow \quad (\tau\mathbf{N}) \,.\, \mathbf{N} = (\mathbf{N}' \times \mathbf{r}') \,.\, \mathbf{N}$$

$$\Rightarrow \quad \tau(\mathbf{N} \,.\, \mathbf{N}) = \mathbf{N} \,.\, (\mathbf{N}' \times \mathbf{r}')$$

$$\Rightarrow \quad \tau = [\mathbf{N},\ \mathbf{N}', \mathbf{r}']. \tag{4}$$

Equation (4) gives the torsion of a geodesic at any point $P(\mathbf{r})$.

Geodesic Tangent. Definition.

The geodesic which touches a curve C at any point is called the geodesic tangent of C at that point.

Corollary 1. The torsion of an asymptotic line is equal to the torsion of its geodesic tangent.

Proof. Let the equation of a surface be given by

$$\mathbf{r} = \mathbf{r}(u, v) \tag{1}$$

and let the equation of an asymptotic line on this surface be given by

$$\mathbf{r} = \mathbf{r}(s). \tag{2}$$

Let a geodesic on the surface (1) touch the asymptotic line (2) at $P(\mathbf{r})$.

We know that the torsion of the asymptotic line (2) at $P(\mathbf{r})$ is given by

$$\tau = [\mathbf{N},\ \mathbf{N}', \mathbf{r}']. \tag{3}$$

But, this is also the expression for the torsion of a geodesic at P ($\mathbf{r}$). Since the asymptotic line and the geodesic touch each other at P, they have the same tangent line at P. Hence the two curves have the same direction coefficients (u', v') at P. Therefore the vectors $\mathbf{N}' = \mathbf{N}_1 u' + \mathbf{N}_2 v'$ and $\mathbf{r}' = \mathbf{r}_1 u' + \mathbf{r}_2 v'$ are the same at P for both the curves. Hence it follows from (3) that the torsion of an asymptotic line is equal to the torsion of its geodesic tangent.

Corollary 2. The torsion of a geodesic vanishes where it touches a line of curvature.

Proof. We know that the expression $[\mathbf{N},\ \mathbf{N}', \mathbf{r}']$ vanishes for a principal direction at a point $P(\mathbf{r})$. But $[\mathbf{N},\ \mathbf{N}', \mathbf{r}']$ is also the expression for the torsion of a geodesic at $P(\mathbf{r})$. Hence it follows that the torsion of a geodesic vanishes at a point where its direction is a principal direction, i.e., the torsion of a geodesic vanishes where it touches a line of curvature.

II. Torsion of a Geodesic in terms of Principal Curvatures.

Let the parametric equation of a surface in vector form be given by

$$\mathbf{r} = \mathbf{r}\,(u, v) \tag{1}$$

and let the equation of a geodesic on this surface be given by

$$\mathbf{r} = \mathbf{r}\,(s). \tag{2}$$

Let $\mathbf{r}$ be the position vector of any point P on the geodesic (2) and let $\mathbf{t}$, $\mathbf{n}$, $\mathbf{b}, \mathbf{N}, \kappa$ and τ have their usual meaning.

Since (2) is a geodesic, we have

$$\mathbf{N} = \mathbf{n}. \tag{3}$$

Now, proceeding as above, it can be proved that

$$\tau = [\mathbf{N},\ \mathbf{N}', \mathbf{r}'] = \mathbf{N} \,.\, (\mathbf{N}' \times \mathbf{r}'). \tag{4}$$

Again, we have

$$\left.\begin{aligned} \mathbf{N}' &= \mathbf{N}_1 u' + \mathbf{N}_2 v', \\ \mathbf{r}' &= \mathbf{r}_1 u' + \mathbf{r}_2 v'. \end{aligned}\right\} \tag{5}$$

Substituting the values of $\mathbf{N}'$ and $\mathbf{r}'$ from (5) in (4), we get

$$\tau = \mathbf{N} \,.\, (\mathbf{N}_1 u' + \mathbf{N}_2 v') \times (\mathbf{r}_1 u' + \mathbf{r}_2 v')$$

or $$\tau = \mathbf{N} \,.\, [(\mathbf{N}_1 \times \mathbf{r}_1)\, u'^2 + (\mathbf{N}_1 \times \mathbf{r}_2 + \mathbf{N}_2 \times \mathbf{r}_1)\, u'v' + (\mathbf{N}_2 \times \mathbf{r}_2)\, v'^2]$$

or $$\tau = [\mathbf{N}, \mathbf{N}_1, \mathbf{r}_1]\, u'^2 + ([\mathbf{N}, \mathbf{N}_1, \mathbf{r}_2] + [\mathbf{N}, \mathbf{N}_2, \mathbf{r}_1])\, u'v' + [\mathbf{N}, \mathbf{N}_2, \mathbf{r}_2]\, v'^2. \tag{6}$$

Also, we know that

$$[\mathbf{N},\ \mathbf{N}_1,\ \mathbf{r}_1]=\frac{EM-FL}{H},$$

$$[\mathbf{N},\ \mathbf{N}_1,\ \mathbf{r}_2]=\frac{FM-GL}{H},$$

$$[\mathbf{N},\ \mathbf{N}_2,\ \mathbf{r}_1]=\frac{EN-FM}{H},$$

$$[\mathbf{N},\ \mathbf{N}_2,\ \mathbf{r}_2]=\frac{FN-GM}{H}.$$

Substituting the above values in (6), we get

$$\tau=\frac{EM-FL}{H}u'^2+\frac{EN-GL}{H}u'v'+\frac{FN-GM}{H}v'^2. \tag{7}$$

Let the lines of curvature be taken as the parametric curves. Then

$$F=0, M=0. \tag{8}$$

Substituting the value of F and M from (8) in (7), we get

$$\tau=\frac{EN-GL}{H}u'v'$$

or $$\tau=\frac{EN-GL}{\sqrt{EG}}u'v' \qquad [\because H^2=EG-F^2=EG-0=EG]$$

or $$\tau=\sqrt{EG}\left(\frac{N}{G}-\frac{L}{E}\right)u'v'. \tag{9}$$

Let κ_a and κ_b be the principal curvatures of the surface (1) at $P(u,v)$.

We know that the normal curvature of the surface (1) at $P(u,v)$ in direction (du, dv) is given by

$$\kappa_n=\frac{L\,du^2+2M\,du\,dv+N\,dv^2}{E\,du^2+2F\,du\,dv+G\,dv^2}$$

or $$\kappa_n=\frac{L\,du^2+N\,dv^2}{E\,du^2+G\,dv^2}. \tag{10}$$

[using (8)]

The principal curvature κ_a, being the normal curvature for the direction $dv=0$, is expressed as

$$\kappa_a=\frac{L}{E}. \tag{11}$$

[using (10)]

Similarly, the principal curvature κ_b, being the normal curvature for the direction $du=0$, is expressed as

$$\kappa_b = \frac{N}{G}. \tag{12}$$

[using (10)]

Let ψ be the inclination of the geodesic to the line of curvature $v =$ constant. Then, we know that

$$\cos\psi = \frac{1}{\sqrt{E}}\left(E\frac{du}{ds} + F\frac{dv}{ds}\right),\ \sin\psi = \frac{H}{\sqrt{E}}\frac{dv}{ds}$$

or $$\cos\psi = \sqrt{E}\frac{du}{ds},\ \sin\psi = \sqrt{G}\frac{dv}{ds}$$ [using (8)]

or $$\cos\psi = \sqrt{E}\,u',\ \sin\psi = \sqrt{G}\,v'. \tag{13}$$

Using (11), (12) and (13), (9) gives

$$\tau = (\kappa_b - \kappa_a)\cos\psi \sin\psi. \tag{14}$$

Equation (14) gives the torsion of a geodesic in terms of principal curvatures.

Corollary 1. Two geodesics at right angles have their torsions equal in magnitude but opposite in sign.

Proof. Let τ_1 and τ_2 be the torsions of two geodesics at right angles. Then we can take ψ and $\frac{\pi}{2}+\psi$ as the inclinations of the two geodesics with the parametric curve $v =$ constant. Therefore

$$\tau_1 = (\kappa_b - \kappa_a)\cos\psi \sin\psi,$$

$$\tau_2 = (\kappa_b - \kappa_a)\cos\left(\frac{\pi}{2}+\psi\right)\sin\left(\frac{\pi}{2}+\psi\right)$$

$$= -(\kappa_b - \kappa_a)\sin\psi\cos\psi.$$

Obviously,

$$\tau_2 = -\tau_1.$$

Hence two geodesics at right angles have their torsions equal in magnitude but opposite in sign.

Corollary 2. The torsion of a geodesic vanishes at an umbilic.

Proof. We know that the torsion of a geodesic is given by

$$\tau = \frac{EM - FL}{H}u'^2 + \frac{EN - GL}{H}u'v' + \frac{FN - GM}{H}\ v'^2. \tag{1}$$

At an umbilic, we have

$$\frac{E}{L} = \frac{F}{M} = \frac{G}{N} = \lambda \text{ (say)}.$$

$$\therefore \qquad E = \lambda L,\ F = \lambda M,\ G = \lambda N. \tag{2}$$

Substituting the values of E, F, G from (2) in (1), we get

$$\tau = 0.$$

Therefore the torsion of a geodesic vanishes at an umbilic.

Corollary 3. Of all geodesics through a given point, those which bisect the angles between the lines of curvature have the greatest torsion.

Proof. We know that if ψ be the inclination of a geodesic to the line of curvature $v =$ constant, then the torsion of the geodesic is given by

$$\tau = (\kappa_b - \kappa_a) \cos\psi \sin\psi$$

i.e., $$\tau = \frac{1}{2}(\kappa_b - \kappa_a) \sin 2\psi.$$

Therefore τ is numerically maximum when sin 2ψ has the maximum value, i.e., when

$$\sin 2\psi = \pm 1$$

or $$2\psi = \frac{\pi}{2} \text{ or } \frac{3\pi}{2}$$

or $$\psi = \frac{\pi}{4} \text{ or } \frac{3\pi}{4}.$$

$\therefore$ Inclination of the geodesic of numerically maximum torsion with the line of curvature $(v = \text{const.}) = \dfrac{\pi}{4}$ or $\dfrac{3\pi}{4}$. (1)

Since the parametric curves are orthogonal and the lines of curvature are taken as the parametric curves, the lines of curvature are orthogonal. Hence it follows from (1) that of all geodesics through a given point, those which bisect the angles between the lines of curvature have the greatest torsion.

III. Curvature of a Geodesic.

Theorem. The curvature of a geodesic at any point is the normal curvature in its direction.

Proof. Let the parametric equation of a surface in vector form be given by

$$\mathbf{r} = \mathbf{r}\,(u, v) \tag{1}$$

and let the equation of a geodesic on this surface be given by

$$\mathbf{r} = \mathbf{r}\,(s). \tag{2}$$

Let $\mathbf{r}$ be the position vector of any point P on the geodesic (2) and let $\mathbf{t}$, $\mathbf{n}$, $\mathbf{b}$, $\mathbf{N}$, κ and τ have their usual meaning.

Since (2) is a geodesic, we have

$$\mathbf{N} = \mathbf{n}. \tag{3}$$

Now,

$$\mathbf{r}'' = \kappa\mathbf{n}$$
$$= \kappa\mathbf{N} \quad \text{[using (3)]}$$

$$\Rightarrow \quad \mathbf{r}''.\mathbf{N} = \kappa\mathbf{N}.\mathbf{N}$$

$$\Rightarrow \quad \kappa_n = \kappa, \tag{4}$$

where $\kappa_n = \mathbf{r}''.\mathbf{N}$ is the normal curvature of surface (1) at P in the direction of (2).

It follows from (4) that the curvature of a geodesic at any point is the normal curvature in its direction.

By Euler's theorem, the normal curvature κ_n at a point P on a surface in any direction is given by

$$\kappa_n = \kappa_a \cos^2\psi + \kappa_b \sin^2\psi, \tag{5}$$

where κ_a and κ_b are the principal curvatures at P and ψ is the angle which the direction (du, dv) of the normal section at P makes with the principal direction through P in which the principal curvature is κ_a.

From (4) and (5), we have

$$\kappa = \kappa_a \cos^2\psi + \kappa_b \sin^2\psi. \tag{6}$$

Equation (6) gives the curvature of a geodesic in terms of principal curvatures.

Solved Examples

Example 1. If κ, τ are the curvature and torsion of a geodesic, prove that

$$\tau^2 = (\kappa - \kappa_a)(\kappa_b - \kappa).$$

Also, if the surface is developable, show that $\kappa = \tau \tan\psi$.

Solution. First Part. We know that if κ and τ are the curvature and torsion of a geodesic at a point P, then

$$\kappa = \kappa_a \cos^2\psi + \kappa_b \sin^2\psi, \tag{1}$$

$$\tau = (\kappa_b - \kappa_a)\cos\psi \sin\psi, \tag{2}$$

where κ_a, κ_b are the principal curvatures of the surface at P and ψ is the inclination of the geodesic to the direction giving κ_a.

From (1), we have

$$\kappa - \kappa_a = (\kappa_a \cos^2\psi + \kappa_b \sin^2\psi) - \kappa_a$$
$$= \kappa_b \sin^2\psi - \kappa_a(1 - \cos^2\psi)$$
$$= \kappa_b \sin^2\psi - \kappa_a \sin^2\psi$$

i.e., $\kappa - \kappa_a = (\kappa_b - \kappa_a)\sin^2\psi.$ (3)

Also, from (1),

$$\kappa_b - \kappa = \kappa_b - (\kappa_a \cos^2\psi + \kappa_b \sin^2\psi)$$
$$= \kappa_b(1-\sin^2\psi) - \kappa_a \cos^2\psi$$
$$= \kappa_b \cos^2\psi - \kappa_a \cos^2\psi$$

i.e., $\kappa_b - \kappa = (\kappa_b - \kappa_a)\cos^2\psi.$ (4)

Multiplying (3) and (4), we get

$$(\kappa - \kappa_a)(\kappa_b - \kappa) = (\kappa_b - \kappa_a)^2 \sin^2\psi \cos^2\psi$$

or $(\kappa - \kappa_a)(\kappa_b - \kappa) = \tau^2.$ (5)

[using (2)]

This proves the first part.

Second Part. If the surface is developable, then the Gaussian curvature $K = \kappa_a \kappa_b = 0.$ Therefore, either $\kappa_a = 0$ or $\kappa_b = 0.$

Let us take $\kappa_a = 0.$ Then, from (1), we have

$$\kappa = \kappa_b \sin^2\psi$$

i.e., $\kappa_b = \kappa \operatorname{cosec}^2\psi.$ (6)

Again, substituting $\kappa_a = 0$ in (5), we get

$$\kappa(\kappa_b - \kappa) = \tau^2$$

or $\kappa(\kappa \operatorname{cosec}^2\psi - \kappa) = \tau^2$ [using (6)]

or $\kappa^2(\operatorname{cosec}^2\psi - 1) = \tau^2$

or $\kappa^2 \cot^2\psi = \tau^2$

or $\kappa \cot\psi = \tau$

or $\kappa = \tau \tan\psi.$

Example 2. Prove that the torsion of a geodesic is equal to

$$\frac{1}{H}\begin{vmatrix} Eu' + Fv' & Fu' + Gv' \\ Lu' + Mv' & Mu' + Nv' \end{vmatrix}.$$

Solution. We know that the torsion of a geodesic is given by

$$\tau = \frac{1}{H}\{(EM - FL)u'^2 + (EN - GL)u'v' + (FN - GM)\}\ v'^2$$

or $$\tau = \frac{1}{H}\{(Eu' + Fv')(Mu' + Nv') - (Fu' + Gv')(Lu' + Mv')\}$$

[adding and subtracting $FMu'v'$]

or $$\tau = \frac{1}{H}\begin{vmatrix} Eu' + Fv' & Fu' + Gv' \\ Lu' + Mv' & Mu' + Nv' \end{vmatrix}.$$

Example 3. Prove that, with usual notation for a geodesic,

$$\kappa \cos\psi - \tau \sin\psi = \kappa_a \cos\psi,$$
$$\kappa \sin\psi + \tau \cos\psi = \kappa_b \sin\psi.$$

Solution. First Part. We know that if κ and τ are the curvature and torsion of a geodesic at a point P, then

$$\kappa = \kappa_a \cos^2\psi + \kappa_b \sin^2\psi, \tag{1}$$
$$\tau = (\kappa_b - \kappa_a)\cos\psi \sin\psi, \tag{2}$$

where the symbols have their usual meaning.

From (1) and (2), we have

$$\begin{aligned} \kappa \cos\psi - \tau \sin\psi &= (\kappa_a \cos^2\psi + \kappa_b \sin^2\psi)\cos\psi \\ &\qquad - (\kappa_b - \kappa_a)\cos\psi \sin^2\psi \\ &= \kappa_a \cos^3\psi + \kappa_a \cos\psi \sin^2\psi \\ &= \kappa_a \cos\psi\,(\cos^2\psi + \sin^2\psi) \\ &= \kappa_a \cos\psi. \end{aligned}$$

Also, from (1) and (2),

$$\begin{aligned} \kappa \sin\psi + \tau \cos\psi &= (\kappa_a \cos^2\psi + \kappa_b \sin^2\psi)\sin\psi \\ &\qquad + (\kappa_b - \kappa_a)\cos^2\psi \sin\psi \\ &= \kappa_b \sin^3\psi + \kappa_b \cos^2\psi \sin\psi \\ &= \kappa_b \sin\psi\,(\sin^2\psi + \cos^2\psi) \\ &= \kappa_b \sin\psi. \end{aligned}$$

Exercises

1. Show that the for a geodesic

$$\frac{1}{\sigma^2} = \left(\frac{1}{\rho} - \frac{1}{\rho_1}\right)\left(\frac{1}{\rho_2} - \frac{1}{\rho}\right).$$

2. Prove that the torsion of a geodesic at any point is given by

$$\frac{1}{\sigma} = \left(\frac{1}{\rho_1} - \frac{1}{\rho_2}\right)\sin\psi \cos\psi,$$

where ψ is the angle between the directions of geodesic and a principal direction, and ρ_1, ρ_2 are the principal radii of curvature.

5.9 CURVES IN RELATION TO GEODESICS

I. Bonnet's Theorem. Let τ be the torsion of a curve at any point and let W be the torsion of the geodesic tangent at this point. Then

$$\frac{d\varpi}{ds}+\tau=W,$$

where ϖ is the normal angle of the curve at the point which is defined as the angle measured positively from the principal normal **n** to the surface normal **N** at the point.

Proof. Let the parametric equation of a surface in vector form be given by

$$\mathbf{r}=\mathbf{r}\,(u,v). \tag{1}$$

Let the vector equation of a curve C on the surface (1) be given by

$$\mathbf{r}=\mathbf{r}\,(s). \tag{2}$$

Let P be any point on C whose position vector is **r**. Let **t**, **n**, **b** denote the fundamental unit vectors of C at P and let **N** be the unit normal vector to the surface (1) at P. Also, let τ be the torsion of the curve C at P and let W be the torsion of the geodesic which touches C at P.

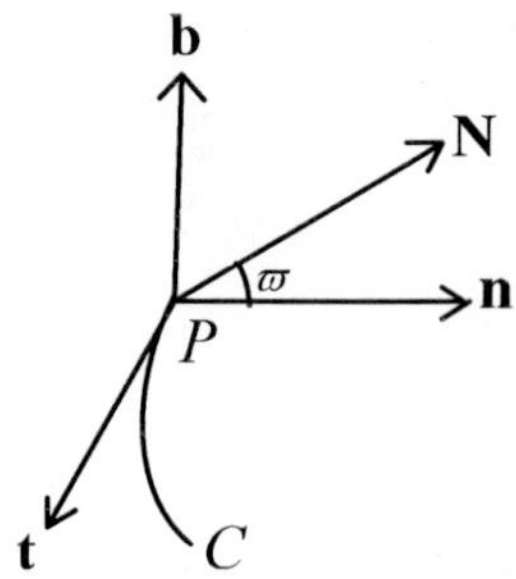

If ϖ is the normal angle of the curve C at P, i.e., the angle measured positively from the principal normal **n** to the surface normal **N**, then

$$\mathbf{n}\,.\,\mathbf{N}=\cos\varpi, \tag{3}$$

$$\mathbf{b}\,.\,\mathbf{N}=\sin\varpi. \tag{4}$$

Differentiating (4) with respect to s, we get

$$\frac{d\mathbf{b}}{ds}\,.\,\mathbf{N}+\mathbf{b}\,.\,\frac{d\mathbf{N}}{ds}=\cos\varpi\frac{d\varpi}{ds}$$

or $$(-\tau\mathbf{n})\,.\,\mathbf{N}+\mathbf{b}\,.\,\frac{d\mathbf{N}}{ds}=\cos\varpi\frac{d\varpi}{ds}$$

or $$-\tau\,(\mathbf{n}\,.\,\mathbf{N})+(\mathbf{t}\times\mathbf{n})\,.\,\mathbf{N}'=\cos\varpi\frac{d\varpi}{ds}$$

or $$-\tau\cos\varpi+(\mathbf{r}'\times\mathbf{n})\,.\,\mathbf{N}'=\cos\varpi\frac{d\varpi}{ds} \qquad \text{[using (3)]}$$

or $$\left(\frac{d\varpi}{ds}+\tau\right)\cos\varpi=(\mathbf{r}'\times\mathbf{n})\,.\,\mathbf{N}'$$

or $$\left(\frac{d\varpi}{ds}+\tau\right)\cos\varpi=[\mathbf{r}',\mathbf{n},\mathbf{N}']$$

$$=[\mathbf{N}',\mathbf{r}',\mathbf{n}]$$

$$=\mathbf{N}'\times\mathbf{r}'.\mathbf{n}$$

$$=(W\mathbf{N}).\mathbf{n} \qquad [\because \mathbf{N}'\times\mathbf{r}'=W\mathbf{N}]$$

$$=W(\mathbf{N}.\mathbf{n})$$

$$=W\cos\varpi.$$

$\therefore$ $$\frac{d\varpi}{ds}+\tau=W.$$

II. Joachimsthal's Theorems.

(i) If a curve on a surface has any two of the following properties, it also has the third :

(a) It is a line of curvature,

(b) It is a plane curve,

(c) Its normal angle is constant.

Proof. Let the parametric equation of a surface in vector form be given by

$$\mathbf{r}=\mathbf{r}(u,v). \tag{1}$$

Let the vector equation of a curve C on the surface (1) be given by

$$\mathbf{r}=\mathbf{r}(s), \tag{2}$$

where s is the arc length of a point on C from a fixed point on it.

Let P be any point on C whose position vector is $\mathbf{r}$.

Let $\mathbf{t}$, $\mathbf{n}$, $\mathbf{b}$ denote the fundamental unit vectors of C at P and let $\mathbf{N}$ be the unit normal vector to the surface (1) at P. Also, let τ be the torsion of the curve C at P and let W be the torsion of the geodesic which touches C at P.

If ϖ is the normal angle of the curve C at P, then, by Bonnet's theorem,

$$\frac{d\varpi}{ds}+\tau=W. \tag{3}$$

Let the curve C satisfy (a) and (b), i.e., we suppose that C is both a line of curvature and a plane curve. Then, we have

$$W=0 \text{ and } \tau=0$$

$\Rightarrow$ $$\frac{d\varpi}{ds}=0 \qquad \text{[using (3)]}$$

$\Rightarrow$ $\varpi=$ constant

$\Rightarrow$ The normal angle of C is constant.

$\Rightarrow$ The curve C satisfies (c).

Now, let curve C satisfy (b) and (c), i.e., we suppose that C is a plane curve and the normal angle of C is constant. Then, we have

$$\tau = 0 \text{ and } \varpi = \text{constant}$$

$$\Rightarrow \quad \tau = 0 \text{ and } \frac{d\varpi}{ds} = 0$$

$$\Rightarrow \quad W = 0 \qquad \text{[using (3)]}$$

$\Rightarrow$ C is a line of curvature.

$\Rightarrow$ The curve C satisfies (a).

Again, let curve C satisfy (c) and (a), i.e., we suppose that the normal angle of C is constant and C is a line of curvature. Then, we have

$$\varpi = \text{constant and } W = 0$$

$$\Rightarrow \quad \frac{d\varpi}{ds} = 0 \text{ and } W = 0$$

$$\Rightarrow \quad \tau = 0 \qquad \text{[using (3)]}$$

$\Rightarrow$ C is a plane curve.

$\Rightarrow$ The curve C satisfies (b).

(ii) If the curve of intersection of a sphere and another surface is a line of curvature on the latter, the two surfaces cut at a constant angle.

Conversely, if a sphere cuts a surface at a constant angle, the curve of intersection is a line of curvature on the surface.

Proof. Let S be a surface and let S_0 be a sphere. Let C be the curve of intersection of S and S_0.

Let P be any point on C whose position vector is **r**.

Let **t**, **n**, **b** denote the fundamental unit vectors of C at P and let τ be the torsion of C at P.

Let **N** and $\mathbf{N}_0$ be unit normals to the surfaces S and S_0 respectively at P. Let W and W_0 be the torsions of the geodesic tangents of C at P corresponding to the surfaces S and S_0 respectively.

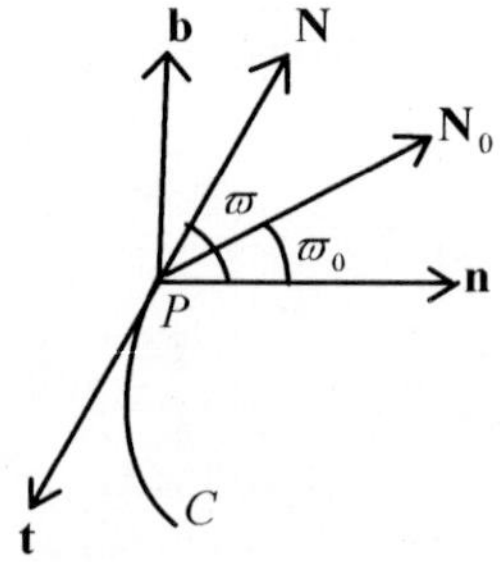

Also, let ϖ and ϖ_0 be the normal angles of C at P for S and S_0 respectively.

Case 1. Let C be a line of curvature on S.

To prove : The surfaces S and S_0 cut at a constant angle.

Since C is a line of curvature on S, we have

$$W = 0. \tag{1}$$

Now, applying Bonnet's theorem to the curve C, regarding it on S, we have

$$\frac{d\varpi}{ds} + \tau = W$$

or $$\frac{d\varpi}{ds} + \tau = 0. \tag{2}$$

[using (1)]

We know that the geodesics on a sphere are great circles of the sphere. Since the great circles on a sphere are plane curves, the geodesics on a sphere are plane curves. Hence it follows that the geodesic tangent of C on the surface S_0 is a plane curve. Therefore

$$W_0 = 0. \tag{3}$$

Now, applying Bonnet's theorem to the curve C, regarding it on S_0, we have

$$\frac{d\varpi_0}{ds} + \tau = W_0$$

or $$\frac{d\varpi_0}{ds} + \tau = 0. \tag{4}$$

[using (3)]

From (2) and (4), we get

$$\frac{d\varpi}{ds} = \frac{d\varpi_0}{ds}$$

$$\Rightarrow \quad \frac{d\varpi}{ds} - \frac{d\varpi_0}{ds} = 0$$

$$\Rightarrow \quad \frac{d}{ds}(\varpi - \varpi_0) = 0$$

$$\Rightarrow \quad \varpi - \varpi_0 = \text{constant}$$

$\Rightarrow$ The angle of intersection of S and S_0 is constant.

Converse. Now, we suppose that the angle of intersection of S and S_0 is constant.

To prove : C is a line of curvature on S.

By supposition, we have

$$\varpi - \varpi_0 = \text{constant}$$

$$\Rightarrow \quad \frac{d}{ds}(\varpi - \varpi_0) = 0$$

$$\Rightarrow \quad \frac{d\varpi}{ds} = \frac{d\varpi_0}{ds}$$

$$\Rightarrow \quad \frac{d\varpi}{ds} + \tau = \frac{d\varpi_0}{ds} + \tau. \tag{5}$$

Now, applying Bonnet's theorem to the curve C, regarding it on S, we have

$$\frac{d\varpi}{ds} + \tau = W$$

and, as in the preceding case, we have

$$\frac{d\varpi_0}{ds} + \tau = 0.$$

Substituting the above values in (5), we get

$$W = 0$$

$\Rightarrow \quad C$ is a line of curvature on S.

5.10 GEODESIC CURVATURE

At a point P of a curve on a surface, there are three non coplanar vectors : $\mathbf{N}$ (the unit normal to the surface) and $\mathbf{r}_1, \mathbf{r}_2$ (the tangential vectors). Therefore the curvature vector $\mathbf{r}''$ of the curve at P can be expressed as

$$\mathbf{r}'' = \nu\mathbf{N} + \lambda\mathbf{r}_1 + \mu\mathbf{r}_2. \tag{1}$$

Taking dot product on both sides of (1) with $\mathbf{N}$, we get

$$\mathbf{r}''.\mathbf{N} = \nu\mathbf{N}.\mathbf{N} + \lambda\mathbf{r}_1.\mathbf{N} + \mu\mathbf{r}_2.\mathbf{N}$$

or $\quad \kappa_n = \nu,$

where κ_n is the normal curvature of the surface at P in the direction of the curve.

Substituting the above value of κ_n in (1), we get

$$\mathbf{r}'' = \kappa_n\mathbf{N} + \lambda\mathbf{r}_1 + \mu\mathbf{r}_2. \tag{2}$$

The part $\lambda\mathbf{r}_1 + \mu\mathbf{r}_2$ on right hand side of (2) is called the geodesic curvature vector of the curve at P and it is denoted by $\mathbf{k}_g$.

The magnitude of $\mathbf{k}_g$, denoted by κ_g is called the geodesic curvature of the curve at P, and $\mathbf{k}_g$ has the direction making an angle $+\frac{\pi}{2}$ or $-\frac{\pi}{2}$ with the tangent.

Formula for Geodesic Curvature

The geodesic curvature vector of a curve at a point P on a surface is given by

$$\mathbf{k}_g = \lambda \mathbf{r}_1 + \mu \mathbf{r}_2, \tag{3}$$

where the curvature vector $\mathbf{r}''$ of the curve at P is given by (2).

Since $\mathbf{r}_1$ and $\mathbf{r}_2$ are tangential to the surface at P, it follows that $\mathbf{k}_g$ is tangential to the surface at P, i.e.,

$$\mathbf{k}_g \text{ is perpendicular to } \mathbf{N}. \tag{4}$$

Also,

$$\mathbf{k}_g \text{ is perpendicular to } \mathbf{t}. \tag{5}$$

From (4) and (5) it follows that

$$\mathbf{k}_g \text{ is parallel to } \mathbf{N}\times\mathbf{t}. \tag{6}$$

Again, we have

$$\mathbf{N}\times\mathbf{t} = |\mathbf{N}|\,|\mathbf{t}|\sin\left(\pm\frac{\pi}{2}\right)\hat{\mathbf{e}} = (\pm 1)\,\hat{\mathbf{e}}$$

$\Rightarrow \quad |\mathbf{N}\times\mathbf{t}| = 1$

$\Rightarrow \quad \mathbf{N}\times\mathbf{t}$ is a unit vector parallel to $\mathbf{k}_g$. [using (6)]

$$\Rightarrow \quad \mathbf{k}_g = \kappa_g(\mathbf{N}\times\mathbf{t}) = \kappa_g(\mathbf{N}\times\mathbf{r}'). \tag{7}$$

Now, from (2), (3) and (7), we have

$$\mathbf{r}'' = \kappa_n \mathbf{N} + \kappa_g(\mathbf{N}\times\mathbf{r}').$$

Taking dot product of both sides with $\mathbf{N}\times\mathbf{r}'$, we get

$$\mathbf{r}''.(\mathbf{N}\times\mathbf{r}') = \kappa_n \mathbf{N}.(\mathbf{N}\times\mathbf{r}') + \kappa_g(\mathbf{N}\times\mathbf{r}').(\mathbf{N}\times\mathbf{r}')$$

$\Rightarrow \quad [\mathbf{r}'', \mathbf{N}, \mathbf{r}'] = \kappa_n[\mathbf{N}, \mathbf{N}, \mathbf{r}'] + \kappa_g|\mathbf{N}\times\mathbf{r}'|^2$

$\Rightarrow \quad [\mathbf{N}, \mathbf{r}', \mathbf{r}''] = \kappa_n(0) + \kappa_g(1) \qquad [\because [\mathbf{N}, \mathbf{N}, \mathbf{r}'] = 0 \text{ and } |\mathbf{N}\times\mathbf{r}'| = 1]$

$$\Rightarrow \quad \kappa_g = [\mathbf{N}, \mathbf{r}', \mathbf{r}'']. \tag{8}$$

The geodesic curvature of a curve at a point P is given by (8).

Substituting $\mathbf{N} = \dfrac{\mathbf{r}_1\times\mathbf{r}_2}{H}$ in (8), we get

$$\kappa_g = \frac{1}{H}(\mathbf{r}_1\times\mathbf{r}_2).(\mathbf{r}'\times\mathbf{r}'')$$

$$= \frac{1}{H}\begin{vmatrix} \mathbf{r}_1.\mathbf{r}' & \mathbf{r}_2.\mathbf{r}' \\ \mathbf{r}_1.\mathbf{r}'' & \mathbf{r}_2.\mathbf{r}'' \end{vmatrix} \qquad \left[\because (\mathbf{a}\times\mathbf{b}).(\mathbf{c}\times\mathbf{d}) = \begin{vmatrix} \mathbf{a}.\mathbf{c} & \mathbf{b}.\mathbf{c} \\ \mathbf{a}.\mathbf{d} & \mathbf{b}.\mathbf{d} \end{vmatrix}\right]$$

or

$$\kappa_g = \frac{1}{H}[(\mathbf{r}_1.\mathbf{r}')(\mathbf{r}_2.\mathbf{r}'') - (\mathbf{r}_2.\mathbf{r}')(\mathbf{r}_1.\mathbf{r}'')]. \tag{9}$$

Equation (9) gives an alternative expression for the geodesic curvature.

Geodesic Curvature in terms of Normal Angle

We have shown that the geodesic curvature κ_g of a curve at a point P on a surface is given by (8), where the curvature vector $\mathbf{r}''$ of the curve at P is given by (2).

Let ϖ be the normal angle of the curve at P, i.e., the angle between the principal normal $\mathbf{n}$ and the surface normal $\mathbf{N}$. Then, by (8), we have

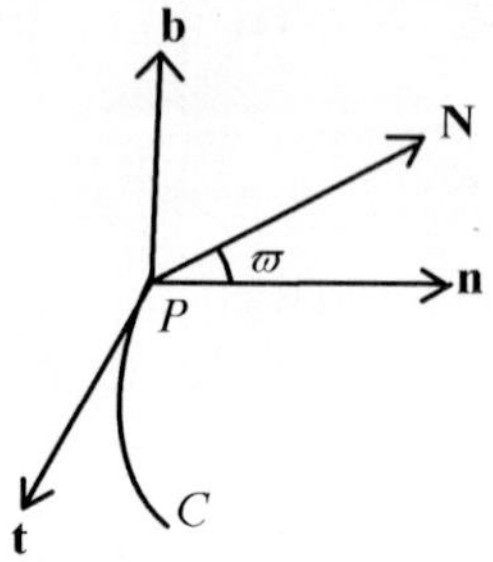

$$\begin{aligned}\kappa_g &= [\mathbf{N},\ \mathbf{r}',\ \mathbf{r}''] \\ &= [\mathbf{N},\ \mathbf{t},\ \kappa\mathbf{n}] \\ &= \kappa[\mathbf{N},\ \mathbf{t},\ \mathbf{n}] \\ &= \kappa\mathbf{N}\,.\,\mathbf{t}\times\mathbf{n} \\ &= \kappa\mathbf{N}\,.\,\mathbf{b} \\ &= \kappa\sin\varpi\end{aligned}$$

i.e., $\kappa_g = \kappa\sin\varpi.$ (10)

Also, from (2) and (3), we have

$$\mathbf{r}'' = \kappa_n\mathbf{N} + \mathbf{k}_g$$

or $\kappa\mathbf{n} = \kappa_n\mathbf{N} + \mathbf{k}_g.$

Taking dot product of both sides with $\mathbf{N}$, we get

$$\kappa\mathbf{n}\,.\,\mathbf{N} = \kappa_n\mathbf{N}\,.\,\mathbf{N} + \mathbf{k}_g\,.\,\mathbf{N}$$

or $\kappa\cos\varpi = \kappa_n + 0$

or $\kappa_n = \kappa\cos\varpi.$ (11)

Squaring and adding (10) and (11), we get

$$\kappa^2 = \kappa_g^2 + \kappa_n^2. \tag{12}$$

Dividing (10) by (11), we get

$$\kappa_g = \kappa_n\tan\varpi. \tag{13}$$

Geodesic Curvature in terms of Gauss Coefficients

We have shown that the geodesic curvature κ_g of a curve at a point P on a surface is given by (9).

Now,

$$\mathbf{r}' = \mathbf{r}_1 u' + \mathbf{r}_2 v'$$

and $\mathbf{r}'' = \mathbf{r}_1 u'' + \mathbf{r}_2 v'' + \mathbf{r}_{11} u'^2 + 2\mathbf{r}_{12} u'v' + \mathbf{r}_{22} v'^2.$

Substituting the values of $\mathbf{r}'$ and $\mathbf{r}''$ in (9), we get

$$\kappa_g = \frac{1}{H}[\{\mathbf{r}_1.(\mathbf{r}_1 u' + \mathbf{r}_2 v')\}\{(\mathbf{r}_2.(\mathbf{r}_1 u'' + \mathbf{r}_2 v'' + \mathbf{r}_{11} u'^2 + 2\mathbf{r}_{12} u'v' + \mathbf{r}_{22} v'^2)\}]$$
$$- \frac{1}{H}[\{\mathbf{r}_2.(\mathbf{r}_1 u' + \mathbf{r}_2 v')\}\{(\mathbf{r}_1.(\mathbf{r}_1 u'' + \mathbf{r}_2 v'' + \mathbf{r}_{11} u'^2 + 2\mathbf{r}_{12} u'v' + \mathbf{r}_{22} v'^2)\}]$$

or
$$\kappa_g = \frac{1}{H}[(Eu' + Fv')(Fu'' + Gv'' + \mathbf{r}_2.\mathbf{r}_{11} u'^2 + 2\mathbf{r}_2.\mathbf{r}_{12} u'v' + \mathbf{r}_2.\mathbf{r}_{22} v'^2)]$$
$$- \frac{1}{H}[(Fu' + Gv')(Eu'' + Fv'' + \mathbf{r}_1.\mathbf{r}_{11} u'^2 + 2\mathbf{r}_1.\mathbf{r}_{12} u'v' + \mathbf{r}_1.\mathbf{r}_{22} v'^2)]. \tag{14}$$

By Gauss's formulae, we have
$$\left.\begin{aligned}\mathbf{r}_{11} &= L\mathbf{N} + l\mathbf{r}_1 + \lambda\mathbf{r}_2,\\ \mathbf{r}_{12} &= M\mathbf{N} + m\mathbf{r}_1 + \mu\mathbf{r}_2,\\ \mathbf{r}_{22} &= N\mathbf{N} + n\mathbf{r}_1 + \nu\mathbf{r}_2.\end{aligned}\right\} \tag{15}$$

Taking the dot product of both sides of the first equation in (15) with $\mathbf{r}_1$ and $\mathbf{r}_2$ successively, we get
$$\left.\begin{aligned}\mathbf{r}_1.\mathbf{r}_{11} &= lE + \lambda F,\\ \mathbf{r}_2.\mathbf{r}_{11} &= lF + \lambda G.\end{aligned}\right\} \tag{16}$$

Similarly, taking the dot product of both sides of the second equation in (15) with $\mathbf{r}_1$ and $\mathbf{r}_2$ successively, we get
$$\left.\begin{aligned}\mathbf{r}_1.\mathbf{r}_{12} &= mE + \mu F,\\ \mathbf{r}_2.\mathbf{r}_{12} &= mF + \mu G.\end{aligned}\right\} \tag{17}$$

Again, taking the dot product of both sides of the third equation in (15) with $\mathbf{r}_1$ and $\mathbf{r}_2$ successively, we get
$$\left.\begin{aligned}\mathbf{r}_1.\mathbf{r}_{22} &= nE + \nu F,\\ \mathbf{r}_2.\mathbf{r}_{22} &= nF + \nu G.\end{aligned}\right\} \tag{18}$$

Using (16), (17) and (18) in (14), we get
$$\kappa_g = \frac{1}{H}[(Eu' + Fv')\{Fu'' + Gv'' + (lF + \lambda G)u'^2 + 2(mF + \mu G)u'v' + (nF + \nu G)v'^2)\}]$$
$$- \frac{1}{H}[(Fu' + Gv')\{Eu'' + Fv'' + (lE + \lambda F)u'^2 + 2(mE + \mu F)u'v' + (nE + \nu F)v'^2\}]$$

or $$\kappa_g = \frac{1}{H}[(EG-F^2)(u'v''+\lambda u'^3+2\mu u'^2v'+\nu u'v'^2]$$
$$-\frac{1}{H}[(EG-F^2)(v'u''+lv'u'^2+2mu'v'^2+nv'^3]$$

or $$\kappa_g = Hu'(v''+\lambda u'^2+2\mu u'v'+\nu v'^2)$$
$$-Hv'(u''+lu'^2+2mu'v'+nv'^2). \tag{19}$$

Liouville's Formula for Geodesic Curvature

If a curve makes an angle θ with the parametric curve $v =$ constant, then its geodesic curvature is given by

$$\kappa_g = \theta' + Pu' + Qv',$$

where

$$P = \frac{2EF_1 - FE_1 - EE_2}{2HE}, \quad Q = \frac{EG_1 - FE_2}{2HE}.$$

Proof. Let θ be the inclination of the given curve C to the parametric curve $v =$ constant and let ϕ be the inclination of the geodesic tangent of C to the parametric curve $v =$ const.

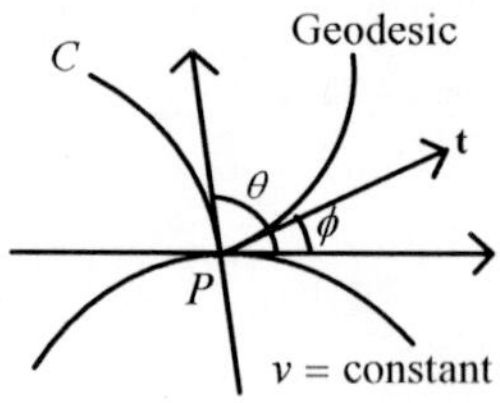

Since the direction coefficients of the geodesic tangent of C are (u', v') and the direction coefficients of the parametric curve $v =$ const. are $\left(\frac{1}{\sqrt{E}}, 0\right)$,

we have

$$\cos\phi = E\frac{1}{\sqrt{E}}u' + F\frac{1}{\sqrt{E}}v'$$

i.e., $$\cos\phi = \frac{1}{\sqrt{E}}(Eu' + Fv'), \tag{1}$$

$$\sin\phi = \frac{H}{\sqrt{E}}v'. \tag{2}$$

From (1), we have

$$Eu' + Fv' = \sqrt{E}\cos\phi.$$

Differentiating this with respect to s, we get

$$\frac{d}{ds}(Eu' + Fv') = \frac{1}{2\sqrt{E}}\frac{dE}{ds}\cos\phi - \sqrt{E}\sin\phi\frac{d\phi}{ds}$$

Using $\dfrac{dE}{ds} = E_1 u' + E_2 v'$ and substituting the values of $\cos\phi$ and $\sin\phi$ from (1) and (2), we get

$$\frac{d}{ds}(Eu' + Fv') = \frac{1}{2E}(E_1 u' + E_2 v')(Eu' + Fv') - Hv'\frac{d\phi}{ds}. \tag{3}$$

Also, for a geodesic, we know that

$$\frac{d}{ds}(Eu' + Fv') = \frac{1}{2}(E_1 u'^2 + 2F_1 u'v' + G_1 v'^2). \tag{4}$$

From (3) and (4), we have

$$\frac{(E_1 u' + E_2 v')(Eu' + Fv')}{2E} - Hv'\frac{d\phi}{ds} = \frac{E_1 u'^2 + 2F_1 u'v' + G_1 v'^2}{2}$$

$$\Rightarrow \qquad Hv'\frac{d\phi}{ds} = \frac{(E_1 u' + E_2 v')(Eu' + Fv')}{2E} - \frac{E_1 u'^2 + 2F_1 u'v' + G_1 v'^2}{2}$$

$$= \frac{(FE_1 + EE_2 - 2EF_1)u'v' + (FE_2 - EG_1)v'^2}{2E}$$

$$= \frac{-[(2EF_1 - FE_1 - EE_2)u' + (EG_1 - FE_2)v']v'}{2E}$$

$$\Rightarrow \qquad \frac{d\phi}{ds} = -\left[\frac{2EF_1 - FE_1 - EE_2}{2HE}u' + \frac{EG_1 - FE_2}{2HE}v'\right]. \tag{5}$$

Now the geodesic curvature of a curve C at a point P is tangential to the surface, and its magnitude is the arc-rate of deviation of C from its geodesic tangent at P. Therefore, we have

$$\kappa_g = \frac{d}{ds}(\theta - \phi) = \frac{d\theta}{ds} - \frac{d\phi}{ds}$$

or $$\kappa_g = \theta' + \frac{2EF_1 - FE_1 - EE_2}{2HE}u' + \frac{EG_1 - FE_2}{2HE}v' \qquad \text{[using (5)]}$$

or $$\kappa_g = \theta' + Pu' + Qv',$$

where

$$P = \frac{2EF_1 - FE_1 - EE_2}{2HE}, \quad Q = \frac{EG_1 - FE_2}{2HE}.$$

Bonnet's Formula for Geodesic Curvature

The geodesic curvature of the curve $\phi(u, v) = \text{constant}$ is given by

$$\kappa_g = \frac{1}{H}\frac{\partial}{\partial u}\left(\frac{F\phi_2 - G\phi_1}{\theta}\right) + \frac{1}{H}\frac{\partial}{\partial v}\left(\frac{F\phi_1 - E\phi_2}{\theta}\right),$$

where $\theta^2 = E\phi_2^{\,2} - 2F\phi_1\phi_2 + G\phi_1^{\,2}$.

Proof. Equation of the given curve is

$$\phi(u, v) = \text{constant}. \tag{1}$$

Differentiating (1) with respect to s, we get

$$\frac{\partial\phi}{\partial u}\frac{du}{ds} + \frac{\partial\phi}{\partial v}\frac{dv}{ds} = 0$$

or $$\phi_1 u' + \phi_2 v' = 0. \tag{2}$$

From (2), we have

$$\frac{u'}{\phi_2} = \frac{v'}{-\phi_1} = \frac{\sqrt{Eu'^2 + 2Fu'v' + Gv'^2}}{\sqrt{E\phi_2^{\,2} - 2F\phi_1\phi_2 + G\phi_1^{\,2}}} = \frac{1}{\theta}$$

or $$\phi_2 = u'\theta,\ \phi_1 = -v'\theta, \tag{3}$$

where

$$\theta^2 = E\phi_2^{\,2} - 2F\phi_1\phi_2 + G\phi_1^{\,2}. \tag{4}$$

Differentiating (2) again with respect to s, we have

$$(\phi_1 u'' + \phi_{11}u'^2 + \phi_{12} u'v') + (\phi_2 v'' + \phi_{21} u'v' + \phi_{22}v'^2) = 0$$

or $$\phi_1 u'' + \phi_2 v'' + \phi_{11}u'^2 + 2\phi_{12} u'v' + \phi_{22}v'^2 = 0$$

or $$(u'v'' - u''v')\theta + \phi_{11}u'^2 + 2\phi_{12} u'v' + \phi_{22}v'^2 = 0. \tag{5}$$

[using (3)]

Now,

$$\frac{\partial}{\partial u}\left(\frac{F\phi_2 - G\phi_1}{\theta}\right) + \frac{\partial}{\partial v}\left(\frac{F\phi_1 - E\phi_2}{\theta}\right)$$

$$= \frac{F_1\phi_2 + F\phi_{12} - G_1\phi_1 - G\phi_{11}}{\theta} - \frac{F\phi_2 - G\phi_1}{\theta^2}\frac{\partial\theta}{\partial u} + \frac{F_2\phi_1 + F\phi_{12} - E_2\phi_2 - E\phi_{22}}{\theta} - \frac{F\phi_1 - E\phi_2}{\theta^2}\frac{\partial\theta}{\partial v}$$

or $$\frac{\partial}{\partial u}\left(\frac{F\phi_2 - G\phi_1}{\theta}\right) + \frac{\partial}{\partial v}\left(\frac{F\phi_1 - E\phi_2}{\theta}\right)$$

$$= (F_1 - E_2)u' + (G_1 - F_2)v' + \frac{2F\phi_{12} - G\phi_{11} - E\phi_{22}}{\theta} - \left(\frac{F\phi_2 - G\phi_1}{\theta^2}\frac{\partial\theta}{\partial u} + \frac{F\phi_1 - E\phi_2}{\theta^2}\frac{\partial\theta}{\partial v}\right). \tag{6}$$

[using (3)]

Differentiating (4) partially with respect to u, we have

$$2\theta\frac{\partial\theta}{\partial u}=E_1\phi_2^{\ 2}+2E\phi_2\phi_{21}-2F_1\phi_1\phi_2-2F\phi_{11}\phi_2$$

$$-2F\phi_1\phi_{21}+G_1\phi_1^{\ 2}+2G\phi_1\phi_{11}$$

$$\Rightarrow\qquad \frac{\partial\theta}{\partial u}=\frac{1}{2\theta}(E_1\phi_2^{\ 2}+2E\phi_2\phi_{21}-2F_1\phi_1\phi_2-2F\phi_{11}\phi_2$$

$$-2F\phi_1\phi_{21}+G_1\phi_1^{\ 2}+2G\phi_1\phi_{11})$$

$$\Rightarrow\qquad \frac{F\phi_2-G\phi_1}{\theta^2}\frac{\partial\theta}{\partial u}=\frac{F\phi_2-G\phi_1}{2\theta^3}(E_1\phi_2^{\ 2}+2E\phi_2\phi_{21}-2F_1\phi_1\phi_2$$

$$-2F\phi_{11}\phi_2-2F\phi_1\phi_{21}+G_1\phi_1^{\ 2}+2G\phi_1\phi_{11}).$$

Substituting the values of ϕ_1, ϕ_2 from (3) in this equation, we get

$$\frac{F\phi_2-G\phi_1}{\theta^2}\frac{\partial\theta}{\partial u}$$

$$=\frac{Fu'+Gv'}{2}(E_1u'^2+2F_1u'v'+G_1v'^2)$$

$$+\frac{Fu'+Gv'}{\theta}[(E\phi_{21}-F\phi_{11})u'+(F\phi_{21}-G\phi_{11})v']. \qquad (7)$$

Again, differentiating (4) partially with respect to v, we have

$$2\theta\frac{\partial\theta}{\partial v}=E_2\phi_2^{\ 2}+2E\phi_2\phi_{22}-2F_2\phi_1\phi_2-2F\phi_{12}\phi_2$$

$$-2F\phi_1\phi_{22}+G_2\phi_1^{\ 2}+2G\phi_1\phi_{12}$$

$$\Rightarrow\qquad \frac{\partial\theta}{\partial v}=\frac{1}{2\theta}(E_2\phi_2^{\ 2}+2E\phi_2\phi_{22}-2F_2\phi_1\phi_2-2F\phi_{12}\phi_2$$

$$-2F\phi_1\phi_{22}+G_2\phi_1^{\ 2}+2G\phi_1\phi_{12})$$

$$\Rightarrow\qquad \frac{F\phi_1-E\phi_2}{\theta^2}\frac{\partial\theta}{\partial v}=\frac{F\phi_1-E\phi_2}{2\theta^3}(E_2\phi_2^{\ 2}+2E\phi_2\phi_{22}-2F_2\phi_1\phi_2$$

$$-2F\phi_{12}\phi_2-2F\phi_1\phi_{22}+G_2\phi_1^{\ 2}+2G\phi_1\phi_{12}).$$

Substituting the values of ϕ_1, ϕ_2 from (3) in this equation, we get

$$\frac{F\phi_1-E\phi_2}{\theta^2}\frac{\partial\theta}{\partial v}$$

$$=-\frac{Eu'+Fv'}{2}(E_2u'^2+2F_2u'v'+G_2v'^2)$$

$$-\frac{Eu'+Fv'}{\theta}[(E\phi_{22}-F\phi_{12})u'+(F\phi_{22}-G\phi_{12})v']. \qquad (8)$$

Adding (7) and (8), we get

$$\frac{F\phi_2 - G\phi_1}{\theta^2}\frac{\partial\theta}{\partial u} + \frac{F\phi_1 - E\phi_2}{\theta^2}\frac{\partial\theta}{\partial v}$$

$$= \frac{u'^3}{2}(FE_1 - EE_2) + \frac{u'^2 v'}{2}(GE_1 + 2FF_1 - FE_2 - 2EF_2)$$

$$+ \frac{u' v'^2}{2}(FG_1 + 2GF_1 - EG_2 - 2FF_2) + \frac{v'^3}{2}(GG_1 - FG_2)$$

$$+ \frac{u'^2}{\theta}(2EF\phi_{12} - F^2\phi_{11} - E^2\phi_{22})$$

$$+ \frac{v'^2}{\theta}(2FG\phi_{12} - G^2\phi_{11} - F^2\phi_{22})$$

$$+ \frac{2u'v'}{\theta}(F^2\phi_{12} + EG\phi_{12} - FG\phi_{11} - EF\phi_{22}). \quad (9)$$

We know that

$$l = \frac{1}{2H^2}(GE_1 - 2FF_1 + FE_2),$$

$$m = \frac{1}{2H^2}(GE_2 - FG_1),$$

$$n = \frac{1}{2H^2}(2GF_2 - GG_1 - FG_2),$$

$$\lambda = \frac{1}{2H^2}(2EF_1 - EE_2 - FE_1),$$

$$\mu = \frac{1}{2H^2}(EG_1 - FE_2),$$

$$\nu = \frac{1}{2H^2}(EG_2 - 2FF_2 + FG_1).$$

Therefore

$$FE_1 - EE_2 = 2(EF_1 - EE_2 - \lambda H^2),$$

$$GE_1 + 2FF_1 - FE_2 - 2EF_2 = 2[2FF_1 - 2FE_2 - EF_2 + EG_1 + (l - 2\mu)H^2],$$

$$FG_1 + 2GF_1 - EG_2 - 2FF_2 = 2[2FG_1 + GF_1 - 2FF_2 - GE_2 + (2m - \nu)H^2],$$

$$GG_1 - FG_2 = 2(GG_1 - GF_2 + nH^2).$$

Substituting the above values in (9), we get

$$\frac{F\phi_2 - G\phi_1}{\theta^2}\frac{\partial\theta}{\partial u} + \frac{F\phi_1 - E\phi_2}{\theta^2}\frac{\partial\theta}{\partial v}$$

$$= u'^3\,(EF_1 - EE_2 - \lambda H^2)$$
$$+ u'^2v'[2FF_1 - 2FE_2 - EF_2 + EG_1 + (l-2\mu)H^2]$$
$$+ u'v'^2[2FG_1 + GF_1 - 2FF_2 - GE_2 + (2m-\nu)H^2]$$
$$+ v'^3(GG_1 - GF_2 + nH^2) - \frac{F^2}{\theta}(\phi_{11}u'^2 - 2\phi_{12}u'v' + \phi_{22}v'^2)$$
$$- \frac{G\phi_{11}}{\theta}(2Fu'v' + Gv'^2) + \frac{2F\phi_{12}}{\theta}(Eu'^2 + Gv'^2)$$
$$- \frac{E\phi_{22}}{\theta}(Eu'^2 + 2Fu'v') + \frac{2EG\,\phi_{12}}{\theta}u'v'$$

$$= -H^2[\lambda u'^3 - (l-2\mu)\,u'^2v' - (2m-\nu)\,u'v'^2 - nv'^3]$$
$$+ F_1u'(Eu'^2 + 2Fu'v' + Gv'^2) - E_2u'(Eu'^2 + 2Fu'v' + Gv'^2)$$
$$+ G_1v'(Eu'^2 + 2Fu'v' + Gv'^2) - F_2v'(Eu'^2 + 2Fu'v' + Gv'^2)$$
$$- \frac{F^2}{\theta}(\phi_{11}u'^2 + 2\phi_{12}u'v' + \phi_{22}v'^2) - \frac{G\phi_{11}}{\theta}(2Fu'v' + Gv'^2)$$
$$+ \frac{2F\phi_{12}}{\theta}(Eu'^2 + 2Fu'v' + Gv'^2) - \frac{E\phi_{22}}{\theta}(Eu'^2 + 2Fu'v')$$
$$+ \frac{2EG\,\phi_{12}}{\theta}u'v'$$

$$= -H^2[\lambda u'^3 - (l-2\mu)\,u'^2v' - (2m-\nu)\,u'v'^2 - nv'^3]$$
$$+ F_1u' - E_2u' + G_1v' - F_2v' - \frac{F^2}{\theta}(\phi_{11}u'^2 + 2\phi_{12}u'v' + \phi_{22}v'^2)$$
$$- \frac{G\phi_{11}}{\theta}(1 - Eu'^2) + \frac{2F\phi_{12}}{\theta} - \frac{E\phi_{22}}{\theta}(1 - Gv'^2) + \frac{2EG\,\phi_{12}}{\theta}u'v'$$
$$[\because Eu'^2 + 2Fu'v' + Gv'^2 = 1]$$

$$= -H^2[\lambda u'^3 - (l-2\mu)\,u'^2v' - (2m-\nu)\,u'v'^2 - nv'^3]$$
$$+ (F_1 - E_2)\,u' + (G_1 - F_2)\,v' - \frac{F^2}{\theta}(\phi_{11}u'^2 + 2\phi_{12}u'v' + \phi_{22}v'^2)$$
$$+ \frac{EG}{\theta}(\phi_{11}u'^2 + 2\phi_{12}u'v' + \phi_{22}v'^2) + \frac{2F\phi_{12} - G\phi_{11} - E\phi_{22}}{\theta}$$

$$= -H^2[\lambda u'^3 - (l-2\mu)u'^2v' - (2m-\nu)u'v'^2 - nv'^3]$$
$$+ (F_1 - E_2)u' + (G_1 - F_2)v' + \frac{2F\phi_{12} - G\phi_{11} - E\phi_{22}}{\theta}$$
$$+ \frac{EG - F^2}{\theta}(\phi_{11}u'^2 + 2\phi_{12}u'v' + \phi_{22}v'^2)$$
$$= -H^2[\lambda u'^3 - (l-2\mu)u'^2v' - (2m-\nu)u'v'^2 - nv'^3]$$
$$+ (F_1 - E_2)u' + (G_1 - F_2)v' + \frac{2F\phi_{12} - G\phi_{11} - E\phi_{22}}{\theta}$$
$$- H^2[u'v'' - v'u'') \qquad \text{[using (5)]}$$
$$= -H^2u'(v'' + \lambda u'^2 + 2\mu u'v' + \nu v'^2)$$
$$+ H^2v'(u'' + lu'^2 + 2mu'v' + nv'^2)$$
$$+ (F_1 - E_2)u' + (G_1 - F_2)v' + \frac{2F\phi_{12} - G\phi_{11} - E\phi_{22}}{\theta}$$
$$= -H[Hu'(v'' + \lambda u'^2 + 2\mu u'v' + \nu v'^2)$$
$$- Hv'(u'' + lu'^2 + 2mu'v' + nv'^2)]$$
$$+ (F_1 - E_2)u' + (G_1 - F_2)v' + \frac{2F\phi_{12} - G\phi_{11} - E\phi_{22}}{\theta}$$
$$= -H\kappa_g + (F_1 - E_2)u' + (G_1 - F_2)v' + \frac{2F\phi_{12} - G\phi_{11} - E\phi_{22}}{\theta}$$

i.e.,
$$\frac{F\phi_2 - G\phi_1}{\theta^2}\frac{\partial\theta}{\partial u} + \frac{F\phi_1 - E\phi_2}{\theta^2}\frac{\partial\theta}{\partial v}$$
$$= -H\kappa_g + (F_1 - E_2)u' + (G_1 - F_2)v' + \frac{2F\phi_{12} - G\phi_{11} - E\phi_{22}}{\theta}. \tag{10}$$

Using (10), (6) gives

$$\frac{\partial}{\partial u}\left(\frac{F\phi_2 - G\phi_1}{\theta}\right) + \frac{\partial}{\partial v}\left(\frac{F\phi_1 - E\phi_2}{\theta}\right) = H\kappa_g$$

or
$$\kappa_g = \frac{1}{H}\frac{\partial}{\partial u}\left(\frac{F\phi_2 - G\phi_1}{\theta}\right) + \frac{1}{H}\frac{\partial}{\partial v}\left(\frac{F\phi_1 - E\phi_2}{\theta}\right).$$

Hence the result.

5.11 GAUSS-BONNET THEOREM

If a curve Γ encloses a simply connected region R and consists of n arcs $A_0A_1, A_1A_2, \ldots, A_{n-1}A_n (A_n = A_0)$, making exterior angles $\alpha_1, \alpha_2, \ldots, \alpha_n$ at the vertices, then

$$2\pi - \sum_{i=1}^{n} \alpha_i - \int_{\Gamma} \kappa_g dS = \int_{R} K\, dS,$$

where κ_g is the geodesic curvature of the arcs, dS is the element of area of the surface and K is Gaussian curvature of the surface.

Proof. Consider a surface of class 3 whose equation is given by

$$\mathbf{r} = \mathbf{r}\,(u, v).$$

Let a closed curve Γ be the boundary of a simply connected region R of the surface (A region R is called simply connected if every closed curve lying in R can be contracted continuously into a point without leaving R). Let the curve Γ consist of n smooth arcs

$$A_0 A_1,\ A_1 A_2,\ \ldots,\ A_{n-1} A_n (A_n = A_0),$$

where n is finite and each arc is of class 2.

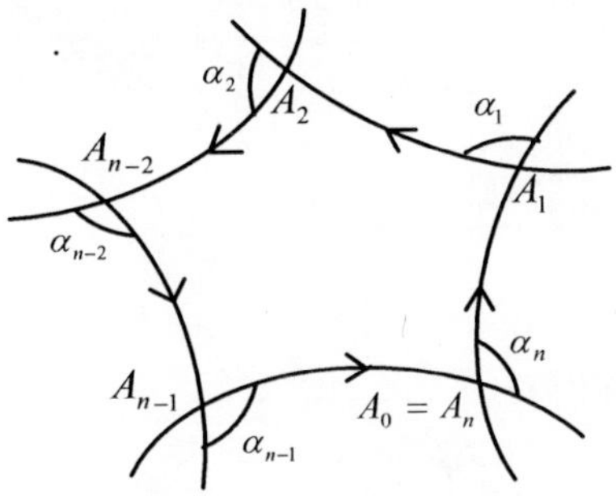

Let α_i be the angle between the tangents to the arcs $A_{i-1}A_i$ and A_iA_{i+1} at the vertex A_i ($i = 1, 2, \ldots, n$) so that $-\pi < \alpha_i < \pi$. At A_n, α_n is the angle between the tangents to the arcs $A_{n-1}A_n$ and A_nA_1.

Then Γ can be regarded as a curvilinear polygon and $\alpha_1, \alpha_2, \ldots, \alpha_n$ as the exterior angles at the vertices $A_1, A_2, \ldots, A_n$.

The geodesic curvature κ_g exists at every point of Γ except possibly at the vertices of the polygon.

From Liouville's formula for geodesic curvature, we have

$$\int_{\Gamma} \kappa_g dS = \int_{\Gamma} (d\theta + P\, du + Q\, dv), \tag{1}$$

where θ is the angle which Γ makes with the parametric curve $v =$ constant and P, Q are functions of u, v given by

$$P = \frac{2EF_1 - FE_1 - EE_2}{2HE},\quad Q = \frac{EG_1 - FE_2}{2HE}.$$

Since the curves $v = \text{constant}$ form a family in the region R enclosed by Γ, the tangent to Γ turns through 2π relatives to these curves, i.e., we have

$$\int_{\Gamma} d\theta + \sum_{i=1}^{n} \alpha_i = 2\pi. \tag{2}$$

Eliminating $\int_{\Gamma} d\theta$ from (1) and (2), we get

$$2\pi - \sum_{i=1}^{n} \alpha_i - \int_{\Gamma} \kappa_g dS = -\int_{C} (P\,du + Q\,dv). \tag{3}$$

Since R is a simply connected region and P, Q are differentiable functions of u, v, we have, by Green's theorem,

$$\int_{\Gamma} (P\,du + Q\,dv) = \iint_{R} \left(\frac{\partial Q}{\partial u} - \frac{\partial P}{\partial v} \right) du\,dv$$

or $$\int_{\Gamma} (P\,du + Q\,dv) = \iint_{R} \left[\frac{1}{H} \left(\frac{\partial Q}{\partial u} - \frac{\partial P}{\partial v} \right) \right] H\,du\,dv.$$

or $$\int_{\Gamma} (P\,du + Q\,dv) = -\int_{R} K dS, \tag{4}$$

where $dS = H\,du\,dv$ is the element of area of the surface and K is a function of u, v independent of the curve Γ, given by

$$K(u, v) = -\frac{1}{H} \left(\frac{\partial Q}{\partial u} - \frac{\partial P}{\partial v} \right). \tag{5}$$

From (3) and (4), we have

$$2\pi - \sum_{i=1}^{n} \alpha_i - \int_{\Gamma} \kappa_g dS = \int_{R} K\,dS. \tag{6}$$

Now, we shall prove that the function $K(u, v)$ given by (5) is determined uniquely.

Suppose, if possible, K_1 is another function of u, v satisfying (6) and independent of the curve Γ. Then, we have

$$2\pi - \sum_{i=1}^{n} \alpha_i - \int_{\Gamma} \kappa_g dS = \int_{R} K_1\,dS. \tag{7}$$

From (6) and (7), we have

$$\int_{R} (K_1 - K)\,dS = 0 \tag{8}$$

for every region R.

Now, let $K_1 \neq K$ at some point P and let $K_1 > K$ at P. Then, since $K_1 - K$ is continuous, there exists a region R containing P such that $K_1 - K > 0$ at every point of that region. Hence for this region R, we have

$$\int_R (K_1 - K)\, dS > 0.$$

But, this contradicts (8). Therefore the possibility $K_1 > K$ at P is ruled out. A similar contradiction exists if we take $K_1 < K$ at P. Therefore the possibility $K_1 < K$ at P is also ruled out.

Hence $K_1 = K$ at each point, i.e., K is determined uniquely.

Remark. The expression on left hand side of (6) is called the excess of Γ, denoted by $ex\,\Gamma$, and the expression on right hand side of (6) is called the total curvature of R. Hence Gauss-Bonnet Theorem can also be stated as follows.

For any curve Γ which encloses a simply connected region R, the excess of Γ is equal to the total curvature of R.

Corollary 1. The total curvature of a geodesic triangle is equal to the excess of the sum of the angles of the triangle over two right angles. In general, the total curvature of a geodesic polygon consisting of n arcs is equal to the excess of the sum of the interior angles over $(n-2)\,\pi$.

Proof. Let ABC be a geodesic triangle consisting of geodesic arcs AB, BC, CA and enclosing a simply connected region R.

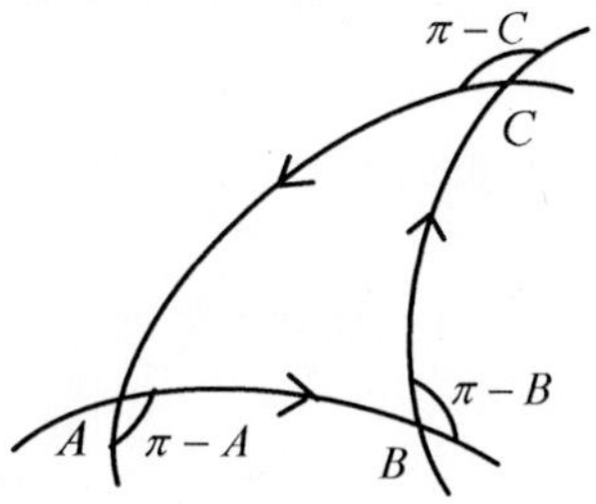

Then the angles between the tangents to the arcs at the vertices A, B and C are $\pi - A$, $\pi - B$ and $\pi - C$ respectively.

Also, we know that the geodesic curvature is zero for a geodesic. Therefore

$$\kappa_g = 0$$

$$\Rightarrow \quad \int_\Gamma \kappa_g\, dS = 0.$$

Hence, by Gauss-Bonnet theorem given by (6), we have

$$\int_R K\, dS = 2\pi - (\pi - A + \pi - B + \pi - C)$$

or

$$\int_R K\, dS = A + B + C - \pi.$$

Similarly, for a geodesic polygon consisting of n arcs,

$$\int_R K\, dS = \text{sum of all interior angles} - (n-2)\,\pi.$$

Corollary 2. When the parametric curves are orthogonal, the formula for the Gaussian curvature K reduces to

$$K = -\frac{1}{2\sqrt{EG}}\left\{\frac{\partial}{\partial u}\left(\frac{G_1}{\sqrt{EG}}\right)+\frac{\partial}{\partial v}\left(\frac{E_2}{\sqrt{EG}}\right)\right\}.$$

Proof. We know that

$$P = \frac{2EF_1 - FE_2 - EE_2}{2HE},\ Q = \frac{EG_1 - FE_2}{2HE}.$$

Substituting the above values of P and Q in (5), we get

$$K = -\frac{1}{H}\left\{\frac{\partial}{\partial u}\left(\frac{EG_1 - FE_2}{2HE}\right)-\frac{\partial}{\partial v}\left(\frac{2EF_1 - FE_1 - EE_2}{2HE}\right)\right\}.$$

When the parametric curves are orthogonal, $F = 0$ and $H = \sqrt{(EG)}$. Therefore, we have

$$K = -\frac{1}{2\sqrt{EG}}\left\{\frac{\partial}{\partial u}\left(\frac{G_1}{\sqrt{EG}}\right)+\frac{\partial}{\partial v}\left(\frac{E_2}{\sqrt{EG}}\right)\right\}.$$

5.12 GEODESIC PARALLELS

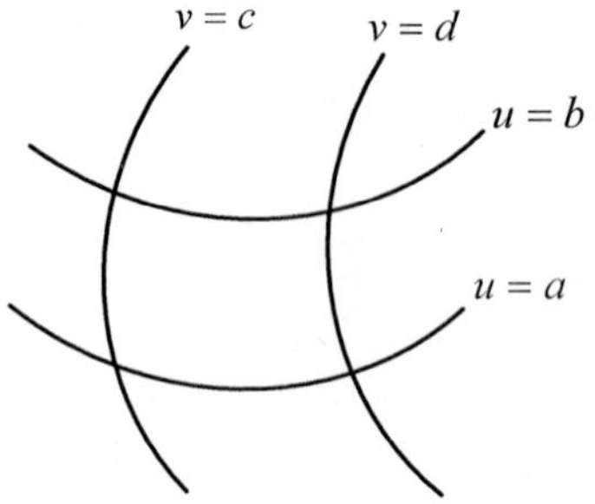

Let a family of geodesics on a surface be taken as parametric curves $v = \text{const.}$ and let their orthogonal trajectories be taken as the curves $u = \text{const.}$ Then $F = 0$. Also, since the parametric curves $v = \text{const.}$ are geodesics, E is a function of u only.

Therefore the square of the linear element is of the form

$$ds^2 = E\,(u)\,du^2 + G\,(u,v)\,dv^2. \tag{1}$$

Now we consider the distance between any two orthogonal trajectories, say $u = a, v = b$, measured along the geodesic $v = c$. Along $v = c$, we have

$$dv = 0$$

$$\Rightarrow \quad ds = \sqrt{E\,(u)}\,du. \qquad \text{[using (1)]}$$

Hence the distance between the curves $u = a$ and $v = b$ measured along the geodesic $v = c$ is given by

$$s = \int_a^b \sqrt{E(u)}\, du.$$

As the distance s given above is independent of c, it follows that it is the same along all geodesics of the family $v = \text{const.}$, and is called the geodesic distance between the two curves. On account of this property, the orthogonal trajectories $u = \text{const.}$ are called geodesic parallels.

Again, since E is a function of u only, we can take $\int \sqrt{E(u)}\, du$ as a new parameter u. Then (1) reduces to

$$ds^2 = du^2 + G(u, v)\, dv^2, \tag{2}$$

where the new parameter u measures the arc length along the geodesics measured from some fixed parallel to the parallel determined by u.

The parameters u and v are called geodesic parameters and (2) is called geodesic form for ds^2.

Deduction. The length of the arc joining two points of the geodesic is least.

Proof. Consider two points P, Q in which a geodesic is cut by the parallels $u = a, v = b$. For any other curve joining the points P, Q, the length of the arc is

$$\int_P^Q ds = \int_P^Q \sqrt{du^2 + G\, dv^2} > \int_P^Q du,$$

where $\int_P^Q du$ is the length of the arc along geodesics in terms of new parameter. Thus the distance is the least in the case of the geodesic.

5.13 GEODESIC POLAR COORDINATES

Let O be an arbitrary point, called the pole on a surface S.

We consider the family of geodesics $v = \text{const.}$ starting from O in all directions, and their orthogonal trajectories $u = \text{const.}$, called geodesic parallels.

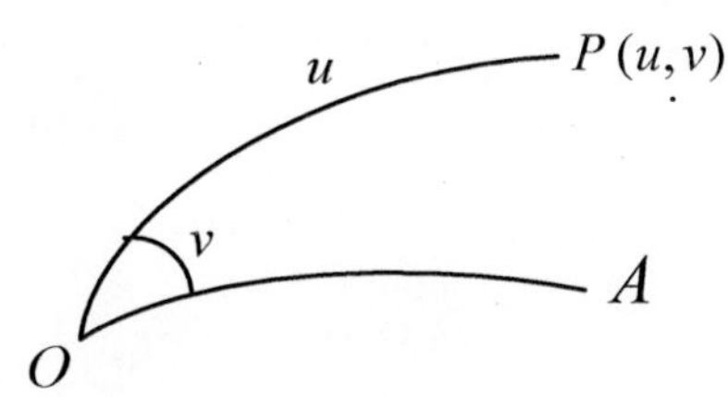

The geodesic parallels can be called geodesic circles when the geodesics are concurrent.

We can choose the parameter u so that relation (2) of Section 5.12 holds, but now u measures the arc length from pole along $v = \text{const}$.

Suppose v is the inclination of the geodesic at O to a fixed geodesic of reference OA.

Then the position of any point P on the surface is determined by the geodesic through O on which it lies, and its distance u from O along that geodesic.

These parameters u, v are called the geodesic polar coordinates of P.

They are analogous to polar coordinates in the plane.

Angle between a Curve on a Surface and the Geodesic through the Pole

Let C be a curve drawn on the surface and let P and Q be two consecutive points (u, v) and $(u + \delta u, v + \delta v)$. Then δv is the angle at O between the geodesics OP and OQ. Let PN be an element of the geodesic circle through P, cutting OQ at N. Then, we have $ON = OP$ and $NQ = \delta u$. Also, $\angle PNQ = \pi / 2$.

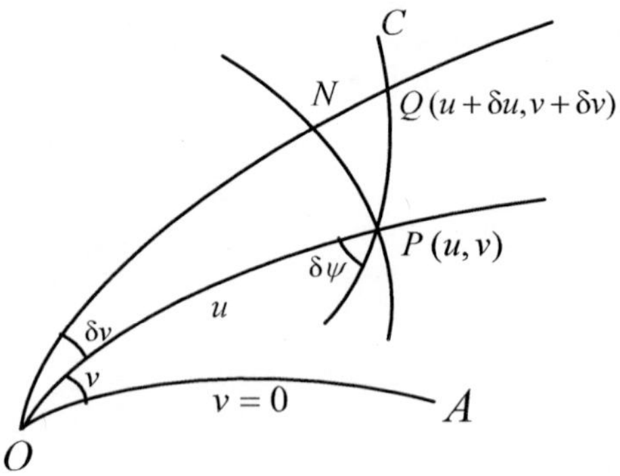

Therefore,

$$PN^2 + NQ^2 = PQ^2$$

or $$PN^2 + \delta u^2 = \delta s^2$$

or $$PN^2 + \delta u^2 = \delta u^2 + G\,\delta v^2. \qquad [\because \delta s^2 = \delta u^2 + G\,\delta v^2]$$

$\therefore$ $$PN^2 = G\,\delta v^2$$

i.e., $$PN = \sqrt{G}\,\delta v.$$

Hence if ψ is the angle between the curve and the geodesic at P, then

$$\sin\psi = \sin\left(\lim_{\delta u \to 0} \angle PQN\right) = \lim_{\delta u \to 0} (\sin \angle PQN)$$

$$= \lim_{\delta u \to 0} \frac{PN}{PQ} = \lim_{\delta u \to 0} \frac{\sqrt{G}\,\delta v}{\delta s} = \sqrt{G}\frac{dv}{ds}.$$

Similarly,

$$\cos\psi = \frac{du}{ds} \text{ and } \tan\psi = \sqrt{G}\frac{dv}{du}.$$

Since G is positive, we may replace it by D^2. Then, we have

$$\sin\psi = D\frac{dv}{ds}, \quad \cos\psi = \frac{du}{ds} \text{ and } \tan\psi = D\frac{dv}{du}.$$

Remark. The area of the element of the surface enclosed by the geodesics $v, v+\delta v$ and the geodesic circles $u, u+\delta u$ is given by

$$\delta S = \sqrt{G}\,\delta u\,\delta v = D\,\delta u\,\delta v.$$

Corollary. When we use geodesic polar parameters, the formula for the Gaussian curvature K reduces to

$$K = -\frac{1}{\sqrt{G}}\frac{\partial^2}{\partial u^2}(\sqrt{G}).$$

Proof. We know that when the parametric curves are orthogonal, the formula for the Gaussian curvature K reduces to

$$K = -\frac{1}{2\sqrt{EG}}\left\{\frac{\partial}{\partial u}\left(\frac{G_1}{\sqrt{EG}}\right) + \frac{\partial}{\partial v}\left(\frac{E_2}{\sqrt{EG}}\right)\right\}.$$

For geodesic polar parameters, we have $F = 0, E = 1$ so that $E_2 = 0$. Therefore, the formula for K takes the form

$$K = -\frac{1}{2\sqrt{G}}\left\{\frac{\partial}{\partial u}\left(\frac{G_1}{\sqrt{G}}\right)\right\}$$

$$= -\frac{1}{2\sqrt{G}}\frac{\partial}{\partial u}\left(\frac{1}{\sqrt{G}}\frac{\partial G}{\partial u}\right)$$

$$= -\frac{1}{2\sqrt{G}}\frac{\partial}{\partial u}\left(\frac{1}{\sqrt{G}}\frac{\partial}{\partial u}(\sqrt{G})^2\right)$$

$$= -\frac{1}{2\sqrt{G}}\frac{\partial}{\partial u}\left(\frac{1}{\sqrt{G}}2\sqrt{G}\frac{\partial}{\partial u}\sqrt{G}\right)$$

$$= -\frac{1}{\sqrt{G}}\frac{\partial^2}{\partial u^2}(\sqrt{G}).$$

Solved Examples

Example 1. If two families of geodesics of a surface intersect at a constant angle, prove that the surface has zero Gaussian curvature.

Solution. Let C be a quadrilateral made up of geodesics on a surface and let P be an arbitrary point in its interior. Let R be the region enclosed by this quadrilateral.

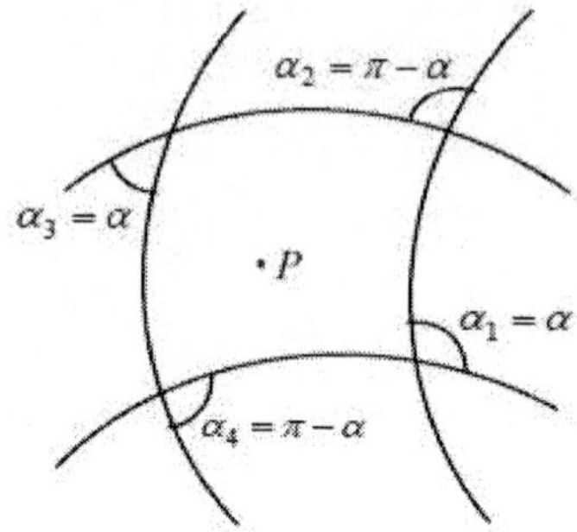

Applying Gauss-Bonnet theorem for this quadrilateral C, we have

$$\int_R K\, dS = 2\pi - \sum_{i=1}^{n} \alpha_i - \int_C \kappa_g dS. \tag{1}$$

Since each side of the quadrilateral C is a geodesic, the geodesic curvature κ_g is zero at each point of C. Therefore,

$$\int_C \kappa_g dS = 0. \tag{2}$$

Since the geodesics intersect at a constant angle, say α, we have

$$\sum_{i=1}^{n} \alpha_i = 2\pi. \tag{3}$$

Using (2) and (3), (1) gives

$$\int_R K\, dS = 0. \tag{4}$$

Since R is at our choice, it can be chosen arbitrarily small. Therefore considering R infinitesimally small, it follows from (4) that K at P is zero. But P is an arbitrary point on the surface. Hence the surface has zero Gaussian curvature.

Example 2. Prove that on a surface of negative curvature, two geodesics cannot meet in two points and enclose a simply connected area.

Solution. We shall prove the result by contradiction. If possible, let two geodesics C_1 and C_2 meet in two points A and B and enclose a simply connected area of the given surface of negative curvature.

Let any other geodesic C cut C_1 and C_2 at the points P and Q respectively. If $\alpha_1, \alpha_2, \alpha_3$ and α_4 are the angles as shown in the figure, we have

$$\alpha_1 + \alpha_2 + \alpha_3 + \alpha_4 = 2\pi. \tag{1}$$

We know that the total curvature of the geodesic triangle AQP is

$$\beta + \alpha_1 + \alpha_2 - \pi$$

and the total curvature of the geodesic triangle QBP is

$$\alpha_4 + \gamma + \alpha_3 - \pi.$$

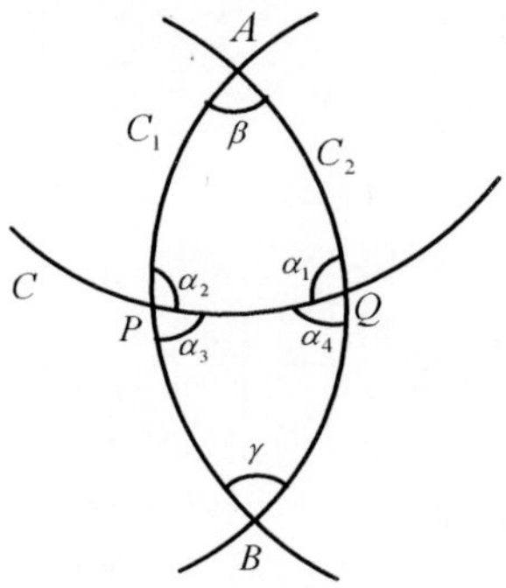

Therefore the total curvature of the area R of the surface is given by

$$\int_R K\,dS = (\beta + \alpha_1 + \alpha_2 - \pi) + (\alpha_4 + \gamma + \alpha_3 - \pi)$$

$$= (\beta + \gamma) + (\alpha_1 + \alpha_2 + \alpha_3 + \alpha_4) - 2\pi$$

$$= \beta + \gamma \qquad \text{[using (1)]}$$

= a positive quantity.

Hence the Gaussian curvature K of the surface is positive at every point of the area $AQBPA$. But this contradicts our assumption.

Hence our supposition that two geodesics C_1 and C_2 meet in two points A and B on a surface of negative curvature is wrong.

Therefore, on a surface of negative curvature, two geodesics cannot meet in two points and enclose a simply connected area.

Exercises

1. State and prove the Gauss-Bonnet Theorem.
2. Prove that the total curvature of a geodesic triangle ABC is $A + B + C - \pi$.

5.14 MINDING THEOREM

Two surfaces of the same constant Gaussian curvature are locally isometric.

Proof. Here, the term 'surface' means a sufficiently small region.

Let S be a surface with constant Gaussian curvature K_0. We shall prove the theorem by proving the following results.

(i) If $K_0 = 0$, then S is isometric with a plane.

(ii) If $K_0 = \frac{1}{a^2}$, then S is isometric with a sphere of radius a.

(iii) If $K_0 = -\frac{1}{a^2}$, then S is isometric with a pseudo-sphere (a certain surface of revolution).

Let P be any point on the surface S of constant Gaussian curvature K_0 and let C be any geodesic through P. We choose a parametric system in the neighbourhood of P as follows :

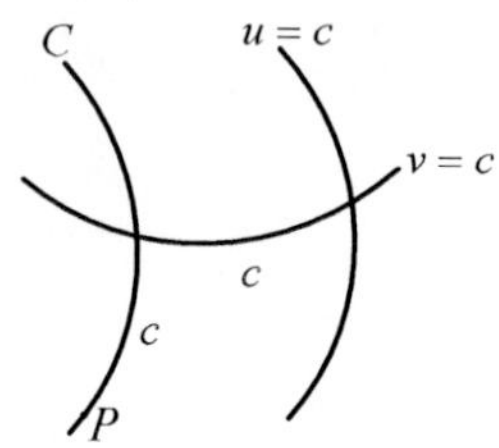

Let the geodesics orthogonal to C and their orthogonal trajectories be taken as the parametric curves. Let the equation of the geodesic orthogonal to C be $v = c$ and let its distance from P measured along C be c. Also, let the orthogonal trajectory of this geodesic be $u = c$ and let its distance from the parallel C measured along the geodesic $v = c$ be c. Then the metric of the surface S is given by

$$ds^2 = du^2 + D^2 dv^2, \tag{1}$$

where $D^2 = G(u, v)$.

We know that a necessary and sufficient condition for the parametric curves $u = c$ to be geodesics is

$$GG_1 + FG_2 - 2GF_2 = 0. \tag{2}$$

In our case, the parametric curve $u = 0$ is a geodesic. Also, $F = 0$.

Hence, by (2), we have

$$GG_1 = 0$$

or $\quad G_1 = 0 \qquad [\because G \neq 0]$

or $\quad \frac{\partial G}{\partial u} = 0$

or $\quad \frac{\partial D^2}{\partial u} = 0$

or $\quad 2D\frac{\partial D}{\partial u} = 0$

or $\qquad \dfrac{\partial D}{\partial u} = 0.$

Also, since v is the arcual distance along C, we have

$$ds = dv,$$

when $u = 0$. Hence, from (1), we have

$$dv^2 = 0 + D^2 dv^2 \text{ when } u = 0.$$

$\Rightarrow \qquad D = 1$ when $u = 0$.

Thus, we have

$$(D)_{u=0} = 1,\ (D_1)_{u=0} = 0. \tag{3}$$

Now, we know that when we use geodesic polar parameters, the formula for the Gaussian curvature K reduces to

$$K = -\frac{1}{\sqrt{G}} \frac{\partial^2}{\partial u^2}(\sqrt{G}) = -\frac{1}{D} \frac{\partial^2 D}{\partial u^2}.$$

Therefore, in our case,

$$\frac{\partial^2 D}{\partial u^2} + K_0 D = 0. \tag{4}$$

Case (i). Let $K_0 = 0$.

In this case, (4) reduces to

$$\frac{\partial^2 D}{\partial u^2} = 0.$$

Integrating, we get

$$D_1 = \frac{\partial D}{\partial u} = \lambda(v)$$

$\Rightarrow \qquad D_1 = \dfrac{\partial D}{\partial u} = 0. \qquad [\because (D_1)_{u=0} = 0, \text{ by } (3)]$

Integrating this again, we get

$$D = f(v)$$

$\Rightarrow \qquad D = 1. \qquad [\because (D)_{u=0} = 1, \text{ by } (3)]$

Therefore the metric (1) of the surface S reduces to

$$ds^2 = du^2 + dv^2,$$

which is the metric of a plane with (u, v) as Cartesian coordinates. Hence the surface S in the neighbourhood of P is mapped isometrically onto a region of the plane.

Case (ii). Let $K_0 = \dfrac{1}{a^2}$.

In this case, (4) reduces to

$$\frac{\partial^2 D}{\partial u^2}+\frac{D}{a^2}=0.$$

The general solution of this differential equation is

$$D(u,v)=A(v)\cos\frac{u}{a}+B(v)\sin\frac{u}{a}. \tag{5}$$

Differentiating (5) partially with respect to u, we get

$$D_1(u,v)=-\frac{A(v)}{a}\sin\frac{u}{a}+\frac{B(v)}{a}\cos\frac{u}{a}. \tag{6}$$

Using (3), (5) and (6) give

$$A(v)=1, B(v)=0.$$

Substituting the values of $A(v)$ and $B(v)$ in (5), we have

$$D(u,v)=\cos\frac{u}{a}.$$

Therefore the metric (1) of the surface S reduces to

$$ds^2=du^2+\cos^2\frac{u}{a}dv^2,$$

which is the metric of a sphere of radius a. Hence the surface S in the neighbourhood of P is mapped isometrically onto a region of a sphere of radius a.

Case (iii). Let $K_0=-\dfrac{1}{a^2}$.

In this case, (4) reduces to

$$\frac{\partial^2 D}{\partial u^2}-\frac{D}{a^2}=0.$$

The general solution of this differential equation is

$$D(u,v)=A(v)\cosh\frac{u}{a}+B(v)\sin\frac{u}{a}. \tag{7}$$

Differentiating (7) partially with respect to u, we get

$$D_1(u,v)=\frac{A(v)}{a}\sinh\frac{u}{a}+\frac{B(v)}{a}\cosh\frac{u}{a}. \tag{8}$$

Using (3), (5) and (6) give

$$A(v)=1, B(v)=0.$$

Substituting the values of $A(v)$ and $B(v)$ in (5), we have

$$D(u,v)=\cosh\frac{u}{a}.$$

Therefore the metric (1) of the surface S reduces to

$$ds^2 = du^2 + \cosh^2 \frac{u}{a} dv^2. \tag{9}$$

In the above metric, E, F, G are functions of u only. Therefore the surface S in the neighbourhood of P is mapped isometrically on a certain surface of revolution.

If we put $u = a\bar{u}, v = a\bar{v}$, then the metric (9) reduces to

$$ds^2 = a^2 (d\bar{u}^2 + \cosh^2 u \, d\bar{v}^2).$$

This is the metric of the surface obtained by revolving the curve

$$x = a \cosh \bar{u}, y = 0, z = a \int_0^{\bar{u}} \sqrt{1 - \sinh^2 \theta} \, d\theta \qquad \left[|\bar{u}| < \log(1 + \sqrt{2}) \right]$$

about z-axis.

This completes the proof of the theorem.

5.15 MAPPING OF SURFACES

Two surfaces S and S^* are said to be mapped onto one another if there exists a one-to-one correspondence between the points of the two surfaces. Let the parametric equations of S and S^* be given by

$$\mathbf{r} = \mathbf{r}(u, v), \mathbf{r}^* = \mathbf{r}(u^*, v^*)$$

respectively. If the mapping of S onto S^* is defined by

$$u^* = \phi(u, v), v^* = \psi(u, v),$$

where ϕ and ψ are analytic functions of u, v, then the corresponding points of S and S^* have the same curvilinear coordinates, and the parametric equations of S and S^* can be written as

$$\mathbf{r} = \mathbf{r}(u, v), \mathbf{r}^* = \mathbf{r}^*(u, v). \tag{1}$$

Therefore the mapping between S and S^* can be defined by (1).

1. Conformal Mapping

Definition. Let S and S^* be two surfaces. If there exists a differentiable homeomorphism, i.e., an analytic one-one mapping of S onto S^* such that the angle between any two curves at an arbitrary point P on S is equal to the angle between the corresponding curves at the corresponding point P^* on S^*, then S is said to be mapped conformally onto S^*. In this case, the mapping is called angle-preserving or a conformal mapping.

Theorem 1. Let S and S^* be two surfaces and let a mapping of S onto S^* be defined by

$$\mathbf{r} = \mathbf{r}(u, v), \mathbf{r}^* = \mathbf{r}^*(u, v).$$

Then a necessary and sufficient condition for this mapping to be conformal is that

$$\frac{E}{E^*} = \frac{F}{F^*} = \frac{G}{G^*} = \left(\frac{ds}{ds^*}\right)^2,$$

where the symbols have their usual meaning.

Proof. Let a surface S be mapped onto a surface S^* by the mapping

$$\mathbf{r} = \mathbf{r}(u, v), \mathbf{r}^* = \mathbf{r}^*(u, v). \tag{1}$$

Let P and P^* be any two corresponding points on the surfaces S and S^* respectively. We suppose that the points P and P^* have the same parametric values. Then, we have

$$ds^2 = E\,du^2 + 2F\,du\,dv + G\,dv^2, \tag{2}$$

$$ds^{*2} = E^*\,du^2 + 2F^*\,du\,dv + G^*\,dv^2. \tag{3}$$

Let (l_1, m_1), (l_2, m_2) be the direction coefficients of any two directions through P and let (l_1^*, m_1^*), (l_2^*, m_2^*) be the direction coefficients of the corresponding two directions through P^*. Then

$$\cos\theta = El_1l_2 + F(l_1m_2 + l_2m_1) + Gm_1m_2, \tag{4}$$

$$\cos\theta^* = E^*l_1^*l_2^* + F^*(l_1^*m_2^* + l_2^*m_1^*) + G^*m_1^*m_2^*. \tag{5}$$

Now,

$$l_1^* = \frac{du}{ds^*} = \frac{du}{ds}\frac{ds}{ds^*} = l_1\frac{ds}{ds^*}.$$

Similarly,

$$m_1^* = m_1\frac{ds}{ds^*}, n_1^* = n_1\frac{ds}{ds^*},$$

$$l_2^* = l_2\frac{ds}{ds^*}, m_2^* = m_2\frac{ds}{ds^*}, n_2^* = n_2\frac{ds}{ds^*}.$$

Substituting the above values in (5), we get

$$\cos\theta^* = \left(\frac{ds}{ds^*}\right)^2 [E^*l_1l_2 + F^*(l_1m_2 + l_2m_1) + G^*m_1m_2]. \tag{6}$$

The condition is necessary. Let the mapping (1) be conformal.

To prove : $\dfrac{E}{E^*} = \dfrac{F}{F^*} = \dfrac{G}{G^*} = \left(\dfrac{ds}{ds^*}\right)^2.$

Since the mapping (1) is conformal, we have

$$\theta = \theta^*$$

$$\Rightarrow \quad \cos\theta = \cos\theta^*$$

$$\Rightarrow \quad El_1l_2 + F(l_1m_2 + l_2m_1) + Gm_1m_2$$
$$= \left(\frac{ds}{ds^*}\right)^2 [E^* l_1l_2 + F^*(l_1m_2 + l_2m_1) + G^* m_1m_2]$$

[using (4) and (6)]

$$\Rightarrow \quad \frac{E}{E^*} = \frac{F}{F^*} = \frac{G}{G^*} = \left(\frac{ds}{ds^*}\right)^2.$$

Hence the condition is necessary.

The condition is sufficient. Now, we suppose that

$$\frac{E}{E^*} = \frac{F}{F^*} = \frac{G}{G^*} = \left(\frac{ds}{ds^*}\right)^2.$$

To prove : The mapping (1) is conformal.

From the given relation, we have

$$E = E^*\left(\frac{ds}{ds^*}\right)^2, F = F^*\left(\frac{ds}{ds^*}\right)^2, G = G^*\left(\frac{ds}{ds^*}\right)^2.$$

Substituting the above values of E, F, G in (4), we get

$$\cos\theta = \left(\frac{ds}{ds^*}\right)^2 [E^* l_1l_2 + F^*(l_1m_2 + l_2m_1) + G^* m_1m_2]$$

$\Rightarrow \quad \cos\theta = \cos\theta^*$ [using (6)]

$\Rightarrow \quad \theta = \theta^*$

$\Rightarrow$ The mapping (1) is conformal.

Hence the condition is sufficient.

Theorem 2. Every point on a surface S has a neighbourhood which can be mapped conformally on a region of the plane.

Proof. We know that the metric of the surface S is given by

$$ds^2 = E\,du^2 + 2F\,du\,dv + G\,dv^2. \quad (1)$$

At any point P of the surface, there are two imaginary directions (called isotropic directions) such that $ds^2 = 0$. Therefore, by (1),

$$E\,du^2 + 2F\,du\,dv + G\,dv^2 = 0. \quad (2)$$

Since $EG - F^2 \neq 0$, it follows from (2) that the two imaginary directions are distinct.

Let the curves along these directions be the parametric curves. For the parametric curve $v = \text{constant}$, (2) reduces to

$$E\,du^2 = 0. \quad (3)$$

Similarly, for the parametric curve $u = \text{constant}$, (2) reduces to

$$G\,dv^2 = 0. \quad (4)$$

Using (3) and (4), (1) becomes

$$ds^2 = 2F\,du\,dv$$

or $$ds^2 = \lambda\,du\,dv, \tag{5}$$

where $\lambda = 2F$.

Now, we consider the transformation

$$u = U + iV, v = U - iV,$$

where U and V are real parameters. Then

$$du = dU + i\,dV, v = dU - i\,dV. \tag{6}$$

Using (6), (5) reduces to

$$ds^2 = \lambda\,(dU^2 + dV^2). \tag{7}$$

We know that the metric of the plane when the parameters $\overline{u}, \overline{v}$ are taken as the Cartesian coordinates is of the form

$$ds^2 = d\overline{u}^2 + d\overline{v}^2. \tag{8}$$

Comparing (7) with (8), we observe that the mapping

$$\overline{u} = U,\ \overline{v} = V$$

gives a conformal mapping of a region of a given surface on a region of the plane. Hence the theorem is proved.

Corollary. Every point on a surface S has a neighbourhood which can be mapped conformally on some neighbourhood of any other surface S^*.

Proof. The proof follows from the above theorem.

2. Isometric Mapping

Definition. Let S and S^* be two surfaces. If there exists a correspondence between the points P on S and P^* on S^* such that as P traces out an arc on S, P^* traces out an arc of equal length on S^*, then S is said to be mapped isometrically onto S^*. In this case, the mapping is called length-preserving or an isometric mapping.

Theorem. Let S and S^* be two surfaces and let a mapping of S onto S^* be defined by

$$\mathbf{r} = \mathbf{r}\,(u, v), \mathbf{r}^* = \mathbf{r}^*(u, v).$$

Then a necessary and sufficient condition for this mapping to be isometric is that

$$E = E^*, F = F^*, G = G^*.$$

Proof. Let a surface S be mapped onto a surface S^* by the mapping

$$\mathbf{r} = \mathbf{r}\,(u, v), \mathbf{r}^* = \mathbf{r}^*(u, v). \tag{1}$$

If the mapping (1) is isometric, then S and S^* have the same linear element, i.e.,

$$ds^2 = ds^{*2}$$

or $$E\,du^2 + 2F\,du\,dv + G\,dv^2 = E^*\,du^2 + 2F^*\,du\,dv + G^*\,dv^2$$

or $$E = E^*, F = F^*, G = G^*.$$

Hence the condition is necessary.

Conversely, if we have

$$E = E^*, F = F^*, G = G^*,$$

then

$$E\,du^2 + 2F\,du\,dv + G\,dv^2 = E^*\,du^2 + 2F^*\,du\,dv + G^*\,dv^2$$

i.e., $$ds^2 = ds^{*2}$$

i.e., S and S^* have the same linear element.

Therefore the mapping (1) is isometric.

Hence the condition is sufficient.

Remark. It is obvious that an isometric mapping is always conformal but a conformal mapping need not be isometric, i.e., an isometric mapping preserves both the angles and the distances whereas a conformal mapping preserves angles only.

3. Geodesic Mapping

Definition. Let S and S^* be two surfaces. If there exists a differentiable homeomorphism, i.e., an analytic one-one mapping of S onto S^* such that the geodesic arcs on S map onto geodesic arcs on S^*, then S is said to be mapped geodesically onto S^*. In this case, the mapping is called a geodesic mapping.

Theorem. If a mapping of a surface S onto a surface S^* is both geodesic and conformal, then it is necessarily an isometry or else a similarity mapping.

Proof. Let S and S^* be two surfaces and let a mapping of S onto S^* be defined by

$$\mathbf{r} = \mathbf{r}(u, v), \mathbf{r}^* = \mathbf{r}^*(u, v). \qquad (1)$$

Let the mapping of S onto S^* defined above be both geodesic and conformal.

Now, consider a system of geodesic coordinates on S so that $v = \text{const.}$ is a family of geodesics on S and the curves $u = \text{const.}$ are their orthogonal trajectories. We know that the metric of the surface S is given by

$$ds^2 = du^2 + G\,dv^2. \tag{2}$$

Since the mapping (1) is conformal, the metric of the surface S^* is given by

$$ds^{*2} = \lambda(u,v)(du^2 + G\,dv^2). \tag{3}$$

Also, since the mapping (1) is geodesic, the curves $v =$ const. must be geodesics with respect to metric (3). But, we know that a necessary and sufficient condition for the parametric curve $v =$ const. to be geodesics is

$$EE_2 + FE_1 - 2EF_1 = 0$$

or $\quad EE_2 = 0 \qquad [\because F = 0]$

or $\quad \lambda\lambda_2 = 0 \qquad [\because E = \lambda \text{ for metric (3)}]$

or $\quad \lambda_2 = 0. \qquad [\because \lambda \neq 0]$

$\therefore \quad \lambda = \lambda(u).$

Now, consider any geodesic on S inclined at an angle θ with the parametric curve $v =$ const. From Liouville's formula, we have

$$\kappa_g = \theta' + Pu' + Qv',$$

where

$$P = \frac{2EF_1 - FE_1 - EE_2}{2HE}, \quad Q = \frac{EG_1 - FE_2}{2HE}.$$

For a geodesic, we have $\kappa_g = 0$. Also, in our case, $F = 0, E = 1$ for metric (2). Therefore, from Liouville's formula, we get

$$0 = d\theta + \frac{1}{2}\frac{G_1}{H}dv. \tag{4}$$

Since the mapping (1) is both geodesic and conformal, the equation for S^* corresponding to (4) is

$$0 = d\theta + \frac{1}{2}\frac{G_1^*}{H^*}dv. \tag{5}$$

Comparing (4) and (5), we have

$$\frac{G_1}{H} = \frac{G_1^*}{H^*}$$

or $\quad \dfrac{G_1}{\sqrt{G}} = \dfrac{(\lambda G)_1}{\lambda\sqrt{G}} \qquad \left[\because H = \sqrt{EG - F^2} = \sqrt{G},\, G^* = \lambda G,\, H^* = \lambda\sqrt{G}\right]$

or $\quad \dfrac{G_1}{\sqrt{G}} = \dfrac{\lambda G_1 + \lambda_1 G}{\lambda\sqrt{G}} = \dfrac{G_1}{\sqrt{G}} + \sqrt{G}\dfrac{\lambda_1}{\lambda}$

or $$0 = \sqrt{G}\frac{\lambda_1}{\lambda}.$$

Therefore $\lambda_1 = 0$, i.e., λ is a constant. Hence the mapping (1) is a similarity mapping. If $\lambda = 1$, the mapping is isometric.

5.16 TISSOT'S THEOREM

Let S and S^* be any two surfaces. Then in any non-conformal mapping of S on S^*, given by a differentiable homeomorphism regular at each point, there exists at each point P of S, a uniquely determined pair of real orthogonal directions such that the corresponding directions on S^* are also orthogonal.

Proof. Let the parameters on S and S^* be so chosen that the corresponding points have the same parameters. Let (l_1, m_1), (l_2, m_2) be the direction coefficients of two orthogonal directions at P on S and let (l_1^*, m_1^*), (l_2^*, m_2^*) be the direction coefficients of the corresponding two directions at P^* on S^*. Then

$$El_1 l_2 + F(l_1 m_2 + l_2 m_1) + G m_1 m_2 = 0, \tag{1}$$

$$E^* l_1^* l_2^* + F^*(l_1^* m_2^* + l_2^* m_1^*) + G^* m_1^* m_2^* = 0. \tag{2}$$

Now,

$$l_1^* = \frac{du}{ds^*} = \frac{du}{ds}\frac{ds}{ds^*} = l_1\frac{ds}{ds^*}.$$

Similarly,

$$m_1^* = m_1\frac{ds}{ds^*},\ n_1^* = n_1\frac{ds}{ds^*},$$

$$l_2^* = l_2\frac{ds}{ds^*},\ m_2^* = m_2\frac{ds}{ds^*},\ n_2^* = n_2\frac{ds}{ds^*}.$$

Substituting the above values in (2), we get

$$E^* l_1 l_2 + F^*(l_1 m_2 + l_2 m_1) + G^* m_1 m_2 = 0. \tag{3}$$

Now, we shall eliminate l_1, m_1 from (1) and (3).

From (1), we have

$$(El_2 + Fm_2)\,l_1 = -(Fl_2 + Gm_2)\,m_1. \tag{4}$$

From (3), we have

$$(E^* l_2 + F^* m_2)\,l_1 = -(F^* l_2 + G^* m_2)m_1. \tag{5}$$

Eliminating l_1, m_1 from (4) and (5), we get

$$(El_2 + Fm_2)(F^* l_2 + G^* m_2) = (E^* l_2 + F^* m_2)(Fl_2 + Gm_2)$$

or $$(EF^* - E^*F)\,l_2^2 + (EG^* - E^*G)\,l_2 m_2 + (FG^* - F^*G)m_2^2 = 0. \tag{6}$$

Equation (6) is a quadratic equation in l_2/m_2. Therefore it determines two directions at each point of the surface S.

We know that the two directions represented by the differential equation

$$P\,du^2 + 2Q\,du\,dv + R\,dv^2 = 0$$

are orthogonal if

$$ER - 2FQ + GP = 0.$$

For the directions given by (6), we have

$$P = EF^* - E^*F,$$
$$2Q = EG^* - E^*G,$$
$$R = FG^* - F^*G.$$

$$\therefore \quad ER - 2FQ + GP = E(FG^* - F^*G) - F(EG^* - E^*G) + G(EF^* - E^*F)$$
$$= 0.$$

Hence (6) determines two orthogonal directions at each point of S.

If Δ denotes discriminant of (6), then

$$\Delta = (EG^* - E^*G)^2 - 4(EF^* - E^*F)(FG^* - F^*G)$$
$$= (EG^* - E^*G)^2 + (EF^* - E^*F)^2\left[\frac{4(EG - F^2)}{E^2} + \frac{4F^2}{E^2}\right] - \frac{4F}{E}(EF^* - E^*F)(EG^* - E^*G)$$
$$[\because E \neq 0]$$

i.e.,
$$\Delta = \frac{4(EG - F^2)}{E^2}(EF^* - FE^*)^2 + \left[(EG^* - GE^*) - \frac{2F}{E}(EF^* - FE^*)\right]^2. \qquad (7)$$

Since the mapping is not conformal, the relations

$$\frac{E}{E^*} = \frac{F}{F^*} = \frac{G}{G^*}$$

do not hold, i.e., $EF^* - FE^*$ and $EG^* - GE^*$ are not both zero. Therefore the discriminant of (6), given by (7) is strictly positive, and so the roots of (6) are real and distinct. Hence at each point P on S, a pair of real orthogonal directions is uniquely determined such that the corresponding directions on S^* are also orthogonal.

5.17 DINI'S THEOREM

If two surfaces S and S^* are mapped geodesically on each other by a non-conformal mapping, then their line elements can be written in the form

$$ds^2 = (U - V)(du^2 + dv^2),$$
$$ds^{*2} = (V^{-1} - U^{-1})(U^{-1}du^2 + V^{-1}dv^2),$$

where U is a function of u only and V is a function of v only.

Proof. Let the parameters on S and S^* be so chosen that the corresponding points have the same parameters u, v.

By Tissot's theorem, in any non-conformal mapping of S on S^*, given by a differentiable homeomorphism regular at each point, there exists at each point P of S, a uniquely determined pair of real orthogonal directions such that the corresponding directions on S^* are also orthogonal. Therefore, $F = F^* = 0$. Consequently, the line elements of S and S^* reduce to the form

$$ds^2 = E\,du^2 + G\,dv^2, \tag{1}$$
$$ds^{*2} = E^*\,du^2 + G^*\,dv^2. \tag{2}$$

Also, since $F = 0$, we have

$$\left.\begin{array}{lll} l = \dfrac{E_1}{2E}, & m = \dfrac{E_2}{2E}, & n = -\dfrac{G_1}{2E}, \\ \lambda = -\dfrac{E_2}{2G}, & \mu = \dfrac{G_1}{2G}, & \nu = \dfrac{G_2}{2G}. \end{array}\right\} \tag{3}$$

Now, we know that the equation of a geodesic on S is given by

$$\frac{d^2v}{du^2} = n\left(\frac{dv}{du}\right)^3 + (2m - \nu)\left(\frac{dv}{du}\right)^2 + (l - 2\mu)\frac{dv}{du} - \lambda$$

or $$\ddot{v} = -\frac{G_1}{2E}\dot{v}^3 + \left(\frac{E_2}{E} - \frac{G_2}{2G}\right)\dot{v}^2 + \left(\frac{E_1}{2E} - \frac{G_1}{G}\right)\dot{v} + \frac{1}{2}\frac{E_2}{G} \qquad \text{[using (3)]}$$

or $$EG\ddot{v} + \frac{1}{2}GG_1\dot{v}^3 + \left(\frac{1}{2}EG_2 - GE_2\right)\dot{v}^2 + \left(EG_1 - \frac{1}{2}GE_1\right)\dot{v} - \frac{1}{2}EE_2 = 0. \tag{4}$$

Similarly, the equation of a geodesic on S^* is

$$E^*G^*\ddot{v} + \frac{1}{2}G^*G_1^*\dot{v}^3 + \left(\frac{1}{2}E^*G_2^* - G^*E_2^*\right)\dot{v}^2 + \left(E^*G_1^* - \frac{1}{2}G^*E_1^*\right)\dot{v} - \frac{1}{2}E^*E_2^* = 0. \tag{5}$$

Since the mapping is geodesic, equations (4) and (5) must be identical. Hence, comparing coefficients, we have

$$\frac{EG}{E^*G^*} = \frac{GG_1}{G^*G_1^*} = \frac{\frac{1}{2}EG_2 - GE_2}{\frac{1}{2}E^*G_2^* - G^*E_2^*}$$

$$= \frac{EG_1 - \frac{1}{2}GE_1}{E^*G_1^* - \frac{1}{2}G^*E_1^*} = \frac{EE_2}{E^*E_2^*}. \tag{6}$$

Taking first two ratios of (6), we get

$$\frac{G_1}{E} = \frac{G_1^*}{E^*}. \tag{7}$$

Taking first and last ratios of (6), we get

$$\frac{E_2}{G} = \frac{E_2^*}{G^*}. \tag{8}$$

Taking third and last ratios of (6), we have

$$\frac{\frac{1}{2}EG_2 - GE_2}{EE_2} = \frac{\frac{1}{2}E^*G_2^* - G^*E_2^*}{E^*E_2^*}$$

or $$\frac{\frac{1}{2}\frac{G_2}{G} - \frac{E_2}{E}}{\frac{E_2}{G}} = \frac{\frac{1}{2}\frac{G_2^*}{G^*} - \frac{E_2^*}{E^*}}{\frac{E_2^*}{G^*}}$$

or $$\frac{1}{2}\frac{G_2}{G} - \frac{E_2}{E} = \frac{1}{2}\frac{G_2^*}{G^*} - \frac{E_2^*}{E^*}$$ [using (8)]

or $$\frac{E_2}{E} - \frac{1}{2}\frac{G_2}{G} = \frac{E_2^*}{E^*} - \frac{1}{2}\frac{G_2^*}{G^*}. \tag{9}$$

Taking second and fourth ratios of (6), we have

$$\frac{EG_1 - \frac{1}{2}GE_1}{GG_1} = \frac{E^*G_1^* - \frac{1}{2}G^*E_1^*}{G^*G_1^*}$$

or $$\frac{\frac{G_1}{G} - \frac{1}{2}\frac{E_1}{E}}{\frac{G_1}{E}} = \frac{\frac{G_1^*}{G^*} - \frac{1}{2}\frac{E_1^*}{E^*}}{\frac{G_1^*}{E^*}}$$

or $$\frac{G_1}{G} - \frac{1}{2}\frac{E_1}{E} = \frac{G_1^*}{G^*} - \frac{1}{2}\frac{E_1^*}{E^*} \tag{10}$$

[using (7)]

Now, $\dfrac{\partial}{\partial v}\left(\dfrac{E^2}{G}\right)=\dfrac{2EE_2G-G_2E^2}{G^2}=\dfrac{2E^2}{G}\left(\dfrac{E_2}{E}-\dfrac{1}{2}\dfrac{G_2}{G}\right).$

$$\therefore \qquad \frac{\dfrac{\partial}{\partial v}\left(\dfrac{E^2}{G}\right)}{\dfrac{E^2}{G}}=2\left(\frac{E_2}{E}-\frac{1}{2}\frac{G_2}{G}\right).$$

Similarly,

$$\frac{\dfrac{\partial}{\partial v}\left(\dfrac{E^{*2}}{G^*}\right)}{\dfrac{E^{*2}}{G^*}}=2\left(\frac{E_2^*}{E^*}-\frac{1}{2}\frac{G_2^*}{G^*}\right).$$

The above two equations with the help of (9) give

$$\frac{\dfrac{\partial}{\partial v}\left(\dfrac{E^2}{G}\right)}{\dfrac{E^2}{G}}=\frac{\dfrac{\partial}{\partial v}\left(\dfrac{E^{*2}}{G^*}\right)}{\dfrac{E^{*2}}{G^*}}.$$

Integrating, we get

$$\log\left(\frac{E^2}{G}\right)=\log\left(\frac{E^{*2}}{G^*}\right)+\log U^3, \text{ where } U \text{ is some function of } u.$$

$$\therefore \qquad \frac{E^2}{G}=\frac{E^{*2}}{G^*}U^3. \tag{11}$$

Again, we have

$$\frac{\partial}{\partial u}\left(\frac{G^2}{E}\right)=\frac{2GG_1E-E_1G^2}{E^2}=\frac{2G^2}{E}\left(\frac{G_1}{G}-\frac{1}{2}\frac{E_1}{E}\right).$$

$$\therefore \qquad \frac{\dfrac{\partial}{\partial u}\left(\dfrac{G^2}{E}\right)}{\dfrac{G^2}{E}}=2\left(\frac{G_1}{G}-\frac{1}{2}\frac{E_1}{E}\right).$$

Similarly,

$$\frac{\dfrac{\partial}{\partial v}\left(\dfrac{G^{*2}}{E^*}\right)}{\dfrac{G^{*2}}{E^*}}=2\left(\frac{G_1^*}{G^*}-\frac{1}{2}\frac{E_1^*}{E^*}\right).$$

The above two equations with the help of (10) give

$$\frac{\dfrac{\partial}{\partial u}\left(\dfrac{G^2}{E}\right)}{\dfrac{G^2}{E}} = \frac{\dfrac{\partial}{\partial v}\left(\dfrac{G^{*2}}{E^*}\right)}{\dfrac{G^{*2}}{E^*}}.$$

Integrating, we get

$$\log\left(\frac{G^2}{E}\right) = \log\left(\frac{G^{*2}}{E^*}\right) + \log V^3,$$

where V is some function of v.

$$\therefore \qquad \frac{G^2}{E} = \frac{G^{*2}}{E^*}V^3. \tag{12}$$

Solving (11) and (12) for E and G, we get

$$E = E^* U^2 V, \; G = G^* U V^2. \tag{13}$$

From (7) and (13), we get

$$\frac{G_1}{E^* U^2 V} = \frac{G_1^*}{E^*}$$

or $\qquad G_1 = U^2 V \dfrac{\partial}{\partial u}(G^*) = U^2 V \dfrac{\partial}{\partial u}\left(\dfrac{G}{UV^2}\right)$

or $\qquad G_1 = U^2 V\left(\dfrac{G_1 UV^2 - U_1 V^2 G}{U^2 V^4}\right)$

or $\qquad VG_1 = G_1 U - U_1 G$

or $\qquad G_1(U - V) = U_1 G$

or $\qquad \dfrac{G_1}{G} = \dfrac{U_1}{U - V}.$

This on integration, when $U > V$ gives

$$\log G = \log(U - V) + \log \bar{V}^2,$$

where $\bar{V}$ is an arbitrary function of v.

$$\therefore \qquad G = (U - V)\bar{V}^2. \tag{14}$$

Similarly, from (8) and (13), we get

$$E = (U - V)\bar{U}^2, \tag{15}$$

where $\bar{U}$ is an arbitrary function of u.

Substituting these values of E and G in (1), the metric of S becomes

$$ds^2 = (U - V)(\bar{U}^2 du^2 + \bar{V}^2 dv^2). \tag{16}$$

Similarly, substituting the values of E^* and G^* from (13) in (2), the metric of S^* becomes

$$ds^{*2} = \frac{E}{U^2 V} du^2 + \frac{G}{UV^2} dv^2$$

or $$ds^{*2} = \frac{U-V}{U^2 V} \overline{U}^2 du^2 + \frac{U-V}{UV^2} \overline{V}^2 dv^2 \qquad \text{[using (14) and (15)]}$$

or $$ds^{*2} = \frac{U-V}{UV} \left[U^{-1} \overline{U}^2 du^2 + V^{-1} \overline{V}^2 dv^2 \right]$$

or $$ds^{*2} = (V^{-1} - U^{-1}) [U^{-1} \overline{U}^2 du^2 + V^{-1} \overline{V}^2 dv^2]. \tag{17}$$

Let us set a transformation of parameters

$$u^* = \int \overline{U}\, du, v^* = \int \overline{V}\, dv.$$

Then (16) and (17) are reduced to the forms

$$ds^2 = (U - V)(du^2 + dv^2),$$

$$ds^{*2} = (V^{-1} - U^{-1})(U^{-1} du^2 + V^{-1} dv^2).$$

Solved Example

Example 1. If a unit-sphere is represented in terms of longitude u and colatitude v by the equation

$$\mathbf{r} = (\cos u \sin v, \sin u \sin v, \cos v),$$

prove the formula

$$x = u, y = \log \tan \frac{1}{2} v.$$

Determine a conformal representation of this sphere on a plane.

Solution. The position vector $\mathbf{r}$ of a current point on the unit sphere is

$$\mathbf{r} = (\cos u \sin v, \sin u \sin v, \cos v).$$

$\therefore$ $$\mathbf{r}_1 = (-\sin u \sin v, \cos u \sin v, 0),$$

$$\mathbf{r}_2 = (\cos u \cos v, \sin u \cos v, -\sin v),$$

$$E = \mathbf{r}_1^2 = \sin^2 v, \; F = \mathbf{r}_1 \cdot \mathbf{r}_2 = 0, G = \mathbf{r}_2^2 = 1.$$

Therefore the metric of the sphere is given by

$$ds^2 = E\, du^2 + 2F\, du\, dv + G\, dv^2 = \sin^2 v\, du^2 + dv^2$$

or $$ds^2 = \sin^2 v\, (du^2 + \operatorname{cosec}^2 v\, dv^2). \tag{1}$$

Now, let us put

$$w = \int \operatorname{cosec} v\, dv = \log \tan \frac{1}{2} v.$$

Then the metric (1) reduces to

$$ds^2 = A^2 (du^2 + dw^2),$$

where $A = \sin v$.

If we represent this sphere on a plane referred to rectangular axes (x, y) by substituting

$$x = u,\ y = \log \tan \frac{1}{2} v = w,$$

we get a conformal mapping (called Mercator's mapping).

Exercises

1. State and prove Tissot's Theorem.
2. State and prove Dini's Theorem.

Index

(Numbers refer to pages)

O

P

R

S

T

U

W